THE INTERNATI(
MONOGRAPHS

THE INTERNATIONAL SERIES OF MONOGRAPHS ON CHEMISTRY

Atoms in Molecules

A Quantum Theory

Richard F. W. Bader

Department of Chemistry,
McMaster University,
Ontario, Canada

CLARENDON PRESS · OXFORD

1994

Oxford University Press, Walton Street, Oxford OX2 6DP
Oxford   New York   Toronto
Delhi   Bombay   Calcutta   Madras   Karachi
Kuala Lumpur   Singapore   Hong Kong   Tokyo
Nairobi   Dar es Salaam   Cape Town
Melbourne   Auckland   Madrid
and associated companies in
Berlin   Ibadan

Oxford is a trade mark of Oxford University Press

Published in the United States
by Oxford University Press Inc., New York

British Library Cataloguing in Publication Data
Bader, Richard F. W.
Atoms in molecules: a quantum theory,
1. Molecules. Structure. Quantum theory
I. Title   II. Series
541.28

Library of Congress Cataloging in Publication Data
Bader, Richard F. W., 1931–
Atoms in molecules: a quantum theory/Richard F. W. Bader.
(The International series of monographs on chemistry; 22)
Includes bibliographical references and index.
1. Quantum chemistry.   I. Title.   II. Series
QD462.2.B33 1994   541.2'8—dc 20   93–46241
ISBN 0 19 855865 1 (Pbk)

Printed in Great Britain by
Bookcraft (Bath) Ltd
Midsomer Norton, Avon

PREFACE

Science is based upon experiment and observation. Its purpose is to enable us to predict the natural events that occur around us and to understand them as best we can. The purpose of theory is to provide the conceptual framework in this quest for the prediction and understanding of observation. Chemistry has been in a unique position in this regard. The molecular structure hypothesis—that a molecule is a collection of atoms linked by a network of bonds—was forged in the crucible of nineteenth century experimental chemistry. It has continued to serve as the principal means of ordering and classifying the observations of chemistry. There can be no question from the experimental point of view but that our ability to understand chemical behaviour is a result of the structure imposed on a system and its properties by the presence of atoms. The difficulty with the molecular structure hypothesis is that it is not related directly to the physics which governs the motions of the nuclei and electrons that make up the atoms and the bonds. We have in chemistry an understanding based on a classification scheme that is both powerful and at the same time, because of its empirical nature, limited. What remains is to show that the atoms in a molecule are real, with properties predicted and defined by the laws of quantum mechanics, and that the structure their presence imparts to a molecule is indeed a consequence of the underlying physics. By demonstrating that the molecular structure hypothesis has a theoretical basis, the classification based upon the concept of atoms in molecules is freed from its empirical constraints and the full predictive power of quantum mechanics can be incorporated into the resulting theory—a theory of atoms in molecules.

It is the purpose of this book to demonstrate to the reader that such a theory is obtained when the definition of a quantum subsystem, as derived from a fundamental principle of physics, is applied to the observed topological properties of a system's distribution of charge. An atom is uniquely defined as an open system in real space, free to exchange charge and momentum with its neighbours. Its properties are described by the same equations of motion and associated theorems of quantum mechanics, such as the virial and Ehrenfest theorems, as apply to the total system. The same topological properties of the charge distribution lead in turn to a definition of the bonds which link the atoms, and to definitions of structure and structural stability for molecular systems. The theory makes possible the identification of all possible structures for a given system and determines the mechanisms of structural change within the system.

The topology of ρ, the charge density, as displayed in the global properties of its gradient vector field, yields a faithful mapping of the chemical concepts of atoms, bonds, and structure. The properties of this field however, provide no indication of maxima in ρ corresponding to the electron pairs of the Lewis model, a model secondary only to the atomic hypothesis itself in our interpretation of chemical reactivity and molecular geometry. The physical basis of this most important model is one level of abstraction above the visible topology of the charge density and appears instead in the topology of the Laplacian of ρ. This function is the scalar derivative of the gradient vector field of the charge density and it determines where electronic charge is locally concentrated and depleted, the local charge concentrations providing a mapping of the electron pairs of the Lewis model. The Laplacian of the charge density plays a dominant role throughout the theory, appearing in the constraint determining the boundary of a quantum subsystem and relating the spatial properties of the charge density to the local contributions to the energy.

The whole of the theory, its description of atoms, bonds, and structure and its stability, together with the discussion of molecular geometry and reactivity in terms of the Laplacian of ρ, is developed in a continuous manner from the initial quantum definition of an atom. At no point in this development is the reader asked to accept the introduction of arbitrary assumptions or personal points of view.

The properties of atoms in molecules can be determined experimentally and quantum mechanics, through the theory of atoms in molecules, predicts these properties just as it predicts the properties of the total system. It comes as no surprise to a chemist to know that there is a carbonyl group, a fragment of real space with a set of definable properties. The energies of individual atoms can be calculated and the mechanism of energy storage in a biological molecule can be determined and understood at the atomic level. There are situations when functional groupings of atoms are transferable between systems, carrying with them essentially constant contributions to the energy and other properties of the systems in which they occur. This is a known experimental result of chemistry and these are the groups identified by the theory of atoms in molecules.

This book presents an account of the theory of atoms in molecules to the scientists who are responsible for performing the experiments and collecting the observations on the properties of matter at the atomic level. This is done in the belief that the transformation of qualitative concepts into a quantitative theory will serve to deepen our understanding of chemistry. This constitutes a wide audience and the book is structured accordingly. The material in each chapter is presented in such a manner that a reader desiring a general understanding of its content may do so by reading less than the complete chapter. The reader desirous of a fuller understanding of the mathematical and physical structure of the theory attains this goal by a complete reading of the chapter. At no point or stage is rigour sacrificed for ease of presentation, only the depth of understanding paralleling the chapter depth. Topics considered to be important but not essential to the first reading of the text are referenced at the appropriate place in the text and placed at the end of the chapter in which they occur.

Tobermory, Ontario   R.F.W.B
October 1989.

This book is dedicated to the members of my research group, past and present, who have contributed to the development of the theory of atoms in molecules.

CONTENTS

LIST OF SYMBOLS

Coordinates

$\mathbf{r}$	position vector
$\mathbf{r}_c$	position vector of critical point in electronic charge density
$\mathbf{x}$	the space and spin coordinates of a set of electrons
$\mathbf{r}_j$	position vector of electron j
$\mathbf{r}_\Omega$	electronic position vector with origin at nucleus of atom Ω
$\mathrm{d}\tau$	volume element for a single electron
∇	gradient with respect to coordinates of an electron
$\mathbf{R}^Q$	nuclear configuration space
$\mathbf{X}$	a point in $\mathbf{R}^Q$ defining a molecular geometry, the set of nuclear coordinates
$\mathbf{X}_\alpha$	position vector of nucleus α
$\mathbf{X}_c$	catastrophe point in $\mathbf{R}^Q$
$\mathbf{X}_b$	bifurcation point in $\mathbf{R}^Q$
∇_x	gradient with respect to nuclear coordinates
$\int \mathrm{d}\tau'$	a summation over all electronic spin coordinates and the integration over the space coordinates of all electrons but one, eqns (1:3) and (1.4)
$\oint \mathrm{d}S(\Omega, \mathbf{r})$	integral over surface of atom Ω

Vectors and dyadics

$\mathbf{i}, \mathbf{j}, \mathbf{k}$	unit vectors in x, y, z directions
$\mathbf{A}$	vector with components A_x, A_y, A_z, eqn (E5.11)
$\mathbf{A} \cdot \mathbf{B}$	scalar product of $\mathbf{A}$ and $\mathbf{B}$, eqn (E5.13)
$\mathbf{A} \times \mathbf{B}$	vector product of $\mathbf{A}$ and $\mathbf{B}$
$\mathbf{A}\mathbf{B}$	dyadic with nine components, eqn (E5.14)

Charge density and related quantities

$\rho(\mathbf{r}; \mathbf{X}), \rho(\mathbf{r})$	electronic charge density, eqn (1.3)
ρ_b	value of $\rho(\mathbf{r})$ at bond critical point
ρ_r	value of $\rho(\mathbf{r})$ at a ring critical point
$\rho^\alpha(\mathbf{r})$	density of α spin electrons
$\nabla\rho(\mathbf{r})$	gradient vector of ρ, eqn (2.1)
s	path parameter of trajectory of $\nabla\rho$
$\mathbf{n}(\mathbf{r})$	unit vector normal to surface
λ_i	eigenvalue of the Hessian matrix of ρ at a critical point

$\mathbf{u}_i$	eigenvector of the Hessian matrix of ρ at a critical point
σ	signature of critical point in ρ
ω	rank of critical point in ρ
ε	bond ellipticity
$\nabla^2\rho(\mathbf{r})$	Laplacian of ρ, eqn (2.3)
α_b	bond path angle
α_e	geometric bond angle
R_b	bond path length
R_e	geometric equilibrium bond length
n	bond order
$\left.\begin{array}{l}\Psi(\mathbf{x}, t), \\ \Psi(\mathbf{x}, \mathbf{X}, t)\end{array}\right\}$	state functions for time-dependent systems
$\psi(\mathbf{x}), \psi(\mathbf{x}; \mathbf{X})$	state functions for stationary states

Functionals

$\mathscr{I}[\psi]$	Schrödinger's energy functional, eqn (5.54)
$\mathscr{G}[\psi]$	Schrödinger's constrained energy functional, eqn (5.58)
$\mathscr{E}[\psi]$	stationary state energy functional, eqn (5.63)
$\mathscr{G}[\psi, \Omega]$	constrained atomic functional, eqn (5.72)
$\mathscr{E}[\psi, \Omega]$	stationary state atomic energy functional, eqn (E5.1)
$\mathscr{H}[\psi, \Omega]$	Hamiltonian energy functional for an atom, eqn (E5.4)
$\mathscr{W}_{12}[\Psi]$	action integral, eqn (8.99)
$\mathscr{L}[\Psi, t]$	Lagrange integral, eqn (8.99)
$\mathscr{H}[\Psi]$	Hamiltonian energy integral
$\mathscr{W}_{12}[\Psi, \Omega]$	atomic action integral, eqn (8.111)
$\mathscr{L}[\Psi, \Omega, t]$	atomic Lagrange integral, eqn (8.110)
$\mathscr{W}_{12}[\Psi, \mathbf{A}, \phi]$	action integral for external fields, eqn (8.210)
$\mathscr{L}[\Psi, \mathbf{A}, \phi, t]$	Lagrange integral for external fields, eqn (8.215)

Operators

$\hat{\mathscr{W}}_{12}$	action integral operator
$\hat{\mathscr{L}}[t]$	Lagrange function operator
$\hat{\mathbf{q}}$	coordinate operator
$\hat{\mathbf{r}}$	electronic position operator
$\hat{\mathbf{p}}$	momentum operator
$\hat{\boldsymbol{\pi}}$	momentum operator in magnetic field
$\hat{F}(t)$	generator of infinitesimal unitary transformation, eqn (8.87)
$\hat{H}$	Hamiltonian operator
$\hat{H}_e$	electronic Hamiltonian, eqn (6.53)
$\hat{\mathbf{F}}_\kappa$	force operator for particle κ

$\hat{T}$	electronic kinetic energy operator
$\hat{T}_\pi$	kinetic energy operator in magnetic field, eqn (8.226)
$\hat{U}$	unitary transformation operator, eqn (8.27)
$\hat{V}$	potential energy operator
$\hat{V}_{en}$	electron–nuclear potential energy operator, eqn (6.9)
$\hat{V}_{ee}$	electron–electron potential energy operator, eqn (6.9)
$\hat{V}_{nn}$	nuclear–nuclear potential energy operator, eqn (6.9)

Classical mechanics

q	position
$\dot{q}$	velocity
p	momentum
W_{12}	action integral, eqn (8.43)
$L(q, \dot{q}, t)$	Lagrangian
$H(p, q, t)$	Hamiltonian
$G(p, q)$	generator of infinitesimal canonical transformation, eqn (8.61)
$\{F, G\}$	Poisson bracket of F and G

Property and energy densities

$\alpha(\mathbf{r})$	polarizability density, eqn (8.258)
$\Gamma^{(1)}(\mathbf{r}, \mathbf{r}')$	first-order density matrix, eqn (E1.1)
$\Gamma^{(2)}(\mathbf{r}_1, \mathbf{r}_2)$	diagonal element of second-order density matrix, eqn (E1.3)
$\rho(\mathbf{r}_1, \mathbf{r}_2)$	pair density, eqn (E7.7)
$E_e(\mathbf{r})$	electronic energy density, eqn (6.70)
$\mathbf{F}(\mathbf{r})$	Ehrenfest force density, eqn (6.16)
$G(\mathbf{r})$	kinetic energy density, eqn (5.49)
$G_\perp(\mathbf{r}_c)$	component of kinetic energy density perpendicular to bond path at $\mathbf{r}_c$
$G_{\|}(\mathbf{r}_c)$	component of kinetic energy density parallel to bond path at $\mathbf{r}_c$
$\mathbf{j}(\mathbf{r})$	vector current density, eqn (5.94)
$\delta_\psi \mathbf{j}(\mathbf{r})$	infinitesimal change in $\mathbf{j}(\mathbf{r})$ caused by variation in ψ, eqn (5.95)
$\mathbf{j}_G(\mathbf{r})$	vector current density for generator $\hat{G}$, eqn (5.98)
$\mathbf{j}_B(\mathbf{r})$	vector current density in presence of magnetic field, eqn (8.220)
$\mathbf{J}_F(\mathbf{r})$	vector current density for property F, eqn (8.140)
$\mathbf{J}_F^d(\mathbf{r})$	diamagnetic current density for property F, eqn (8.222)
$K(\mathbf{r})$	kinetic energy density, eqn (5.49)
$L(\Psi, \nabla\Psi, \dot{\Psi}, t)$	many-particle Lagrange density, eqn (8.100)
$\mathscr{L}^0(\mathbf{r}, t)$	Lagrange density at point of variation, eqn (8.107)
$\mathbf{m}(\mathbf{r})$	magnetization density, eqn (8.272)

$\rho_A(\mathbf{r})$	density for property A, eqn (6.35)
$\overset{\leftrightarrow}{\sigma}(\mathbf{r})$	quantum mechanical stress tensor, eqn (5.28), (6.12)
$\mathrm{Tr}\,\overset{\leftrightarrow}{\sigma}(\mathbf{r})$	trace of the stress tensor, eqns (5.32) and (6.26)
$\mathscr{V}(\mathbf{r})$	virial density, eqn (6.30)

Atomic properties

$\alpha_b(\Omega)$	basin contribution to atomic polarizability, eqn (8.260)
$\alpha_s(\Omega)$	surface contribution to atomic polarizability, eqn (8.259)
$A(\Omega\|\Omega')$	contribution to $\Delta A(\Omega)$ from interatomic surface between atoms Ω and Ω', the corresponding bond contribution, eqn (6.102)
$D_2(\Omega,\Omega)$	average number of electron pairs in atom Ω, eqn (E7.17)
$E_e(\Omega)$	electronic energy, eqn (6.73), (6.74)
$\mathbf{F}(\Omega)$	Ehrenfest force, eqn (6.14)
$\mathbf{F}_\alpha(\Omega)$	Hellmann–Feynman force exerted on electronic charge in atom Ω by nucleus α
$F(\Omega,\Omega)$	electron correlation in Ω
$G(\Omega)$	average kinetic energy, eqn (5.50)
$K(\Omega)$	average kinetic energy, eqn (5.50)
$L(\Omega)$	atomic Lagrangian, eqn (6.88)
$l(\Omega)$	percent localization of electrons, eqn (E7.23)
$\Lambda(\bar{N},\Omega)$	fluctuation in average electron population of Ω, eqn (E7.9)
$\mathbf{M}(\Omega)$	atomic first moment, eqn (6.45)
$N(\Omega)$	average electron population, eqn (6.42)
$P_n(\Omega)$	probability of n electrons being in Ω, eqn (E7.1)
$Q_{zz}(\Omega)$	component of quadrupole moment tensor, eqn (6.48)
$q(\Omega)$	charge on atom Ω, eqn (6.43)
$r_b(\Omega)$	bonded radius
$r_n(\Omega)$	non-bonded radius
$S(\Omega), S(\Omega,\mathbf{r})$	atomic surface
$S(\Omega\|\Omega')$	interatomic surface between atoms Ω and Ω'
$T(\Omega)$	average electronic kinetic energy
$T_\pi(\Omega)$	average electronic kinetic energy in magnetic field
$V_{ee}(\Omega,\Omega)$	contribution to $V_{ee}(\Omega)$ from electrons within atom Ω
$V_{ee}(\Omega,\Omega')$	contribution to $V_{ee}(\Omega)$ from repulsion of electrons in Ω with those in Ω'
$V_{en}(\Omega), V_a(\Omega)$	electron–nuclear attractive energy for atom Ω, eqn (6.80)
$V_{en}^0(\Omega), V_a^0(\Omega)$	contribution to $V_{en}(\Omega)$ from interaction of the electronic charge density in the basin of atom Ω with its own nucleus, eqn (6.81)
$V_{nn}(\Omega)$	nuclear–nuclear repulsion energy for atom Ω, eqn (6.86)

$v(\Omega)$	atomic volume
$\mathscr{V}(\Omega)$	atomic virial, eqn (6.24)
$\mathscr{V}_b(\Omega)$	basin virial
$\mathscr{V}_s(\Omega)$	surface virial, eqn (6.21)
$\mathscr{V}(\Omega, \mathbf{B}, \phi)$	atomic virial in external fields
$\chi_b(\Omega)$	basin contribution to atomic magnetic susceptibility, eqn (8.277)
$\chi_S(\Omega)$	surface contribution to atomic magnetic susceptibility, eqn (8.277)

Total properties

$\boldsymbol{\alpha}$	polarizability tensor
$\mathbf{A}$	vector potential of electromagnetic field
$\mathbf{B}$	magnetic field vector
$\mathscr{E}$	electric field vector
e	elementary charge
ϕ	scalar potential of electromagnetic field
D_e	dissociation energy
E	total energy, eqn (6.49)
E_e	total electronic energy, eqn (6.53)
E_n	total nuclear energy, eqn (6.55)
$\mathbf{F}_\alpha$	Hellmann–Feynman force on nucleus α, eqn (5.37)
$\mathbf{F}_{\alpha e}$	electronic contribution to force on nucleus α
$\mathbf{F}_{\alpha n}$	nuclear contribution to force on nucleus α
$\mathbf{F}_{e\alpha}$	Hellmann–Feynman force exerted on electrons by nucleus α
m	mass of electron
$\mathbf{m}$	magnetic dipole moment vector
N	total number of electrons
T	average kinetic energy
$T_{\parallel}$	component of kinetic energy parallel to chosen axis
$T_\perp$	component of kinetic energy perpendicular to chosen axis
V_e	electronic potential energy, eqn (6.54)
$\boldsymbol{\mu}$	dipole moment vector
χ	magnetic susceptibility, eqn (8.273)
$W_{\mu\nu}$	component of energy–momentum tensor, eqn (8.176)
Z	nuclear charge

1

ATOMS IN CHEMISTRY

I have been enabled . . . to exhibit a new view of the first principles or elements of bodies and their combinations, which, if established, as I doubt not it will in time, will produce the most important changes in the system of chemistry.

John Dalton (1807)

1.1 Atoms and the molecular structure hypothesis

Chemistry is the study of the properties of substances and their transformations. Since substances are identified and distinguished one from another by the elements they contain and since the properties of each element are determined by the atoms unique to that element, chemistry is the study of matter at the atomic level. The realization of this fact by Dalton in 1807 marked the birth of chemistry as a branch of science. Dalton's atomic hypothesis accounted for the existence of the observed relative combining weights of the elements and predicted a new relationship among them, the law of multiple proportions. Dalton did not simply resurrect the Greek idea of atoms; he postulated that atoms retained their identity even when in chemical combination with other atoms—that atoms exist in molecules. As a result of the work of van't Hoff and Le Bel in 1874 the geometrical aspect of molecular structure was incorporated into the hypothesis by postulating the existence of preferred geometrical arrangements of bonds terminating at a single atom. Before the close of his century, there had evolved from Dalton's atomic hypothesis the concept of molecular structure—that a molecule is a collection of atoms linked by a network of bonds, the bonds imparting the structure.

The primary purpose in postulating the existence of atoms in molecules is a consequence of the observation that atoms or groupings of atoms appear to exhibit characteristic sets of properties (static, reactive, and spectroscopic), which, in general, vary between relatively narrow limits. In some series of molecules, the variation in the properties is so slight that a group additivity scheme for certain properties, including the energy, can be established. Dalton defined the first of these atomic properties by assigning to each element a relative combining weight. As the body of descriptive chemistry grew, it was realized that the chemistry of some total system could be rationalized by assigning separate properties to each type of atom or functional group in the system. Thus, the knowledge of chemistry is ordered, classified, and understood by assigning properties to atoms and to functional groupings of atoms, properties which can then used to identify the presence of a given group or to understand and predict the behaviour of some total system.

It was a bold step that Dalton took in assuming that each type of atom possessed a particular mass whose value was independent of its chemical state. Why should an atom when in combination with any other atom always possess the same mass? The vindication of Dalton's assumption had to await the results of Rutherford's alpha-scattering experiments and his proposal of the nuclear atom. In this model, essentially all of the atomic mass is found in the chemically stable nucleus. It also places all of the positive charge in the nucleus, thereby stamping each atom with a chemical identity. Because the positive charge is concentrated at a point within the atom, the nuclear–electron attractive force is the single most dominant force in a molecular system. As a consequence, a nucleus plays a unique role, that of an attractor, in determining the topological properties of a molecular charge distribution—a role which leads directly to the definition of an atom. In this manner the chemical identity of an atom is imposed on the electronic charge distribution.

It is the purpose of this book to demonstrate that the existence of atoms with definable properties and the associated concepts of the molecular structure hypothesis are a consequence of the quantum description of matter as applied to the properties of the electronic charge distribution. In so doing, the hypothesis is transformed into a theory and the complete description of matter afforded by quantum mechanics can be applied to the study of atoms in molecules.

1.2   Necessary criteria for a theory of atoms in molecules

The definition of a bound atom—an atom in a molecule—must be such that it enables one to define all of its average properties. For reasons of physical continuity, the definition of these properties must reduce to the quantum mechanical definitions of the corresponding properties for an isolated atom. The atomic values for a given property should, when summed over all the atoms in a molecule, yield the molecular average for that property. The atomic properties must be additive in the above sense to account for the observation that, in certain series of molecules, the atoms and their properties are transferable between molecules, leading to what are known as additivity schemes. An additivity scheme requires both that the property be additive over the atoms in a molecule and that the atoms be essentially transferable between molecules.

Two identical pieces of matter possess identical properties. This elementary fact should extend down to the atomic domain and requires that atoms be defined in real space so that, if an atom is identical in two different systems or at different sites within a given system (e.g. in a solid), then it must contribute identical amounts to the total properties of the systems in which it occurs. Atoms are objects in real space. Theory defines them through a partitioning of real space as determined by the topological properties of a molecular charge distribution, that is, by its form in real space. The constancy in the properties of an atom of theory, including its contribution to the total energy of a system, is observed to be directly determined by a corresponding constancy in its distribution of charge. When the distribution of charge over an atom is the same in two different molecules, i.e. when the atom or some functional grouping of atoms is the same in the real space of two systems, then it makes the same contribution to the total energy and other properties in both systems (Bader and Beddall 1972). It is because of the direct relationship between the spatial form of an atom and its properties that we are able to identify them in different systems. The reader is referred to Fig. 1.1 for an illustration of such transferable functional groups.

There is in this discussion no implication that atoms are always essentially transferable. It is simply that, in the limit of an atom being transferable between systems, the relationship between the form and properties of an atom is most evident. Of equal importance, this limit provides a stringent test of the theory. The properties of atoms in molecules can be experimentally measured and, when the atoms are transferable, these properties can be determined with impressive precision. Rossini and co-workers at the National Bureau of Standards have determined that the incremental increase in the standard heat of formation of a normal hydrocarbon per methylene group is -20.6 ± 1.3 kJ/mol. The theory of atoms in molecules not only defines a methylene group whose distribution of charge in real space up to its atomic boundaries is transferable between hydrocarbon molecules, but, in addition, defines the energy of this group. This energy is as constant and transferable as is the group itself and it accounts for the observed additivity of the energy in this series of molecules.

Following on this observation, if, as is usually the case, the charge distribution of a group does change when it is transferred to another system, then its contribution to the total energy will be different and one may relate its change in energy to this change in its distribution of charge as induced by the new environment. The ultimate justification for identifying the atoms of theory with the atoms of chemistry is this very observation—that their properties change in direct proportion to their change in form. Thus, whether the charge distribution of an atom changes by a little or by a lot, its energy and other properties change by corresponding amounts. Having found the transferable methylene group, for example, one may inquire as to how its properties change when it is introduced into the cyclopropane molecule and subjected to geometric strain. As a result of the geometric constraints present in the cyclopropane molecule, a small amount of electronic charge is transferred

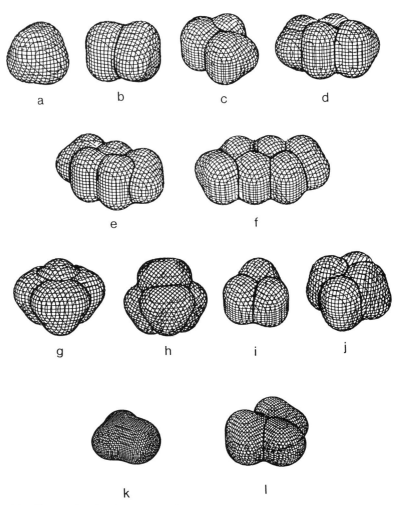

FIG. 1.1. Shapes of molecules represented by envelopes of constant electronic charge density. The envelope shown has the value of 0.001 au. The molecules are: (a)–(f) the normal alkanes from methane to hexane; (g) isobutane; (h) neopentane; (i) cyclopropane; (j) cyclobutane; (k) formaldehyde, $H_2C=O$; (l) acetone, $(CH_3)_2C=O$. The intersections of the 'zero-flux' interatomic surfaces with the envelope are shown in some cases. They define the methyl, methylene, hydrogen, and carbonyl groups. The isobutane molecule (g), for example, exhibits three methyl groups topped by a hydrogen atom.

from hydrogen to carbon. This relatively small change in the distribution of charge of the standard methylene group causes a correspondingly small increase of ~ 38 kJ/mol in its energy, the so-called strain energy. The experimental heat of formation of cyclopropane when compared to three times the energy of the standard methylene group shows an energy increase which equals the theoretically predicted strain energy. This is another example of the atoms of theory recovering the experimental value of a chemically important energy.

It is a well-established procedure to use quantum mechanics to relate a spectroscopically measured energy to the theoretically defined difference in energies between two states of a system. In a less direct but no less rigorous manner, the same quantum mechanics relates the difference in the heats of formation of butane and pentane to the theoretically defined energy of the methylene group. It is the purpose of the atoms in molecules concept to relate molecular properties to those of its constituent atoms, whether these are strictly transferable contributions or, as is more generally the case, physical or spectroscopic properties or modes of reaction that are characteristic of a given grouping of atoms within the molecule. What was lacking in chemistry was the theoretical framework necessary for the incorporation of these chemical observations into the fabric of quantum mechanics so that they could be treated along with the properties of the total system by a single theory. There is only one test of the validity of any theory and that is comparison with observation. The ability of the theory of atoms in molecules to recover measured properties is proper justification that atoms do indeed exist in molecules as separate definable pieces of real space and that their properties are predicted by quantum mechanics as described in this treatise.

1.3 The role of the charge density

The question 'are there atoms in molecules?' is equivalent to asking two equally necessary questions of quantum mechanics: does the state function $\Psi(\mathbf{x}, t)$, which contains all of the information that can be known about a system, predict a unique partitioning of a molecule into subsystems and, if so, can one define the observables, their average values, and their equations of motion for the subsystem? Affirmative answers to both these questions are what is required to establish the physics of an atom in a molecule within the framework of quantum mechanics. Such questions must be posed within a completely general development of quantum mechanics and, as demonstrated in Chapters 5 and 8, affirmative answers are indeed obtained. A most important result of this demonstration is that the quantum subsystems are open systems defined in real space, their boundaries being determined by a particular property of the electronic charge density (Bader and Beddall 1972). Thus, while the theory has its origin in quantum mechanics, its vehicle of expression is the charge density. It is for this reason that we begin with a study of the electronic charge density. Accepting from theory the statement of the quantum boundary condition, this initial study will lead directly to the definition of atoms, bonds, molecular structure, and the mechanisms of structural change and will provide the reader with the knowledge required to apply the theory to chemical problems.

The state function Ψ determines all of the information that can be known about a quantum system. If one desires a theory of molecular structure that is free of arbitrary or subjective assumptions, no information beyond that contained in Ψ should be used in its development. The state function for a molecular system is a function of the electronic and nuclear coordinates and of the time t, $\Psi(\mathbf{x}, \mathbf{X}, t)$, where $\mathbf{x}$ denotes the collection of electronic space and spin coordinates and $\mathbf{X}$ the collection of nuclear coordinates. While the general theory applies to the time-dependent case, we are primarily interested in systems in stationary states whose properties do not change with time. Denoting a solution to Schrödinger's stationary-state equation for a fixed arrangement of the nuclei by $\psi(\mathbf{x}; \mathbf{X})$, the probability of finding each one of N electrons in a particular elemental volume $d\tau_i = dx_i dy_i dz_i$ with spin co-ordinate σ_i (equal to either the α or β spin coordinate) for a given configuration of the nuclei $\mathbf{X}$ is

$$\psi^*(\mathbf{x}; \mathbf{X})\psi(\mathbf{x}; \mathbf{X})d\mathbf{x}_1 d\mathbf{x}_2 \ldots d\mathbf{x}_N \qquad (1.1)$$

where $d\mathbf{x}_i = d\tau_i \sigma_i$. The corresponding probability independent of spin is obtained by summing eqn (1.1) over the spin coordinates. If the summing over all spins is followed by an integration over the spatial coordinates of all the electrons but one (it matters not which electron is chosen, since ψ is an antisymmetrized function and hence all electrons are equivalent), the resulting expression gives the probability of finding one electron in some elemental volume, independent of the instantaneous positions of the remaining electrons,

Probability of one electron being in $d\tau_1$

$$= \sum(\text{spins})\left\{\int d\tau_2 \int d\tau_3 \ldots \int d\tau_N \psi^*(\mathbf{x}; \mathbf{X})\psi(\mathbf{x}; \mathbf{X})\right\} d\tau_1. \qquad (1.2)$$

Multiplication of this probability by the number of electrons N gives the probability of finding any of the electrons in $d\tau_1$, that is, the total probability of finding electronic charge in $d\tau_1$. The corresponding probability density, the probability per unit volume, is called the electron density or the electronic charge density. It is denoted by the symbol $\rho(\mathbf{r}; \mathbf{X})$, the space coordinate of a single electron being given by the position vector $\mathbf{r} = \mathbf{i}x + \mathbf{j}y + \mathbf{k}z$,

$$\rho(\mathbf{r}; \mathbf{X}) = N\sum(\text{spins})\left\{\int d\tau_2 \int d\tau_3 \ldots \int d\tau_N \psi^*(\mathbf{x}; \mathbf{X})\psi(\mathbf{x}; \mathbf{X})\right\}. \qquad (1.3)$$

The subscript 1 is suppressed in this equation since the result refers to any of the electrons. The form of integration given in eqn (1.3), which yields the density of electronic charge in real space, recurs throughout the theory and is designated in an abbreviated manner as

$$\rho(\mathbf{r}; \mathbf{X}) = N\int d\tau' \psi^*(\mathbf{x}; \mathbf{X})\psi(\mathbf{x}; \mathbf{X}) \qquad (1.4)$$

where τ' denotes the spin coordinates of all the electrons and the Cartesian coordinates of all electrons but one. If one inserts an operator $\hat{M}(\mathbf{r})$ for some property between the state functions in eqn (1.4), the result is a real space density distribution function for that property. All atomic properties are defined in terms of such density distributions, the atomic average of a property being obtained by an integration of the corresponding density over the region of space assigned to the atom. The charge density is a special case of this more general situation with the operator $\hat{M}$ equalling the unit operator and the atomic integration yielding the average number of electrons for the atom in question. It should be borne in mind that, while one integrates over the coordinates of the other electrons to obtain the charge density or property-weighted density, all of the electrons contribute to the form of the density so obtained. The integration denoted by $\int d\tau'$ yields a density of one electron determined by the average motion of the remaining electrons.

The analysis of the topological properties of the charge density which follows can also be applied to the density obtained by an averaging of ρ over the motions of the nuclei

$$\rho(\mathbf{r}; \bar{\mathbf{X}}) = \int d\mathbf{X} \rho(\mathbf{r}; \mathbf{X}) / \int d\mathbf{X}. \tag{1.5}$$

It is the charge density averaged over the thermal motions of the nuclei that is determined in X-ray and electron diffraction experiments of solids and solid surfaces. The charge density, whether determined experimentally or calculated from theory for a single configuration of the nuclei, is a representation of the time-independent distribution of negative charge throughout three-dimensional space and this is an alternative and equally valid interpretation of the quantity defined in eqn (1.4) or (1.5).

One can, in principle, perform what are necessarily inelastic scattering experiments which would determine the probability of finding an electron in some relatively small region of space, small relative to the dimensions of the system to which the electron is bound. This must result in the excitation or loss of the electron as a consequence of the uncertainty principle and the electronic state of the system is thus changed. Such experiments, leading to a change in state of the system under observation, are probabilistic in their outcome and ρ in this case is justifiably interpreted as a probability distribution function.

It is equally possible, however, to measure ρ in an experiment in which the electronic state of the system is not changed as the scattering is elastic and the recoil is to the entire system, the whole of the crystal lattice, for example. There is no need here for a probabilistic interpretation of the results. These experiments may be interpreted as providing a measurement of a spatial distribution of charge. The properties determined directly by the charge density, such as the electronic moments of a system, are obtained by averaging the corresponding operator over the charge density, an operation equivalent to the assumption that ρ provides a description of the static distribution of electronic charge throughout three-dimensional space. This is the interpretation given to $\rho(\mathbf{r}; \mathbf{X})$ or $\rho(\mathbf{r}; \bar{\mathbf{X}})$ in this work. Schrödinger (1926) defined the 'electric density' in his fourth paper as a quantity which, aside from a summation over the spin coordinates, is identical with the definition of the electronic charge density given in eqn (1.3). It is interesting to note that Schrödinger regarded $\psi^*\psi$ as providing a description of the spreading out of electronic charge throughout space. To simplify notation, the symbol $\rho(\mathbf{r})$ will normally be used for the charge density, it being understood from the context that it denotes a charge distribution for a fixed or averaged nuclear configuration.

The charge density $\rho(\mathbf{r})$ is the fundamental property measured in a coherent X-ray scattering experiment, as is commonly employed in the determination of a crystal structure. This follows since the operator which describes the interactions of X-rays with the electrons is, to a very good approximation (the Born approximation), given by a sum of one-electron interactions

$$\sum_i \exp\left[i(\mathbf{k} - \mathbf{k}_0)\cdot\mathbf{r}_i\right] \qquad (1.6)$$

where $\mathbf{k}_0$ and $\mathbf{k}$ are the wave vectors of the X-rays incident upon and scattered by the electron at $\mathbf{r}_i$. Because of its one-electron nature, the averaging of this operator over $\psi^*\psi$ reduces to N times the average of the operator for a single electron over the charge density (eqn (1.4)). This procedure yields the so-called X-ray scattering factor

$$f(s) = \int \rho(\mathbf{r}) \exp\left[i(\mathbf{k} - \mathbf{k}_0)\cdot\mathbf{r}\right] d\tau \qquad (1.7)$$

where $s = 4\pi \sin\theta/\lambda$. Experimentally, one measures the ratio of the intensities of the scattered to the incident radiation of wavelength λ as a function of the scattering angle θ. Such a measurement is proportional to f^2. Thus, from measurements of scattered X-ray intensities, one may obtain the scattering factors and, by their Fourier transform, the charge density, $\rho(\mathbf{r})$. In X-ray scattering experiments on crystals, which are the most common sort, the diffraction pattern is determined by the repeating unit cell, characteristic of the crystal, and by the particular crystallographic plane (as denoted by the Miller indices h, k, l) from which the scattering occurs. Thus for crystals one defines a structure factor $F(hkl)$ which is expressed as

$$F(hkl) = \sum_j f_j \exp(2\pi i\mathbf{h}\cdot\mathbf{r}_j) \qquad (1.8)$$

where f_j is the atomic scattering factor of the jth atom in the unit cell and $\mathbf{r}_j$ is its position. The components of the vector $\mathbf{h}$ are h/a, k/b, and l/c where a, b, and c denote the dimensions of the unit cell. The charge distribution at a point $\mathbf{r}$ within a unit cell is then expressed as

$$\rho(\mathbf{r}) = V^{-1}\sum_h\sum_k\sum_l F(hkl)\exp(-2\pi i\mathbf{h}\cdot\mathbf{r}) \qquad (1.9)$$

where V is the volume of the unit cell. Experimentally, the maxima in $\rho(\mathbf{r})$ are found to occur at the positions of the atomic nuclei and eqn (1.9) can be used to determine the arrangement of the atoms in a crystal. The mathematical description of a crystal structure is, therefore, given by a Fourier series which represents the electron density distribution. The techniques needed to overcome associated problems, such as thermal smearing of the nuclear positions, are being constantly refined and eqn (1.9) is being increasingly used to determine the charge density $\rho(\mathbf{r})$ itself from experimental X-ray scattering results; see, for example, Kappkhan *et al.* (1989). Stewart (1979) has given relations that allow for the direct mapping of electrostatic properties of the charge density, together with its gradient vector field and its associated Laplacian density, directly from the experimental X-ray structure factors.

Few chemical observations are dependent upon a nucleus possessing a small but finite radius and, for questions of molecular structure and chemical reactivity, nuclei may be regarded as point charges. Thus, the model of matter employed here is one wherein the distribution of negative charge is measurably finite over a relatively large region of space and is most highly concentrated at the positions of the point-like nuclei which are imbedded in it.

Figure 1.1 is a representation of a number of molecules in terms of an outer envelope of their charge distributions. Also indicated are the intersections of the atomic boundaries as determined by the theory of atoms in molecules with this envelope. These representations look familiar, resembling models presently employed to aid in the visualization of molecules. They are also similar to the pictures of atoms obtained experimentally by the scattering of electrons in super microscopes or from the scanning tunnelling electron microscope (Fig. 1.2). It is the distribution of charge that scatters the X-rays or electrons in these experiments and it is the distribution of charge that determines the form of matter in real space. Theory enables us to identify atoms in terms of the morphology of the charge distribution. The functional groups of chemistry, as represented by the methyl, methylene, and carbonyl groups shown in Fig. 1.1, are objects in real space and their properties are predicted by quantum mechanics.

If the state function is approximated by a single Slater determinant, an antisymmetrized sum of products of one-electron spin-orbital states ϕ_i, the expression for the charge density is given by a sum over the products of the one-electron states

$$\rho(\mathbf{r}) = \sum_i \phi_i^*(\mathbf{r})\phi_i(\mathbf{r}). \tag{1.10}$$

If the state function is determined beyond the one-electron approximation through some form of configuration interaction, an expression corresponding to eqn (1.10) still obtains, the spin-orbitals being replaced by a set of one-

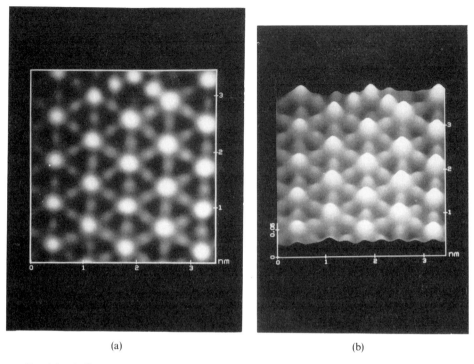

(a)   (b)

FIG. 1.2. Iodine atoms absorbed on the (111) face of platinum. The data have been filtered in order to remove spatial frequencies below 0.25 nm. Image B is a three-dimensional, height shaded, surface plot of the same image data shown in A.
(Courtesy of B. C. Schardt, from Science, (1989), **243**, 1050. Copyright 1989 by the AAAS.)

electron functions which diagonalize the one-electron density matrix, the so-called natural orbitals η_i,

$$\rho(\mathbf{r}) = \sum_i \lambda_i \eta_i^*(\mathbf{r})\eta_i(\mathbf{r}) = \sum_i \lambda_i |\eta_i(\mathbf{r})|^2 \qquad (1.11)$$

where the natural orbital occupation numbers λ_i possess values between zero and one (Löwdin 1955).

There is necessarily a great reduction in the amount of accessible information in passing from the state function ψ, a vector in the infinite-dimensional Hilbert space, via eqn (1.4) to the charge density, a distribution function in real space. However, from a purely operational point of view there is too much information in the state function, some of it being redundant because of the indistinguishability of the electrons and because of the symmetry of their interactions, and some of it being unnecessary as a result of the two-body nature of the Coulombic interaction. In choosing to describe a system in terms of ρ rather than ψ, one is not just discarding physically irrelevant information. Instead one obtains a description of the system in real space and through this, as demonstrated in the following chapters, a basis for the definition of structure in this space.

E1.1 Density matrices

Not all properties of a system in a given state, notably its energies, total, kinetic, and potential, are determined by the charge density, at least not in an operational form—density functional theory not withstanding. All these properties are, however, determined by the one-electron density matrix or one-matrix $\Gamma^{(1)}(\mathbf{r}, \mathbf{r}')$, which differs from the charge density only in that one saves separately the information regarding the coordinates of one electron from both ψ^* and ψ. Thus,

$$\Gamma^{(1)}(\mathbf{r}, \mathbf{r}') = N \int d\tau' \psi^*(\mathbf{x}_1, \mathbf{x}_2, \ldots \mathbf{x}_N) \psi(\mathbf{x}_1', \mathbf{x}_2, \ldots \mathbf{x}_N). \tag{E1.1}$$

Expressed in terms of the natural orbitals η_i,

$$\Gamma^{(1)}(\mathbf{r}, \mathbf{r}') = \sum_i \lambda_i \eta_i^*(\mathbf{r}) \eta_i(\mathbf{r}'). \tag{E1.2}$$

When $\mathbf{r} = \mathbf{r}'$, eqn (E1.2) becomes eqn (1.11); hence, $\rho(\mathbf{r})$ is said to be a diagonal element of $\Gamma^{(1)}(\mathbf{r}, \mathbf{r}')$. While eqns (1.11) and (E1.2) are formally alike, one can calculate the kinetic energy from the latter but not from the former, for only in the latter can one insert the operator between the natural orbitals and let it act separately on η_i or η_i^*. The average value of a two-electron property can be expressed in terms of the diagonal elements of the second-order density matrix $\Gamma^{(2)}(\mathbf{r}_1, \mathbf{r}_2)$. Assuming a summation over electron spins, its definition is

$$\Gamma^{(2)}(\mathbf{r}_1, \mathbf{r}_2) = \{N(N-1)/2\} \int d\tau_3 \int d\tau_4 \ldots \int d\tau_N \psi^* \psi. \tag{E1.3}$$

Just as the diagonal element of $\Gamma^{(1)} = \rho(\mathbf{r})$ gives the total density of electrons at $\mathbf{r}$ (it is normalized to N), so $\Gamma^{(2)}(\mathbf{r}_1, \mathbf{r}_2)$ gives the total density of pairs at $\mathbf{r}_1$ and $\mathbf{r}_2$ (it is normalized to the number of distinct pairs $N(N-1)/2$). The expression for $\Gamma^{(2)}$ obtained from a determinantal wave function expressed in terms of spin orbitals ϕ_i is

$$\Gamma^{(2)}(\mathbf{r}_1, \mathbf{r}_2) = \tfrac{1}{2} \sum_i \sum_j \{\phi_i^*(\mathbf{r}_1) \phi_i(\mathbf{r}_1) \phi_j^*(\mathbf{r}_2) \phi_j(\mathbf{r}_2)$$
$$- \phi_i^*(\mathbf{r}_1) \phi_i(\mathbf{r}_2) \phi_j^*(\mathbf{r}_2) \phi_j(\mathbf{r}_1)\}. \tag{E1.4}$$

For discussions of density matrices see McWeeny (1954, 1960), McWeeny and Sutcliffe (1969, pp. 76–109), and Löwdin (1955).

References

Bader, R. F. W. and Beddall, P. M. (1972). *J. Chem. Phys.* **56**, 3320.

Kappkhan, M., Tsirel'son, V. G., and Ozerov, R. P. (1989). *Doklady Phys. Chem.* **303**, 1025.

Löwdin, P.-O. (1955). *Phys. Rev.* **97**, 1474.

McWeeny, R. (1954). *Proc. r. Soc.* **A223**, 63.

McWeeny, R. (1960). *Rev. mod. Phys.* **32**, 335.

McWeeny, R. and Sutcliffe, B. T. (1969). *Methods of molecular quantum mechanics*, Academic Press, London.

Schardt, B. C., Yau, S.-H., and Rinaldi, F. (1989). *Science* **243**, 1050.

Schrödinger, E. (1926). *Ann. d. Phys.* **81**, 109. Referenced in *Collected papers on wave mechanics*, *E. Schrödinger* (3rd edn) (1982), together with his four lectures on wave mechanics. (Chelsea Publishing Co, New York) See in particular the fourth lecture, p. 205.

Stewart, R. F. (1979). *Chem. Phys. Lett.* **65**, 335.

ATOMS AND THE TOPOLOGY OF THE CHARGE DENSITY

It would be a voluntary and unnecessary abandonment of most valuable aid, if an experimentalist, who chooses to consider magnetic power as represented by lines of magnetic force, were to deny himself the use of iron filings. By their employment he may make many conditions of the power, even in complicated cases, visible to the eye at once; may trace the varying direction of the lines of force and determine the relative polarity; may observe in which direction the power is increasing or diminishing; and in complex systems may determine the neutral points or places where there is neither polarity nor power, even when they occur in the midst of powerful magnets. By their use probable results may be seen at once, and many a valuable suggestion gained for future leading experiments.

Michael Faraday (1851)

2.1 Introduction

This chapter describes the spatial properties of the electronic charge distribution and how its dominant morphological feature defines an atom in a molecule. The charge density, $\rho(\mathbf{r})$, is a physical quantity which has a definite value at each point in space. It is a scalar field defined over three-dimensional space. The topological properties of such a scalar field are conveniently summarized in terms of the number and kind of its critical points. These are points where the first derivatives of $\rho(\mathbf{r})$ vanish and they thus determine the positions of extrema in the charge density—maxima, minima, or saddles. Since the charge density is not an arbitrary field, but one whose form is dominated by the forces exerted on it by the nuclei, its topological structure is relatively simple. This structure is made more evident and more amenable to analysis through the study of the associated gradient vector field of the charge density, $\nabla\rho(\mathbf{r})$. The properties of a vector field are characterized by associating a direction as well as a magnitude with each point in space. Such fields are frequently encountered in physics. A representation of lines of force is a display of a vector field. Hence the opening quotation from Faraday, who used iron filings to obtain a display of a magnetic field in terms of its lines of force. The null points referred to by Faraday are what we have referred to as critical points, points where the magnetic field or the $\nabla\rho(\mathbf{r})$ field vanishes. While the field we wish to study is different from a magnetic field in that it is derivable from a scalar field (it is a gradient vector field), the reason for studying its 'lines of force' is the same as that given by Faraday—'to make visible to the eye at once' its important features, features which we shall relate back to the scalar field $\rho(\mathbf{r})$.

A display of $\nabla\rho(\mathbf{r})$ for a molecule will make visible to the eye, without further mathematical analysis, the definitions of its atoms and of a particular set of lines linking certain pairs of nuclei within the molecule—its molecular graph. There are four possible kinds of stable critical points in $\rho(\mathbf{r})$ and each will be associated with a particular element of structure. From this study of the general topological properties of the charge density, we shall conclude that atoms exist in molecules and that they may be linked together to form structures consisting of chains, rings, and cages (Bader and Beddall 1972; Bader et al. 1979a,b, 1981).

2.2   Topological properties of the charge density

2.2.1   The dominant form in the charge density

The form assumed by the distribution of charge in a molecular system is the physical manifestation of the forces acting within the system. Dominant among these is the attractive force exerted by the nuclei, a consequence of the localized nature of the nuclear charge. This interaction is responsible for the single most important topological property exhibited by a molecular charge distribution of a many-electron system—that, in general, $\rho(\mathbf{r}; \mathbf{X})$ exhibits local maxima only at the positions of the nuclei. This is an observation based on experimental results obtained from X-ray diffraction studies on crystals and on the results of theoretical calculations on a large number of systems. A more detailed discussion of this topic, both theoretical and experimental, is given in the end-note, Section E2.1.

Because the charge density exhibits local maxima at the nuclei, recognizable atomic forms are created within a molecular charge distribution. These forms are so dominant in determining the structure of the charge density that their individual properties form recognizable contributions to the properties of the total system. Thus, the atomic basis for the classification of chemical properties could and did evolve as the working model of chemistry before its underlying physical basis was known. The spherical atom model introduced by Bragg to model crystal structures and the more recent model of overlapping van der Waals spheres or the space-filling models, commonly used by chemists and employed to determine the shapes and sizes of molecules, are all reflections of the basic fact that the charge density is a maximum at a nucleus and decays in a nearly spherical manner away from this point.

The local maxima in a charge distribution are illustrated in Fig. 2.1. The charge density in the ethylene molecule, C_2H_4, is displayed in two ways for each of three planes: as a projection in the third dimension above the geometric plane and as a contour map, in which each contour represents a

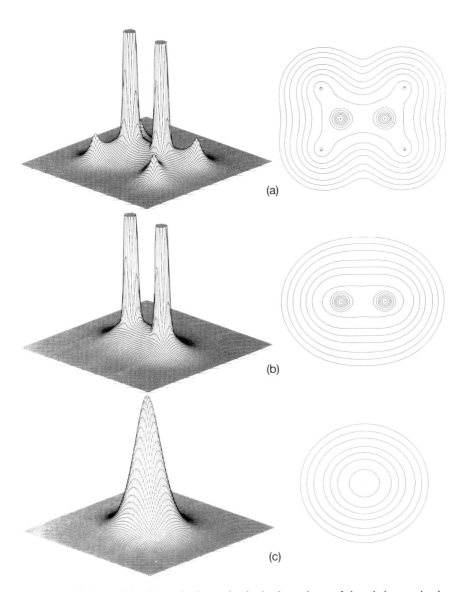

FIG. 2.1.   Displays of the electronic charge density in three planes of the ethylene molecule, C_2H_4. In the diagrams on the left, the value of the density is shown as a projection above the geometric plane; the diagrams on the right are corresponding contour plots of $\rho(\mathbf{r})$. (a) The plane containing the nuclei. Values of $\rho(\mathbf{r})$ above an arbitrarily chosen value are not shown. (b) The plane obtained by a 90° rotation about the C–C axis of the plane shown in (a). (c) The plane perpendicular to the C–C axis at its midpoint. The values of the contours increase in magnitude from the outer contour inwards with the values given in the Appendix, Table A2. The projected positions of out-of-plane nuclei are indicated by open crosses.
(From Bader *et al* (1981).)

line of constant density. The charge distribution exhibits a maximum at the position of each nucleus in Fig. 2.1(a), the plane of the nuclei. Figure 2.1(b) is for a plane perpendicular to that shown in Fig. 2.1(a), obtained by a rotation about the carbon–carbon axis and containing these nuclei. The density again exhibits maxima at the positions of the carbon nuclei. The observation of a maximum in $\rho(\mathbf{r})$ at a nuclear position is true when the distribution is viewed in any plane containing the nucleus. It is this property that is classified as a local maximum in $\rho(\mathbf{r})$.

This behaviour of $\rho(\mathbf{r})$ is to be contrasted with that displayed at the midpoint of the carbon–carbon axis. In the planes shown in Fig. 2.1(a) and (b), $\rho(\mathbf{r})$ has the appearance of a saddle at this point, while in the plane perpendicular to and bisecting the carbon–carbon axis, $\rho(\mathbf{r})$ is a maximum at this same point. In this case $\rho(\mathbf{r})$ is a maximum in one particular plane. Knowledge of $\rho(\mathbf{r})$ in one or two dimensions is insufficient to characterize its three-dimensional form. What is needed is a method of summarizing in a precise manner the principal topological features of a charge distribution. This information is provided by the curvatures of $\rho(\mathbf{r})$ at its critical points (Collard and Hall 1977; Smith *et al.* 1977).

2.2.2   Critical points and their classification

Each topological feature of $\rho(\mathbf{r})$, whether it be a maximum, a minimum, or a saddle, has associated with it a point in space called a *critical point*, where the first derivatives of $\rho(\mathbf{r})$ vanish. Thus, at such a point denoted by the position vector $\mathbf{r}_c$, $\nabla\rho(\mathbf{r}_c) = 0$ where $\nabla\rho$ denotes the operation

$$\nabla\rho = \mathbf{i}\,\partial\rho/\partial x + \mathbf{j}\,\partial\rho/\partial y + \mathbf{k}\,\partial\rho/\partial z. \tag{2.1}$$

Whether a function is a maximum or a minimum at an extremum is, of course, determined by the sign of its second derivative or curvature at this point. The second derivative of a function $f(x)$ at x (illustrated diagrammatically in Fig. 2.2) is the limiting difference between its two first derivatives or tangent lines which bracket that point

$$d^2 f(x)/dx^2 = \lim([\lim\{[f(x + \Delta x) - f(x)]/\Delta x\}$$
$$- \lim\{[f(x) - f(x - \Delta x)]/\Delta x\}]/\Delta x). \tag{2.2}$$

At a point where $f(x)$ is a minimum, the second derivative is the difference between a positive and a negative slope and is, therefore, greater than zero, while, at a maximum in $f(x)$, the second derivative is the difference between a negative and a positive slope and the result is a value less than zero. For values of x lying between such extrema, both slopes are either positive or negative and the curvature can be of either sign depending upon whether x is in the region of a maximum or a minimum. The second

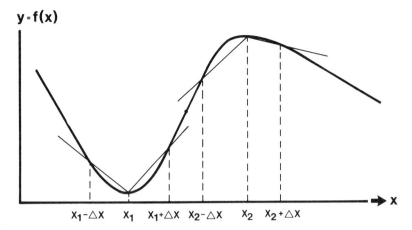

FIG. 2.2. Definition of curvature as the limiting difference ($\Delta x \to 0$) in the tangent lines which bracket a given point; at x_1 where $f(x)$ is a minimum and the curvature is positive and at x_2 where $f(x)$ is a maximum and the curvature is negative.

derivative undergoes a change in sign at the point of inflexion where its value is zero. It is also clear from its defining equation that, when the curvature is negative at x, the value of $f(x)$ is greater than the average of its values at the neighbouring points $x + \mathrm{d}x$ and $x - \mathrm{d}x$, with the reverse being true when the curvature is positive. These considerations carry over into three dimensions.

It is clear from Fig. 2.1 that the critical point at the centre of the ethylene molecule between the two carbon nuclei has one positive curvature, along the internuclear axis, and a negative curvature along each of the two perpendicular axes. Thus the central saddles in the charge density appearing in Fig. 2.1(a), (b) exhibit the positive curvature along the carbon–carbon axis and one each of the two negative curvatures, while the maximum shown in Fig. 2.1(c) exhibits just the two negative curvatures.

In general, for an arbitrary choice of coordinate axes, one will encounter nine second derivatives of the form $\partial^2 \rho / \partial x \partial y$ in the determination of the curvatures of ρ at a point in space. Their ordered 3×3 array is called the *Hessian matrix* of the charge density, or simply, the Hessian of ρ. This is a real, symmetric matrix and as such it can be diagonalized. This is an eigenvalue problem and its solution corresponds to finding a rotation of the coordinate axes to a new set such that all of the off-diagonal elements—the mixed second derivatives of ρ—vanish. This general problem is ubiquitous throughout chemistry and physics and its solution is outlined in Section E2.2 for the reader unfamiliar with it. The new coordinate axes are called the principal axes of curvature because the magnitudes of the three second derivatives of ρ calculated with respect to these axes are extremized. The principal axes will correspond to symmetry axes if the critical point is at the origin of such a set of axes, as it is for the central critical point in ethylene discussed above. In the case of symmetrically equivalent axes, the corresponding curvatures are equal and any linear combination of the degenerate set of axes will serve as a principal axis of curvature. The trace of the Hessian matrix, the sum of its diagonal elements, is invariant to a rotation of the coordinate system. Thus, the value of the quantity $\nabla^2\rho$, called the Laplacian of ρ,

$$\nabla^2\rho = \mathbf{V}\cdot\mathbf{V}\rho = \partial^2\rho/\partial x^2 + \partial^2\rho/\partial y^2 + \partial^2\rho/\partial z^2, \qquad (2.3)$$

is invariant to the choice of coordinate axes. The principal axes and their corresponding curvatures at a critical point in ρ are obtained as the eigenvectors and corresponding eigenvalues in the diagonalization of the Hessian matrix of $\rho(\mathbf{r}_c)$. Thus the pairs of names 'curvature and eigenvalue' and 'axes of curvature and eigenvectors' can be used interchangeably in describing the properties of a critical point in ρ.

While all of the eigenvalues of the Hessian matrix of ρ at a critical point are real, they may equal zero. The *rank* of a critical point, denoted by ω, is equal to the number of non-zero eigenvalues or non-zero curvatures of ρ at the critical point. The *signature*, denoted by σ, is simply the algebraic sum of the signs of the eigenvalues, i.e. of the signs of the curvatures of ρ at the critical point. The critical point is labelled by giving the duo of values (ω, σ). Thus the central critical point in ethylene with three non-zero curvatures, one positive and two negative, is a $(3, -1)$ critical point.

With relatively few exceptions, the critical points of charge distributions for molecules at or in the neighbourhood of energetically stable geometrical configurations of the nuclei are all of rank three. The near ubiquitous occurrence of critical points with $\omega = 3$ in such cases is another general observation regarding the topological behaviour of molecular charge distributions. It is in terms of the properties of critical points with $\omega = 3$ that the elements of molecular structure are defined. A critical point with $\omega < 3$, i.e. with at least one zero curvature, is said to be degenerate. Such a critical point is unstable in the sense that a small change in the charge density, as caused by a displacement of the nuclei, causes it to either vanish or to bifurcate into a number of non-degenerate or stable $(\omega = 3)$ critical points. Since structure is generic in the sense that a given structure or arrangement of bonds persists over a range of nuclear configurations, the observed limited occurrence of degenerate critical points is not surprising. One correctly anticipates that the appearance of a degenerate critical point in a molecular charge distribution denotes the onset of structural change, the subject of Chapter 3.

There are just four possible signature values for critical points of rank three.

$(3, -3)$ All curvatures are negative and ρ is a local maximum at $\mathbf{r}_c$.

(3, − 1)   Two curvatures are negative and ρ is a maximum at $\mathbf{r}_c$ in the
               plane defined by their corresponding axes. ρ is a minimum at
               $\mathbf{r}_c$ along the third axis which is perpendicular to this plane.

(3, + 1)   Two curvatures are positive and ρ is a minimum at $\mathbf{r}_c$ in the
               plane defined by their corresponding axes. ρ is a maximum at
               $\mathbf{r}_c$ along the third axis which is perpendicular to this plane.

(3, + 3)   All curvatures are positive and ρ is a local minimum at $\mathbf{r}_c$.

2.2.3   Critical points of molecular charge distributions

The Coulombic potential becomes infinitely negative when an electron and a nucleus coalesce and, because of this, the state function for an atom or molecule must exhibit a cusp at a nuclear position. That is, as shown by Kato (1957), the first derivative of the function is discontinuous at the position of a nucleus. Thus, while the charge density is a maximum at the position of a nucleus, this point is not a true critical point because $\nabla\rho$, like $\nabla\psi$, is discontinuous there. However, as discussed in Section E2.1, this is not a problem of practical import and the nuclear positions behave topologically as do (3, − 3) critical points in the charge distribution and hereafter they will be referred to as such.

All of the saddle points shown in Fig. 2.1(a) are (3, − 1) critical points. A (3, − 1) critical point is found between every pair of nuclei which are considered to be linked by a chemical bond in the ethylene molecule. Figure 2.3 shows representations of the charge distribution, similar to those shown in Fig. 2.1, for the three symmetry planes of the diborane molecule. While the topology of the charge distribution for the plane of the four terminal hydrogen nuclei and the two boron nuclei shown in Fig. 2.3(a) is indistinguishable from that shown in Fig. 2.1(a) for ethylene, the central critical point in the former molecule is not a (3, − 1) critical point. This is made clear from Fig. 2.3(b) which shows the charge density in the plane of the bridging hydrogens obtained by a 90° rotation about the boron–boron axis. The charge density is a minimum at the central point in this plane showing that it is a (3, + 1) critical point in the charge density. The final view of this critical point is shown in Fig. 2.3(c), in the plane perpendicular to the boron–boron axis and containing the bridging protons, where it again appears as a saddle in ρ. The axis of the single negative curvature of this critical point is perpendicular to the plane shown in Fig. 2.3(b). Thus Fig. 2.3(a), (c) exhibits this one negative and one each of the two positive curvatures and ρ has the appearance of a saddle point in these planes. While (3, − 1) and (3, + 1) critical points can both appear as (2, 0) critical points when viewed in specific planes where every critical point is of rank two, their full three-dimensional behaviour is quite different. The other critical points,

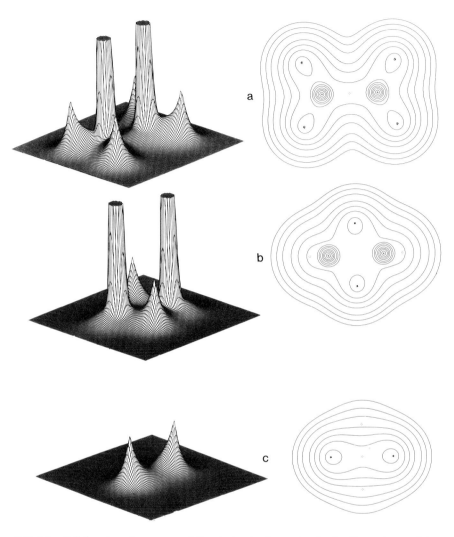

FIG. 2.3.   Relief and contour maps of the electronic charge density for three planes of the diborane molecule, B_2H_6. (a) The plane containing the boron nuclei and the four terminal protons. (b) The plane obtained by a 90° rotation about the B–B axis of the plane shown in (a). (c) The plane perpendicular to the B–B axis at its midpoint and containing the bridging hydrogen atoms. The axis linking the two bridging hydrogens has been rotated by 90° relative to this same axis in (b). The projected positions of out-of-plane nuclei are indicated by open crosses.

appearing as saddles connecting neighbouring nuclei in Fig. 2.3(a), (b), are indeed $(3, -1)$ critical points in agreement with the chemical structure usually assigned to this molecule, i.e. terminal BH_2 groups linked by two bridging protons. One notes for future reference that the minimum of the

(3, + 1) central critical point in Fig. 2.3(b) is bounded by a ring of four (3, − 1) critical points which link the bridging protons to the boron nuclei.

The fourth and final kind of stable critical point is illustrated in Fig. 2.4 which gives a representation of the charge distribution in the tetrahedrane molecule, C_4H_4. The chemical structure assigned to this molecule is that of a cage. The critical point in the centre of the molecule appears as a minimum in Fig. 2.4 and, because of the symmetry possessed by this molecule, this critical point will have the same appearance when viewed in any plane. It is a (3, + 3) critical point and the charge density is a local minimum at the centre of the cage structure. Figure 2.4 is interesting as it gives two-dimensional views of all four kinds of stable critical points. In terms of the structural properties associated with the critical points illustrated above, the reader can satisfy herself or himself that the (2, 0) saddles between the carbon nuclei and between each carbon and its neighbouring hydrogen are (3, − 1) critical points, that the (2, − 2) maximum is in the unique plane of the (3, − 1) critical point between the two out-of-plane carbon nuclei, and that the remaining two (2, 0) saddles are (3, + 1) critical points in the triangular faces formed by the carbon nuclei joined by (3, − 1) critical points.

This discussion has shown that the principal topological features of a charge distribution can be summarized using the rank and signature classification scheme of its critical points. It has further demonstrated the existence

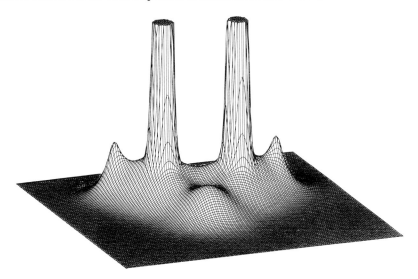

FIG. 2.4.   Relief map of the electronic charge density in the tetrahedrane molecule, C_4H_4. The plane shown is a σ_d symmetry plane and contains two carbon nuclei and their associated protons. The charge density at the central critical point is a local minimum with a value of 0.165 au. The two-dimensional maximum in the foreground is the (2, − 2) maximum in ρ in the interatomic surface of the out-of-plane carbon nuclei. The value of ρ at this point is 0.246 au. A contour map of the charge density in the same plane is shown in Fig. 2.10.

of a connection between the number and kind of critical points appearing in a charge distribution and its conventional chemical structure. The next stage of the development is to show that these qualitative associations of topological features of ρ with elements of molecular structure can be replaced with a complete theory, one which recovers all of the elements of structure in a manner that is totally independent of any information other than that contained in the charge density. The topology of the charge density will define molecular structure.

There is in the charge density an underlying structure which is brought to the fore in its associated gradient vector field. We have observed above that the only local maxima in a many-electron charge distribution occur at the positions of the nuclei. This observation confers upon a nucleus the special role of attractor in the gradient vector field of the charge density. This identification, which is itself a reflection of the dominance of the nuclear force in determining the form of ρ, provides the basis for the definition of an atom and the associated elements of molecular structure. These observations aside, one would still proceed with a study of the gradient vector field of the charge density, for the boundary condition of a quantum subsystem is stated in terms of this field. It is this most remarkable coincidence of reasons for studying the gradient vector field—one from the quantum definition of a subsystem, the other from the quite independent demonstration that its form yields a mapping of the elements of molecular structure—that gives the theory of atoms in molecules its unified structure.

2.3   Gradient vector field of the charge density

2.3.1   Trajectories of the gradient vector field

The gradient vector field of the charge density is represented through a display of the trajectories traced out by the vector $\nabla\rho$. A trajectory of $\nabla\rho$, also called a gradient path, starting at some arbitrary point r_0 is obtained by calculating $\nabla\rho(r_0)$, moving a distance Δr away from this point in the direction indicated by the vector $\nabla\rho(r_0)$ and repeating this procedure until the path so generated terminates. This operation is the three-dimensional analogue of approximating a function $f(x)$ in terms of its tangent line at x, $f(x + \Delta x) = f(x) + (df/dx)\Delta x$, an expression which, as a consequence of the definition of a derivative, becomes exact in the limit $\Delta x \to dx$.

Figure 2.5 is a contour diagram of the charge density of the sodium chloride diatomic molecule overlaid with trajectories of $\nabla\rho$ to illustrate the following general properties.

1. Since the gradient vector of a scalar points in the direction of greatest

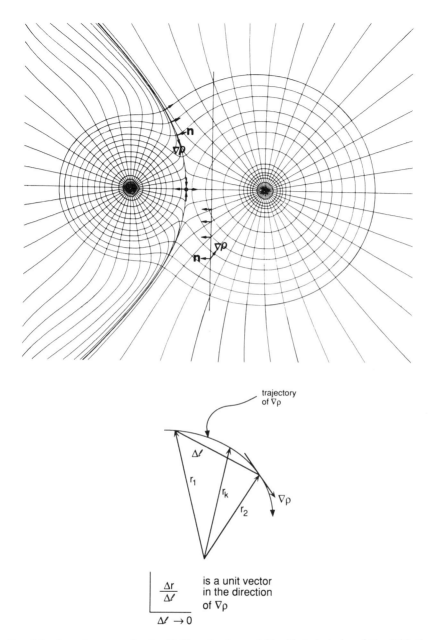

FIG. 2.5. A contour map for the NaCl molecule overlaid with trajectories of $\nabla\rho$. With the exception of the four trajectories associated with the $(3, -1)$ critical point (denoted by a dot), the trajectories originate at infinity and terminate at one of the two nuclei. Two trajectories originate at infinity and terminate at the $(3, -1)$ critical point, while two others originate at this point and terminate, one each, at the nuclei. The property of zero flux in the gradient vectors of ρ is illustrated for the interatomic surface whose intersection with this plane is given by the two trajectories which terminate at the critical point. An arbitrarily drawn surface is shown not to have this property of zero flux. The definition of the derivative $d\mathbf{r}/dl$ as the limit of $\Delta\mathbf{r}/\Delta l$ is also shown.

increase in the scalar, the trajectories of $\nabla\rho$ are perpendicular to lines of constant density—to contour lines of ρ.

2. The vector $\nabla\rho(\mathbf{r})$ is tangent to its trajectory at each point $\mathbf{r}$.

3. Every trajectory must originate or terminate at a point where $\nabla\rho(\mathbf{r})$ vanishes, i.e. at a critical point in ρ.

4. Trajectories cannot cross since $\nabla\rho(\mathbf{r})$ defines but one direction at each point $\mathbf{r}$.

All but two of the trajectories shown in the Fig. 2.5 originate at infinity where $\nabla\rho$ vanishes. The remaining two originate at the $(3, -1)$ critical point located between the nuclei and these, along with all but another two of the trajectories, terminate at one of the two nuclei. The remaining two trajectories in this plane terminate at the $(3, -1)$ critical point.

An understanding of these properties of trajectories is sufficient for a reader to proceed with the discussion of their use in the definition of structure given in the following section. The present section is concluded with the derivation of the differential equation for $\nabla\rho(\mathbf{r})$ and the corresponding integral curve which defines the points on its trajectory.

Consider the two points defined by the position vectors $\mathbf{r}_1$ and $\mathbf{r}_2$ on a gradient path and separated by the distance Δl as shown in Fig. 2.5. By taking the limit, $\Delta l \to 0$, one obtains the dimensionless derivative

$$d\mathbf{r}/dl = \lim_{\Delta l \to 0}\{(\mathbf{r}_2 - \mathbf{r}_1)/\Delta l\} = \lim_{\Delta l \to 0}\{\Delta\mathbf{r}/\Delta l\}. \tag{2.4}$$

This derivative is the unit vector tangent to the curve defined by a succession of points $\mathbf{r}_k$ at the point $\mathbf{r}$, and hence it must give the direction of the vector $\nabla\rho(\mathbf{r})$ at this point. Thus

$$d\mathbf{r}/dl = \nabla\rho(\mathbf{r})/|\nabla\rho(\mathbf{r})|. \tag{2.5}$$

What is desired is a differential equation for $\nabla\rho(\mathbf{r})$ itself. Such a result can be obtained from eqn (2.5) by defining a parameter s as a weighted measure of distance along a given path and related to the true path length l by the property that

$$\lim_{\Delta l \to 0}\{\Delta s/\Delta l\} = 1/|\nabla\rho|$$

or

$$ds = dl/|\nabla\rho|. \tag{2.6}$$

Substitution of ds for dl in eqn (2.5) yields the desired differential equation

$$d\mathbf{r}(s)/ds = \nabla\rho(\mathbf{r}(s)) \tag{2.7}$$

where the notation $\mathbf{r}(s)$ implies that a point $\mathbf{r}$ on a given trajectory is dependent upon the path parameter s.

Equation (2.7) represents three first-order differential equations and it yields unique solutions only when particular values are assigned to three constants of integration. This corresponds to fixing some initial point on a trajectory, at $s = s_1$, for example. Then every other point on the trajectory that passes through the point $\mathbf{r}(s_1)$ is obtained by integrating eqn (2.7) with the three constants of integration given by the components of $\mathbf{r}(s_1)$,

$$\mathbf{r}(s) = \mathbf{r}(s_1) + \int_{s_1}^{s} \mathbf{\nabla}\rho(\mathbf{r}(t))\mathrm{d}t. \qquad (2.8)$$

A trajectory of the gradient vector field of $\rho(\mathbf{r})$ is, therefore, a parametrized integral curve, a solution curve, of the differential equation for $\mathbf{\nabla}\rho(\mathbf{r})$. By fixing a point on a given trajectory all other points which lie on the same trajectory can be obtained by solving eqn (2.8). It is clear from eqn (2.6) that, since $|\mathbf{\nabla}\rho|$ becomes increasingly smaller as a critical point is approached, the value of the path parameter s must, correspondingly, become increasingly larger, equalling infinity at the critical point where $\mathbf{\nabla}\rho$ vanishes. Whether the actual path is of finite or infinite length, the path parameter varies between the limits $-\infty$, the origin of the path, and $+\infty$, its terminus. Mathematicians refer to the origin of a gradient path as its α-limit set, and to its terminus as its ω-limit set.

2.3.2 Phase portraits of the gradient vector field

First-order differential equations, such as eqn (2.7) for $\mathbf{\nabla}\rho$, are common throughout classical physics. In the nineteenth century, the French mathematician Poincaré suggested that one could discern the qualitative behaviour of the solutions to such equations in the pattern of the trajectories representing the vector field defined by the differential equation, its so-called phase portrait. We are also interested in the global arrangement of the gradient paths rather than in their individual behaviour. Our primary concern is to look for and to describe any regularity or structure that the gradient paths exhibit. Since the critical points in a charge distribution serve as the origins and/or termini of the gradient paths, they emerge as fundamental in the study of the topology of the charge density, and the elements of structure are to be found in the properties exhibited by the phase portraits of the gradient vector field in the vicinities of the four stable critical points.

An eigenvalue and its associated eigenvector of the Hessian of ρ (a principal curvature and its associated axis) at a critical point define a one-dimensional system. If the eigenvalue or curvature is negative, then ρ is a maximum at the critical point on this axis and a gradient vector will approach and terminate at this point from both its left- and right-hand side as illustrated in Fig. 2.6 for the case $(1, -1)$, a system of rank 1 and signature

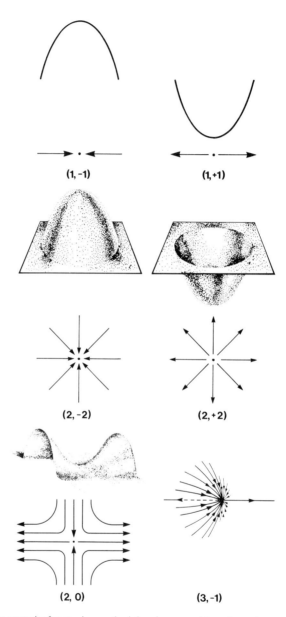

FIG. 2.6.   Phase portraits for maxima and minima in one and two dimensions and for a saddle. The final diagram is a display of the trajectories, which terminate at a (3, − 1) critical point and define the interatomic surface, and of the two trajectories which originate at this point and define the bond path.

$- 1$. If the eigenvalue is positive, then ρ is a minimum at the critical point on this axis and two gradient vectors will originate at this point as illustrated for the case $(1, + 1)$. In two dimensions, if both eigenvalues are negative, ρ is a maximum at the critical point and all trajectories of $\nabla\rho$ will terminate at the critical point as illustrated for the case $(2, - 2)$. This set of trajectories is defined by all possible linear combinations of the two associated eigenvectors which span a two-dimensional space. Similarly, if both eigenvalues are positive and ρ is a minimum at the critical point, all trajectories will originate at the critical point and again define a surface as illustrated for the case $(2, + 2)$. A more interesting situation is obtained when the eigenvalues are of opposite sign (the signature is zero) and the charge density in a plane has the form of a saddle as illustrated for the case $(2, 0)$. In this situation the two trajectories associated with the axis of the negative curvature terminate at the critical point while the two associated with the positive curvature originate there. The trajectories formed by linear combinations of the two associated eigenvectors neither terminate nor originate at the critical point but instead avoid this point as indicated for the case labelled $(2, 0)$.

In three dimensions, the pairs of eigenvectors associated with the two negative or two positive eigenvalues of a $(3, - 1)$ or $(3, + 1)$ critical point, respectively, will again define a surface. Unlike the two-dimensional examples discussed above, these surfaces will be planar only if the critical point $\mathbf{r}_c$ lies in a symmetry plane. If not, the surface will be planar only in the immediate neighbourhood of $\mathbf{r}_c$ and will, in general, be curved beyond this region but still defined by the unique set of trajectories which terminate at a $(3, - 1)$ critical point or originate at a $(3, + 1)$ critical point. The final diagram in Fig. 2.6 is a display of the three-dimensional form of the phase portrait of a $(3, - 1)$ critical point. The unique pair of trajectories associated with the single positive eigenvalue originate at the critical point, as in the $(1, + 1)$ case illustrated above. The set of trajectories defined by linear combinations of the pair of eigenvectors associated with the two negative eigenvalues terminate at the critical point and define a surface. The charge density is a maximum in the surface at the critical point and a minimum at this same point along the perpendicular axis. The behaviour of the charge density at a $(3, + 1)$ critical point is just the opposite of this and its phase portrait is obtained by reversing all of the arrows shown for the $(3, - 1)$ case.

An analytical expression for a trajectory in the vicinity of a critical point is given in Section E2.2. While this expression is necessary for numerical work, a knowledge of the phase portraits given above is all that is required for the definition of the elements of molecular structure which follows.

2.4 Elements of molecular structure

2.4.1 Equivalence of the topological and quantum definitions
of an atom

The gradient vector field of the charge density in the plane containing the
nuclei of the ethylene molecule is illustrated in Fig. 2.7. A $(3, -3)$ critical
point, such as occurs at each of the nuclear positions, serves as the terminus,
the ω-limit set, of all the paths starting from and contained in some
neighbourhood of the critical point. A $(3, -3)$ critical point exhibits the
property which defines a point attractor in the gradient vector field of the
charge distribution: there exists an open neighbourhood B of the attractor
which is invariant to the flow of $\nabla\rho$ such that any trajectory originating in B
terminates at the attractor. The largest neighbourhood satisfying these
conditions is called the basin of the attractor.

Since $(3, -3)$ critical points in a many-electron charge distribution are
generally found only at the positions of the nuclei, the nuclei act as the
attractors of the gradient vector field of $\rho(\mathbf{r}; \mathbf{X})$. The result of this identification is that the space of a molecular charge distribution, real space, is
partitioned into disjoint regions, the basins, each of which contains one point
attractor or nucleus. This fundamental topological property of a molecular
charge distribution is illustrated in Fig. 2.7(a), which depicts only those
gradient paths of the charge density which terminate at each of the nuclear
attractors in the molecule. While Fig. 2.7(a) illustrates this property for only
one plane, it is to be emphasized that, because ρ is a local maximum at a
nucleus, a $(3, -3)$ critical point, the basin of an attractor is a region of three-
dimensional space and the partitioning so clearly indicated in Fig. 2.7 extends
throughout all of space. *An atom, free or bound, is defined as the union of an
attractor and its associated basin.*

Alternatively, an atom can be defined in terms of its boundary. The basin of
the single nuclear attractor in an isolated atom covers the entire three-
dimensional space R^3. For an atom in a molecule the atomic basin is an open
subset of R^3. It is separated from neighbouring atoms by interatomic
surfaces. The existence of an interatomic surface S_{AB} denotes the presence of a
$(3, -1)$ critical point between neighbouring nuclei A and B. The presence of
such a critical point between certain pairs of nuclei was noted above as being
a general topological property of molecular charge distributions. Their
presence now appears as providing the boundaries between the basins of
neighbouring atoms. As discussed above and illustrated in Fig. 2.6, the
trajectories which terminate at a $(3, -1)$ critical point define a surface, the
interatomic surface S_{AB}. The reader will recall Fig. 2.1(c) showing that the
charge density is a maximum at a $(3, -1)$ critical point in this surface. Figure
2.7(b) shows, in addition to the trajectories that terminate at each of the nuclei

in ethylene, the trajectories associated with each of the $(3, -1)$ critical points. Two trajectories terminate at each such critical point in the plane of the diagram. They denote the intersection of the interatomic surfaces with this plane. It should be borne in mind that each such pair of trajectories are but two of an infinity of gradient paths, all of which terminate at a $(3, -1)$ critical point and define a surface in three dimensions as illustrated in Fig. 2.6. The atomic surface S_A of atom A is defined as the boundary of its basin. In set-theoretic language $S_A \supseteq \cup_B S_{AB}$. (A summary of symbols used in set theory is given in Section E3.1.) Generally, this boundary comprises the union of a number of interatomic surfaces, separating two neighbouring basins, and some portions which may be infinitely distant from the attractor. The atomic surface of a carbon atom in ethylene, as seen in Fig. 2.7 for example, consists of three interatomic surfaces, two with hydrogens and another with a carbon atom.

If the topological property which defines an atom is also one of physical significance, then it should be possible to obtain from quantum mechanics an equivalent mechanical definition. As demonstrated in Chapters 5 and 8, this can be accomplished through a generalization of the quantum action principle to obtain a statement of this principle which applies equally to the total system or to an atom within the system. The result is a *single* variational principle which defines the observables, their equations of motion, and their average values for the total system or for an atom within the system.

The generalization of the action principle to a subsystem of some total system is unique, as it applies only to a region that satisfies a particular constraint on the variation of its action integral. The constraint requires that the subsystem be bounded by a surface of zero flux in the gradient vectors of the charge density, i.e.

$$\nabla \rho(\mathbf{r}) \cdot \mathbf{n}(\mathbf{r}) = 0 \qquad \text{for all points on the surface } S(\mathbf{r}). \qquad (2.9)$$

In order for the scalar product of $\mathbf{n}$, the vector normal to the surface (see Fig. 2.5), with $\nabla \rho$ to vanish, it is necessary that the atomic surface not be crossed by any trajectories of $\nabla \rho$ and as such it is referred to as a *zero-flux surface*. The state function ψ and $\nabla \psi \cdot \mathbf{n}$, where the gradient is taken with respect to the coordinates of any one of the electrons, vanish on the boundary of a bound system at infinity. Thus, ρ and $\nabla \rho$ vanish there as well and a total isolated system is also bounded by a surface satisfying eqn (2.9). Since the generalized statement of the action principle applies to any region bounded by such a surface, the zero-flux surface condition places the description of the total system and the atoms which comprise it on an equal footing.

Because of the dominant topological property exhibited by a molecular charge distribution—that it exhibits local maxima at the positions of the nuclei—the imposition of the quantum boundary condition of zero flux leads directly to the topological definition of an atom. Indeed, the interatomic

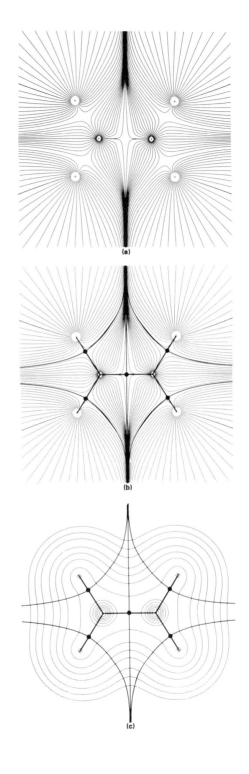

(a)

(b)

(c)

surfaces, along with the surfaces found at infinity, are the only closed surfaces of R^3 which satisfy the zero-flux surface condition of eqn (2.9). This is a natural result of the property of an atomic basin as shown in Fig. 2.7(a)—all the trajectories in the vicinity of a given nucleus terminate at that nucleus and no trajectories cross from one basin to another. Since trajectories of $\mathbf{V}\rho$ never cross, the zero-flux surface condition follows directly from the definition of an interatomic surface in terms of the set of trajectories which terminate at a $(3, -1)$ critical point. In terms of this same definition, it follows that the vector $\mathbf{V}\rho(\mathbf{r})$ will be tangent to the surface $S(\mathbf{r})$ of an atom at every point $\mathbf{r}$.

It is clear from Fig. 2.5 that a surface, arbitrarily drawn through a charge distribution, will be crossed by gradient vectors of ρ and will not satisfy the zero-flux condition. Any surface including a nuclear position coordinate is also excluded as $\mathbf{V}\rho$ is not defined there and again the zero-flux condition, eqn (2.9), will be violated.

The coincidence of the topological and quantum definitions of an atom means that the topological atom is an open quantum subsystem, free to exchange charge and momentum with its environment across boundaries which are defined in real space and which, in general, change with time. It should be emphasized that the zero-flux surface condition is universal—it applies equally to an isolated atom or to an atom bound in a molecule. The approach of two initially free atoms causes a portion of their surfaces to be shared in the creation of an interatomic surface. Atomic surfaces undergo continuous deformations as atoms move relative to one another. They are, however, not destroyed as atoms separate.

A linked grouping of neighbouring atoms (a concept defined more fully in the following Section) as well as individual atoms are bounded by a zero-flux surface. Such a surface may be used to define a Wigner–Seitz cell in a solid, a solute in a solution, or a molecule in a molecular crystal. The zero-flux surface, eqn (2.9), is the natural boundary condition for a system defined in real space and such a surface can always be used to define the physically

---

FIG. 2.7.   Maps of the gradient vector field of the charge density for the plane containing the nuclei of the ethylene molecule. Each line represents a trajectory of $\mathbf{V}\rho(\mathbf{r})$. In (a) only those trajectories which terminate at the positions of the nuclei are shown. Each trajectory is arbitrarily terminated at the surface of a small circle about a nucleus. The set of trajectories which terminate at a given nucleus or attractor define the basin of the attractor. (b) The same as (a), but including the trajectories which terminate and originate at the $(3, -1)$ critical points in the charge distribution. The position of a $(3, -1)$ critical point in these and succeeding diagrams is denoted by a full circle. The pair of trajectories which, in this plane, terminate at each $(3, -1)$ critical point mark the intersection of an interatomic surface with the plane of the figure. The gradient paths which originate at the $(3, -1)$ critical points and define the bond paths are shown by the heavy lines. (c) A superposition of the trajectories associated with the $(3, -1)$ critical points on a contour map of the charge density. These trajectories define the boundaries of the atoms and the molecular graph.

(From Bader et al (1981).)

relevant pieces of a total system whether they be in intimate contact or essentially isolated. There is, of course, no such thing as a truly isolated system. Each is bounded by a zero-flux surface, and systems so defined differ only in the degree of their interaction with neighbouring systems. One should also anticipate at this point, that the average properties of a subsystem and their equations of motion are determined by the same quantum mechanical expressions as apply to an 'isolated' system. The major difference between the properties of a subsystem and those of a system considered to be isolated is to be found in the fluctuation of their average values. Such fluctuations vanish for an isolated system, thereby ensuring, for example, that an integer number of electrons are present. The fluctuation in the population of an interacting subsystem is, however, not zero and, correspondingly, its average population is not necessarily an integer.

This book is primarily concerned with atoms in molecules but the more general nature of the theory regarding its ability to provide a description of the properties of any physically relevant piece of the universe and of its interaction with the remainder of the universe could prove to be of equal if not greater usefulness. Further discussion of this point is postponed until Chapter 8 where the proper quantum mechanical framework is established, particularly the equation of motion for a subsystem.

2.4.2   Chemical bonds and molecular graphs

Also shown in Fig. 2.7(b) are the pairs of gradient paths which originate at each $(3, -1)$ critical point and terminate at the neighbouring attractors. As previously discussed and illustrated in Fig. 2.6, each such pair of trajectories is defined by the eigenvector associated with the unique positive eigenvalue of a $(3, -1)$ critical point. These two unique gradient paths define a line through the charge distribution linking the neighbouring nuclei along which $\rho(\mathbf{r})$ is a maximum with respect to any neighbouring line. Such a line is found between every pair of nuclei whose atomic basins share a common inter-atomic surface and, in the general case, it is referred to as an *atomic interaction line* (Bader 1975; Runtz *et al.* 1977; Bader and Essen 1984).

The existence of a $(3, -1)$ critical point and its associated atomic inter-action line indicates that electronic charge density is accumulated between the nuclei that are so linked. This is made clear by reference to the displays of the charge density for such a critical point, as given in Fig. 2.1 for example, particularly Fig. 2.1(c) which shows that the charge density is a maximum in an interatomic surface at the position of the critical point. This is the point where the atomic interaction line intersects the interatomic surface and charge is so accumulated between the nuclei along the length of this line. Both theory and observation concur that the accumulation of electronic charge between a pair of nuclei is a necessary condition if two atoms are to be bonded to one another. This accumulation of charge is also a sufficient condition when the forces on the nuclei are balanced and the system possesses a minimum energy equilibrium internuclear separation. Thus, the presence of an atomic interaction line in such an equilibrium geometry satisfies both the necessary and sufficient conditions that the atoms be bonded to one another. In this case the line of maximum charge density linking the nuclei is called a *bond path* and the $(3, -1)$ critical point referred to as a *bond critical point* (Bader and Essén 1984).

For a given configuration **X** of the nuclei, *a molecular graph* is defined as the union of the closures of the bond paths or atomic interaction lines. Pictorially, the molecular graph is the network of bond paths linking pairs of neighbouring nuclear attractors. The molecular graph isolates the pairwise interactions present in an assembly of atoms which dominate and characterize the properties of the system be it at equilibrium or in a state of change.

A molecular graph is the direct result of the principal topological properties of a system's charge distribution: that the local maxima, $(3, -3)$ critical points, occur at the positions of the nuclei thereby defining the atoms, and that $(3, -1)$ critical points are found to link certain, but not all pairs of nuclei in a molecule. The network of bond paths thus obtained is found to coincide with the network generated by linking together those pairs of atoms which are assumed to be bonded to one another on the basis of chemical considerations. Molecular graphs for a sampling of molecules in equilibrium geometries with widely different bonding properties are illustrated in Fig. 2.8. The existence and position of a bond or $(3, -1)$ critical point in this and other figures is indicated by a black dot. More examples of molecular graphs are given throughout the remainder of the book. The recovery of a chemical structure in terms of a property of the system's charge distribution is a most remarkable and important result. The representation of a chemical structure by an assumed network of lines has evolved through a synthesis of observations on elemental combination and models of how atoms combine, particularly models of chemical valency—a model which states that the ability of a given type of atom to form bonds, its valency, can be saturated and that valency is determined by the number of valence electrons. A great deal of chemical knowledge goes into the formulation of a chemical structure and, correspondingly, the same information is successfully and succinctly summarized by such structures. The demonstration that a molecular structure can be faithfully mapped on to a molecular graph imparts new information—that nuclei joined by a line in the structure are linked by a line through space along which electronic charge density, the glue of chemistry, is maximally accumulated. Finding the physical basis for a molecular structure also leads to a broadening of the concept that the dominant interactions between atoms, be they attractive or repulsive, have a common physical representation. This is not an entirely surprising result since the ever present

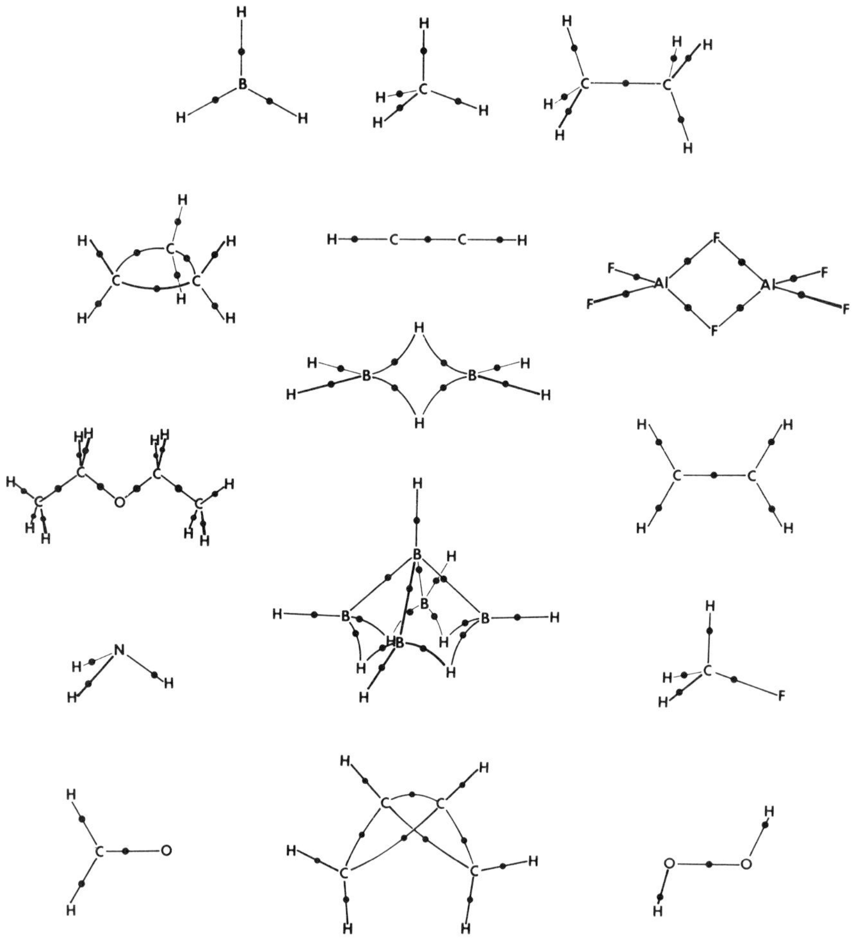

FIG. 2.8. Molecular graphs for some molecules in their equilibrium geometries. A bond critical point is denoted by a black dot.
(From Bader *et al.* (1981).)

nuclear excursions from an equilibrium separation between a pair of atoms force a sampling of these same portions of a potential surface even though the atoms are considered to be bonded to one another. It is the closely related questions of what is meant by the making and breaking of chemical bonds that lead one to consider the most important extension of the molecular structure concept. As discussed in Chapter 3, the dynamic behaviour of the molecular graphs, as caused by the relative motions of the nuclei, forms the basis for the definition of structural stability and the analytical description of the mechanisms of structural change.

It is to be stressed that a bond path is not to be understood as representing a 'bond'. The presence of a bond path linking a pair of nuclei implies that the corresponding atoms are bonded to one another. As demonstrated later, the interaction can be characterized and classified in terms of the properties of the charge density at its associated $(3, -1)$ critical point. The complete description of the interaction, however, requires the evaluation of operators over the associated interatomic surface. These are the topics of later discussions. At this point we continue with the identification of the elements of molecular structure with the topological properties of the remaining stable critical points, $(3, +1)$ and $(3, +3)$ critical points.

2.4.3   Rings and cages

The remaining critical points of rank three occur as consequences of particular geometrical arrangements of bond paths and they define the remaining elements of molecular structure—rings and cages. If the bond paths are linked so as to form a ring of bonded atoms, as in the bridging plane of the diborane molecule, for example, whose molecular graph is shown in Fig. 2.8, then a $(3, +1)$ critical point is found in the interior of the ring. As discussed above and illustrated in Fig. 2.6, the eigenvectors associated with the two positive eigenvalues of the Hessian matrix of ρ at this critical point generate an infinite set of gradient paths which originate at the critical point and define a surface, called the ring surface. This behaviour is illustrated here by the gradient paths in the bridging plane of the diborane molecule as shown in Fig. 2.9. All of the trajectories which originate at the critical point at the centre of the ring of nuclei, the $(3, +1)$ or ring critical point, terminate at the ring nuclei, but for the set of single trajectories, each of which terminates at one of the bond critical points whose bond paths form the perimeter of the ring. These bond paths are noticeably inwardly curved away from the geometrical perimeter of the ring, a behaviour characteristic of systems which are electron-deficient. The remaining eigenvector of a ring critical point, its single negative eigenvector, generates a pair of gradient paths which terminate at the critical point and define a unique axis perpendicular to the ring surface at the critical point. In diborane this axis is perpendicular to the plane shown in Fig. 2.9. It represents the intersection of the boundaries of the basins of the hydrogen and boron atoms forming the ring. *A ring*, as an element of structure, *is defined as a part of a molecular graph which bounds a ring surface.*

If the bond paths are so arranged as to enclose the interior of a molecule with ring surfaces, then a $(3, +3)$ or cage critical point is found in the interior of the resulting cage. The charge density is a local minimum at a cage critical point. The phase portrait in the vicinity of a cage critical point is shown in Fig. 2.10 for the tetrahedrane molecule. Trajectories only originate at such a critical point and terminate at nuclei, and at bond and ring critical points,

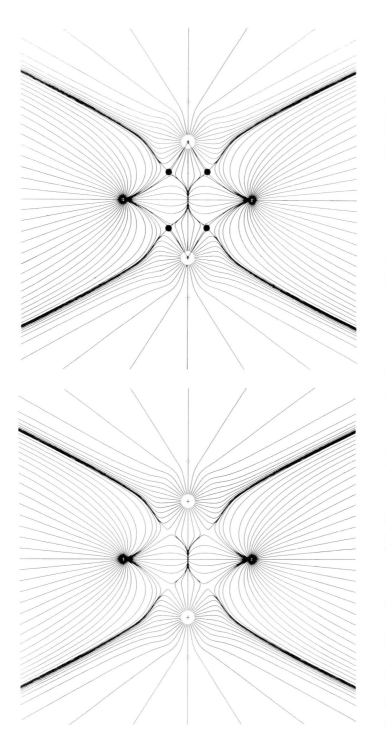

FIG. 2.9. Maps of the gradient vector field of the electronic charge density in the bridging plane of the diborane molecule (see Fig. 2.3(b)). The diagram on the left shows only those trajectories which originate either at infinity or at the central $(3, +1)$ critical point, the ring critical point. The diagram on the right shows, in addition, those trajectories which terminate and originate at the four $(3, -1)$ critical points. The former define the positions of the interatomic surfaces in the plane while the latter define the bond paths.

thereby defining a bounded region of space. *A cage*, as the final element of molecular structure, *is a part of a molecular graph which contains at least two rings, such that the union of the ring surfaces bounds a region of* R^3 *which contains a* (3, + 3) *critical point.* While it is mathematically possible for a cage to be bounded by only two ring surfaces, the minimum number found in an actual molecule so far is three, as in bicyclo[1.1.1]pentane, for example.

The representation of the gradient vector field of the tetrahedrane molecule for the plane shown in Fig. 2.10 is of some interest as all four critical points of rank three are present. As a consequence of the symmetry present in this molecule, the bond paths that are shown indicate that each carbon is linked to a single hydrogen and to three other carbons to form a cage. The carbon–carbon and carbon–hydrogen bond critical points linking the nuclei in this plane appear as saddles in the charge density (see Fig. 2.4) and as (2, 0) critical points in terms of their phase portraits. The two trajectories terminating at each such point and defining the intersection of an interatomic surface with this plane are shown, as are the two trajectories that originate at each of these points and define a bond path. The critical point at the top of the diagram is also a bond critical point, the one whose bond path links the two out-of-plane carbon nuclei. The interatomic surface for this pair of nuclei coincides with the plane of the diagram and hence this bond critical point appears as a (2, − 2) critical point, a maximum in ρ. Indeed, in this particular plane the phase portrait of this critical point is topologically equivalent to that for a nucleus. A bond critical point behaves as an attractor in this one special surface, the interatomic surface. The two remaining critical points are ring critical points and in this plane they also appear as saddles in ρ and as (2, 0) critical points in terms of their phase portraits. Of the two trajectories which originate at each of these ring critical points, one terminates at a carbon nucleus, the other at the upper bond critical point. These pairs of trajectories denote the intersection of two corresponding ring surfaces with the plane of the diagram. One notes that the ring surfaces, as well as the carbon–carbon bond paths which together bound the cage, are all outwardly curved from the geometrical lines and faces of the tetrahedral structure. This behaviour is, in general, characteristic of organic molecules with geometrical strain. Of the two trajectories that terminate at each of the ring critical points and define the unique axis perpendicular to the ring surface, one originates at the cage critical point, the other at infinity. These two ring axes, together with the two trajectories terminating at the carbon–carbon bond critical point, mark the intersections of the interatomic surfaces with the plane of the diagram. The two ring axes are themselves intersections of the boundaries of the basins of three carbon atoms.

The number and type of critical points which can coexist in a system with a finite number of nuclei are governed by the Poincaré–Hopf relationship. With the above association of each type of critical point with an element of

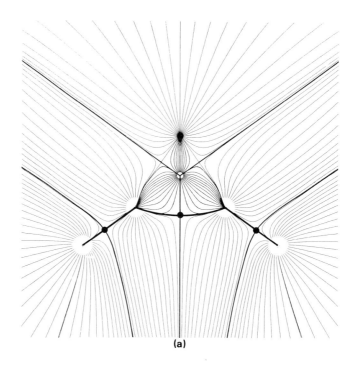

(a)

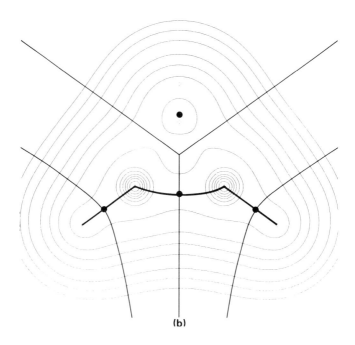

(b)

molecular structure, this relationship states that

$$n - b + r - c = 1 \tag{2.10}$$

where n is the number of nuclei, b is the number of bond paths (or atomic interaction lines), r is the number of rings, and c is the number of cages (Collard and Hall 1977). The collection of numbers (n, b, r, c) is called the *characteristic set* of the molecule.

This chapter has demonstrated that elements of the molecular structure hypothesis, atoms, and bonds and the linking together of the atoms to form chains, rings, and cages, as evidenced by the forms of the molecular graphs, follow directly from the topological properties of the electronic charge distribution. The experimental development of the structural aspects of this hypothesis thus appears as an inevitable consequence of the form and properties possessed by the charge distribution, properties which themselves are a reflection of the forces acting within the system.

The finding that the charge distribution contains the information required to define a molecular graph for every geometrical arrangement of the nuclei allows for much more than a pedestrian recovery of the notion of structure as a set of atoms linked by a network of bonds. Instead, the dynamical behaviour of the molecular graphs, as induced by a motion of the system through nuclear configuration space, is shown to be describable within the framework of a relatively new field of mathematics, one which demonstrates that the notions of structure and structural stability are inseparable. The application of these ideas to the dynamical properties of the molecular graphs defined by the charge density leads to a complete theory of structure and structural stability, one which predicts that a change in structure as caused by the making or breaking of chemical bonds is an abrupt and discontinuous process. It is this important aspect of the theory of atoms in molecules that we develop in the next chapter.

FIG. 2.10. (a) A map of the gradient vector field of the electronic charge density for the tetrahedrane molecule in a σ_d symmetry plane (see Fig. 2.4). (b) A contour map of the electronic charge density in the same plane showing the bond critical points, and overlaid with the bond paths and the intersections made with the interatomic surfaces. The diagram shows the phase portrait for the $(3, -1)$ critical point between the two out-of-plane carbon nuclei (denoted by the upper dot) in the plane of its interatomic surface where it behaves as a $(2, -2)$ critical point (all trajectories terminate at the critical point). The dot directly beneath this point denotes the $(3, -1)$ critical point between the in-plane carbon nuclei and the phase portrait in this case is for a plane containing the bond path where the bond critical point behaves as a $(2, 0)$ critical point. The same plane is illustrated for the two C–H bond critical points. The central critical point is a $(3, +3)$ or cage critical point. In this plane it behaves as does a $(2, +2)$ critical point, with all the trajectories originating there.

E2.1   Local properties of ρ and associated theorems

E2.1.1   Local maxima in ρ

With relatively few exceptions, the electronic charge density exhibits a local maximum only at the position of a nucleus in both the ground and excited states of many-electron systems, molecules, or solids. The value of the charge density at a nuclear position, a quantity denoted by $\rho(0)$, is, with the exception of a proton, many times larger than its value at any other of its extrema. The value of $\rho(0)$ at the nucleus of a free atom in the Hartree–Fock approximation is roughly proportional to the cube power of the nuclear charge, the expression $\rho(0) = 0.4798Z^{3.1027}$ au, having a mean deviation of 2.6 per cent for $1 < Z < 55$. The symbol au denotes atomic units. Atomic units for a number of properties are defined in Table A1 of the Appendix which also gives the factors for their conversion into other units. On the other hand the values of ρ at the saddle or bond critical points found between certain pairs of nuclei over the range of chemically significant internuclear separations, lie in the range $0 \leqslant \rho_b \leqslant 1.0$ au. It is the presence and dominance of the nuclear maxima in a charge distribution that imposes the atomic form on the observed properties of a system.

As mentioned in the text, the local maxima in ρ at the positions of nuclei are not true $(3, -3)$ critical points because the gradient vector of the charge density is discontinuous at the nuclear cusp that is present in both the state function and the density (Kato 1957). However, there always exists a function homeomorphic to $\rho(\mathbf{r}; \mathbf{X})$ which coincides with ρ almost everywhere and for which the nuclear positions are $(3, -3)$ critical points. In this sense, the nuclear positions behave topologically as do $(3, -3)$ critical points in the charge distribution.

The assumption that the charge distribution possesses local maxima at the positions of the nuclei is the basis for the determination of crystal structures by X-ray diffraction techniques. A crystal structure is determined by a least squares fitting of the amplitudes of calculated structure factors to the experimentally observed ones. The calculated structure factor is based on a model which assumes a Hartree–Fock spherical free-atom charge density centred on each nucleus in a unit cell and convoluted with a Debye–Waller thermal factor. By assigning the nuclear positions to the centroids of such thermally averaged atomic charge densities in the fitted structures, one in general obtains good agreement with the geometrical structures determined by neutron diffraction techniques. This agreement clearly supports the assertion that local maxima in $\rho(\mathbf{r}; \mathbf{X})$ occur at the positions of the nuclei but it does not preclude the existence of maxima, of considerably smaller magnitude, at other positions.

The principal justification for the observation that local maxima in $\rho(\mathbf{r}; \mathbf{X})$ generally occur only at the positions of the nuclei comes from extensive studies of theoretically determined charge distributions calculated from state functions close to the Hartree–Fock limit. There is well-documented evidence that molecular Hartree–Fock charge densities (charge densities obtained from single determinant SCF calculations carried essentially to the Hartree–Fock limit) are accurate to within 2–5 per cent of the total charge density at any point in space, except perhaps for very large distances from the nuclei (Cade 1972; Bader 1975; Smith *et al.* 1977). This degree of accuracy, and the correspondingly good reputation which Hartree–Fock wave functions enjoy in the prediction of related one-electron properties, are accounted for in terms of Brillouin's theorem (Brillouin 1933–34; Moller and Plesset 1934). Cade and co-workers generated a set of wave functions for diatomic molecules all close to the Hartree–Fock limit using an SCF method with a carefully selected set of Slater basis functions together with optimization of the orbital exponents (Cade and Huo 1973, 1975; Cade and Wahl 1974). McLean and Yoshimine (1967) have published wave functions of slightly lesser quality for a number of linear polyatomic molecules. The charge densities derived from these functions have been analysed using a program which locates all of their critical points and classifies them in terms of the eigenvalues of their Hessian matrices. The results of this topological analysis, which included charge densities of ground and excited states of neutral as well as of positively and negatively charged species, have been described previously (Bader *et al.* 1981). All 50 second- and third-row diatomic hydrides AH, all 83 heteronuclear diatomics AB, and all 13 linear polyatomics exhibit local maxima only at the positions of the nuclei. In all these molecules, save two excited states, a single $(3, -1)$ critical point is found on the internuclear axis between neighbouring nuclear maxima. The only exceptions to these findings were for a small number of the 47 homonuclear diatomic molecules which exhibited a small maximum in ρ at the midpoint of the internuclear axis. The values of the charge density at these exceptional maxima exceeded the values at the neighbouring minima on the internuclear axis by very small amounts, within the error of the Hartree–Fock predictions. For example, in $C_2(X^1\Sigma_g^+)$ the value of ρ at the midpoint of the internuclear axis is 0.3076 au, a value which exceeds that of the neighbouring minima by 0.0029 au, i.e. approximately 1 per cent. Thus the maxima that have been found at other than the nuclear positions are so small as to be within the limits of error placed on the calculated densities and are trivially small compared to the maxima found at the nuclear positions. In the above state of C_2, the value of ρ at a carbon nucleus is 127.323 au. The values of ρ at the $(3, -1)$ critical points for the ground state charge distributions of the diatomics are given in the Appendix, Table A3.

The topological properties of the charge distributions of a very large number of polyatomic molecules derived from basis sets of varying quality have now been analysed in many laboratories. These systems include many different examples of what are conventionally called bonded pairs of nuclei. In almost all cases the only maxima in $\rho(\mathbf{r}; \mathbf{X})$ are found at the nuclei and, between each such pair of nuclei, a $(3, -1)$ critical point is observed. Occasionally, spurious maxima, again small in magnitude, are observed for charge distributions obtained from approximate wave functions expanded in terms of small or unbalanced basis sets or basis sets without polarizing functions. These spurious maxima are not present in charge distributions derived from a calculation using a larger or improved set of basis functions. Multiple bonds and other bonds containing a diffuse density component are difficult to describe using Gaussian basis sets. Large Gaussian sets yield a small maximum in $\rho(\mathbf{r}; \mathbf{X})$ at the midpoint of the C–C axis of acetylene, for example. This spurious maximum is replaced by a proper $(3, -1)$ critical point when a singles and doubles configuration interaction calculation using the same basis set is performed. This spurious maximum in $\rho(\mathbf{r}; \mathbf{X})$ is not present in the density for acetylene or in substituted acetylenes obtained from a single determinant self-consistent field (SCF) calculation using a large Slater basis set which includes f functions.

A number of studies on the effects of Coulomb electron correlation on the charge density obtained in a single determinant SCF calculation have shown them to be relatively small. This result is not unexpected since Hartree–Fock charge distributions are, because of Brillouin's theorem, correct to second-order. It is essential that electron correlation computed by a configuration interaction (CI) method include single as well as double excitations with respect to suitable zero-order reference wave functions. Bender and Davidson (1968) have pointed out that, with the inclusion of only double excitations, the highly occupied natural spin orbitals will always be simply a transformation of the original occupied SCF orbitals. Thus double excitations, while accounting for an important fraction of the correlation energy, play only a minor role in correcting the charge density. Only the inclusion of single excitations brings about a significant mixing of the virtual orbitals with the original occupied SCF set to yield improvements in the SCF charge density. The one-electron correlation functions of Sinanoğlu correspond to the sum of single excitations and, as pointed out by Sinanoğlu and Tuan (1963), they represent the corrections to the Hartree–Fock orbitals.

A study, using such CI methods together with the generalized valence bond and multi-configuration SCF methods for introducing Coulomb correlation, demonstrates that the correlated charge distributions possess the same number and kind of critical points in both the $\rho(\mathbf{r})$ and $\nabla^2\rho(\mathbf{r})$ fields as are found for the SCF distributions (Gatti *et al.* 1988). Thus, the topology of a charge distribution and the structure it defines are unaffected by the addition of Coulomb correlation. The study covered a wide range of chemical structures, and the quantitative changes in the properties of the charge density at the critical points induced by correlation are found to be small in magnitude and to be more pronounced for shared or covalent interactions than for systems with pronounced charge transfer between the atoms. This study and earlier ones (Bader 1975; Smith *et al.* 1977) shows that, in general, a Hartree–Fock charge density overestimates, by a few per cent, the value of $\rho(\mathbf{r})$ in the neighbourhood of a bond critical point relative to that predicted by a correlated wave function. The effect is greatest for interactions with a significant shared density and a relatively small value for the curvature of ρ along the bond axis.

Excited states of one-electron atoms do exhibit non-nuclear maxima as can excited states of one-electron molecular ions, as demonstrated for H_2^+ by Bates *et al.* (1953). There is no doubt that the success of the atomic concept in the classification and prediction of chemical knowledge is a reflection of the fact that, in most systems, nuclei are the sole attractors of the charge density and the atomic form is dominant. There is, however, the possibility of observing non-nuclear maxima that could be responsible for particular physical effects. An apparent example of such behaviour has been found. The Li_2 molecule is among the homonuclear diatomics that exhibit a small relative maximum in the charge density at the bond midpoint at the Hartree–Fock level. It has now been demonstrated by Gatti *et al.* (1987) in a study of charge densities of clusters of Li atoms, including electron correlation, that the central non-nuclear maximum in the ground-state charge distribution of Li_2 is not an artefact of the single determinantal state function but is instead a property not only of this charge distribution but of all lithium clusters they investigated up to and including Li_6. The same behaviour has also been found by Cao *et al.* (1987) for clusters of sodium atoms. The value of ρ at a non-nuclear attractor is small, 0.0078 au, and exceeds that at the $(3, -1)$ critical points to which it is linked by a very small amount— $\approx 1 \times 10^{-4}$ au for the sodium clusters and only slightly greater than this for the lithium clusters. The non-nuclear attractors in ρ in these clusters behave as pseudoatoms and it is found that the metal atoms are not linked directly to one another but only indirectly through an intervening pseudoatom (Fig. 2.11). It appears that the optimum structure for planar lithium and sodium clusters corresponds to a network of linked pseudoatoms, each of which is bonded to a singly bonded metal atom and to two bridging metal atoms. The metal atoms bear substantial positive charges and the clusters correspond to positively charged metal atoms with localized charge distributions bound by an intermeshed network of negatively charged pseudoatoms. The loosely bound and delocalized electronic charge of the pseudoatoms is

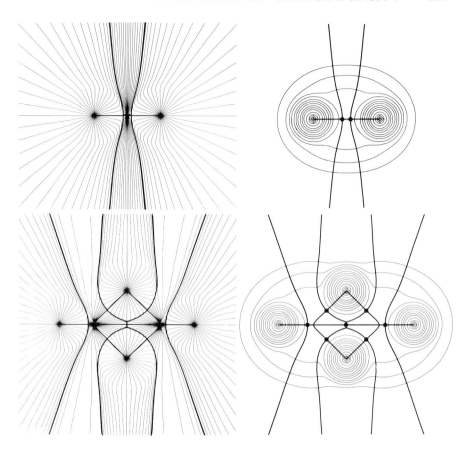

FIG. 2.11.   Maps of the gradient vector fields of the charge density for Na_2 and Na_4, displayed in terms of trajectories of $\nabla\rho$, and contour maps of the corresponding charge densities overlaid with bond paths and lines denoting the intersections of the interatomic surfaces with the plane of the diagrams. The latter trajectories of $\nabla\rho$ are denoted by heavy lines in both sets of maps. There are two bonds in Na_2 and seven bonds in Na_4. The positions of the non-nuclear attractors are denoted by stars on the gradient vector field maps while the positions of the bond critical points are denoted by dots. Each (3, + 1) or ring critical point in Na_4 is located at an intersection of three gradient paths as occurs immediately above and below the central bond critical point in the gradient vector field map.

responsible for the binding in these systems and for the conducting properties, assuming the same structure exists in their bulk metals. The properties of
the pseudoatoms are futher discussed in Section 7.4.2.

Another change in the usual topology of ρ can occur when two nuclei with
disparate charges approach very close to one another. Under these conditions the charge density will continue to exhibit the required cusp at the
position of each nucleus but, at some small separation, the local maximum at

the position of the lesser of the two charges will disappear. (This is an example of what is termed a fold catastrophe, a subject to be discussed in Chapter 3. In this case the $(3, -1)$ critical point initially between the nuclei approaches and eventually merges with the $(3, -3)$ nuclear maximum. Both critical points vanish upon their coalescence but the cusp condition at the nucleus is still obeyed.) This behaviour is not observed for energies normally encountered in chemical reactions for an atom that possesses a core density, as the value of $\rho(0)$ is then very great and extremely short internuclear separations would be required to increase the value of ρ at the intervening $(3, -1)$ critical point to where it equalled $\rho(0)$. Hydrogen, however, does not possess a core and, when in combination with a very electronegative species such as F^+, its nuclear maximum is absent (Bader *et al.* 1980). In this molecule the total charge distribution is dominated by the forces exerted by the partially unshielded F nucleus and the properties of the molecule FH^+ approach those of the corresponding united atom limit, Ne^+. This behaviour is illustrated in Fig. 2.12.

E2.1.2   Theorems concerning the properties of ρ

Hoffmann-Ostenhof and Morgan (1981) were able to prove that the ground-state charge distribution of a one-electron homonuclear diatomic molecule can exhibit maxima in ρ only at the positions of the nuclei. In this proof an important inequality is used (Hoffmann-Ostenhof and Hoffman-Ostenhof 1977),

$$-\tfrac{1}{2}\nabla^2 \rho^{1/2}(\mathbf{r}) + (\varepsilon - Z/|\mathbf{r}|)\rho^{1/2}(\mathbf{r}) \leqslant 0 \qquad (E2.1)$$

where ε is the first ionization potential. Unfortunately, only a few theorems concerning the local behaviour of the exact charge density are known. One, derived from the nuclear–electron cusp condition on the state function (Kato 1957), relates the value of the charge density at a nucleus, $\rho(0)$, to the spherical average of the derivative of $\rho(\mathbf{r})$ (Steiner 1963; Bingel 1963)

$$\lim_{r \to 0}[d\rho(\mathbf{r})/d\mathbf{r}]_{av} = -2Z\rho(0). \qquad (E2.2)$$

An upper bound on the value of $\rho(0)$ in an atom, given by Hoffmann-Ostenhof *et al.* (1978) is

$$\rho(0) \leqslant (Z/2\pi)\int r^{-2}\rho(\mathbf{r})d\tau. \qquad (E2.3)$$

The long-range behaviour of $\rho(\mathbf{r})$ for both atoms and molecules has been discussed by a number of authors (Ahlrichs 1972; Hoffmann-Ostenhof and Hoffmann-Ostenhof 1977; Tal 1978; Katriel and Davidson 1980). The results

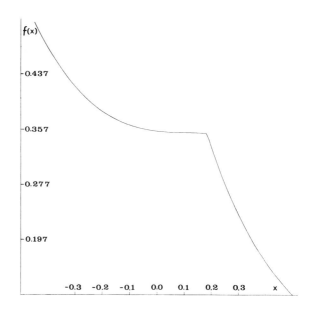

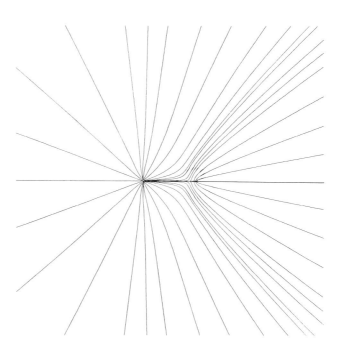

FIG. 2.12. Profile of the charge density along the internuclear axis of the HF$^+$ molecule and the gradient vector field of the charge density for a plane containing the two nuclei. A trajectory of ρ originates at infinity, passes through the proton, and terminates at the F nucleus.

of these studies show that the charge density, at a sufficiently large distance from all nuclei, decays exponentially according to

$$\rho(\mathbf{r}) \sim \exp[-(2\varepsilon)^{1/2}\mathbf{r}] \tag{E2.4}$$

where ε is the first ionization potential of the system.

E2.2 Mathematical properties of ρ at a critical point

E2.2.1 Eigenvalues and eigenvectors of Hessian of ρ

In general, there are nine second derivatives of $\rho(\mathbf{r})$ with respect to an arbitrarily chosen set of coordinate axes. Their ordered array is called the Hessian matrix, or simply the Hessian of ρ. This matrix, when evaluated at the position of the critical point $\mathbf{r}_c$ and denoted by $\mathbf{A}(\mathbf{r}_c)$, is given by

$$\mathbf{A}(\mathbf{r}_c) = \begin{pmatrix} \dfrac{\partial^2\rho}{\partial x^2} & \dfrac{\partial^2\rho}{\partial x\partial y} & \dfrac{\partial^2\rho}{\partial x\partial z} \\[2mm] \dfrac{\partial^2\rho}{\partial y\partial x} & \dfrac{\partial^2\rho}{\partial y^2} & \dfrac{\partial^2\rho}{\partial y\partial z} \\[2mm] \dfrac{\partial^2\rho}{\partial z\partial x} & \dfrac{\partial^2\rho}{\partial z\partial y} & \dfrac{\partial^2\rho}{\partial z^2} \end{pmatrix}_{\mathbf{r}=\mathbf{r}_c} \tag{E2.5}$$

Since it is real and symmetric, $\mathbf{A}(\mathbf{r}_c)$ can always be put in a diagonal form. This corresponds to finding a rotation of the original coordinate system $\mathbf{r}(x, y, z)$ which aligns the new coordinate axes $\mathbf{r}'(x', y', z')$ with the principal axes of the critical point. The diagonalized form of $\mathbf{A}$ is denoted by the matrix $\mathbf{\Lambda}$

$$\mathbf{\Lambda} = \begin{pmatrix} \dfrac{\partial^2\rho}{\partial x'^2} & 0 & 0 \\[2mm] 0 & \dfrac{\partial^2\rho}{\partial y'^2} & 0 \\[2mm] 0 & 0 & \dfrac{\partial^2\rho}{\partial z'^2} \end{pmatrix}_{\mathbf{r}'=\mathbf{r}_c} = \begin{pmatrix} \lambda_1 & 0 & 0 \\ 0 & \lambda_2 & 0 \\ 0 & 0 & \lambda_3 \end{pmatrix}, \tag{E2.6}$$

in which each element is the curvature of $\rho(\mathbf{r}_c)$ with respect to a principal axis. The matrix which generates this coordinate transformation is a unitary matrix $\mathbf{U}$ and, if $\mathbf{r}$ and $\mathbf{r}'$ denote row vectors, one has

$$\mathbf{r}' = \mathbf{r}\mathbf{U}. \tag{E2.7}$$

The matrix $\mathbf{U}$ is obtained by assuming that the matrix $\mathbf{A}$ satisfies three eigenvalue equations of the form

$$\mathbf{A}\mathbf{u}_i = \lambda_i\mathbf{u}_i \qquad (E2.8)$$

where $\mathbf{u}_i$ is a column vector in the matrix $\mathbf{U}$ which expresses the ith component of the vector $\mathbf{r}'$ in terms of the components of $\mathbf{r}$, eqn (E2.7). Thus,

$$\mathbf{U} = \begin{pmatrix} u_{11} & u_{12} & u_{13} \\ u_{21} & u_{22} & u_{23} \\ u_{31} & u_{32} & u_{33} \end{pmatrix} \qquad (E2.9)$$

where $\mathbf{U}^{-1} = \mathbf{U}^{\mathrm{T}}$ since $\mathbf{U}$ is unitary. All three equations for the $\mathbf{u}_i$ can be expressed in a single matrix equation

$$\mathbf{A}\mathbf{U} = \mathbf{U}\mathbf{\Lambda}. \qquad (E2.10)$$

Multiplication of eqn (E2.10) from the left by $\mathbf{U}^{-1}$ subjects $\mathbf{A}$ to a similarity transformation which places it in its diagonalized form (eqn (E2.6)),

$$\mathbf{U}^{-1}\mathbf{A}\mathbf{U} = \mathbf{U}^{-1}\mathbf{U}\mathbf{\Lambda} = \mathbf{\Lambda}. \qquad (E2.11)$$

Thus, if $\mathbf{A}$ solves an eigenvalue equation of the form (E2.8), then a coordinate transformation can be found which will yield a diagonalized form of the matrix. The matrix $\mathbf{A}$ is itself a tensor (see Section E5.2) which can be constructed by taking the product of a column and a row vector

$$\nabla_r\nabla_r\rho = \begin{pmatrix} \dfrac{\partial}{\partial x} \\[2mm] \dfrac{\partial}{\partial y} \\[2mm] \dfrac{\partial}{\partial z} \end{pmatrix} \begin{bmatrix} \dfrac{\partial\rho}{\partial x} & \dfrac{\partial\rho}{\partial y} & \dfrac{\partial\rho}{\partial z} \end{bmatrix} = \mathbf{A}(\mathbf{r})$$

where the derivatives are taken at $\mathbf{r}_c$. Thus, the similarity transformation can be expressed as

$$\mathbf{U}^{-1}\mathbf{A}\mathbf{U} = \mathbf{U}^{-1}\nabla_r\nabla_r\rho\mathbf{U} = \nabla_r\nabla_r\rho = \mathbf{\Lambda},$$

making clear that the diagonalization results from a transformation of the coordinates used in the evaluation of the second derivatives.

The eigenvalues λ_i and eigenvectors $\mathbf{u}_i$ of eqn (E2.8) are obtained through the solution of the corresponding determinantal equation

$$\det|\mathbf{A} - \lambda_i\mathbf{I}| = 0 \qquad (E2.12)$$

where **I** is the unit matrix. The associated eigenvectors $\mathbf{u}_i$ are obtained by solving the equation for each value of λ_i

$$(\mathbf{A} - \lambda_i \mathbf{I})\mathbf{u}_i = 0. \tag{E2.13}$$

E2.2.2   Analytical expression for a trajectory in the vicinity of a critical point

It is possible to obtain an analytical expression for the points $\mathbf{r}(s)$ on a given trajectory of $\nabla\rho$ in the neighbourhood of a critical point. This corresponds to solving the differential equation (2.7) and the solution may be expressed in terms of the eigenvalues λ_i and eigenvectors $\mathbf{u}_i$ of the Hessian matrix of $\rho(\mathbf{r})$ at $\mathbf{r}_c$ as

$$\mathbf{r}(s) = c_1 \mathbf{u}_1 e^{\lambda_1 s} + c_2 \mathbf{u}_2 e^{\lambda_2 s} + c_3 \mathbf{u}_3 e^{\lambda_3 s} \tag{E2.14}$$

where c_1, c_2, and c_3 are arbitrary constants. Equation (E2.14) states that the vectors $\mathbf{r}(s)$, which for varying s define the succession of points lying on a given trajectory, are expressible as a *linear* combination of the three mutually orthogonal eigenvectors, $\mathbf{u}_i$. Because of this simple linear relationship between $\mathbf{r}(s)$ and the $\mathbf{u}_i$, one may use eqn (E2.14) to predict the form of the phase portraits shown in Fig. 2.6.

Equation (E2.14) is obtained by first expressing $\rho(\mathbf{r})$ in the form of a Taylor series expansion about the position of the critical point, $\mathbf{r}_c$. In what follows it is to be understood that all derivatives of $\rho(\mathbf{r}_c)$ are evaluated at $\mathbf{r} = \mathbf{r}_c$. The components of the position vector $\mathbf{r}$, denoting a point in the vicinity of the critical point $\mathbf{r}_c$, are given by (x, y, z).

Writing out explicitly only the first few terms in the expansion and recalling that the first derivatives of $\rho(\mathbf{r})$ vanish at $\mathbf{r}_c$, one obtains

$$\rho(\mathbf{r}) = \rho(\mathbf{r}_c) + 1/2(x - x_c)(\partial^2\rho/\partial x^2)(x - x_c)$$

$$+ 1/2(x - x_c)(\partial^2\rho/\partial x\partial y)(y - y_c)$$

$$+ 1/2(x - x_c)(\partial^2\rho/\partial x\partial z)(z - z_c) + \cdots$$

$$+ \text{remaining second-order terms and third-}$$

$$\text{and higher-order terms.} \tag{E2.15}$$

The differential expression for $\nabla\rho(\mathbf{r})$ in the neighbourhood of $\mathbf{r}_c$ (see eqn (2.7)) is obtained by taking the gradient of equation (E2.15),

$$\frac{\mathrm{d}\mathbf{r}(s)}{\mathrm{d}s} \equiv \nabla\rho(\mathbf{r}(s)) = \mathbf{i}\{(x - x_c)(\partial^2\rho/\partial x^2) + (y - y_c)(\partial^2\rho/\partial x\partial y)$$

$$+ (z - z_c)(\partial^2\rho/\partial x\partial z)\}$$

$$+ \mathbf{j}\{(x - x_c)(\partial^2\rho/\partial y\partial x) + (y - y_c)(\partial^2\rho/\partial y^2)$$

$$+ (z - z_c)(\partial^2\rho/\partial y\partial z) + \ldots . \tag{E2.16}$$

By retaining only the first-order terms in $(\mathbf{r} - \mathbf{r}_c)$ eqn (E.2.16) may be expressed as

$$\nabla\rho(\mathbf{r}(s)) = [x - x_c \quad y - y_c \quad z - z_c] \begin{pmatrix} \dfrac{\partial^2\rho}{\partial x^2} & \dfrac{\partial^2\rho}{\partial x\,\partial y} & \cdots \\[2mm] \dfrac{\partial^2\rho}{\partial y\,\partial x} & \dfrac{\partial^2\rho}{\partial y^2} & \cdots \\[2mm] \cdots & \cdots & \cdots \end{pmatrix}.$$

Thus, the differential eqn (2.7) for a point $\mathbf{r}$ in the neighbourhood of a critical point is given by

$$d\mathbf{r}(s)/ds = (\mathbf{r}(s) - \mathbf{r}_c)\cdot\mathbf{A} \qquad (E2.17)$$

where $\mathbf{A}$ is the Hessian matrix defined in eqn (E2.5).

Eqn (E2.17) may be re-expressed in terms of the diagonal form of $\mathbf{A}$ and its eigenvectors as

$$\frac{d\mathbf{r}'(s)}{ds} = (\mathbf{r}(s) - \mathbf{r}_c)UU^{\mathsf{T}}\mathbf{A}U$$

or

$$\frac{d\mathbf{r}'(s)}{ds} = [x' - x'_c \quad y' - y'_c \quad z' - z'_c] \begin{pmatrix} \lambda_1 & & \\ & \lambda_2 & \\ & & \lambda_3 \end{pmatrix}. \qquad (E2.18)$$

Each component on eqn (E2.18) defines a first-order differential equation of the form

$$\frac{dx'(s)}{ds} = \lambda_1 x'(s)$$

whose solution is of the form

$$x'(s) = c_1 e^{\lambda_1 s}.$$

Hence the general solution to the differential equation for $\nabla\rho$ in the neighbourhood of a critical point is as given in eqn (E2.14).

The eigenvectors $\mathbf{u}_i$ span the space in the neighbourhood of a critical point and, by fixing the values of the constants c_i, one chooses some initial point on a particular trajectory. One may then follow the succession of points $\mathbf{r}(s)$ and determine the trajectory by varying the path parameter s. If $c_2 = c_3 = 0$, for example, then $\mathbf{r}(s)$ lies on the axis defined by $\mathbf{u}_1$. If the corresponding eigenvalue $\lambda_1 < 0$, then the coefficient $e^{\lambda_1 s}$ decreases as s increases. Hence the trajectory of the points $\mathbf{r}(s)$ approaches $\mathbf{r}_c$ along $\mathbf{u}_1$ and the trajectory terminates at $\mathbf{r}_c$ when $s = +\infty$ as illustrated in Fig. 2.6 for the case $(1, -1)$.

Correspondingly, if $\lambda_1 > 0$, the points $\mathbf{r}(s)$ move away from $\mathbf{r}_c$ as s increases, the trajectory is directed away from $\mathbf{r}_c$, and it terminates at some other critical point when $s = +\infty$.

Any trajectory described by a combination of the eigenvectors associated with both the positive and negative eigenvalues will neither originate nor terminate at the critical point but instead will curve so as to avoid $\mathbf{r}_c$. This is illustrated in Fig. 2.6 for the case (2, 0). The two eigenvectors associated with the negative eigenvalues of a $(3, -1)$ critical point define a surface, planar in the immediate neighbourhood of $\mathbf{r}_c$, and all of the trajectories in this surface terminate at $\mathbf{r}_c$; $\rho(\mathbf{r}_c)$ is a maximum in this plane. The eigenvector associated with the single positive eigenvalue of a $(3, -1)$ critical point defines a unique axis, perpendicular to the plane described above, and the two trajectories lying on this axis originate at $\mathbf{r}_c$. If the plane defined by the eigenvectors associated with the two negative eigenvalues is not a symmetry plane of the molecule, then the linear nature of eqn (E2.10) does not obtain beyond the immediate neighbourhood of $\mathbf{r}_c$. In this case the surface associated with the $(3, -1)$ critical point will be curved, but still well-defined by the unique set of trajectories which terminate at the critical point as illustrated in Fig. 2.6.

The two eigenvectors associated with the two positive eigenvalues of a $(3, +1)$ critical point also define a unique surface but in this case all of the trajectories in this surface originate at the critical point and $\rho(\mathbf{r})$ is a minimum at $\mathbf{r}_c$. The third eigenvector defines a unique axis and the two trajectories on this axis terminate at $\mathbf{r}_c$. Thus the phase portraits of a $(3, +1)$ critical point are the reverse of those found for a $(3, -1)$ critical point.

References

Ahlrichs, R. (1972). *Chem. Phys. Lett.* **15**, 609.

Bader, R. F. W. (1975). *MTP International Review of Science. Physical Chemistry*, Ser. 2, Vol. I. *Theoretical chemistry* (ed. A. D. Buckingham and C. A. Coulson), p. 43. Butterworth, London.

Bader, R. F. W. and Beddall, P. M. (1972). *J. Chem. Phys.* **56**, 3320.

Bader, R. F. W. and Essén, H. (1984). *J. Chem. Phys.* **80**, 1943.

Bader, R. F. W., Anderson, S. G., and Duke, A. J. (1979a). *J. Am. chem. Soc.* **101**, 1389.

Bader, R. F. W., Nguyen-Dang, T. T., and Tal, Y. (1979b). *J. Chem. Phys.* **70**, 4316.

Bader, R. F. W., Tal, Y., Anderson, S. G., and Nguyen-Dang, T. T. (1980). *Israel J. Chem.* **19**, 8.

Bader, R. F. W., Nguyen-Dang, T. T., and Tal, Y. (1981). *Rep. Prog. Phys.* **44**, 893.

Bates, D. R., Ledsham, K., and Stewart, A. L. (1953). *Phil. Trans. r. Soc., London* **246A**, 215.

Bender, C. F. and Davidson, E. R. (1968). *J. Chem. Phys.* **49**, 4222.

Bingel, W. A. (1963). *Z. Naturforsch.* **A18**, 1249.

Brillouin, L. (1933–34). *Actualities Sci. Ind.* **71**, Sections 159, 160.

Cade, P. E. (1972). *Trans. Am. crystallogr. Assoc.* **8**, 1.

Cade, P. E. and Huo, W. M. (1973). *Atom. Data nucl. Data Tables* **12**, 415.

Cade, P. E. and Huo, W. M. (1975). *Atom. Data nucl. Data Tables* **15**, 1.

Cade, P. E. and Wahl, A. C. (1974). *Atom. Data nucl. Data Tables* **13**, 339.

Cao, W. L., Gatti, C., MacDougall, P. J., and Bader, R. F. W. (1987). *Chem. Phys. Lett.* **141**, 380.

Collard, K. and Hall, G. G. (1977). *Int. J. Quantum Chem.* **12**, 623.

Faraday, M. (1855). *Experimental researches in electricity*, Vol. III, paragraph 3234, Dec. 1851, Taylor and Francis. Reprinted in 1965, Dover, New York, N.Y.

Gatti, C., Fantucci, P., and Pacchioni, G. (1987). *Theor. Chim. Acta* **72**, 433.

Gatti, C., MacDougall, P. J., and Bader, R. F. W. (1988). *J. Chem. Phys.* **88**, 3792.

Hoffmann-Ostenhof, M. and Hoffmann-Ostenhof, T. (1977). *Phys. Rev.* **A16**, 1782.

Hoffmann-Ostenhof, T. and Morgan III, J. D. (1981). *J. Chem. Phys.* **75**, 843.

Hoffmann-Ostenhof, M., Hoffmann-Ostenhof, T., and Thirring, W. (1978). *J. Phys. B. Atom. Mol. Phys.* **11**, L571.

Kato, W. A. (1957), *Commun. pure appl. Math.* **10**, 151.

Katriel, J. and Davidson, E. R. (1980). *Proc. nat. Acad. Sci., USA* **77**, 4403.

McLean, A. D. and Yoshimine, M. (1967). *Tables of linear molecule wave functions, IBM J. Res. Dev.* **12**, 206.

Moller, C. and Plesset, M. S. (1934). *Phys. Rev.* **46**, 618.

Runtz, G. R., Bader, R. F. W., and Messer, R. (1977). *Can. J. Chem.* **55**, 3040.

Sinanoğlu, O. and Tuan, Fu-Tai, D. (1963). *J. Chem. Phys.* **38**, 1740.

Smith, V. H., Price, P. F., and Absar, I. (1977). *Israel J. Chem.* **16**, 187.

Steiner, E. (1963). *J. Chem. Phys.* **39**, 2365.

Tal, Y. (1978). *Phys. Rev.* **A18**, 1781.

3

MOLECULAR STRUCTURE AND ITS CHANGE

No one really understands the behaviour of a molecule until he knows its structure . . .

C. A. Coulson (1972)

3.1 The notion of structure in chemistry

The essential understanding and original intent associated with the notion of structure in chemistry is that it be a generic property of a system. Structure originally implied the existence of a particular network of bonds, which was presumed to persist over a range of nuclear displacements until some geometrical parameter attained a critical value at which point bonds were assumed to be broken and/or formed to yield a new structure. However, the word 'structure' has over the years acquired a duality of meanings. This has occurred for two reasons. The first was a result of our inability to unambiguously assign a network of bonds to a given system. The second was a result of our ever-increasing ability to experimentally measure the geometrical parameters which characterize the minimum energy nuclear configuration of a system within the Born–Oppenheimer model. Thus, if one references the heading 'molecular structure' in early books on descriptive chemistry, one finds a discussion of valency and bonding and of models for assigning a network of bonds. Referenced under the same heading in present-day physical chemistry or spectroscopy texts are discussions of the experimental methods used to determine the bond lengths and bond angles which are used to characterize an equilibrium geometrical structure. Frequently, one encounters discussions of molecular structure where the two interpretations of structure are used interchangeably, which is a most unfortunate practice, particularly when the resulting confusion of ideas is used to criticize the physical relevance of the very notion of molecular structure.

It is most important to distinguish clearly between molecular geometry and the original intent and use of the notion of structure in the molecular structure hypothesis. Geometry is a non-generic property since any infinitesimal change in a set of nuclear coordinates, denoted collectively by $\mathbf{X}$, results in a different geometry. Molecular structure, on the other hand, was assumed to be a generic property of a system. Any configuration of the nuclei $\mathbf{X}'$ in the neighbourhood of a given configuration $\mathbf{X}$, while it has a different geometry, should possess the same structure, that is, the same nuclei should be linked by the same network of bonds in both $\mathbf{X}$ and $\mathbf{X}'$. Difficulties ascribed to the notion of molecular structure are the inabilities to assign a single geometrical structure, average or otherwise, to rotation- or inversion-related isomers, to a molecule in an excited vibrational state with geometrical parameters very different from those for the same molecule in its ground state, or to a molecule in a 'floppy' state wherein the nuclear excursions cover a wide range of geometrical parameters. In reality, these are shortcomings of attempts to impose the classical idea of a geometry on a quantum system. The nuclei, like the electrons, cannot be localized in space and instead are described by a corresponding distribution function. The definition of structure proposed here recognizes this essential point and associates a molecular structure with an open neighbourhood of nuclear configuration space, with a corresponding average being taken over the nuclear distribution function.

It is demonstrated here that the topological properties of a system's charge distribution enable one to assign a molecular graph to each point X in the nuclear configuration space of a system. This assignment corresponds to defining a unique network of atomic interaction lines for each molecular geometry. A molecular structure is then defined as an equivalence class of molecular graphs. This definition associates a given structure with an open neighbourhood of the most probable nuclear geometry and removes the need of invoking the Born–Oppenheimer approximation for the justification or rationalization of structure in a molecular system (Bader *et al.* 1979, 1981). By defining all possible structures for a given system, the theory shows that a change in structure must be an abrupt and discontinuous process, one which is described in terms of the mathematical theory of dynamical systems and their stabilities.

3.2   The definition of molecular structure

3.2.1   The equivalence relation—a qualitative discussion

We have so far discussed and illustrated the structural elements of molecules for a fixed nuclear geometry, that is, at isolated points X in nuclear configuration space. This structural information is summarized in the molecular graph which indicates how the nuclei are linked for a given nuclear geometry. As changes in the nuclear configuration occur, one anticipates changes in structure and, correspondingly, changes in the molecular graph. A quantitative, detailed description of changes in the molecular graph is uninteresting. Instead, we seek to identify in these changes an invariant property of the system which embodies the notion and properties of molecular structure. This aim is accomplished through the use of a mathematical device known as an equivalence relation as applied to the gradient vector fields of the charge density. The definition of molecular structure is made precise through the language of mathematics. However, the underlying ideas are easily grasped in their qualitative form and to this end we first illustrate them in terms of the changes observed in the topology of the charge density and in its associated molecular graphs for a simple chemical change in structure. The concise and more precise statement of structure and structural stability is given in Section 3.3.

It is important to realize from the outset that two different spaces are involved in the discussion of structure. First is real space, R^3, the space wherein the behaviour of the system is determined and observed through the properties of the charge density. This space may be referred to as *behaviour space*. The topology of the charge density as displayed in behaviour space is determined and controlled by the positions of the nuclei. Changes in the relative positions of the nuclei cause corresponding changes in the distribution of electronic charge, and, as a result, structures are created, they evolve, and are destroyed. Chemical changes are said to have occurred. The relative positions of the nuclei are determined by fixing a point in nuclear configuration space, R^Q, by assigning values to Q internal nuclear coordinates. This is a Q-dimensional mathematical space which may, alternatively, be referred to as the *control space*.

Atoms exist in real space and they are defined by a partitioning of real space. The notion of structure is abstract and it is defined through a partitioning of control space in a manner determined by the behaviour of the system in real space. In this way structure is associated with an open region of nuclear configuration space and necessarily satisfies the condition that it be a generic property of the system.

Suppose a water molecule is at or in the neighbourhood of its equilibrium geometry denoted by $\mathbf{X}_e$, to give the first of the molecular graphs shown in Fig. 3.1. One assumes that the network of bonds denoted by this molecular graph will persist as the nuclei execute vibrational motions in the neighbourhood of the equilibrium geometry. In agreement with this essential idea of the meaning of structure, the molecular graph remains unchanged—each hydrogen nucleus remains linked to the oxygen nucleus by a bond path—for all configurations $\mathbf{X}$ in the neighbourhood of $\mathbf{X}_e$, even those corresponding to considerable excursions of the nuclei from their equilibrium positions as illustrated, for example, by the second of the molecular graphs shown in Fig. 3.1. All of these molecular graphs are said to be equivalent. One obtains an equivalent molecular graph for any arbitrary $\mathbf{X}$ in the vicinity of $\mathbf{X}_e$, that is, all nuclear configurations in the open neighbourhood of $\mathbf{X}_e$ have associated with them charge distributions which yield equivalent molecular graphs. The structure denoted by any one of these graphs is said to be stable.

This is the general idea behind the definition of a stable structure—an open region of nuclear configuration space, all points of which have the same molecular graph. A molecular graph is determined by the gradient vector

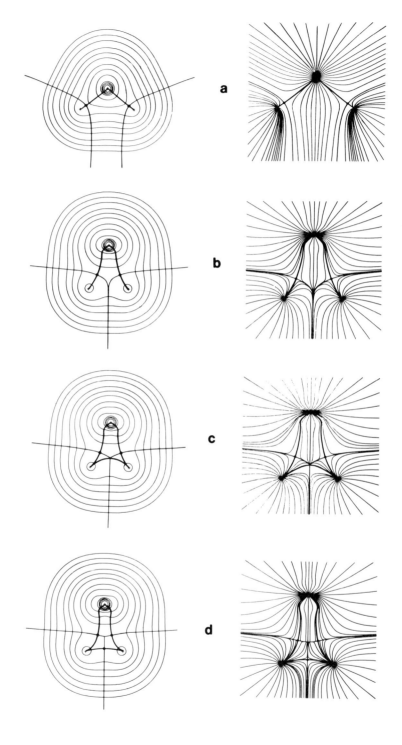

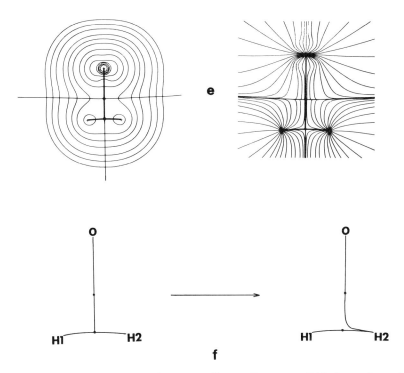

FIG. 3.1.   Contour plots of $\rho(\mathbf{r})$ and corresponding gradient vector fields for configurations along the C_{2v} dissociative path of ground-state water. The molecular graphs and the interatomic surfaces are superposed on the contour plots. (c) corresponds to a catastrophe point of the bifurcation type. At this geometry a singularity in $\rho(\mathbf{r})$ is found between the protons. It is at this geometry that a bond path is formed between the protons in the dissociation of water into $O(^1D) + H_2(^1\Sigma_g^+)$. (e) corresponds to a catastrophe point of the conflict type. The molecular graphs in (f) indicate how an asymmetric motion of the system in this configuration causes the bond path from oxygen to switch to one of the protons.
(From Bader *et al.* (1981).)

field and, in the general case, the equivalence relationship must be established first in terms of this field. Two vector fields are said to be equivalent if one can map every trajectory of one field on to a corresponding trajectory of the other. By applying this definition to the gradient vector fields of the charge density we obtain an equivalence relation between points in the nuclear configuration space. For every arrangement of the nuclei which corresponds to some point $\mathbf{X}$ in nuclear configuration space, there is a charge density $\rho(\mathbf{r}; \mathbf{X})$ and an associated gradient vector field $\nabla\rho(\mathbf{r}; \mathbf{X})$. If we use the symbols $\nabla\rho(\mathbf{r}; \mathbf{X})$ and $\nabla\rho(\mathbf{r}; \mathbf{X}')$ to denote such fields for two configurations $\mathbf{X}$ and $\mathbf{X}'$, say those depicted in Fig. 3.1(a), (b), then we can state that the two nuclear configurations are equivalent only if their associated gradient vector fields are equivalent. The nuclear configuration **X** is structurally stable if **X** is an interior point of its equivalence class. That is, one can always find a neighbourhood of a structurally stable configuration **X**, such that the neighbourhood is totally contained within the region of structures equivalent to **X**.

The equivalence of gradient vector fields for two configurations **X** and **X**′ ensures that a critical point (ω, σ) in $\rho(\mathbf{r}; \mathbf{X})$ is mapped on to a critical point of the same type in $\rho(\mathbf{r}; \mathbf{X}')$ and that a gradient path connecting a pair of critical points in $\rho(\mathbf{r}; \mathbf{X})$ is transformed into a corresponding path in $\nabla\rho(\mathbf{r}; \mathbf{X}')$, linking the images of these two critical points. Thus, the molecular graphs associated with the set of equivalent gradient vector fields consist of the same number of bond paths linking the same nuclei. These molecular graphs represent a single structure and the region of nuclear configuration space containing **X** and all configurations equivalent to it is called the structural region associated with **X**.

Two molecular graphs are defined to be equivalent if and only if they belong to the same structural region. *An equivalence class of molecular graphs is called a molecular structure.* Thus, a unique molecular structure is associated with a given structural region, and molecular structure, as defined above through the equivalence of molecular graphs, necessarily fulfils the requirement of being generic.

Clearly, one could average the electronic charge distribution over the nuclear distribution function generated by the zero-point vibrational motions. This latter distribution peaks around the configuration $\mathbf{X}_e$ and the nuclear averaged molecular graph would be equivalent to the one obtained for this configuration, as it is obtained at all points in the neighbourhood of $\mathbf{X}_e$. In this manner one can dissociate the notion of structure from molecular geometry and define a molecular graph and a corresponding molecular structure for the water molecule in its electronic and vibrational ground state.

3.2.2   Changes in structure

There are a number of dissociative pathways for ground-state water and again, within the notion of structure, one anticipates that, for sufficiently large nuclear excursions, bonds will be broken and/or formed to yield the possible dissociation products. How are these anticipated changes in structure reflected in the topology of the charge distribution as monitored by the molecular graph? Consider the particular case of the minimum-energy adiabatic pathway which maintains C_{2v} symmetry and leads to the formation of an oxygen atom and a hydrogen molecule,

$$H_2O(^1A_1) \rightarrow O(^1D) + H_2(^1\Sigma_g^+).$$

The gradient vector fields and molecular graphs for points along this pathway are illustrated in Fig. 3.1. In Fig. 3.1(b) the oxygen nucleus has receded from the protons which in turn have approached one another. While the atomic surfaces of the two hydrogen atoms approach one another, they remain distinct as no $(3, -1)$ critical point has appeared between the protons. Thus, as noted above, the gradient vector field and its molecular graph for the configuration in Fig. 3.1(b) are equivalent to those in Fig. 3.1(a) and they represent a unique structure, the open structure associated with the equilibrium geometry of water. This region does not, however, contain the remaining configurations shown in the figure whose molecular graphs define other structures and other structural regions for the system consisting of an oxygen and two hydrogen atoms. The result of applying the equivalence relation to the gradient vector fields of the charge density is, therefore, a partitioning of nuclear configuration space into a number of non-overlapping structural regions, each of which is characterized by a unique molecular structure. What is demonstrated next is that a change in structure, that is, passage from one structural region to another, is necessarily an abrupt and discontinuous process.

The gradient vector field and the structure it defines shown in Fig. 3.1(c) are not equivalent to those which preceded it. At this point an abrupt change occurs in the topology of the charge density. A new critical point in the charge density appears between the protons. It is at this configuration along the dissociative pathway that the bond is formed between the hydrogen atoms. The reader should note that the change in the molecular graph monitors the important change in the charge distribution which occurs at this critical nuclear configuration X_c—the accumulation of sufficient electronic charge between the protons to define a line which links them and along which the charge density is a maximum with respect to any neighbouring line.

The molecular graph for X_c is not equivalent to those which preceded it and it represents a new structure. Unlike the structure from which it evolved, however, it is unstable. A further infinitesimal displacement along the reaction pathway causes the new critical point to change through its bifurcation into two new critical points creating a new pattern for the gradient vector field and another structure. This new gradient vector field, shown in Fig. 3.1(d), defines a ring structure and this one, like the structure which preceded the unstable structure at X_c, is stable—it persists for arbitrary motions of the nuclei in the vicinity of the reaction pathway and for some further motion of the system along the pathway.

The new critical point formed between the hydrogens for the nuclear configuration X_c is unstable because it has one zero curvature—the curvature along the C_2 symmetry axis. It is a critical point of rank two. Of the two non-vanishing curvatures of the charge density at this point, one is positive and is associated with the line linking the protons, while the second is negative, lying along the axis perpendicular to the plane containing the nuclei. When the system is extended past X_c, the form of the charge density with respect to these two curvatures does not change but the flat portion—the zero curvature—is replaced by a ripple in ρ causing the critical point to split in two. This behaviour is illustrated in Fig. 3.2 which shows a series of profiles of the charge density along the C_2 symmetry axis. Each profile is for a particular value of R_0, the distance of the oxygen nucleus from the midpoint of the proton–proton internuclear axis. For configurations preceding the nuclear configuration X_c in which the separation R_0 attains its critical value, the charge density falls off monotonically from the oxygen nucleus along the C_2 axis out to infinity. The only critical point on this axis for these configurations is the maximum at the oxygen nucleus. There is an abrupt and discontinuous change in this behaviour for the nuclear configuration X_c. A point of inflexion appears in the profile of the density when R_0 attains its critical value, a point where both the first and second derivatives of ρ with respect to the coordinate along the C_2 axis vanish. An increase in the separation R_0 causes the inflexion point to be replaced by two critical points, one with a positive curvature and the other, between the protons, with a negative curvature. This situation is again a stable one for it persists with further increase in the parameter R_0.

In this one particular example, all of the changes in the topology of the charge density occur along this one axis and the changes occurring in three dimensions are easily obtained from those indicated in Fig. 3.2 by adding one negative and one positive curvature to each of the critical points obtained in the bifurcation. Thus, the new critical point between the protons is a $(3, -1)$

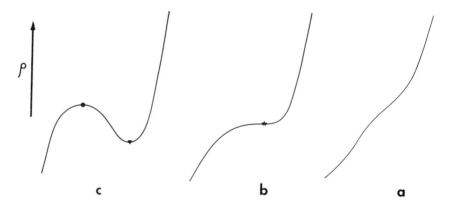

FIG. 3.2. Profiles of the charge density along the C_2 symmetry axis for the dissociation of the water molecule. In (a) ρ decreases monotonically from a maximum value at the oxygen nucleus. As R_0 is increased, an abrupt change occurs in the form of the curve when a point of inflexion appears (b). This unstable critical point exists only for this single configuration X_c of the nuclei and a further increase in R_0 results in its bifurcation to yield a maximum, corresponding to the formation of a bond critical point and a minimum, corresponding to the formation of a ring critical point as typified by (c). The system has entered the region of the ring structure.

or bond critical point and the second is a $(3, +1)$ or ring critical point. If one reverses the motion of the system starting from the ring structure, the same bond and ring critical points approach one another as the distance R_0 is decreased, until at its critical value they coalesce, a negative curvature of one annihilating a positive curvature of the other along their line of approach to produce the inflexion point in ρ between the protons for the critical config- uration $\mathbf{X}_c$, as illustrated in Fig. 3.2.

The above example has illustrated one way of bringing about a change in structure, through the creation of a critical point in the charge density of rank less than three, a so-called *degenerate critical point*. Such a critical point is not stable with respect to the changes induced in the charge distribution by arbitrary displacements of the nuclei away from the critical configuration $\mathbf{X}_c$. Thus, the structure defined by the molecular graph associated with the critical nuclear configuration $\mathbf{X}_c$ is an unstable structure and exists for just one particular configuration of the nuclei along the dissociative pathway. It will be referred to as a bifurcation structure and the point in nuclear configuration space for the critical configuration as a *bifurcation point*.

A further extension of the parameter R_0 leads to the destruction of the ring structure and to the formation of another degenerate critical point. This one results from the simultaneous coalescence of the two O–H bond critical points with the ring critical point. Passage of the system through this bifurcation point in nuclear configuration space causes the unstable critical point in the charge density to degenerate into a $(3, -1)$ critical point, thereby yielding the new structure illustrated by the molecular graph in Fig. 3.1(e). This structure, in which the oxygen nucleus is no longer linked to either proton, persists for continued motion along the C_{2v} pathway and eventually yields an oxygen atom and a hydrogen molecule.

The structure shown in Fig. 3.1(e) is not a stable structure. It is abruptly changed into the structure represented by the second of the molecular graphs in Fig. 3.1(f), or its mirror image, for any motion of the nuclei which destroys the twofold symmetry axis. Thus, as is characteristic of all unstable structures, the structure in Fig. 3.1(e) is stable to some but not to all nuclear displace- ments. It represents the second and final type of instability that can be exhibited by a molecular charge distribution. While all of the critical points in the molecular graph in Fig. 3.1(e) are of rank three and are themselves stable, the graph itself is unstable. There are two $(3, -1)$ or bond critical points, the bond path of one linking the protons, that of the second linking the oxygen nucleus not to either proton but to the surface manifold of the first bond critical point. Unlike a nuclear position where ρ is a local maximum, which therefore acts as an attractor of gradient paths in three-dimensional space, a bond critical point is a maximum only in the interatomic surface and can act as an attractor of only those gradient paths which lie in this surface. In this particular system, the surface includes one of the two gradient paths originating at the upper $(3, -1)$ critical point. Any motion of the nuclei which causes this trajectory to leave the surface manifold, that is, to move in three dimensions rather than in just the two-dimensional space of the surface, causes it to abruptly switch to one or the other of the three-dimensional attractors whose basins share the interatomic surface, the protons.

The ability of a $(3, -1)$ critical point to act as an attractor in the two-dimensional manifold of its associated interatomic surface can be seen in Fig. 3.3 which illustrates the gradient vector field for the unstable structure shown in Fig. 3.1(e) for a plane perpendicular to the plane of the nuclei, along

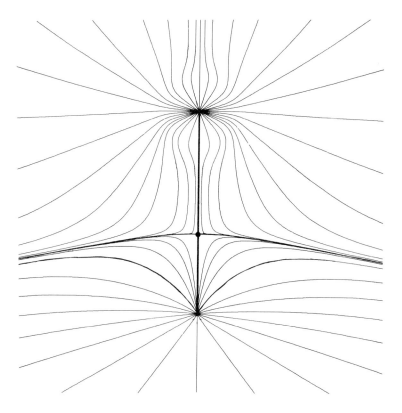

FIG. 3.3.   Gradient vector field of the electronic charge density for the unstable conflict structure shown in Fig. 3.1(e). The plane shown contains the C_2 symmetry axis and is perpendicular to the plane of the nuclei. In the lower portion of the diagram, trajectories terminate at the $(3, -1)$ critical point between the hydrogen nuclei. In the upper portion, trajectories terminate at the pseudo $(3, -3)$ critical point at the oxygen nucleus. These two critical points are linked by the pair of trajectories which originate at the central $(3, -1)$ critical point indicated by the dot. This is an unstable intersection of the one-dimensional manifold of this $(3, -1)$ critical point with the two-dimensional manifold of the $(3, -1)$ critical point between the protons.

the C_2 axis. It shows, in the lower portion, the trajectories which terminate at the H–H bond critical point and define the interatomic surface between the hydrogens. Both this critical point and the one at the oxygen nucleus appear as $(2, -2)$ critical points, or two-dimensional attractors in this plane. The second $(3, -1)$ critical point appears as a $(2, 0)$ critical point and its bond path links the oxygen nucleus and the critical point of the H–H bond. The gradient vector field of this system for the one particular plane shown in Fig. 3.3 is indistinguishable in form from that for a diatomic molecule, the critical point of the bond between the hydrogens serving as an attractor. The phase portrait in the vicinity of the H–H bond critical point in this case, however, is unstable and is obtained only when the bond path trajectory lies within its surface manifold, which, in this particular case, is a symmetry plane.

The type of unstable structure shown in Fig. 3.1(e) is called a *conflict structure*—the protons are in conflict as they compete for the terminus of the line of maximum negative charge density linked to the oxygen nucleus. In the region of the energy surface where the conflict structure is found, which corresponds to a relatively large separation between the oxygen nucleus and the midpoint of the hydrogen molecule, there is a small energy barrier along the length of the exit channel for the separation of the oxygen along the C_2 symmetry axis, it being energetically more favourable for the oxygen to recede from the hydrogen molecule when it is on either side of this axis. Thus, the conflict structure in this system is both energetically and topologically unstable, and perpendicular motions of the exiting oxygen nucleus will cause the structure to oscillate from one stable form to the other.

This analysis of the quantum mechanical calculations for various geometries along a reaction pathway illustrates how structural information can be obtained from the state function. Table 3.1 summarizes the values of the charge density at the bond critical points, values denoted by the symbol $\rho(\mathbf{r}_c)$, and the total integrated electron populations of the atoms for the various structures along the reaction pathway. (An atomic population, obtained by integration of the charge density over the basin of the atom, is an obvious atomic property which is used here in advance of the formal definition of the properties of atoms.) It is clear from Fig. 3.1(a) that the oxygen nucleus dominates the equilibrium charge distribution of the water molecule. The bond critical points and interatomic surfaces are close to the protons and there is a substantial transfer of charge to the oxygen atom, each hydrogen bearing a net charge of $+0.58\,e$. The value of $\rho(\mathbf{r}_c)$ for the two O–H bonds in the equilibrium geometry is typical. This value undergoes a continuous decrease as the oxygen nucleus recedes from the hydrogens and, accompanying this, there is a transfer of charge from the oxygen to the hydrogens. Correspondingly, the O–H bond critical points and their interatomic surfaces move towards the oxygen nucleus as the hydrogens grow in size. Electronic

Table 3.1

Charge densities at bond critical points and atomic charges for C_{2v} dissociative pathway of H_2O X $^1A_1$

Nuclear geometry*		$\rho(\mathbf{r}_c)$			Net charge	
R_O (au)	R_H (au)	O–•–H	H–•–H	O–•–H_2	$q(O)$	$q(H)$
(a) 1.109	2.868	0.3677			− 1.16	+ 0.58
(b) 2.300	2.160	0.1428			− 0.32	+ 0.16
(c)† 2.390	2.100	0.1308	0.1230		− 0.28	+ 0.14
(d) 2.500	2.000	0.1187	0.1312‡		− 0.24	+ 0.12
(e) 2.750	1.800		0.1623	0.0930	− 0.14	+ 0.07

*These data, all in au, refer to the five configurations shown in Fig. 3.1 and are correspondingly labelled.
†The value of R_O for the catastrophe point lies between 2.40 and 2.39 au.
‡The value of $\rho(\mathbf{r}_c)$ at the (3, + 1) or ring critical point in this structure is 0.1176 au.

charge transferred to the hydrogens is accumulated between their nuclei and, as indicated by Fig. 3.1(c), a bond is formed between the hydrogens before the O–H bonds are broken, resulting in the formation of a ring structure. The value of ρ at the H–H bond critical point at the instant of its formation is only slightly less than the value of $\rho(\mathbf{r}_c)$ found for the two O–H bonds. The value of $\rho(\mathbf{r}_c)$ for the H–H bond undergoes a further continuous increase as the oxygen recedes, eventually attaining the value of 0.27 au found in an isolated H_2 molecule. The formation of the conflict structure indicates the presence of only a weak interaction remaining between the oxygen and the forming hydrogen molecule. This is reflected in the relatively low value of ρ at the corresponding critical point, a value which decays to zero for the eventual separation of the oxygen atom from the hydrogen molecule.

3.2.3   Structure diagrams

The discussion so far has demonstrated that the definition of structure is inextricably bound up with the definition of structural stability. We now show that the two definitions taken together lead to a partitioning of nuclear configuration space into a finite number of structural regions.

Associated with each point in nuclear configuration space is a real space charge distribution and its associated gradient vector field. By applying the equivalence relation to the gradient vector fields defined in real space we are able to define equivalent points in nuclear configuration space. The result is a partitioning of nuclear configuration space R^Q into a finite number of non-overlapping regions, each of which is characterized by a unique molecular structure. These structurally stable, open regions are separated by boundaries, hypersurfaces in the space R^Q. A point on a boundary possesses a structure which is different from that characteristic of either of the structural regions it separates. Since a boundary is of dimension less than R^Q, arbitrary motions of the nuclei will carry a point on the boundary into neighbouring stable structural regions and its structure will undergo corresponding changes. The boundaries are the loci of the structurally unstable configurations of a system. In general, the trajectory representing the motion of a system point in R^Q will carry it from one stable structural region through a boundary to a neighbouring stable structural region. The result is an abrupt and discontinuous change in structure as illustrated in Fig. 3.1. In this sense a change in structure is catastrophic (Thom 1975) and the set of unstable structures is called the catastrophe set. We shall henceforth refer to a point in a structurally stable region of nuclear configuration space as a *regular point*, and to a point on one of the structurally unstable boundaries as a *catastrophe point*.

A knowledge of the stable structural regions and their boundaries as defined by the catastrophe set enables one to construct a structure diagram, a diagram which determines all possible structures and all mechanisms of structural change for a given chemical system. Figure 3.4 is a two-dimensional cross-section of the structure diagram for an ABC system. (The reason for the particular form shown there is justified in Section 4.3.2.) The letters may stand for atoms or functional groupings of atoms. The full lines, denoting the catastrophe set, partition nuclear configuration space into its structural regions. The hypocycloid portion denotes the loci of the bifurcation catastrophes, of the type illustrated in Fig. 3.1(c), and the three semiaxes, the loci of the conflict catastrophes as illustrated in Fig. 3.1(e).

All of the essential information concerning the possible structures and the mechanisms of their change for an ABC system are represented in this structure diagram. The reaction of an oxygen atom with a hydrogen molecule in the formation of a water molecule along the path represented by the molecular graphs (e) to (a) in Fig. 3.1 and, with A = O, the oxygen atom, is represented by motion beginning at the top of the vertical semiaxis in Fig. 3.4 with the conflict structure. The conflict structure represents the average structure obtained as the bond path from the oxygen switches between the hydrogens as depicted by the molecular graphs for the stable structural regions separated by this axis. At the point where the axis coincides with the bifurcation set, the system undergoes a change in structure and enters the region of the stable ring structure. The ring critical point when first formed will lie immediately between the critical points of the O–H bonds. As the system progresses down through the ring region, the charge density must change in such a way that the ring critical point migrates towards the critical point of the H–H bond so that, as previously described, they coalesce to form

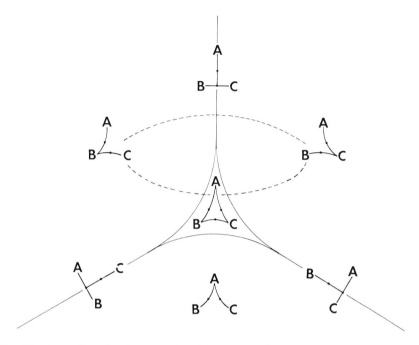

FIG. 3.4.   A two-dimensional cross-section of the structure diagram for an ABC system. The full lines, denoting the catastrophe set, partition nuclear configuration space into its structural regions. The structure associated with each region is indicated by a representative molecular graph. The two broken lines indicate two possible reaction mechanisms for changing the structure A–B–C into the structure B–C–A.

a degenerate critical point and a second bifurcation catastrophe point, a point on the boundary separating the ring from the open region of the water system.

The structure diagram predicts the existence of two possible mechanisms for a reaction of the type

$$A–B–C \rightarrow B–C–A.$$

The conflict mechanism, indicated by the upper path in Fig. 3.4, corresponds to the passage of the system through a conflict catastrophe point. In this case the change in structure is obtained by an abrupt switching of attractors, the bond path from A switching from B to C. This change in structure is illustrated by the sequence of molecular graphs shown in Fig. 3.5(a) for the case of the migration of a methyl group in the carbonium ion $C_3H_7^+$. A structure of the conflict type is frequently found in systems where one invokes an interaction between a sigma bond of the migrating group with the 'π' bond of the ethylenic fragment.

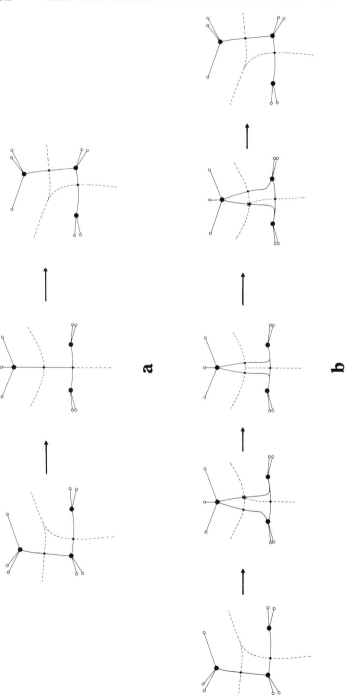

FIG. 3.5.   Molecular graphs, determined by SCF calculations, illustrating the two mechanisms for structural change indicated in Fig. 3.4 for the migration of a CH_3 group across a bonded pair of carbon atoms in $C_3H_7^+$. The conflict mechanism is shown in (a), the bifurcation mechanism in (b). The hydrogen nuclei are denoted by open circles, the carbon nuclei by (large) full circles. Also shown are the carbon–carbon interatomic surfaces. A singularity in $\rho(\mathbf{r})$ for the configurations in (b) which correspond to bifurcation catastrophe points is denoted by a star. At the first of these configurations, a new bond path linking two of the carbon nuclei is formed and, at the second, the original carbon–carbon bond path to the methyl group is destroyed.

(From Bader et al. (1981).)

The second of the two possible mechanisms, the bifurcation mechanism, involves the formation of an intermediate ring structure. This mechanism is again illustrated by the methyl group migration in $C_3H_7^+$, by the second sequence of molecular graphs shown in Fig. 3.5(b). At the catastrophe point separating the open from the ring structure, a new, degenerate critical point appears in the charge distribution. Associated with this critical point is a pair of trajectories which link the methyl group to the second methylene group and a new C–C bond is formed. Further motion of the system point results in the bifurcation of this unstable critical point into two non-degenerate critical points, a bond critical point connecting the methyl with the second methylene group and a ring critical point, in the interior of the newly formed ring structure. Continued motion of the methyl group across the ethylenic fragment causes the ring critical point to migrate towards the bond critical point which formed the original bond path to the methyl group. When these two critical points coalesce, the system is at a second bifurcation catastrophe point. At this point, the original bond to the methyl group is broken and the system enters the structural region of the product molecule.

What have just been described are the topological changes in the charge density that must occur in the formation or destruction of a bonded ring of atoms. As the nuclei are displaced to increase the separation between a bonded pair of atoms forming a perimeter of the ring, the charge density must undergo a corresponding change so that the minimum in the ring surface migrates towards the critical point of the extended bond. The curvatures of the charge density at both the ring and bond critical points are progressively softened in their direction of approach until, at the point of coalescence, their common curvature vanishes and the bond is ruptured. These are new windows on the essential changes in the charge density which accompany changes in structure and the views so obtained lead to new questions concerning the mechanics governing a change in structure. How can one facilitate or hinder the formation of a degenerate critical point in a charge distribution? Critical points in ρ as well as the nuclei must move to achieve a change in structure. Could one develop a liquid model of the charge density, a principal parameter being the viscosity experienced by a migrating critical point? As demonstrated in Chapter 6, one can relate the curvatures of the charge density at a point in space to the local potential and kinetic energy densities of the electrons. While there is a mechanics of the charge density, it is not complete and the theory of atoms in molecules offers opportunities for the formulation of new models in the search for answers to old problems.

One final example of a change in structure is illustrated by the succession of gradient vector fields shown in Fig. 3.6 for the thermal isomerization of HCN to CNH. The calculated minimum energy path for this reaction exhibits a single energy maximum and, correspondingly, the change in structure occurs via the conflict mechanism, the geometry of the unique topologically unstable structure coinciding with that of the transition state. Motion along the

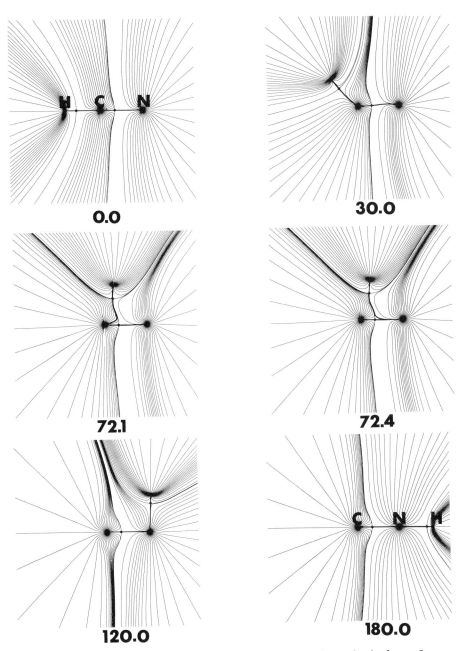

FIG. 3.6. Displays of the gradient vector fields of the electronic charge density for configurations along the isomerization reaction coordinate for HCN. The label refers to the angle θ formed between the C–N axis and the vector from the proton to the CN centre of mass. The bond path from the proton switches attractors, from the carbon nucleus to the nitrogen nucleus for some configuration lying between $\theta = 72.1°$ and $72.4°$.

reaction pathway is monitored by the value of the angle θ formed between the C–N axis and the vector to the proton with origin at the CN centre of mass. The molecular geometries, lying on the reaction coordinate for values of $0 \leqslant \theta \leqslant 72.1°$, all possess equivalent gradient vector fields, hence, they are all represented by an equivalent molecular graph, one which defines the stable structure H–C–N. Similarly, all geometries occurring in the range $72.4° \leqslant \theta \leqslant 180°$ exhibit a different but again equivalent set of molecular graphs which define the stable structure C–N–H. Thus, in the range $72.1° < \theta < 72.4°$, a switching of attractors occurs, as is characteristic of the conflict mechanism.

It would be extremely difficult to determine the exact geometry of the unstable conflict structure wherein the proton is linked to the surface manifold of the C–N bond critical point. Unlike the previous examples, the conflict point is not fixed by symmetry. The conflict structure is very nearly realized in the molecular graphs for $\theta = 72.1°$ and $72.4°$ and is clearly bracketed by them. The exact geometry of the transition state is equally difficult to determine and, within the accuracy of the SCF calculations, its geometry is bracketed by the same values of the parameter θ. The conflict mechanism is also found to occur in the corresponding isomerization of methyl cyanide to methyl isocyanide and again, to within computational accuracy, the transition state and topologically unstable structure possess the same geometry. These are examples wherein one finds a coincidence between an energetically unstable geometry and a topologically unstable structure. The general discussion of the extent to which these two definitions of instability coincide is given in Section 3.4.1.

The set of molecular graphs in Fig. 3.6 exemplify the principal ideas of structure and structural stability. The initial and final structures are each represented by a unique molecular graph, one which persists over a range of nuclear motions. These structures are stable. They are separated by a structure whose molecular graph is intermediate in form and, since it exists for only a single nuclear geometry along the reaction path, is unstable.

3.2.4   Bonds and structure

Hydrocarbon molecules are used to illustrate the definitions of structure and structural stability introduced above and also the manner in which the properties of the electronic charge density at a bond critical point can be used to summarize the important physical characteristics of a bonded interaction between atoms (Bader *et al.* 1983). Figure 3.7 gives the molecular graphs for saturated hydrocarbon molecules with widely varying equilibrium structures, beginning with acyclic molecules both branched and unbranched, followed by cyclic and bicyclic molecules, and ending with a number of so-called propellane molecules and other highly strained systems, tetrahedrane, cubane, and spiropentane (Wiberg *et al.* 1987). These structures are determined in their entirety by the information contained in the corresponding state functions as employed in the definition of the electronic charge density (eqn (1.3)).

With the exception of the propellanes, for which existing models are inadequate, the structures in this figure are in complete accord with those that have been assigned to these molecules on the basis of chemical information and models of directed valency, augmented with the concept of bent bonds. The agreement of structures, predicted in this manner by quantum mechanics for a wide variety of systems with chemically accepted structures, shows that the many models used to rationalize the network of bonds in a molecule may be replaced with a single theory of molecular structure. The small ring propellanes, such as [1.1.1]propellane **20**, possess unusual structures with 'inverted' geometries at the bridgehead carbon atoms. A bond path links the bridgehead nuclei in each of the propellane molecules **20**–**23**, even though this results in structures that have four bond paths, all to one side of a plane, terminating at each of the bridgehead nuclei in **20**, **21** and **22**. While the model of hybridized orbitals cannot describe such a situation, even assuming bent bonds, the molecular graphs for the propellanes demonstrate that the charge distribution of a carbon atom can be so arranged as to yield bond paths—lines of maximum charge density—which correspond to an inverted structure.

The structures of the propellanes consist of three rings, either three- or four-membered, sharing a common bond, the bridgehead bond. Addition of a molecule of hydrogen to a propellane breaks the bridgehead bond and yields the corresponding bicyclic molecule, **16**–**19**. Molecules **16** and **19** possess cage structures, the interior of each molecule being bounded by three curved ring surfaces and containing a cage critical point. The propellanes and their bicyclic congeners exemplify the contrasting behaviour exhibited by the charge density between a pair of nuclei in situations where the atoms are and are not bonded to one another. Figure 3.8 gives relief maps of the charge density in the plane which bisects and is perpendicular to the bridgehead internuclear axis for each propellane and its bicyclic analogue. Such a plane contains the critical point between the bridgehead atoms. It also contains the critical point in the interatomic surface between the methylenic carbon atoms linked by a peripheral bond and/or the C and H nuclei of a peripheral CH_2 group in the case of a three-membered ring in a propellane or the corresponding group in the bicyclic molecule. The charge density is a maximum at a bond critical point in the interatomic surface. Thus, the diagrams for the [2.2.2] systems shown in Fig. 3.8(a) both exhibit three maxima in the charge density corresponding to the three peripheral bonds between the methylene groups. These maxima in bonded density are successively replaced by the

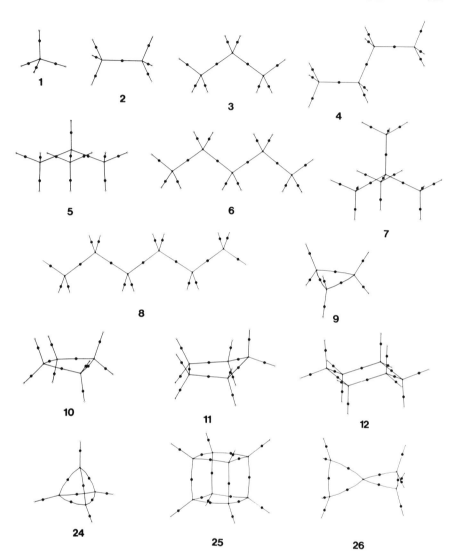

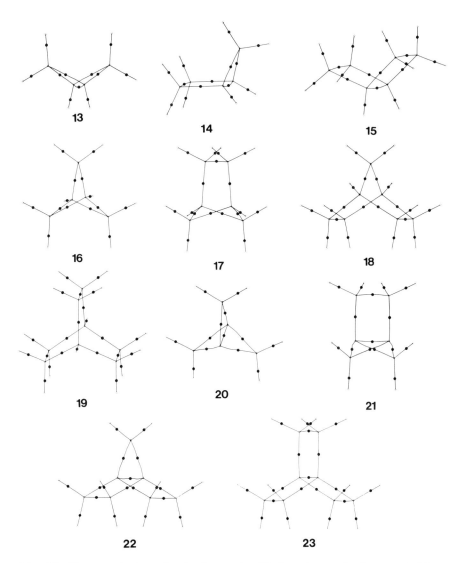

FIG. 3.7.   Planar projections of molecular graphs of hydrocarbon molecules generated from theoretical charge distributions. Bond critical points are denoted by black dots. Structures **1** to **4** are normal hydrocarbons from methane to butane, **5** is isobutane, **6** is pentane, **7** is neopentane, and **8** is hexane. The remaining structures are identified in Table 3.2. The structures depicted in these diagrams are determined entirely by information contained in the electronic charge density.

nuclear maxima of a CH_2 group through the series to the [1.1.1] systems. The purpose of the diagram is to contrast the behaviour of the charge density at the bridgehead critical point for the propellanes with that found for the bicyclic molecules. *In the propellanes, which possess a bridgehead bond, the*

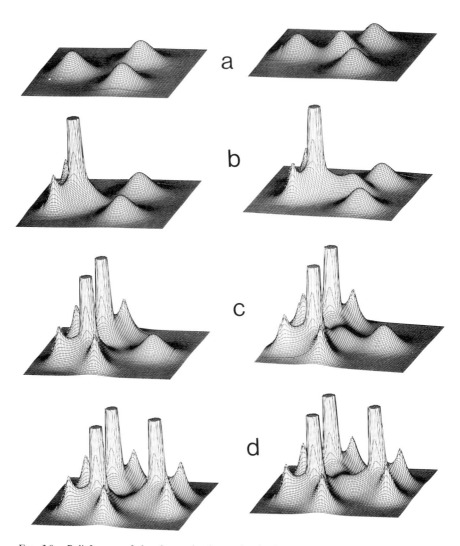

FIG. 3.8. Relief maps of the electronic charge density in the symmetry plane bisecting the bridgehead bond in the propellanes, shown on the right, and in the corresponding symmetry plane in each of the related bicyclic structures, shown on the left. (a) [2.2.2]propellane and bicyclooctane; (b) [2.2.1]propellane and norborane; (c) [2.1.1]propellane and bicyclohexane; (d) [1.1.1]propellane and bicyclopentane. Contrast the presence of a central maximum in ρ in [2.2.2]propellane with its absence in the bicyclic compound. There is a bridgehead bond in the former, but not in the latter.

charge density is a maximum at this point, while in the bicyclic molecules, which do not possess a bridgehead bond, the charge density is a minimum at this same point. In the former molecules there is a line of maximum charge density linking the nuclei, while in the latter such a line is absent and the charge density is instead a local minimum at the central critical point in the first and final of the bicyclic molecules. There is, therefore, an essential, qualitative difference in the manner in which electronic charge is distributed along a line linking a pair of bonded nuclei (as between the bridgehead nuclei in the propellanes) and along a line linking two nuclei that are not bonded (as between the nuclei in the corresponding bicyclic structures).

In addition to the fundamental difference in the form of the bridgehead charge density between the two sets of molecules, there are also substantial quantitative differences in the values of ρ. Thus, its value at the bond critical point in [1.1.1]propellane is 0.203 au, four-fifths of the value for a normal C–C bond, while the value of ρ at the corresponding cage critical point in the bicyclic molecule is 0.098 au. It is also clear from Fig. 3.8(d) that the values of ρ at the bond critical point and the three adjacent ring critical points are almost equal, giving rise to a very broad bonded maximum in ρ in the bridgehead interatomic surface in [1.1.1]propellane.

The value of the charge density at a bond critical point can be used to define a bond order (Bader *et al.* 1983; Cremer and Kraka 1984). The molecular graphs for ethane, ethylene, and acetylene are shown in Fig. 2.8. In each case the unique pair of trajectories associated with a single $(3, -1)$ critical point is found to link the carbon nuclei to one another. Multiple bonds do not appear as such in the topology of the charge density. Instead, one finds that the extent of charge accumulation between the nuclei increases with the assumed number of electron pair bonds and this increase is faithfully monitored by the value of ρ at the bond critical point, a value labelled ρ_b. For carbon–carbon bonds, one can define a bond order n in terms of the values of ρ_b using a relationship of the form

$$n = \exp\left[A(\rho_b - B)\right]$$

where A and B are constants which vary slightly with basis set. B is set equal to the value of ρ_b for ethane and a value for A can be found such that the relationship yields bond orders of 1.0, 1.6, 2.0, and 3.0 for ethane, benzene, ethylene, and acetylene, respectively. The bond order–ρ_b relationship is distinct for each pair of atoms A–B, since the values of ρ_b for bonds of given formal order can vary markedly between different pairs of atoms. Thus, one cannot estimate relative bond strengths for bonds between dissimilar atoms by comparing their respective ρ_b values. The orders of nitrogen–nitrogen bonds are well described by a linear relationship with ρ_b. The original bond order scheme proposed by Coulson (1951) was based on a bond order–bond length relationship. The value of ρ_b is found to increase as a bond length decreases and it is possible to obtain useful bond length–ρ_b value relationships (Boyd *et al.* 1988) for a wide range of A–B systems. While these schemes are empirical, they serve to relate familiar models and concepts to a property of the charge density. However, they also underscore an important point— that the strength of the bonding between a given pair of atoms increases as the amount of charge in the interatomic surface for the equilibrium bond length, as monitored by its maximum value, ρ_b, increases. This point is well illustrated by showing that the behaviour of the value of ρ_b on ionization can be predicted on the basis of the bonding or antibonding nature of the orbital from which the electron is removed.

There is a notion of simple molecular orbital theory that may be a cause for concern in employing ρ_b as a measure of bond order. One may be led to conclude that, since a bond critical point can lie in the nodal surface of a π orbital, the value of ρ_b may not properly reflect the π orbital occupancy. As stressed by Mulliken, essential to the orbital theory of electronic structure is the property of self-consistency—that each orbital be determined by its interaction with the average Coulomb and exchange potential generated by electrons in the other occupied orbitals. Thus, the density distributions derived from σ and π orbitals are not independent of one another. This coupling is very pronounced and is best illustrated by considering the change in ρ_b values for a given bond in vertical (no change in internuclear separation) ionization and electron attachment. Table 3.2 lists such values for a number of diatomic molecules whose ground electronic configuration involves occupation of the 2π antibonding orbital ($1\pi_g$ of homonuclear diatomics). It is a general observation that the vertical ionization of a π electron causes an increase in the σ density distribution along the internuclear axis, while the addition of a π electron has the reverse effect (Cade *et al.* 1971). As a result, the value of ρ_b increases with the removal of, and decreases with the addition of, a 2π electron. Just as the progressive filling of a π bonding orbital causes ρ_b and hence n to increase in the hydrocarbons, the occupation of an antibonding π orbital leads to a decrease in ρ_b. These observations are in accord with the predictions of molecular orbital theory regarding the opposite changes in bond length by a change in the occupation of a bonding or antibonding orbital (Mulliken 1932, 1939). Indeed, the effects of changing the occupation of the 2π orbital are accentuated when the accompanying change in equilibrium bond length is allowed for. For example, in CF^+, ρ_b increases to 0.3636 au at $R_e = 1.229$ Å and in NO^+, ρ_b increases to 0.7608 au at its $R_e = 1.062$ Å.

The bond paths are noticeably curved in those structures in Fig. 3.7 where the geometry of the molecule precludes the possibility of tetrahedrally directed bonds between the carbon nuclei. The differences between the bond path angle α_b, the limiting value of the angle subtended at a nucleus by two bond paths, and the geometrical or so-called bond angle α_e is important in

Table 3.2

Change in ρ at bond critical point for vertical electron ionization and addition

Occupation of 2π and state	Molecule and value of ρ_b (au)									
$2\pi^0$ ($^1\Sigma^+$)	CF^+	0.3333	NO^+	0.6130	NF^+	0.3765	OF^+	0.3890	F_2^+	0.3752
$2\pi^1$ ($^2\Pi$)	CF	0.2924	NO	0.5933	NF	0.3420	OF	0.3691	F_2	0.3696
$2\pi^2$ ($^3\Sigma^-$)			NO^-	0.5755	NF^-	0.3029	OF^-	0.3373		
$2\pi^3$ ($^2\Pi$)										
$2\pi^4$ ($^1\Sigma^+$)										
R_e of neutral (Å)	1.271		1.151		1.317		1.321		1.336	

*Calculated from wave functions using STO basis sets close to the Hartree–Fock limit: Cade and Huo (1975); Cade and Wahl (1974); O'Hare and Wahl (1970).

quantifying the concept of strain in these molecules. Another related quantity is the bond path length, R_b, which for a curved bond path exceeds the corresponding internuclear separation or so-called bond length, R_e. The values of the bond path angle and the difference $\Delta\alpha = \alpha_b - \alpha_e$ are listed in Table 3.3 for the cyclic, bicyclic, and propellane molecules together with their strain energies as calculated using Franklin's group equivalents. For most of these molecules $\Delta\alpha > 0$ and, in these cases, the bonds are less strained than the geometrical angles α_e would suggest. In general, $\Delta\alpha$ provides a measure of the degree of relaxation of the charge density away from the geometrical constraints imposed by the nuclear framework. The reader is referred to the original publication of these results for their full discussion. Here we note a few interesting results for the propellane systems. Because of the additional constraints resulting from the fusion of three small rings to a single bridge-head bond, the relaxation of the apex bond paths in these latter molecules is inhibited relative to that found for cyclopropane. Thus $\Delta\alpha$ for the apex angle in [1.1.1]propellane is actually slightly negative, while the relatively large values found for the three angles formed by the bond paths of the three-membered rings that terminate at a bridgehead carbon nucleus result in bond path angles that differ by only $\sim 1°$ from the tetrahedral angle.

   The charge density along a bond path attains its minimum value at the bond critical point and the associated curvature or eigenvalue of the Hessian of ρ at $\mathbf{r}_c$, λ_3, is thus positive. The charge density in an interatomic surface, on the other hand, attains its maximum value at the bond critical point and the two associated curvatures of ρ at $\mathbf{r}_c$, λ_1 and λ_2, those directed along axes perpendicular to the bond path, are thus negative. In a bond with cylindrical symmetry, these two negative curvatures of ρ at the bond critical point are of equal magnitude. However, if electronic charge is preferentially accumulated in a given plane along the bond path (as it is for a bond with π-character, for example), then the rate of fall-off in ρ is less along the axis lying in this plane than along the one perpendicular to it, and the magnitude of the corresponding curvature of ρ is smaller. If λ_2 is the value of the curvature of smallest magnitude, then the quantity $\varepsilon = [\lambda_1/\lambda_2 - 1]$, the ellipticity of the bond, provides a measure of the extent to which charge is preferentially accumulated in a given plane (Bader *et al.* 1983). The axis of the curvature λ_2, the major axis, determines the relative orientation of this plane within the molecule. The elliptical nature of the charge distribution in the C–C inter-atomic surface in ethylene is illustrated in Fig. 3.9. The ellipticities of the C–C bonds in ethane, benzene, and ethylene are 0.0, 0.23, and 0.45, respectively, for densities calculated from basis sets containing proper polarizing functions and the major axis of the ellipticity in each of the latter two molecules is perpendicular to the plane of the nuclei. The bond ellipticities faithfully recover the anticipated consequences of the conjugation and hyperconjugation models of electron delocalization. A classic example of delocalization through the conjugative effect is provided by trans-1,3-butadiene. The central bond in this molecule exhibits partial double bond character as evidenced by $n > 1$ and $\varepsilon > 0$. Correspondingly, as a result of the delocalization of charge, the n and ε values of the terminal C–C bonds are decreased from the values found for ethylene. Of equal importance is the observation illustrated in Fig. 3.9, that the major axes of the ellipticity induced in the central bond and of the terminal bonds are parallel to one another.

The increase in ellipticity of the C–C bond from benzene to ethylene is primarily a result of an increased contraction of the density towards the bond path, as seen from the magnitude of λ_1 increasing from 0.70 to 0.81 au, rather than of a decrease in the magnitude of λ_2, the curvature of ρ along the principal axis, which changes only by 0.01 au from 0.57 to 0.56 au. The magnitudes of both curvatures are increased over their value in ethane where they degenerate to the value -0.48 au. As discussed more fully below, it is by this mechanism that more charge is concentrated along the bond path, reducing the associated positive curvature λ_3 from 0.27 to 0.20 au, decreasing $\nabla^2 \rho$ and increasing the bond order from 1.6 to 2.0. Thus, while the ellipticity does provide a measure of the π character of a bond, it does not follow that an increase in its value results from a simple decrease in the magnitude of λ_2, as one might anticipate for an increase in a π population. Such a pedestrian change in the charge density, cannot account for the associated increases in bond order and in the magnitude of $\nabla^2 \rho$. Rather, it is the relative proportioning of the charge density between the two planes containing the minor and major axes as measured by the ratio λ_1 / λ_2 that determines ε and correctly summarizes the topological changes in ρ which accompany what, in the orbital model, is described as an increase in π population.

The chemistry of a three-membered ring is very much a consequence of the high concentration of charge in the interior of the ring relative to that along its bond paths, a fact which is reflected in substantial bond ellipticities (Cremer et al. 1983). The values of ρ_r, the value of the charge density at a ring critical point, is generally only slightly less than, and in some cases almost equal to, the values of ρ_b for the peripheral bonds in the case of a three-membered ring of carbon atoms. In four-membered and larger rings the values of ρ_r are considerably smaller, as the geometrical distance between the bond and ring critical points is greater than in a three-membered ring. Because electronic charge is concentrated to an appreciable extent over the entire surface of a three-membered ring, the rate of fall-off in the charge density from its maximum value along the bond path toward the interior of the ring is much less than its rate of decline in directions perpendicular to the ring surface. Thus the C–C bonds have substantial ellipticities, and their major axes lie in the plane of the ring. In hydrocarbons, this property is unique to a three-membered ring. The ellipticity of a C–C bond in cyclopropane is actually slightly greater than that for the 'double bond' in ethylene,

Table 3.3
Geometry and bond path angles

Molecule	Structure	Angle	Geometric angle α_c (°)	Bond path angle α_b (°)	$\Delta\alpha$ $\alpha_b - \alpha_c$ (°)	Strain energy (kJ/mol)
Cyclopropane	9	1	60.0	78.84	18.84	115
Cyclobutane	10	1	89.01	95.73	6.72	111
Cyclopentane	11	1	104.55	104.02	− 0.53	26
		2	105.36	104.43	− 0.93	
		3	106.50	105.03	− 1.54	
Cyclohexane	12	1	111.41	110.08	− 1.34	0.0
Bicyclo[1.1.0]butane	13	1	58.99	72.78	13.79	267
		2	60.50	76.62	16.12	
		3	97.91	105.07	7.16	
Bicyclo[2.1.0]pentane	14	1	60.86	79.28	18.42	229
		2	59.57	78.30	18.72	
		3	90.84	96.76	5.92	
		4	89.16	95.77	6.61	
		5	110.03	109.72	− 0.31	

Compound						
Bicyclo[2.2.0]hexane	15	1	90.17	97.01	6.85	217
		2	89.91	96.59	6.67	
		3	114.43	112.64	−1.79	
Bicyclo[1.1.1]pentane	16	1	74.44	84.72	10.27	285
		2	87.20	95.85	8.65	
Bicyclo[2.1.1]hexane	17	1	99.20	100.69	1.48	155
		2	101.77	103.46	1.69	
		3	86.11	94.21	8.10	
		4	82.60	90.95	8.35	
Bicyclo[2.2.1]heptane	18	1	94.37	97.42	3.05	60
		2	101.51	102.90	1.39	
		3	108.43	108.69	0.26	
		4	103.13	103.57	0.44	
Bicyclo[2.2.2]octane	19	1	109.30	108.68	−0.62	31
		2	109.64	108.56	−1.08	
[1.1.1]propellane	20	1	61.81	59.37	−2.44	410
		2	95.98	107.99	12.01	
		3	59.09	69.09	9.99	
[2.1.1]propellane	21	1	88.97	93.65	4.69	435
		2	91.03	97.91	6.88	
		3	112.30	116.3	4.00	
		4	57.72	67.96	10.25	
		5	97.27	111.45	14.18	
		6	64.57	65.27	0.71	

Table 3.3 (*Contd*)

Molecule	Structure	Angle	Geometric angle α_c (°)	Bond path angle α_b (°)	$\Delta\alpha$ $\alpha_b - \alpha_c$ (°)	Strain energy (kJ/mol)
[2.2.1]propellane	22	1	59.18	75.40	16.22	439
		2	61.64	74.32	12.68	
		3	112.38	116.43	4.06	
		4	90.86	92.97	2.11	
		5	89.14	96.43	7.30	
		6	128.49	126.57	−1.91	
[2.2.2]propellane	23	1	91.15	97.05	5.90	372
		2	119.96	118.51	−1.44	
		3	88.85	96.81	7.97	
Tetrahedrane	24	1	60.00	81.36	21.36	586
Cubane	25	1	90.00	97.43	7.43	647
Spiropentane	26	1	59.24	79.04	19.80	264
		2	61.51	84.84	23.33	
		3	137.6	123.02	−14.58	

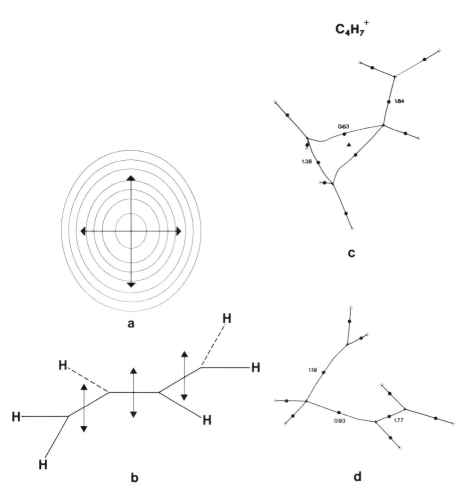

FIG. 3.9.   (a) Contour map of the electronic charge density for ethylene in the plane perpendicular to the plane of the nuclei and containing the C–C bond critical point. The charge density is a
maximum at this point in this plane. The magnitude of the curvature of ρ at the critical point is a
minimum in the direction of the major axis, the axis of the curvature λ_2. This axis is
perpendicular to the plane of the nuclei. The diagram illustrates the elliptical nature of the
distribution of charge associated with the presence of a π bond. (b) The major axes of the C–C
bond ellipticities in trans-butadiene are parallel to one another and perpendicular to the plane
containing the nuclei. This observation is consistent with the orbital model of a delocalization of
π charge density through conjugation. (c) The molecular graph of the $C_4H_7^+$ molecular ion. The
bond orders are given in this and in the following structure. Note the proximity of the critical
points of the two long bonds to the ring critical point. (d) The resulting open structure of $C_4H_7^+$.

indicating that the extent to which charge is preferentially accumulated in the plane of the ring is greater than that accumulated in the π-plane of ethylene. This property accounts for the well-documented ability of three-membered rings to act as an unsaturated system with the charge distribution in the plane of the ring exhibiting properties characteristic of a π-like system, one that is able to conjugate with a neighbouring unsaturated system. Such conjugation is illustrated by the interaction of the cyclopropyl group with the (formally) vacant 2p orbital in CH_2^+ as depicted in Fig. 3.9(c). The major axis of the ellipticity induced in the $C-CH_2$ bond has an overlap of 0.97 with the corresponding axes of the neighbouring C–C bonds of the cyclopropyl group. (The overlap is determined simply by taking the scalar product of the eigenvectors defining the major axes of the two bond critical points.) Such conjugating ability of the cyclopropyl group is rationalized using molecular orbital models through the choice of a particular set of orbitals, the so-called Walsh orbitals. Theory shows that the 'π-like' nature of a three-membered ring is a property of its total charge distribution, one that results from the proximity of its ring and bond critical points. Understanding the physical basis of this effect enables one to predict its appearance and consequences in other systems.

In Section 3.2.2 it was shown that the opening of a ring structure resulted from the coalescence of the ring and a bond critical point, the positive curvature of the ring point annihilating the in-plane negative curvature λ_2 of the bond point to yield a zero curvature characteristic of an unstable or degenerate critical point (Fig. 3.2). The decrease in the magnitude of λ_2 and its eventual disappearance means that the ellipticity of the bond which is to be broken increases dramatically and becomes infinite at the geometry of the bifurcation point. Thus, a structure possessing a bond with an unusually large ellipticity is potentially unstable. The two equivalent ring bonds of the cyclopropylcarbinyl cation, $C_4H_7^+$ (Fig. 3.9), provide an example of this behaviour. The two long bonds of the structure in Fig. 3.9(c) are of order 0.6 and exhibit ellipticities equal to 6.7. Their corresponding paths are very inwardly curved and their bond path length exceeds the internuclear separation by 0.20 Å. The structure verges on instability since either of the bond critical points of the long bonds can be annihilated by a coalescence with the ring critical point. The curvature of ρ at the ring critical point, whose position as indicated in Fig. 3.9 lies almost on the line joining the two bond critical points, is close to zero and, correspondingly, the associated negative curvature, λ_2, of each of the neighbouring bond critical points is equally small in magnitude. As anticipated for long bonds, the positive curvature of ρ along their bond paths is relatively large, as is the second and parallel positive curvature of ρ at the ring critical point. The values of ρ at the bond and ring critical points differ by only 0.001 au. Thus, there is a nearly flat-bottomed trough in the distribution of charge linking these three critical points and little energy is required to cause a migration of the ring point along the trough to coalesce with a bond point and yield the open structure shown in Fig. 3.9(d). It is a general observation that little energy is required for the nuclear motions which result in a migration of a critical point along an axis associated with vanishingly small curvature of the charge density. Thus, the energy surface in the neighbourhood of the structure in Fig. 3.9(c) is very flat for such a motion of the nuclei and the structure in Fig. 3.9(d) differs from it in energy by less than 1 kcal/mol (Cremer *et al.* 1983).

Further examples of potentially unstable structures being revealed through exceptionally high bond ellipticities are provided by the propellanes, particularly [2.1.1]propellane, **21** in Fig. 3.7. The bridgehead bond critical point and each of the ring critical points of the two three-membered rings in this molecule are separated by only 0.07 Å and the value of $\rho_b = 0.197$ au exceeds that of ρ_r by only 0.001 au. Consequently, as is evident from the relief map of the charge density shown in Fig. 3.8(c), the close proximity and nearly equal values for the bond and ring critical points result in a near zero value for the curvature of the density at the bond critical point in the direction of the three-membered ring critical points. (See also Fig. 3.8(b) for the [2.2.1] case which shows the interaction with a single three-membered ring critical point in profile more clearly.) The result is a very large ellipticity, equal to 7.21, for the bridgehead bond in this molecule. The bridgehead bonds in both [2.2.1] and [2.1.1]propellane are predicted to be the most susceptible to rupture by the bifurcation mechanism and both molecules readily undergo polymerization at 50 K. The bifurcation catastrophe undergone by the [1.1.1]propellane molecule is used to illustrate the mathematical modelling of a structural instability in Chapter 4.

These properties of a charge distribution have been applied to a study of the position of the equilibrium dinorcaradiene $\rightleftharpoons$ [10]-annulene as a function of the substituents R. While X-ray diffraction studies yield the geometries of the relevant species, and in particular the C1–C6 internuclear separation, they do not enable one to determine whether or not carbons C1 and C6 are bonded to one another and hence to determine which of the above two structures is the correct one for a given set of substituents R and R'. Gatti *et al.* (1985) determined the topological properties of the theoretically determined charge distributions at the experimentally measured geometries for combinations of substituents R and R' = CN, CH$_3$, H, and F yielding C1–C6 separations ranging from 1.543 Å for R = R' = CN to 2.269 Å for R = R' = F. The study demonstrated that the dicyano derivative and the methyl-cyano derivative (in the β crystallographic phase) possess a C1–C6 bond, whereas, of the two crystallographically distinct molecules in a unit cell of the dimethyl derivative, one possesses a C1–C6 bond, the other, with the annulenic structure, does not. The remaining systems also exhibit the annul-enic structure. The dimethyl derivative with the shortest C1–C6 separation, equal to 1.770 Å, is the last number of the series to possess a dinorcaradiene structure. The bond order of the C1–C6 bond and the separation between its critical point and that of the three-membered ring undergo a continuous decrease and, correspondingly, the C1–C6 bond ellipticity exhibits a contin-uous increase for the three 'closed' molecules. The display of the gradient vector field for the next member in the plane of the three-membered ring, which possesses a C1–C6 separation only slightly greater, at 1.783 Å, indi-cates that this separation is past the geometry of the bifurcation point created by the coalescence of the ring and bond critical point, as both critical points are absent from the display.

   The sum of the three curvatures of ρ at a bond critical point, the quantity $\nabla^2\rho_b$, provides a useful characterization of the manner in which the electronic charge density is distributed in the internuclear region. As pointed out in Section 2.2.2, the Laplacian of the charge density, the quantity $\nabla^2\rho(\mathbf{r})$, determines where electronic charge is locally concentrated, where $\nabla^2\rho < 0$, and where it is locally depleted, where $\nabla^2\rho > 0$. A chemical bond requires the creation of a (3, − 1) or bond critical point in the charge density which has two negative and one positive curvature. Thus, mechanisms involving both charge concentrations and charge depletions are bought into play in the formation of a chemical bond. The charge density is a maximum at the bond critical point $\mathbf{r}_c$ in the interatomic surface, its two associated curvatures, those perpendicular to the bond path, are negative, and charge is locally concen-trated at $\mathbf{r}_c$ in the interatomic surface. The charge density is a minimum at the bond critical point along the bond path; the associated curvature of ρ is positive and charge is locally depleted at this point relative to neighbouring points on the bond path. Thus, the formation of a chemical bond is the resultant of two competing effects: the radial contraction of the charge density towards the bond path, leading to its concentration at $\mathbf{r}_c$ and in the interatomic surface, as opposed by a parallel expansion of ρ which removes charge density from $\mathbf{r}_c$ and from the interatomic surface and results in its separate accumulation in each of the atomic basins. Which of the two effects dominates a given interaction is determined by the sign of the Laplacian of ρ at $\mathbf{r}_c$. The complete spectrum of values from positive, as is characteristic of the interaction between two closed-shell systems, to negative, as found for interactions with a sharing of charge accumulated in the internuclear region, is observed.

All of the C–H bonds and all of the C–C bonds shown in Fig. 3.7, save the bridgehead bonds in [1.1.1] and [2.1.1]propellane, have negative values for $\nabla^2 \rho_b$. In these bonds the perpendicular contractions of ρ dominate the interaction and electronic charge is concentrated between the nuclei along the bond path. The result is a sharing of charge between the atoms similar to that one associates with a covalent interaction. The very strained nature of the bridgehead bonds in [1.1.1] and [2.1.1]propellane result in local concentrations of charge in their non-bonded as well as in their bonded regions, a result which, as discussed by Wiberg *et al.*, 1987, accounts for their susceptibility to acetolysis. When there is substantial charge transfer across an interatomic surface, electronic charge is separately accumulated in the basins of the two atoms and, as a result, λ_3, the parallel curvature of ρ at $\mathbf{r}_c$, is large and its sign dominates the value of $\nabla^2 \rho_b$. The presence of a large positive curvature along the bond path imparts a 'stiffness' to the bond, making it resistant to change by changing substituent effects. On the other hand, the large accumulation of charge in the internuclear region obtained when $\nabla^2 \rho_b$ is negative, results in a softening of the parallel curvature and λ_3 is relatively small in value. These bonds are more easily perturbed, the shift in the interatomic surface and the associated changes in the other bond parameters increasing with a decrease in the value of λ_3 (Slee 1986).

The full usefulness of the classification using $\nabla^2 \rho_b$ must await the development of the quantum mechanical aspects of the theory. The Laplacian of the charge density appears in the local expression of the virial theorem and it is shown that its sign determines the relative importance of the local contributions of the potential and kinetic energies to the total energy of the system. A full discussion of this topic is given in Section 7.4.

These examples demonstrate that molecular structure and its stability are predicted by a theory which uses only the information contained in the quantum mechanical state function and that the static and dynamic properties of a bond can be characterized in terms of the properties of the charge density at the bond critical point. The values of ρ_b, α_b, ε, and $\nabla^2 \rho_b$ enable one to translate the predicted electronic effects of orbital models into observable consequences in the charge distribution.

3.3 A theory of molecular structure

3.3.1 A coming together of mathematics and chemistry

The definition of structure and structural stability for a molecular system as presented above is an example of the application of a general mathematical theory of structural stability. This work has evolved, under the general headings of differential topology and qualitative dynamics, a theory of

dynamical systems and their stabilities. Contributions of particular importance to the present work are the theory of elementary catastrophes developed
by R. Thom (1975) and the general statement of the theorem of structural
stability given by J. Palis and S. Smale (1970). What must surely strike the
reader during the unfolding of this application is the remarkable ease with
which the abstract mathematical ideas are mapped on to the topological
properties of the charge distribution, a real physical object.

A central problem of catastrophe theory is to model the discontinuous
changes in the observed behaviour of a system that are caused by smooth,
continuous variations in the values of the parameters which control it. In the
light of the discussion already presented, this is the central problem of
molecular structure—to predict the discontinuous changes in structure that
are caused by a continuous change in the nuclear coordinates, the parameters
which control the behaviour of the charge density as observed in real space.
This aspect of the theory—the mathematical modelling of the structural
changes occurring in the neighbourhood of a bifurcation point in nuclear
configuration space—is developed in Chapter 4. A very good exposition of
catastrophe theory is given by Poston and Stewart (1978). In the next section
we restate the equivalence relationship and its application to the gradient
vector fields of the charge density in the concise language of set theory and
linear mappings. It is demonstrated that the two mechanisms of structural
instability encountered in a molecular system, the bifurcation and conflict
mechanisms, are predicted as a corollary of the theorem of structural stability
proposed by Palis and Smale (1970). A brief summary of the elementary
notions of set theory appears in Section E3.1. A reader, disinterested in the
more mathematical as opposed to the applied aspects of the theory, may
peruse the examples and conclusions and then proceed to Chapter 5.

3.3.2   The equivalence relationship and structural stability

The idea of studying the properties of the gradient vector field of the charge
density has its basis in quantum mechanics. The boundary condition for the
definition of a quantum subsystem, eqn (2.9), is stated in terms of this field—
that the surface of the subsystem be one of zero flux in the gradient vectors of
the charge density. The resulting study of the topology of the charge density
reveals the existence not only of atoms as defined by the zero-flux surface
condition, but also of a molecular graph as defined through the pairs of
gradient paths associated with the positive eigenvector of the $(3, -1)$ critical
points. In fact the two manifolds, as defined by the trajectories that terminate
and originate at such a critical point, define, respectively, the interatomic
surface shared by the basins of neighbouring attractors and the bond path
which links them. Thus, the topology of the charge density is such that the

quantum definition of an atom is inseparable from the definition of a molecular graph. It is good fortune or design that this is so, for the idea underlying the elementary theory of structure is that it is to be found in the dynamical properties of the gradient vector field of the system's scalar realization in behaviour space. It is important to remember that, while the structural aspects of the theory of atoms in molecules are described by the dynamical changes in the topology of the charge density, the theory is rooted in quantum mechanics. It is the atom and its properties which are defined by quantum mechanics. The bond paths and the structure they define just mirror and summarize in a convenient way what the atoms are doing, performing the same role here as does the assignment of a set of bonds in the molecular structure hypothesis.

To define structure and structural stability, use is made of an equivalence relation of vector fields over R^3. The equivalence relation is defined as follows: two vector fields $\mathbf{v}$ and $\mathbf{v}'$ over R^3 are said to be equivalent if and only if there exists a homeomorphism, i.e. a bijective and bicontinuous mapping of R^3 into R^3, which maps the trajectories of $\mathbf{v}$ into the trajectories of $\mathbf{v}'$. By applying this definition to the gradient vector fields $\nabla\rho(\mathbf{r}; \mathbf{X})$, $\mathbf{X} \in R^Q$, one obtains an equivalence relation operating in the nuclear configuration space R^Q which states: two nuclear configurations $\mathbf{X}$ and $\mathbf{X}' \in R^Q$ are equivalent if and only if their associated gradient vector fields $\nabla\rho(\mathbf{r}; \mathbf{X})$ and $\nabla\rho(\mathbf{r}; \mathbf{X}')$ are equivalent. We further say that the nuclear configuration $\mathbf{X} \in R^Q$ is structurally stable if $\mathbf{X}$ is an interior point of its equivalence class. In other words, one can always find a neighbourhood V of a structurally stable configuration $\mathbf{X}$, such that V is totally contained in the equivalence class of $\mathbf{X}$. All configurations in V possess the same molecular graph as does the stable configuration $\mathbf{X}$.

Indeed for any point $\mathbf{X}'$ of V one can find a homeomorphism, $h_{\mathbf{XX}'}$, which maps the gradient paths of $\nabla\rho(\mathbf{r}; \mathbf{X})$ into those of $\nabla\rho(\mathbf{r}; \mathbf{X}')$. It is easily seen that through $h_{\mathbf{XX}'}$ a critical point of type (ω, σ) in $\rho(\mathbf{r}; \mathbf{X})$ is mapped on to a critical point of the same type in $\rho(\mathbf{r}; \mathbf{X}')$. Moreover, a gradient path connecting a pair of critical points $(\mathbf{r}_c, \mathbf{r}'_c)$ in $\rho(\mathbf{r}; \mathbf{X})$ is transformed into a corresponding path of $\nabla\rho(\mathbf{r}; \mathbf{X}')$, which connects the images of $\mathbf{r}_c$ and $\mathbf{r}'_c$. Since the distance between $\mathbf{X}'$ and $\mathbf{X}$ can be made arbitrarily small and since all nuclear configurations in V are equivalent, one finds that the molecular graphs associated with the points of V consist of the same number of bond paths linking the same nuclei. These molecular graphs represent a single structure, and the maximal neighbourhood contained in the equivalence class of $\mathbf{X}$ is called the structural region associated with $\mathbf{X}$.

An equivalence relation for molecular graphs is defined as follows: two molecular graphs are equivalent if and only if they are associated with two points of the same structural region. An equivalence class of molecular graphs is called a molecular structure. It is then seen that a unique molecular structure is associated with a given structural region and that molecular structure, as defined above through the equivalence of molecular graphs, necessarily fulfils the requirement of being generic.

These definitions were previously illustrated in terms of the gradient paths shown in Fig. 3.1 for successive stages along a dissociative pathway of the water molecule. As discussed in terms of these examples, the application of the notion of structural stability to the topological study of the molecular charge distribution leads to a partitioning of the nuclear configuration space into a finite number, l, of non-overlapping regions, the structural regions, W_i, $i = 1, \ldots, l$, each of which is characterized by a unique molecular structure. These structural regions form a dense open subset of the nuclear configuration space, i.e.

$$\overline{\left(\bigcup_{i=1}^{l} W_i \right)} = R^Q \tag{3.1}$$

where the symbol $\bar{A}$ denotes the closure of the set A and $\bigcup$ denotes the set-theoretic union. A point which belongs to the union of the W_i belongs to some structural region and is called a regular point. A nuclear configuration belonging to the complementary of the set of regular points is called a catastrophe point. The catastrophe set, C, is the collection of all structurally unstable points of nuclear configuration space. Let ∂W_i denote the boundary of the structural region W_i. Since $\bigcup_{i=1}^{l} W_i$ is dense in R^Q, we have

$$C = \bigcup_{i=1}^{l} (\partial W_i), \tag{3.2}$$

i.e. the catastrophe set is the union of the boundaries of all l structural regions W_i. Equation (3.2) denotes the catastrophe set C as the loci of structural changes. Indeed, according to eqn (3.2), a catastrophe point $Y \in C$ belongs to ∂W_i, for some $i \in \{1, \ldots, l\}$. Any neighbourhood of Y in R^Q thus has non-empty intersections with W_i and at least one structural region W_j, with $j \neq i$. Consequently, the slightest displacement of the system from the nuclear configuration Y will cause the molecular graph to change from the graph associated with Y to one which represents either the structure over W_i or that over W_j. Thus, the molecular graph associated with a catastrophe point denotes a discontinuous change in structure which results from a continuous variation in the set of control parameters, the nuclear coordinates.

As discussed above and illustrated in Fig. 3.4, the partitioning of nuclear configuration space obtained as a result of the definition of molecular structure leads to the concept of a structure diagram. The space R^Q is partitioned into a finite number of structural regions with their boundaries, as defined by the catastrophe set, denoting the configurations of unstable structures. This information constitutes a system's structure diagram, a diagram which determines all possible structures and all mechanisms of structural change for a given chemical system.

The role of a structure diagram is equivalent to that of a phase diagram in thermodynamics. In both cases, the space of the control parameters, the nuclear coordinates in one case, P, V, and T in the other, is partitioned through the assignment of some property to each point in this space. This property is generic since it assumes a unique value over each structural region or phase. The boundaries between such open regions correspond to instabilities, i.e. to discontinuous changes in the generic property. Thus, an important consequence of the definition of molecular structure is that a change in structure, like a change in phase, is predicted to be an abrupt, discontinuous process resulting from a continuous change in the control parameters.

The examples previously discussed with reference to the structure diagram demonstrated the existence of two kinds of catastrophe points, called bifurcation and conflict points. Both types of instabilities were illustrated in terms of the behaviour observed for molecular charge distributions. What we now show is that the existence of these two kinds of catastrophes and just these two, is a direct consequence of a theorem of structural stability stated by Palis and Smale in 1970. This theorem predicts what are the two basic mechanisms for structural change in a chemical system.

3.3.3   Stable and unstable intersections of lines and surfaces in R^3

In order to state the conditions under which one will observe a catastrophe of the conflict type, it is necessary to introduce a number of definitions and the notion of a transversal or stable intersection of two submanifolds of three-dimensional space.

Atoms and bonds are defined, respectively, by surfaces and lines embedded in three-dimensional space. A surface and a line are submanifolds of dimensions two and one, respectively, of R^3. It is necessary that the surface or line be embedded smoothly in R^3, i.e. that it possess a unique tangent hyperplane at each point. This condition of smoothness is certainly satisfied in the present case since an interatomic surface and a bond path are defined in terms of the trajectories of the gradient vectors of the charge density associated with a $(3, -1)$ critical point, as the surface and axis of a ring structure are defined by the trajectories associated with a $(3, +1)$ critical point. One can picture the tangent plane to a point of the curved interatomic surface illustrated in Fig. 2.6, a plane defined by the gradient vector of ρ at that point on the surface.

The problem we wish to consider is how to judge whether or not a given intersection of the manifolds of two critical points is stable or unstable to a slight jiggling of the nuclei—does the nature of their intersection remain unchanged or is it radically altered? The two kinds of intersections are termed transversal or non-transversal, respectively. As a simple example, consider the possible interactions between a plane and a curve in R^3 as pictured in Fig. 3.10. When the curve intersects the plane at some finite number of points, or does not intersect the plane at all, we say that their intersection is transversal. Such an intersection will obviously not be affected by small changes in the relative positions of the curve and the plane. If, on the other hand, the curve happens to be tangent to the plane, the intersection is non-transversal. A small change in their relative positions will result in a transition from a non-transversal to a transversal one.

Some necessary definitions are first introduced. Consider a critical point r_c, labelled by its rank and signature as (ω, σ). The stable manifold of r_c is defined to be the $(1/2)(\omega - \sigma)$-dimensional manifold generated by the eigenvectors of the Hessian matrix at r_c associated with its negative eigenvalues. Its unstable manifold is the $(1/2)(\omega + \sigma)$-dimensional manifold generated by the eigenvectors associated with the positive eigenvalues of the Hessian matrix. Thus, for example, the interatomic surface and the bond path are the stable and unstable manifolds respectively, of a $(3, -1)$ critical point.

The general statements regarding the transversality of the intersections of submanifolds are stated in terms of the properties of the intersections of their associated tangent hyperplanes. These are vector subspaces, a plane or line spanned by a corresponding set of gradient vectors, not necessarily at the origin of the coordinate system. Such a vector subspace is called an affine subspace.

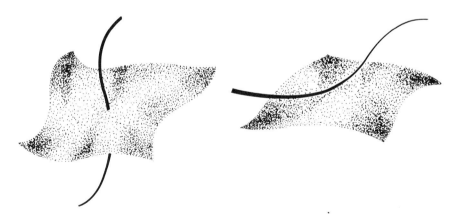

FIG. 3.10.   The diagram on the left illustrates a transversal intersection of a line with a surface. A motion of either manifold relative to the other will not change the nature of this intersection. The diagram on the right illustrates a non-transversal intersection. The line is tangent to the surface at one point. This situation is not stable to arbitrary relative motions of the two manifolds. Such motions will cause the line to either intersect the surface or to not graze the surface.

Let E and F be such affine subspaces in R^3 of dimensions s and t, respectively. The intersection of E and F is said to be transversal if either

(1) $E \cap F = 0$, that is, they do not intersect, or

(2) $s + t \geqslant 3$ and $\dim(E \cap F) = s + t - 3$.

One can now state the conditions for the intersection of two submanifolds M and N in R^3 to be transversal:

(1) $M \cap N = 0$, or

(2) the tangent affine hyperplanes of M and N intersect transversely.

Examples of non-transversal intersections of bond paths, ring axes, and interatomic and ring surfaces are illustrated in Fig. 3.11. They represent the conflict structures that can be found in the topologies of molecular charge distributions. The first arises from the intersection of a bond path of one $(3, -1)$ critical point (its unstable manifold) with the interatomic surface (the stable manifold) of a second such critical point. Examples of the resulting unstable structure have been given previously in the dissociation of water and in the isomerization of hydrogen cyanide (Figs 3.1 and 3.6). The example given in Fig. 3.11 is for the minimum-energy structure of the CH_5^+ molecular ion. The molecular graph suggests that the structure of this ion corresponds to a CH_3^+ ion interacting with a hydrogen molecule. The second example of a non-transversal intersection results from the intersection of the stable and unstable manifolds of $(3, +1)$ critical points. In this case the ring axis of one ring critical point intersects with the surface of the other. The resulting structure is exemplified by the molecular graph shown for the [1.1.1]propellane molecule in which the bond path between the bridgehead carbons has been previously annihilated in a bifurcation catastrophe. (The

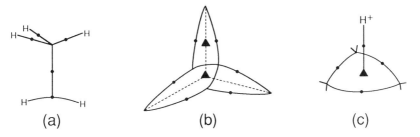

FIG. 3.11. Examples of the possible non-transversal intersections in a molecular system: (a) between the stable and unstable manifolds of two bond critical points in CH_5^+; (b) between the stable and unstable manifolds of two ring critical points in a distorted structure of [1.1.1]propellane; (c) between the unstable manifold of a bond critical point with the stable manifold of a ring critical point (the ring axis) in face-protonated cyclopropane.

structures obtained for the [1.1.1]propellane molecule in the neighbourhood of this bifurcation catastrophe point are discussed in some detail in Chapter 4.) The final example concerns the intersection of the bond path of a $(3, -1)$ critical point with the ring axis of a $(3, +1)$ critical point. The example shown in Fig. 3.11(c) is obtained for what is referred to as face-protonated cyclopropane. This unstable structure is transformed into the corner-protonated species when the bond path from the proton switches from the one-dimensional ring attractor to any one of the three nuclear attractors.

The instability arising from the intersection of the stable and unstable manifolds of two $(3, +1)$ critical points does not lead to a change in the network of bond paths, but only to the formation of a new, stable two-ringed structure. The intersection of a bond path of a $(3, -1)$ critical point with either a ring axis or an interatomic surface, however, leads to instabilities whose removal results in a change in the bonded structure of a molecule. From the examples of the interactions that have been found so far, unstable structures of the conflict type appear to be characteristic of relatively weak interactions associated with longer than normal interatomic separations or with relatively high-energy transition-state structures.

3.3.4   Criteria for structural stability

We are now in a position to adopt Palis and Smale's theorem on structural stability (1970) to describe structural changes in a molecular system. A configuration $\mathbf{X} \in R^Q$ is structurally stable if $\rho(\mathbf{r}; \mathbf{X})$ has a finite number of critical points such that:

(1) each critical point is non-degenerate, and

(2) the stable and unstable manifolds of any pair of critical points intersect transversely.

The immediate consequence of the theorem is that a structural instability can be established through one of two mechanisms which correspond to the bifurcation and conflict catastrophes previously described. A change in molecular structure can only be caused by the formation of a degenerate critical point in the electronic charge distribution or by the creation of an unstable intersection of the submanifolds of $(3, -1)$ and/or $(3, +1)$ critical points. The mechanisms of the making and breaking of chemical bonds are thus seen to be the predicted consequences of changes in the morphology of a molecular charge distribution.

3.4   Comparison of a structure diagram with other partitionings of R^Q

3.4.1   Relation between topological and energetic stabilities of
            molecular structures

A structure diagram partitions nuclear configuration space into structural regions. The stable structural regions are separated by hypersurfaces which are the loci of the unstable structures. The Born–Oppenheimer potential energy is also defined over nuclear configuration space to yield an energy hypersurface $E(\mathbf{X})$ for a molecular system. In terms of the critical points of $E(\mathbf{X})$ one may define and distinguish between energetically stable and unstable geometries. Thus, there are two definitions of stability associated with the nuclear configuration space of a system. The first defines regions of this space as determined by the topological properties of the system's charge distribution. The other defines points within the same space as determined by the properties of the system's energy hypersurface. One may inquire as to the extent to which these two definitions of stability may be juxtaposed (Tal $et\ al.$ 1981).

The reader is reminded that it is important to distinguish between molecular geometry and molecular structure. Geometry is a non-generic property since any infinitesimal change in $\mathbf{X}$ results in a different geometry. Molecular structure, on the other hand, is a generic property. Any configuration $\mathbf{X}'$ in the neighbourhood of a given regular point $\mathbf{X}$, while it has a different geometry, possesses the same structure. Only in this way can one account for the invariance of the network of bonds to the nuclear motions of a system.

The critical points of $E(\mathbf{X})$ enable one to distinguish between energetically stable and unstable geometries. Energetically stable geometries are points $\mathbf{X} \in R^Q$ for which $E(\mathbf{X})$ is at a local minimum with respect to all internal motions of a system. Energetically unstable geometries are saddle points in $E(\mathbf{X})$, geometries which are at a maximum with respect to one or more of the system's internal coordinates. One may also define unique 'reaction co-ordinates' (Tachibana and Fukui 1978) in terms of trajectories of $-\nabla_\mathbf{X} E(\mathbf{X})$. A reaction coordinate consists of a pair of trajectories of $-\nabla_\mathbf{X} E(\mathbf{X})$ originating at a saddle point and terminating at two different local minima. The associated energetically unstable geometry is the transition state for the conversion of one stable geometry into the other.

Mezey (1981) has proposed a partitioning of nuclear configuration space R^Q into catchment regions as determined by the gradient vector field, $-\nabla_\mathbf{X} E(\mathbf{X})$. This latter partitioning is similar to the partitioning of a molecular charge distribution into atomic basins. Corresponding to the definition of an atomic basin as the open region of R^3 traversed by all the trajectories of $\nabla\rho$ which terminate a given nucleus, Mezey's catchment region may be simply defined as the open region of R^Q traversed by all the trajectories of

$- \nabla_X E(\mathbf{X})$ which terminate at a local maximum in $- E(\mathbf{X})$ (an energetically stable geometry). We have found that a molecular graph is, in general, invariant to the nuclear motions of a molecule in the neighbourhood of an energetically stable geometry. Thus, the definition of molecular structure as an equivalence class of molecular graphs associates a given structure with a neighbourhood of an energetically stable geometry.

The boundaries of a catchment region or basin of $E(\mathbf{X})$ are the surfaces of zero flux in the vector $- \nabla_X E(\mathbf{X})$. These boundaries represent the loci of unstable geometries, as $E(\mathbf{X})$ on a boundary is a maximum with respect to one or more of the nuclear motions of the system. Such an energetically unstable geometry, which represents the transition state on a reaction coordinate, may or may not lie on the boundary between two structural regions of the structure diagram defined by the topological properties of a system's charge distribution. It is not possible to give a general answer to the question 'Do the geometries of the energetically unstable configurations of a molecular system correspond to the topologically unstable structures of the system as defined by its catastrophe set?' since the functional relationship between the charge density and the energy is not known. We have discussed a number of examples above where this coincidence of energetic and topological instabilities is indeed found, but it is clear from the same set of examples that a catastrophe point will not always have an energetically unstable geometry associated with it. The C_{2v} approach of an oxygen atom to a hydrogen molecule, described above (Fig. 3.1), is energetically downhill and no maxima in $E(\mathbf{X})$ along the reaction coordinate are found at the catastrophe points denoting the formation and destruction of the ring structure. That is, the stable ring structure does not appear as an energy intermediate along the reaction coordinate. It is also true that not all reactions involving an energy barrier involve a change in the topological structure of a system. The transition-state geometries for reactions involving internal rotations or inversions cannot belong to the catastrophe set as both reactant and product possess equivalent molecular graphs in these cases.

The gradient vector field of $- E(\mathbf{X})$ which determines the partitioning of the energy hypersurface is determined in turn by the Hellmann–Feynman force exerted on the nuclei. The electronic contribution to this force is given by an integral over the charge density; thus, the topology of $E(\mathbf{X})$ is not totally unrelated to the form of the charge density. One can, in fact, re-express this electronic contribution as an integral over the gradient vector field of the charge density $\nabla \rho(\mathbf{r}; \mathbf{X})$, weighed by the nuclear–electron potentials. It has been shown (Tal et al. 1981) that one may obtain an understanding of the correspondence between an energetically unstable critical point as determined by the field $- \nabla E(\mathbf{X})$ and an unstable structure as determined by the field $\nabla \rho(\mathbf{r}; \mathbf{X})$ when this correspondence is indeed observed, as in the isomerizations of HCN and its methyl derivative, for example.

3.4.2   Structural homeomorphism between ρ and the nuclear potential

We have frequently commented on the fundamental role played by the nuclear–electron potential in the determination of the topological properties of the electronic charge distribution. Because of the general observation that the only local maxima of charge distributions of many-electron systems occur at the positions of the nuclei, the nuclei are the point attractors of the gradient vector field of the charge density as well as of the gradient vector field derived from the nuclear potential. To what extent is the form exhibited by the charge density determined by the properties of the nuclear potential $V(\mathbf{r}; \mathbf{X})$? The similarity of two fields is best judged by comparing their topological properties. A considerable similarity is shown to exist between the scalar fields $\rho(\mathbf{r}; \mathbf{X})$ and $V(\mathbf{r}; \mathbf{X})$ through a comparison of the structure which is definable in terms of their associated gradient vector fields. Indeed, it has been argued that the structure diagrams generated by these two scalar fields will, in general, be found to be homeomorphic (Tal *et al.* 1980).

   The topology of the nuclear potential

$$V(\mathbf{r}; \mathbf{X}) = \sum_\alpha Z_\alpha (|\mathbf{r} - \mathbf{X}_\alpha|)^{-1}, \qquad \mathbf{X} = \{\mathbf{X}_\alpha\}, \tag{3.3}$$

is displayed and characterized by its gradient vector field $\nabla V(\mathbf{r}; \mathbf{X})$, where the gradient is taken with respect to the electronic variable $\mathbf{r}$ in a fashion totally analogous to that followed in the analysis of the topology of $\rho(\mathbf{r}; \mathbf{X})$. The field of eqn (3.3) becomes infinite if and only if $\mathbf{r} = \mathbf{X}_\alpha$, the coordinates of one of the nuclei. At such a point ∇V is both discontinuous and infinite. Thus, the maxima in $V(\mathbf{r}; \mathbf{X})$, like the corresponding maxima in $\rho(\mathbf{r}; \mathbf{X})$, are not true $(3, -3)$ critical points. However, again as for $\rho(\mathbf{r}; \mathbf{X})$, the phase portrait for this point is indistinguishable from that for a true $(3, -3)$ critical point. This behaviour is made evident in Fig. 3.12 which portrays maps of the trajectories of $\nabla V(\mathbf{r}; \mathbf{X})$ and also of $\nabla\rho(\mathbf{r}; \mathbf{X})$ for the C_{2v} approach of an oxygen atom to a hydrogen molecule. All of the trajectories of ∇V in the neighbourhood of a given nucleus terminate at that nucleus. Thus, a nucleus acts as an attractor in both $V(\mathbf{r}; \mathbf{X})$ and $\rho(\mathbf{r}; \mathbf{X})$ fields. Moreover, nuclei are the only attractors in $V(\mathbf{r}; \mathbf{X})$ and they are observed to be attractors in many-electron charge distributions. Since topological structure and its stability are determined through the attainment of a state of balance in the competition between the attractors of a system, one anticipates a significant degree of similarity in the topological properties of V and ρ. Thus, one finds that the basins of neighbouring attractors in both fields are separated by the surface generated by the trajectories which terminate at a $(3, -1)$ critical point and the attractors themselves are linked by the unique pair of trajectories which originate at this critical point.

   Figure 3.12 shows that the same sequence of elementary graphs for the H_2O system is obtained for $V(\mathbf{r}; \mathbf{X})$ as is obtained for $\rho(\mathbf{r}; \mathbf{X})$, including a

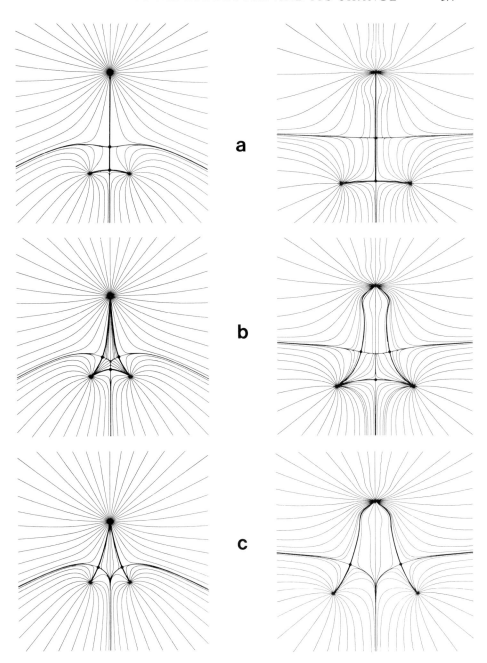

FIG. 3.12.   Gradient maps of the nuclear potential $V(\mathbf{r}; \mathbf{X})$ on the left and of the electronic charge density $\rho(\mathbf{r}; \mathbf{X})$ on the right for symmetric structures of the H_2O system. The pairs of maps demonstrate the existence of equivalent structures.

graph corresponding to the bonded ring of nuclei, Fig. 3.12(b). However, as a consequence of Poisson's equation

$$\nabla^2 V(\mathbf{r}; \mathbf{X}) = -4\pi \sum_\alpha Z_\alpha \delta(\mathbf{r} - \mathbf{X}_\alpha), \tag{3.4}$$

the topological properties of $V(\mathbf{r}; \mathbf{X})$ are constrained relative to those for $\rho(\mathbf{r}; \mathbf{X})$. Since

$$\nabla^2 V(\mathbf{r}; \mathbf{X}) = \sum_i \lambda_i, \tag{3.5}$$

the sum of the eigenvalues λ_i of the Hessian matrix for a critical point in $V(\mathbf{r}; \mathbf{X})$ other than at a $(3, -3)$ critical point must equal zero. No such restriction exists for critical points in $\rho(\mathbf{r}; \mathbf{X})$. Thus $V(\mathbf{r}; \mathbf{X})$ may possess both $(3, -1)$ and $(3, +1)$ critical points but with the restriction that $|\lambda_1 + \lambda_2| = |\lambda_3|$, where λ_3 is the eigenvalue of unique sign. It cannot possess a $(3, +3)$ critical point and hence it cannot exhibit a graph corresponding to a cage structure. However, as discussed later, even in this case $V(\mathbf{r}; \mathbf{X})$ may possess a graph which is topologically equivalent to a bonded cage of $\rho(\mathbf{r}; \mathbf{X})$.

A gradient path of V has a simple physical interpretation. It is a line of force—the path traversed by a test charge moving under the influence of the potential $V(\mathbf{r}; \mathbf{X})$. At a critical point other than a $(3, -3)$ critical point, the force vanishes. Thus a critical point in the field $V(\mathbf{r}; \mathbf{X})$ denotes a point of electrostatic balance between the attractors of the system. Since trajectories defining the surface which separates neighbouring basins satisfy the 'zero-flux' condition

$$\nabla V(\mathbf{r}; \mathbf{X}) \cdot \mathbf{n}(\mathbf{r}) = 0 \qquad \text{for all } \mathbf{r} \in S(\mathbf{r}), \tag{3.6}$$

a test charge on such a surface is not drawn to either attractor. On the contrary, a test charge lying within a basin is drawn to the attractor (the nucleus) and, thus, the boundary of a given basin defines the region of space dominated by the field of its own attractor. With a common set of elements as attractors, $V(\mathbf{r}; \mathbf{X})$ exhibits the same basic elements of structure as does $\rho(\mathbf{r}; \mathbf{X})$, with the atoms and bond paths of the latter being replaced by the topologically equivalent nuclear basins and nuclear–nuclear interaction lines of the former field. The simple and direct physical interpretation one may give to a nuclear basin of $V(\mathbf{r}; \mathbf{X})$ as a region of space dominated by the electrostatic force exerted by its nucleus on an element of electronic charge is the embodiment of the idea of what an atomic basin of $\rho(\mathbf{r}; \mathbf{X})$ represents—a region of space whose chemistry is dominated by its contained nucleus.

One finds that the structure diagrams obtained for $V(\mathbf{r}; \mathbf{X})$ and $\rho(\mathbf{r}; \mathbf{X})$ for the H_2O system are homeomorphic in the sense that both exhibit an identical partitioning of the control space yielding the same sets of structures, both stable and, as illustrated in Fig. 3.13, unstable. In addition to finding the same sets of structures for both fields, it has been found that their mechanisms of structural change are also the same. Thus, the bifurcation and conflict

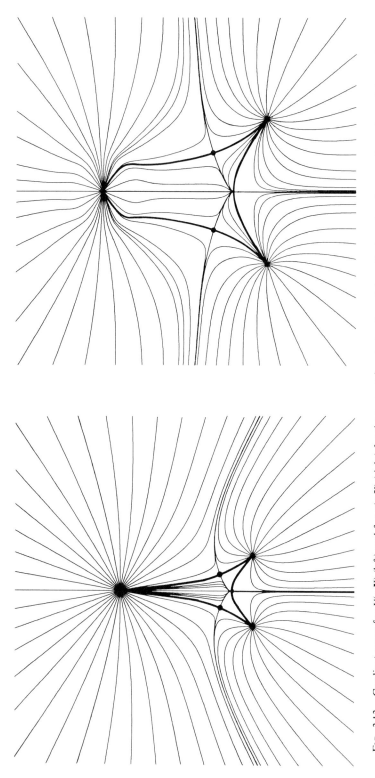

FIG. 3.13. Gradient maps for $V(\mathbf{r}; \mathbf{X})$ (left) and for $\rho(\mathbf{r}; \mathbf{X})$ (right) for the symmetric catastrophe point on the boundary between the regions of ring and open structures of H_2O. The singularities in both $V(\mathbf{r}; \mathbf{X})$ and $\rho(\mathbf{r}; \mathbf{X})$ result from the coalescing of the ring critical point with the H–H critical point.

mechanisms describe all possible structural changes in both fields. The formal statement of this homeomorphism is given in the paper by Tal *et al.* (1980). Other systems have not been studied over such a broad range of nuclear configurations as has been done for the water system, but comparisons of the structures determined by $V(\mathbf{r}; \mathbf{X})$ and $\rho(\mathbf{r}; \mathbf{X})$ have been made for many molecules at their equilibrium geometries. In general, equivalent structures are obtained. In some instances the graph for $V(\mathbf{r}; \mathbf{X})$ must be determined at a geometry other than the equilibrium one to obtain an equivalent structure. It seems reasonable to conjecture that the structure diagrams for $V(\mathbf{r}; \mathbf{X})$ and the ground state $\rho(\mathbf{r}; \mathbf{X})$ will, in general, be homeomorphic for a particular system. The primary reason for making this conjecture is that $\rho(\mathbf{r}; \mathbf{X})$ is observed to have the same set of attractors as does $V(\mathbf{r}; \mathbf{X})$; the topological properties of both fields are controlled and determined by the forces exerted by the nuclei.

There are a few limiting cases where the mapping between the two structure diagrams is not possible because of constraint imposed on $V(\mathbf{r}; \mathbf{X})$ by Poisson's equation (eqns (3.4) and (3.5)). Thus $V(\mathbf{r}; \mathbf{x})$ cannot possess a $(3, +3)$ critical point, for in such a case each $\lambda_i > 0$. Thus, in the interior of a bonded cage of nuclei, where $\nabla\rho$ possesses such a critical point, the corresponding point in V must be a degenerate critical point with three zero roots. In tetrahedrane, for example, such a critical point results from the coalescing of a $(3, +3)$ critical point with the six $(3, -1)$ critical points whose bond paths define the bonds of the cage and with the four ring critical points whose surfaces enclose the cage critical point. This singularity in V behaves as does a sixfold degenerate bond critical point, and the resulting molecular graph is characterized by six bond paths linking the four nuclei as in a regular cage structure. Thus the physical equivalence in structure observed for $\rho(\mathbf{r}; \mathbf{X})$ and $V(\mathbf{r}; \mathbf{X})$ as determined by the equality in the number of their interaction lines and the nuclei they link is maintained even in the case where the constraint of the Poisson equation prevents a one-to-one mapping of the two structure diagrams.

3.4.3   The mechanics of ρ in the one-electron case

The potential energy operator which determines the electron–nuclear attractive energy for a single electron is $\mathscr{V}(\mathbf{r}; \mathbf{X}) = -eV(\mathbf{r}; \mathbf{X})$. Thus for a fixed configuration of the nuclei, an interaction line in an elementary graph of $V(\mathbf{r}; \mathbf{X})$ represents a minimum-energy path linking two neighbouring nuclei in the potential energy surface $\mathscr{V}$. If V were the sole potential acting in a molecular system, one might expect the electronic charge density to be concentrated along corresponding lines, i.e. the interaction lines of V would correspond with the bond paths of ρ. In a many-electron system however, there are electron–electron repulsive forces acting in addition to the force determined by the external potential. Thus, the homeomorphism between the structure diagrams of V and ρ does not, in general, map a point $\mathbf{X} \in R^Q$ on to itself, but rather on to another point $\mathbf{X}' \neq \mathbf{X}$. For a given $\mathbf{X}$ the critical points of V do not coincide with those of ρ. This is true even for a one-electron system where V is the sole Coulombic potential as we now show. One may obtain from Schrödinger's equation for a one-electron case the expression

$$E\rho(\mathbf{r}) = -(1/4)\nabla^2\rho(\mathbf{r}) + (1/8)[\nabla\rho(\mathbf{r})\cdot\nabla\rho(\mathbf{r})]/\rho(\mathbf{r}) + \mathscr{V}(\mathbf{r})\rho(\mathbf{r}) \quad (3.7)$$

where E is the total energy and $\rho(\mathbf{r}) = \psi^*(\mathbf{r})\psi(\mathbf{r})$. This equation may be used to obtain an expression for $\nabla V(\mathbf{r})$ at $\mathbf{r}_c$, a critical point in $\rho(\mathbf{r})$, where $\nabla\rho(\mathbf{r}) = 0$,

$$\nabla V(\mathbf{r}_c) = -\nabla\mathscr{V}(\mathbf{r}_c) = -(1/4)\nabla[\nabla^2\rho(\mathbf{r}_c)]/\rho(\mathbf{r}_c). \quad (3.8)$$

Unless demanded by symmetry, the value of $\nabla^2\rho$ will not be an extremum at a critical point in ρ. Thus, the critical points in V and ρ will not, in general, coincide and the distribution of electronic charge, even in a one-electron system, is not determined entirely by the external force $-\nabla V(\mathbf{r})$. Bohm (1952) ascribed the stability of a stationary state in a quantum system to the balance between the classical force $-\nabla V$ and the quantum mechanical force which is given by the gradient of the 'quantum potential'. For a one-electron system at a critical point in $\rho(\mathbf{r})$, Bohm's quantum mechanical force is just the right-hand side of eqn (3.8). The Laplacian of the charge density, the quantity $\nabla^2\rho$ appearing in the quantum potential, is an important local property of a system and is the subject of Chapter 7.

3.5 Summary

This chapter has demonstrated that the chemical notion of structure finds physical expression in the topological properties exhibited by a molecular charge distribution. The question as to what is the structure of a given molecule can be unambiguously answered. Such questions have been central to many of the protracted and vigorous discussions of chemistry in the past quarter century, the structures of non-classical ions being one example. As is typical of a theory replacing a model, however, new questions can be both asked and answered. Not only can a structure be assigned, but its stability predicted and the mechanisms of a change in structure understood. Structure and structural stability are inseparable concepts. The following chapter demonstrates how Thom's theory of elementary catastrophes can be used to obtain a quantitative description of the structural changes encountered in the neighbourhood of a bifurcation catastrophe point. As will be demonstrated, an abstract mathematical theory can predict all of the topological changes and the corresponding structural changes exhibited by a molecular charge distribution for motion of a system in the neighbourhood of a bifurcation point.

E3.1 Summary of set theoretic language

$x \in X$ x is a member of the set X

$X \subseteq Y$ X is a subset of Y if every element of X is an element of Y. Y contains X.

$\{x|P(x)\}$ the set of all elements x for which a specific property $P(x)$ holds.

$X \cup Y = \{x|x \in X \text{ or } x \in Y\}$ union of two sets X and Y.

$X \cap Y = |x|x \in X \text{ and } x \in Y\}$ intersection of two sets X and Y.

$R^n = \{(x_1, \ldots, x_n)|x_i \in R \text{ for } i = 1, \ldots, n\}$ R^n is an n-dimensional Euclidean space.

Open disc: if $x \in R^n$ and $r > 0$, then the open disc of radius r, centre x is defined to be the set $\{y \in R^n|\,\|x - y\| < r\}$ where $\|x - y\|$ is the distance between x and y.

Open set: a subset X of R^n is open if for each $x \in X$ there is some open disc centre x lying inside X.

Neighbourhood of x: a subset Y is a neighbourhood of x if x lies in some open set X which is contained in Y.

E3.2 Natural coordinate system for an atomic basin

The definition of an atom, as the union of an attractor and its basin, serves to define a natural coordinate system for an atom in a molecule (Biegler–König *et al.* 1981). The new system is obtained by a mapping of the coordinates (x, y, z) of a point $\mathbf{r}$ in an atomic basin into the triple (s, θ, ϕ). The parameter s (eqn (2.6)) determines the position of the point $\mathbf{r}$ along the gradient path that is defined by some initial set of angular coordinates θ and ϕ. This local coordinate system resembles the spherical polar coordinate system with the radial coordinate r replaced by the path parameter s, and the rays replaced by the generally curved gradient paths that traverse the basin of a given atom. Since the limits of s are $\pm\infty$, the mapping sends the bounded space of each atom into a complete space homeomorphic to R^3. Thus, in terms of this coordinate system, every atom, whether free or bound, is assigned a complete space. The space of a bound atom, while 'curved' because of its interactions with its neighbours, is still complete. The collection of atomic basins and the associated coordinate transformations yield an atlas for the molecule.

A gradient path $g(\mathbf{r}_m)$ through a given point $\mathbf{r}_m \in R^3$ is defined by

$$g(\mathbf{r}_m) = \left\{ \mathbf{r}(s) \in R^3 \left| \frac{d\mathbf{r}(s)}{ds} = \nabla\rho(\mathbf{r}(s), \mathbf{X}); \mathbf{r}(0) = \mathbf{r}_m \right. \right\}. \tag{E3.1}$$

We define the α-limit set of $g(\mathbf{r}_m)$, the origin of the path, by

$$\alpha(\mathbf{r}_m) = \lim_{s \to -\infty} \mathbf{r}(s) \qquad \text{(E.3.2a)}$$

and its ω-limit set, the terminus of the path, by

$$\omega(\mathbf{r}_m) = \lim_{s \to +\infty} \mathbf{r}(s). \qquad \text{(E3.2b)}$$

For a given set of initial values $\mathbf{r}(0) = \mathbf{r}_0$, the differential equation

$$d\mathbf{r}(s)/ds = \nabla\rho(\mathbf{r}(s)) \qquad \text{(E3.3a)}$$

defines a unique gradient path. By virtue of the definition of an attractor A of $\nabla\rho$, this path will be contained in the basin Ω of A, if $\mathbf{r}_0 \in \Omega$. Equation (E.3.3a) is equivalent to the integral equation

$$v(s)(\mathbf{r}_0) = \mathbf{r}(s) = \mathbf{r}_0 + \int_0^s \nabla\rho(\mathbf{r}(t))dt \qquad \text{(E3.3b)}$$

which defines a one-parameter group of diffeomorphisms $\{v(s), s \in R\}$.

In the following, eqn (E3.3b) will be regarded as defining a coordinate transformation within Ω. The basic ideas behind the construction of this transformation are as follows. The parameter s will be one of the new coordinates; to obtain a proper coordinate transformation from R^3 to R^3, we need two additional coordinates to specify the initial value of eqn (E3.3a). Thus $\mathbf{r}(s = 0) = \mathbf{r}_0$ must belong to a two-dimensional differentiable manifold, and the two additional coordinates mentioned above are the local co-ordinates of $\mathbf{r}_0$ in the manifold. We make the following general observations: let M be a two-manifold contained in Ω, with an atlas $\{(U_\alpha, \mathscr{E}_\alpha)\}$, where the open sets U_α cover M, and the one-to-one, continuous maps $\mathscr{E}_\alpha: U_\alpha \to R^2$ define local coordinate systems in M. Let $\bar{M} V_\alpha$ be the images of M and U_α, respectively, under the group of diffeomorphisms $\{v(s); s \in R\}$, that is

$$\bar{M} = \bigcup_{s \in R} v(s)(M),$$

$$V_\alpha = \bigcup_{s \in R} v(s)(U_\alpha).$$

For each α, define the map $h_\alpha: V_\alpha \to R^3$ by $h_\alpha(\mathbf{r}) = (s, \mathscr{E}_\alpha(\mathbf{r}_m))$, $\forall \mathbf{r} \in V_\alpha$, with $\mathbf{r}_m = g(\mathbf{r}) \cap M$, and $\mathbf{r} = \mathbf{r}_m + \int_0^s \nabla\rho(\mathbf{r}(t))dt$. Then the h_α define local coordinate systems over M. In particular, let M be a sphere of radius β centred on A, so that $\bar{M} = \Omega$. An appropriate atlas over the sphere M consists of the single chart $(M, \mathscr{E})$, where

$$\mathbf{r} = (x_0, y_0, z_0) \in M, \qquad \mathscr{E}(\mathbf{r}_0) = (\theta, \phi), \qquad 0 < \theta < \pi, \qquad 0 < \phi < 2\pi$$

with

$$x_0 = \beta \sin\theta \cos\phi, \qquad y_0 = \beta \sin\theta \sin\phi, \qquad z_0 = \beta \cos\theta. \qquad \text{(E3.4)}$$

Correspondingly, we obtain a single coordinate system over Ω, illustrated in Fig. E3.1, through the mapping h defined by

$$h(\mathbf{r}) = (s, \theta, \phi) \qquad \forall \mathbf{r} \in \Omega,$$

$$\mathscr{E}^{-1}(\theta, \phi) = g(\mathbf{r}) \cap M = \mathbf{r}_0(\mathbf{r}) \qquad \text{(E.3.5)}$$

and

$$\mathbf{r} = \mathbf{r}_0 + \int_0^s \nabla\rho(\mathbf{r}(t))\,dt.$$

Equation (E3.5) implies that, through the one-to-one map h, we can uniquely associate with a point $\mathbf{r}$ contained in the basin Ω, a triple (s, θ, ϕ) where (θ, ϕ) determines the path through $\mathbf{r}$ and s is the coordinate of $\mathbf{r}$ along that path.

This natural coordinate system is of particular use in the determination of atomic properties. As discussed in Chapter 6, the atomic value of a property F is given by the average over the atomic basin of an effective single-particle density $f(\mathbf{r})$. Thus the value of the property F for atom Ω is

$$F(\Omega) = \int_\Omega f(\mathbf{r})\,d\mathbf{r}. \qquad \text{(E3.6)}$$

The evaluation of $F(\Omega)$ is a non-trivial problem since the atomic surface which bounds the region Ω does not in general have a local definition, much less a simple geometrical structure. The integration method, developed here

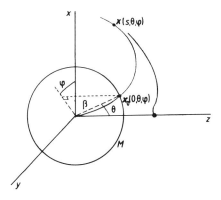

FIG. E3.1. The coordinate system (s, θ, ϕ) with M being a sphere of radius β. Also shown is a single trajectory of $\nabla\rho$ in the interatomic surface which terminates at the $(3, -1)$ critical point on the z-axis.

in terms of the natural coordinate system, avoids the direct and often difficult determination of the atomic surface by implementing the definition of an atom as the union of an attractor and its basin. By integrating along the trajectories of $\nabla\rho(\mathbf{r})$ which terminate at a given nucleus, one must necessarily cover the basin of an atom. In addition, because of the zero-flux surface condition, it is impossible to cross an interatomic surface into the basin of a neighbouring atom.

Since the mapping h is one-to-one, we can use its inverse

$$h^{-1}: R^3 \to \Omega$$

$$\begin{pmatrix} s \\ \theta \\ \phi \end{pmatrix} + \begin{pmatrix} x \\ y \\ z \end{pmatrix} \in \Omega \qquad\qquad \text{(E3.7a)}$$

defined by

$$h^{-1}\begin{pmatrix} s \\ \theta \\ \phi \end{pmatrix} = \begin{pmatrix} x \\ y \\ z \end{pmatrix} = \begin{pmatrix} \beta \sin\theta \cos\phi \\ \beta \sin\theta \sin\phi \\ \beta \cos\theta \end{pmatrix} + \int_0^s \nabla\rho(\mathbf{r}(t))\,dt \qquad \text{(E3.7b)}$$

to rewrite eqn (E3.6) in the form

$$F(\Omega) = \int_0^{2\pi} d\phi \int_0^{\pi} d\theta \int_{-\infty}^{+\infty} ds\, f(\mathbf{r}(s,\theta,\phi))|\mathscr{J}(h^{-1}(s,\theta,\phi))|. \qquad \text{(E3.8)}$$

In eqn (E3.8) $|\mathscr{J}| \equiv |\mathscr{J}(h^{-1}(s,\theta,\phi))|$ denotes the determinant of the Jacobian matrix of h^{-1},

$$\mathscr{J}(h^{-1}) = \begin{pmatrix} \partial x/\partial s & \partial x/\partial\theta & \partial x/\partial\phi \\ \partial y/\partial s & \partial y/\partial\theta & \partial y/\partial\phi \\ \partial z/\partial s & \partial z/\partial\theta & \partial z/\partial\phi \end{pmatrix} = \begin{pmatrix} x_s & x_\theta & x_\phi \\ y_s & y_\theta & y_\phi \\ z_s & z_\theta & z_\phi \end{pmatrix}, \qquad \text{(E3.9)}$$

where

$$x_s = \partial x/\partial s, \qquad x_\theta = \partial x/\partial\theta, \qquad \text{etc.}$$

To evaluate $F(\Omega)$ using eqn (E3.8) we need the values of the Jacobian determinant $|\mathscr{J}(h^{-1})|$ for all s, θ, ϕ. Equation (E3.7b) reveals that a direct computation of this determinant would involve products of integrals of the form

$$\int_0^s \frac{\partial^2\rho(t)}{\partial\theta\,\partial x}\,dt.$$

Instead, as will be shown below, we can obtain $|\mathscr{J}(h^{-1})|$ at all points (s, θ, ϕ) as the solution to an initial-value problem. We first obtain a differential equation for the matrix $\mathscr{J}(h^{-1})$ by differentiating eqn (E3.9) with respect to s. Writing eqn (E3.9) in the form

$$\mathscr{J}(h^{-1}) = (\mathbf{r}_s, \mathbf{r}_\theta, \mathbf{r}_\phi) \tag{E3.10}$$

where, for example, $\mathbf{r}_s$ is the vector $\partial \mathbf{r}/\partial s$, we have

$$\partial \mathscr{J}(h^{-1})/\partial s = (\partial \mathbf{r}_s/\partial s, \partial \mathbf{r}_\theta/\partial s, \partial \mathbf{r}_\phi/\partial s). \tag{E3.11}$$

From $\mathbf{r}_s = \nabla \rho$, we get, using the chain rule,

$$\partial \mathbf{r}_s/\partial s = \partial(\nabla \rho)/\partial s = \mathscr{H}(\rho) \cdot \mathbf{r}_s \tag{E3.12}$$

where $\mathscr{H}(\rho)$ denotes the Hessian matrix of ρ. From eqn (E3.2a), we have

$$\mathbf{r}_0 = \begin{pmatrix} \beta \cos \theta \cos \phi \\ \beta \cos \theta \sin \phi \\ \beta \sin \theta \end{pmatrix} + \frac{\partial}{\partial \theta} \int_0^s \nabla \rho \, dt$$

so that

$$\frac{\partial \mathbf{r}_\theta}{\partial s} = \frac{\partial}{\partial s} \left(\frac{\partial}{\partial \theta} \int_0^s \nabla \rho \, dt \right) = \frac{\partial}{\partial \theta} \left(\frac{\partial}{\partial s} \int_0^s \nabla \rho \, dt \right) = \frac{\partial}{\partial \theta} (\nabla \rho).$$

Using the chain rule, we obtain

$$\partial \mathbf{r}_\theta/\partial s = \mathscr{H}(\rho) \cdot \mathbf{r}_\theta. \tag{E3.12b}$$

Similarly, we find

$$\partial \mathbf{r}_\phi/\partial s = \mathscr{H}(\rho) \cdot \mathbf{r}_\phi. \tag{E3.12c}$$

Substituting equations (E3.12a–c) into eqn (E3.11), we obtain

$$\partial \mathscr{J}(h^{-1})/\partial s = \mathscr{H}(\rho) \mathscr{J}(h^{-1}). \tag{E3.13}$$

Equation (E3.13) defines an initial-value problem, with initial values at $s = 0$,

$$\mathscr{J}_0 = \begin{pmatrix} \rho_x(\mathbf{r}_0) & \beta \cos \theta \cos \phi & -\beta \sin \theta \sin \phi \\ \rho_y(\mathbf{r}_0) & \beta \cos \theta \sin \phi & \beta \sin \theta \cos \phi \\ \rho_z(\mathbf{r}_0) & \beta \sin \theta & 0 \end{pmatrix} \tag{E3.14}$$

where $\mathbf{r}_0$ is given by eqn (E3.4).

The system of differential equations (E3.13) can be reduced to a single differential equation for the determinant of $\mathscr{J}(h^{-1})$ through the application of Davidenko's (1960) trace theorem: let $\mathscr{P}(s)$ be an $n \times n$ matrix whose elements are differentiable functions in s. Let $|\mathscr{P}(s)|$ be the characteristic polynomial associated with $\mathscr{P}(s)$. Then, for all s for which $|\mathscr{P}(s)| \neq 0$,

$$\frac{d}{ds}|\mathscr{P}(s)| = \mathrm{Tr}(\mathscr{P}(s)^{-1}) \frac{d}{ds} (\mathscr{P}(s)) |\mathscr{P}(s)|. \tag{E3.15}$$

Since the characteristic polynomial $\mathscr{P} = \mathscr{J}(h^{-1})$ is the determinant $|\mathscr{J}(h^{-1})|$, application of the above theorem yields, using eqn (E3.13),

$$\partial|\mathscr{J}(h^{-1})|/\partial s = \mathrm{Tr}(\mathscr{J}(h^{-1})^{-1}\mathscr{H}(\rho)\mathscr{J}(h^{-1}))|\mathscr{J}(h^{-1})|$$
$$= \mathrm{Tr}(\mathscr{H}(\rho))|\mathscr{J}(h^{-1})|. \qquad (E3.16)$$

Noting that $\mathrm{Tr}(\mathscr{H}(\rho)) = \nabla^2\rho$, we obtain the differential equation

$$\partial|\mathscr{J}(h^{-1})|/\partial s = \nabla^2\rho(\mathbf{r})|\mathscr{J}(h^{-1})|, \qquad (E3.17a)$$

with initial value at $s = 0$,

$$|\mathscr{J}(h^{-1})|(0, \theta, \phi) = |\mathscr{J}_0| \qquad (E3.17b)$$

where the matrix $\mathscr{J}_0$ is given in eqn (E3.14).

Equation (E3.8) together with eqns (E3.17) form the basis of the algorithm described by Biegler–König et al. (1981) for the evaluation of the atomic property $F(\Omega)$.

The author wishes to offer his thanks to Professor Zeeman for bringing to his attention the theorem on structural stability by Palis and Smale and for a stimulating and most useful discussion of our use of catastrophe theory, all of which occurred during a visit to the University of Warwick in the spring of 1980.

References

Bader, R. F. W., Nguyen-Dang, T. T., and Tal, Y. (1979). *J. Chem. Phys.* **70**, 4316.

Bader, R. F. W., Nguyen-Dang, T. T., and Tal, Y. (1981). *Rep. Prog. Phys.* **44**, 893.

Bader, R. F. W., Slee, T. S., Cremer, D., and Kraka, E. (1983). *J. Am. chem. Soc.* **105**, 5061.

Biegler-König, F. W., Nguyen-Dang, T. T., Tal, Y., Bader, R. F. W., and Duke, A. J. (1981). *J. Phys. B: At. Mol. Phys.* **14**, 2739.

Bohm, D. (1952). *Phys. Rev.* **85**, 166.

Boyd, R., Knop, O., and Choi, S. C. (1988). *J. Am. chem. Soc.* **110**, 7299.

Cade, P. E., Bader, R. F. W., and Pelletier, J. (1971). *J. Chem. Phys.* **54**, 3517.

Cade, P. E. and Huo, W. M. (1975). *Atom. Data nucl. Data Tables* **15**, 1.

Cade, P. E. and Wahl, A. C. (1974). *Atom. Data nucl. Data Tables* **13**, 339.

Coulson, C. A. (1951). *Proc. r. Soc.* **A207**, 91.

Coulson, C. A. (1972). From Preface of first edition of *The shape and structure of molecules* (2nd edn) (ed. R. McWeeny). Oxford University Press, Oxford.

Cremer, D. and Kraka, E. (1984). *Croatica Chem. Acta* **57**, 1259.

Cremer, D., Kraka, E., Slee, T. S., Bader, R. F. W., Lau, C. D. H., Nguyen-Dang, T. T., and MacDougall, P. J. (1983). *J. Am. chem. Soc.* **105**, 5069.

Davidenko, D. F. (1960). *Sov. J. Math.* **1**, 316.

Gatti, C., Barzaghi, M., and Simonetta, M. (1985). *J. Am. chem. Soc.* **107**, 878.

Mezey, P. G. (1981). *Theor. Chim. Acta* **58**, 309.

Mulliken, R. S. (1932). *Rev. mod. Phys.* **4**, 1.

Mulliken, R. S. (1939). *Phys. Rev.* **56**, 778.

O'Hare, P. A. G. and Wahl, A. C. (1970). *J. Chem. Phys.* **53**, 2469.

Palis, J. and Smale, S. (1970). *Pure Math.* **14**, 223.

Poston, T. and Stewart, I. (1978). *Catastrophe theory and its applications.* Pitman, London.

Slee, T. (1986). *J. Am. chem. Soc.* **108**, 606.

Tachibana, A. and Fukui, K. (1978). *Theor. Chim. Acta* **49**, 321.

Tal, Y., Bader, R. F. W., and Erkku, J. (1980). *Phys. Rev.* **A21**, 1.

Tal, Y., Bader, R. F. W., Nguyen-Dang, T. T., Ojha, M., and Anderson, S. G. (1981). *J. Chem. Phys.* **74**, 5162.

Thom, R. (1975). *Structural stability and morphogenesis.* W.A. Benjamin, Reading, Massachusetts.

Wiberg, K. B., Bader, R. F. W., and Lau, C. D. H. (1987). *J. Am. chem. Soc.* **109**, 985.

MATHEMATICAL MODELS OF STRUCTURAL CHANGE

Next we must concede that the universe we see is a ceaseless creation, evolution and destruction of forms and that the purpose of science is to foresee this change of form and, if possible, explain it.

René Thom (1975)

4.1 Introduction

The definition of structure obtained through the application of the equivalence relation to the gradient vector fields of the charge density leads to the partitioning of nuclear configuration space into a finite number of structural regions. The boundaries of these regions denote the configurations of the unstable structures. This information enables one to construct a system's structure diagram, a diagram which details all of its stable structures and all of the unstable structures that are transitional in form between those of neighbouring structural regimes. The collection of the structurally unstable points in nuclear configuration space R^Q, is called the catastrophe set C. As a corollary of the theorem of structural stability stated by Palis and Smale (1970) one finds that only two types of catastrophe points are possible: a configuration $X \in R^Q$ is structurally unstable if and only if its associated charge density $\rho(\mathbf{r}; X)$ exhibits (a) at least one degenerate critical point, also known as a singularity in the charge density or (b) a non-transversal intersection of the stable and unstable manifolds of a pair of its critical points. A configuration satisfying either (a) or (b) is called a bifurcation or conflict point, respectively.

This is a concise summary of a theory of structure which, as demonstrated in the preceding chapters, contains the ideas essential to the notion of molecular structure and, in addition, expands that notion to include a definition of structural stability (Bader *et al.* 1981; Nguyen-Dang and Bader 1982). It is the instability of certain structures and the associated mechanisms of structural change which are the subject of this chapter. It will be demonstrated that Thom's theory of elementary catastrophes (1975) provides a quantitative basis for the description and prediction of the behaviour of unstable structures. The theory provides a mathematical model for structural changes in the neighbourhood of a bifurcation point through the analysis of what are called the universal unfoldings associated with singularities of particular kinds. The possibility of using the theory of elementary catastrophes to describe changes in molecular structure was first pointed out by Collard and Hall (1977).

4.2 Catastrophe theory

It is the purpose of catastrophe theory to account for and explain the observation of sudden, discontinuous changes in the behaviour of a system that are brought about by smooth continuous changes in the variables which control it. Change of this kind is ubiquitous throughout nature and is usually monitored through a change in form. In biological systems, this is the problem of morphogenesis. In chemistry, the structural behaviour of a molecule is determined by the form of its distribution of charge in real space; the changes in this form as caused by nuclear motions are conveniently monitored and summarized by corresponding changes in its molecular graph. This is the physical reality and the mathematical formalism underlying the notion of the making and breaking of chemical bonds, processes which, as we now appreciate, are necessarily discontinuous.

One cannot, of course, develop or even comprehensively review catastrophe theory in the brief space afforded the subject in this book. The theory has in any event been admirably presented by Poston and Stewart (1978) and the reader desirous of more detail is referred to this text. What we do here is illustrate the principal ideas of the theory and its method of application to obtain useful results in the area of molecular structure. The knowledge of the theory presented here is sufficient to enable the reader to make similar applications and it will serve as a suitable introduction to the subject for an interested novice. To one familiar with the method, the examples given are further evidence of the ability of abstract mathematics to describe and predict the events which occur around us. All readers will observe that the theory applies in a direct and natural manner to the study of changes in chemical structure. There is a behaviour space, the real space R^3 occupied by the atoms in a molecule, and there is a control space, nuclear configuration space R^Q, as the interactions between the atoms are altered by the relative motions of their nuclei.

4.2.1   Isolating the unstable piece of a critical point

We have observed in the discussions given in the two preceding chapters that a non-degenerate critical point in the charge distribution is stable, i.e. its form as summarized by its rank and signature persists during nuclear motions which cause changes in the charge density. Such a critical point is called a Morse critical point. There is a theorem called the Morse theorem which states that every non-degenerate critical point in an n-dimensional space may be transformed, by a smooth reversible coordinate change, into a Morse-l saddle which is a function of the form $\sum_{i=1}^{n} c_i x_i^2$, where the first $n - l$ coefficients equal $+1$ and the remaining coefficients equal -1. In this convention, when $l = n$, the critical point is a local maximum. It can be proven that a Morse critical point is structurally stable, i.e. its type is unchanged by the addition of a small perturbation to the function in the neighbourhood of the critical point.

We have also observed that a singularity in the charge distribution, a degenerate critical point, is not stable to a motion of the nuclei. It is a corollary of the above theorem that a degenerate critical point is always structurally unstable. If the rank of a degenerate critical point in n-dimensional space is ω, then its co-rank is defined to be $n - \omega$. The co-rank is the number of zero eigenvalues of the Hessian matrix of the function at the critical point. A most important theorem regarding the description of the behaviour of a function in the neighbourhood of a degenerate critical point is the splitting theorem. This theorem isolates the degenerate critical point of co-rank $k = n - \omega$ into a Morse piece of the form $\sum_{i} (\pm x_i^2)$, with $i = 1$ to ω, and a degenerate part which is expressed as a function of the k remaining variables, called the essential variables.

4.2.2 Elementary catastrophes

Consider a bifurcation point $\mathbf{X}_b$ for which the charge density $\rho(\mathbf{r}; \mathbf{X}_b)$ exhibits a singularity at $\mathbf{r}_c$ of rank $\omega < 3$. Then, as a consequence of the above splitting theorem, the charge density may be expanded in a Taylor series in a sufficiently small neighbourhood of $\mathbf{r}_c$, in which only ω components of $\mathbf{r}$ appear up to second-order,

$$\rho(\mathbf{r}; \mathbf{X}_b) = \rho(\mathbf{r}_c) + Q + f. \tag{4.1}$$

In eqn (4.1), Q is a quadratic expression in ω components of $\mathbf{r}$ with respect to a local coordinate system centred at $\mathbf{r}_c$, and f denotes a polynomial in the remaining components of $\mathbf{r}$, the essential variables, starting with a cubic term.

As we consider configurations $\mathbf{X} \neq \mathbf{X}_b$, lying in a small neighbourhood in R^Q of $\mathbf{X}_b$, the topological properties of the charge distribution in the region of the original singularity in $\rho(\mathbf{r})$ vary, thereby inducing global changes in the entire ρ field. New structures are generated. To discuss these structural changes in the neighbourhood of $\mathbf{X}_b$, we need only consider the perturbation in the term f of eqn (4.1) as caused by nuclear displacements from the point $\mathbf{X}_b$. The configuration $\mathbf{X}_b$ is structurally stable with respect to the Morse part as described by the quadratic term Q. This term, therefore, does not contribute to the generation of new structures and is omitted from any further consideration. Only the term f leads to changes in structure and Thom has shown that a sufficiently small but otherwise arbitrary perturbation of f can always be described in terms of a member of a family of functions which contains f and which is termed the universal unfolding of f.

A function f of the essential variables, is said to be of finite co-dimension if a small perturbation of f gives rise to only a finite number of topological types. It is in this case that a map f of co-rank k can be embedded in a family of deformations ($f_\mu : R^k \times R^q \to R$), parametrized by q variables which form the control space $R^q \subset R^Q$. By definition, q, the dimension of the control space, is the co-dimension of the singularity represented by f. The family of functions ($f_\mu : R^k \times R^q \to R$) thus defined, describes the universal unfolding of f, in which the function f itself is the member associated with the origin of control space. In our applications the control space is a subset of the nuclear configuration space.

The number of elementary catastrophe types depends upon q. It has been shown that only for values of $q \leqslant 5$ is the number of catastrophe types finite. Thom has classified these types by their co-rank k and co-dimension q for values of $q \leqslant 4$. The concept of an unfolding and the accompanying definitions are illustrated first in terms of the simplest of all catastrophe types, the so-called fold catastrophe for which both the co-rank and co-dimension equal one.

4.3   Catastrophes in molecular structures

4.3.1   Opening a ring structure—the fold catastrophe

It was noted in the discussion of the dissociation of the water molecule that the creation of a bond between the hydrogen atoms occurs via the formation of a singularity in the charge density with a single zero eigenvalue. That is, the curvature of the charge density at the singularity formed between the protons vanishes only along the C_2 symmetry axis. This behaviour is illustrated by the series of profiles of the charge density illustrated in Fig. 3.2. We shall consider the reverse process, that of an oxygen approaching a hydrogen molecule, breaking the H–H bond, and forming the open structure of the water molecule. This is pictured by the gradient vector maps and molecular structures shown in the sequence (d), (c), (b) of Fig. 3.1.

The catastrophe associated with the breaking of the H–H bond may be described by a simple unfolding with one essential variable and one control parameter. The essential variable or behaviour coordinate lies along the C_2 axis which is taken to be the z-axis. The control parameter corresponds to the location of the system along the reaction coordinate and this is given by R_0, the distance of the oxygen nucleus from the H–H midpoint. This so-called fold catastrophe is illustrated again in Fig. 4.1. In this diagram the three characteristic curves are plotted relative to the value of ρ at the singularity, i.e. at the inflexion point in curve c. The labelling of the curves corresponds to that shown for the molecular graphs in Fig. 3.1. Thus, the profile labelled d is characteristic of the ring structure. The secondary maximum in this profile is found at the position of the H–H bond critical point and the minimum at the ring saddle point. As oxygen approaches H_2, these two extrema in ρ approach one another in value and in position. At $R_0 = 2.39$ au, profile c, they merge yielding a singularity in ρ and a catastrophe point in configuration (or control) space. This configuration defines a point on the boundary separating the ring structure from the open structure associated with the equilibrium geometry of water. For values of $R_0 < 2.39$ au, ρ exhibits a simple monotonic decrease (profile b) indicating the loss of the H–H bond critical point and the absence of an H–H bond.

The analytical description of this fold catastrophe is obtained by replacing $\rho(\mathbf{r})$ by a function of the single behaviour coordinate ξ and parametrized by a single control parameter v, denoting displacements from the catastrophe point along the reaction coordinate. The topological behaviour of $\rho(\mathbf{r})$ in the

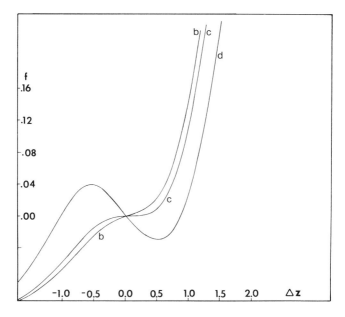

FIG. 4.1. Fold catastrophe in H_2O. Each curve is a profile of $\rho(\mathbf{r})$ along the symmetry axis, the z-axis, relative to the value at the inflexion point z_i; $f = \rho(z) - \rho(z_i)$ and $\Delta z = z - z_i$. The values of R_0 are 2.75 au for curve d, 2.39 au for curve c (where the singularity in $\rho(\mathbf{r})$ occurs), and 2.30 au for curve b.

neighbourhood of the catastrophe point is then described by the expression

$$f(\xi) = (1/3)\xi^3 + v\xi. \tag{4.2}$$

The behaviour variable ξ contains implicitly the third derivative of ρ with respect to the actual coordinate z. The germ of this unfolding is the term $(1/3)\xi^3$, which is unstable as a critical point: both the first and second derivatives of this term are zero for $\xi = 0$. If we perturb it through the addition of a small term $v\xi$ then its character is changed. This is seen by setting $\partial f/\partial \xi = 0$ which yields two roots for

$$\xi = \pm \sqrt{-v}.$$

For $v < 0$, corresponding to $R_0 > 2.39$ au, f exhibits two critical points, equally spaced on either side of the position of the original singularity at $\xi = 0$, thereby predicting a molecular graph of the type shown in Fig. 3.1(d) or (e). As $|v|$ approaches zero, i.e. as the system approaches the catastrophe point at $R_0 = 2.39$ au, the two critical points approach one another and eventually merge at $v = 0$ giving rise to the singularity in ρ. Beyond this point the two critical points have vanished to yield the open structure of water.

The complete description of the catastrophes involved in the dissociation of the water molecule must clearly involve two behaviour variables and at least two control parameters. We show in the following section that the most general description of the formation or destruction of a three-membered ring structure is described by such an unfolding, the elliptic umbilic.

4.3.2   A general analysis of three-centre systems

The function f, called the unfolding of the elliptic umbilic,

$$f(x, y; \mu) = x^2 y - (1/3)y^3 + wx^2 - ux - vy, \qquad \mu = (w, u, v) \in M, \tag{4.3}$$

accounts for the structural changes observed for a three-membered system. These changes, by symmetry, are confined to a plane of the behaviour space. The behaviour variables (x, y) of eqn (4.3) refer to a local coordinate system on this plane. Variations in the control parameters u, v, and w within a subset M of R^Q generate all the structures attainable by a perturbation of the germ of this unfolding defined by

$$f(x, y; 0) = x^2 y - (1/3)y^3. \tag{4.4}$$

The singularity described by eqn (4.4) is exhibited by the charge distribution of the H_3^+ molecular-ion in its equilibrium D_{3h} geometry and the unfolding of this germ, as given by eqn (4.3) with $w = 0$, predicts all structures of this ion observed in the neighbourhood of this geometry (Bader *et al.* 1979).

Figure 4.2 illustrates the sequence of structures and associated phase portraits found in the C_{2v} approach of H to H_2^+. The initial and final

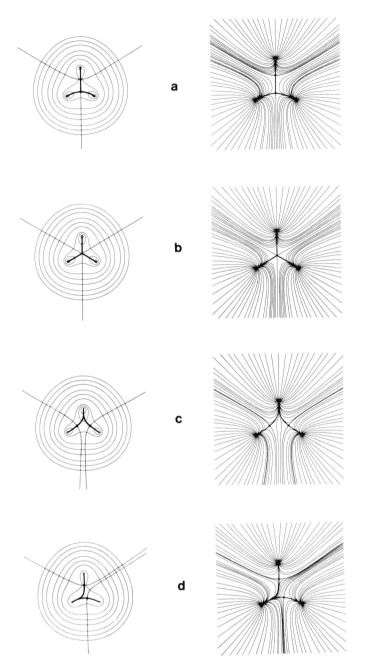

Fig. 4.2.   Contour plots of $\rho(\mathbf{r})$ and their corresponding gradient vector fields for the ground state of the H_3^+ system for configurations in the neighbourhood of the bifurcation catastrophe point shown in diagram (b). The structures illustrated by the molecular graphs in (a), (c), and (d), and the corresponding structures derived from permutations of the protons, are all obtained from the unstable structure in (b) by suitable displacements of the nuclei.

structures shown, $(H–H_2)^+$ and $(H–H–H)^+$, are the same as those found in the $O + H_2$ reaction to form H–O–H. There is, however, apparently no intervening ring formation in the H_3^+ reaction. The molecular graph and phase portrait for the equilibrium geometry of H_3^+ is given by Fig. 4.2(b). This phase portrait is characteristic of the elliptic umbilic. The reader may compare this portrait with that given by Berry and Mackley (1977) for the streamlines of a fluid flowing in a six-roll mill or with the form of the three-dimensional pattern decorating the geometrical optics caustic surface obtained by Berry *et al.* (1979) to gain some idea of its universality. The (x, y) coordinate system of eqn (4.3) is in the molecular plane with origin at the singularity with the y-axis coincident with a C_2 axis. The dimensionless control parameters u and v are directly related to the nuclear displacement coordinates. A displacement from the equilibrium configuration which preserves a single C_2 axis along y is denoted by $(v \neq 0, u = 0)$. A motion denoted by $(u \neq 0, v = 0)$ is asymmetric and reduces the symmetry to that of the C_s point group.

Differentiation of eqn (4.3) for $w = 0$ yields, for the critical points in f,

$$\partial f/\partial x = 2xy - u = 0, \qquad \partial f/\partial y = x^2 - y^2 - v = 0. \qquad (4.5)$$

In general, for any value of u, these equations admit two solutions (see Table 4.1 (p. 128)). The solutions obtained for $u = 0$ correctly predict the number and relative positions of the critical points observed when the system is displaced along the v-axis in control space, i.e. displacements of A_1 symmetry. Thus, for $(u = 0, v < 0)$ it predicts a structure characterized by the molecular graph in Fig. 4.2(a), while for $(u = 0, v > 0)$ it predicts a molecular graph corresponding to a different structure, as displayed in Fig. 4.2(c). For an asymmetric displacement from the bifurcation point, $u \neq 0$, the two critical points no longer lie along either of the behaviour axes, but are located in two quadrants of the behaviour plane, symmetrically placed with respect to the original singularity. The molecular graph obtained for such a displacement corresponds to the same structure as that obtained for a motion, $v > 0, u = 0$, i.e. a structure in which one proton is bonded to the two remaining protons. These predictions are again in agreement with the observed behaviour as illustrated by the phase portrait and molecular graph given in Fig. 4.2(d).

Since the three protons are symmetrically equivalent in the unique structure found at the bifurcation point, an equivalent molecular graph, and hence the same structure, is obtained for an outward displacement of any one of the protons along its C_2 axis. Thus, a motion along any one of the three semiaxes, $(u = 0, v < 0), (u = \pm \sqrt{3}v, v > 0)$, in configuration space results in a structure of the type represented by the molecular graph in Fig. 4.2(a). Any other displacement of the system from the bifurcation point yields a structure of the form $(H–H–H)^+$ (Fig. 4.2(c) or (d)). Thus, the unfolding of the elliptic umbilic with $w = 0$, in the neighbourhood of the catastrophe point, leads to a partitioning of the control or configuration space of the H_3^+ system into regions, to each of which a definite structure can be assigned, the structure diagram, Fig. 4.3(b). The three semiaxes, which originate at the catastrophe point where $u = v = 0$, define the boundaries of three regions of configuration space. Within each such region only a single structure is found, of the type $(H–H–H)^+$. The existence of three such regions results from the permutational symmetry of the protons. An exchange of the form

$$[H(1)–H(2)–H(3)]^+ \rightarrow [H(1)–H(3)–H(2)]^+$$

can occur only by passage from one such region to another. To do so, the system must cross one of the semiaxes partitioning configuration space, i.e. it must pass through the conflict structure of the type shown in Fig. 4.2(a). Thus each of the semiaxes is itself a subregion of structure space along which the molecular graphs are of the form $[H–(H_2)]^+$. These latter structures are necessarily unstable as they exist in a space of dimension less than the full dimension of the control space. The union of the three strata of conflict structures with the bifurcation point forms the catastrophe set describing the structural changes in H_3^+ and this union forms a closed subset of the full configuration space.

Finally, the bifurcation catastrophe point itself defines a third structure in which all three protons are bonded to one another via the gradient paths which originate at the singularity and terminate one at each nucleus (Fig. 4.2(b)). In this electron-deficient molecule, all three nuclei are bound by the charge accumulated in the centre of the pseudoring. The molecule is also partitioned into three atoms by three surfaces symmetrically disposed between the bond paths and generated by the gradient vectors which terminate at the singularity. Thus, all three atoms are defined and linked by gradient paths associated with the singularity in ρ. In this sense, a singularity in ρ resulting from the merger of three $(3, -1)$ bond critical points and a ring critical point behaves as a degenerate bond critical point.

In the more general case, where the control parameter w assumes non-zero values, the unfolding of the elliptic umbilic predicts a more complex partitioning of control space into regions of different structures. In particular, the bifurcation set, a singleton for the case $w = 0$, in which all three nuclei are bonded to one another in a pseudoring structure, expands to define a region of finite size within which a true ring structure exists. The bifurcation set predicted by the unfolding of the elliptic umbilic in the (u, v, w) control space is pictured in Fig. 4.3(a). It is seen that, for a case in which $w \neq 0$, this set in the corresponding u, v-plane defines a hypocycloid-shaped region whose corners are the points $(u, v) = (0, +w^2), (\pm(3/8)\sqrt{3}w^2, -w^2/8)$. Thus, in the full three-dimensional control space (u, v, w), the bifurcation set comprises two tapered cones which are joined at the umbilic, the origin of the control space.

At any point in configuration space lying outside the hypocycloid, at a given $w \neq 0$, only two critical points, which are $(3, -1)$ or bond critical points may exist. When applied to an A_3 system this unfolding predicts that the only structures which can exist outside the hypocycloid are of the types $A–(A_2)$ and $(A–A–A)$. Inside the hypocycloid, the unfolding establishes the existence of four critical points. Three of these are bond critical points, and the fourth may be either a (local) minimum or a (local) maximum in the scalar field $f(x, y)$ depending on whether $w > 0$ or $w < 0$, respectively. Since ρ is a maximum in the (x, y) behaviour plane for any perpendicular displacement and since local maxima in ρ generally occur only at nuclear positions, we may exclude negative values of w. The same argument shows that, for a positive value of w, the unfolding will yield a $(3, +1)$ critical point in ρ as required to obtain a ring structure.

An example of a triatomic system which exhibits a bonded ring structure over a non-empty set of nuclear configuration space is the water molecule. We now demonstrate that the elliptic umbilic unfolding eqn (4.3) can be used to describe the formation and destruction of this ring structure. The origin of the control plane $(u = 0, v = 0, w > 0)$ is placed at the bifurcation catastrophe point illustrated in Fig. 3.1(c). The control coordinates u and v denote nuclear displacements of B_2 and A_1 symmetry, respectively. In the symmetrical bifurcation configuration, the singularity in ρ is located on the C_2 axis between the protons. The defining equations of the critical points derived from the unfolding in eqn (4.3) for $w > 0$ are

$$\partial f/\partial x = 2x(y + w) - u = 0 \quad \text{and} \quad \partial f/\partial y = (x^2 - y^2) - v = 0$$

and, in general, they lead to a quartic equation. For some special values of u and v, however, the solutions are simple and they suffice to give an overall description of the dynamics of a three-atom system. These solutions are presented in detail by Bader et $al.$ (1979). Here we summarize the structural predictions so obtained.

For a displacement $(u = 0, v > 0)$ from the bifurcation point illustrated in Fig. 3.1(c), the singularity between the protons is transformed into a $(3, -1)$ H–H bond critical point and a $(3, +1)$ critical point. The resulting molecular graph is characteristic of a ring structural region, the region bounded by the hypocycloid in Fig. 4.3(a). For the opposite displacement from the catastrophe point $(u = 0, v < 0)$ corresponding to the oxygen nucleus approaching the H_2 molecule, the singularity in ρ vanishes. The H–H bond path is broken and the molecular graph generated is typical of the open structure found for H_2O in its equilibrium geometry (Fig. 3.1(a), (b)). The same type of singularity occurs for nuclear configurations of the two remaining sides of the hypocycloid except that the singularity occurs at either of the O–H bond critical points. A motion across either of these boundaries of the hypocycloid corresponds to an opening of the ring structure by the breaking of an O–H

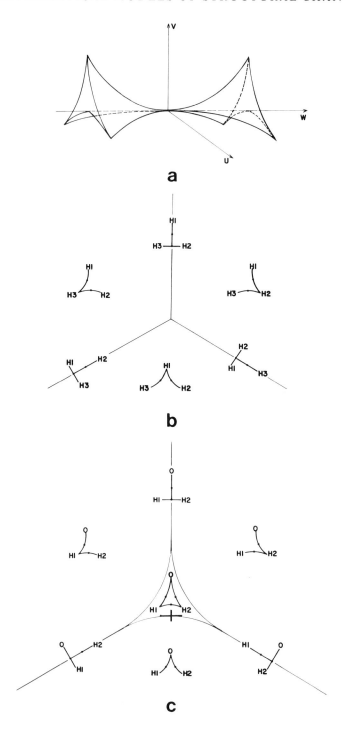

bond to yield structures of the type O–H–H. Three open structures are thus found outside of the ring region. Their associated structural regions are separated from one another by three semiaxes denoting the set of conflict catastrophe points characterized by the structures O–(H$_2$) and H–(OH).

Thus the unfolding of the elliptic umbilic for positive values of the control parameter w predicts the structure diagrams for the H$_3^+$ and H$_2$O systems and these diagrams agree with the theoretically determined behaviour of these two systems.

4.3.3   Formation of a cage structure

We have demonstrated that Thom's theory of elementary catastrophes finds a direct application in the analysis of structural instabilities which correspond to the making and/or opening of a ring structure. The usefulness of Thom's classification theorem is a consequence of the fact that all the changes in $\nabla\rho$ that are involved in such a process occur on a two-dimensional submanifold of the behaviour space of the electronic coordinates. Clearly, more complex cases of structural changes are to be expected, cases whose complete description will necessitate the use of the full three-dimensional behaviour space. Such a case is illustrated by the formation of a cage structure.

The system we use to exemplify this process, the [1.1.1]propellane molecule C$_5$H$_6$, belongs to a class of systems which chemically are regarded as highly strained ring systems. In the equilibrium geometry X$_e$ of D$_{3h}$ symmetry, the charge distribution of this molecule exhibits three, three-membered rings (Fig. 4.4). Each of these rings contains the two bridgehead carbons and an apical carbon atom. Because of the symmetry of the system, the associated ring surfaces are planar. The molecular graph for the equilibrium structure is shown in Fig. 4.4(a).

This structure of three fused rings is transformed into a cage structure by breaking the bond between the bridgehead carbon atoms. The bridgehead bond length of 1.59 Å in this molecule is longer by ~ 0.06 Å than a normal C–C bond. This increased equilibrium separation has two effects on the charge distribution: (1) the value of $\rho(\mathbf{r})$ at the carbon–carbon bond critical point is reduced in value from that in a normal C–C bond; (2) the (3, + 1) ring critical points are in the immediate proximity of the carbon–carbon bond critical point and the value of the charge density at these two points are of almost equal value. This proximity of the (3, + 1) ring and (3, − 1) bond

---

Fig. 4.3.   (a) A plot in the (u, v, w) control space of the bifurcation set obtained from the unfolding of the elliptic umbilic. The structure diagrams for H$_3^+$ in (b) and for H$_2$O in (c) correspond to sections in the (u, v)-plane of the bifurcation set shown in (a) for $w = 0$ and $w > 0$, respectively.

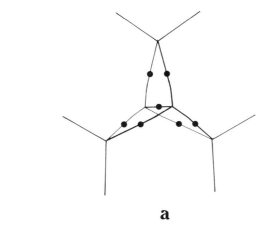

a

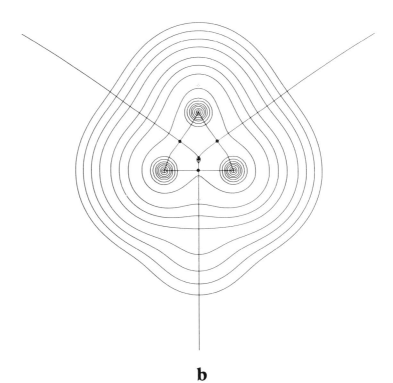

b

FIG. 4.4. (a) A molecular graph illustrating the equilibrium structure of [1.1.1]propellane and, in (b), a plot of $\rho(\mathbf{r})$ in the plane of one of its three-membered rings. In (b) and in succeeding diagrams, the position of a $(3, +1)$ critical point is denoted by a triangle. Note the proximity of the $(3, +1)$ ring critical point to the $(3, -1)$ bridgehead bond critical point.

critical points portends the instability of this system. An increase in the separation between the bridgehead carbon atoms results in the coalescence of the $(3, -1)$ bond critical point with all three of the $(3, +1)$ ring critical points to create a singularity in $\rho(\mathbf{r})$ and an unstable structure (Fig. 4.5(b)). The phase portrait of this singularity in the plane of the apical carbons is similar to that found for the singularity in the H_3^+ system (Fig. 4.2(b)). The singularity is thus of the elliptic umbilic type. As illustrated in Fig. 4.5(c), a further increase in the bridgehead separation causes this singularity to bifurcate, to yield the elements of a cage structure, i.e. a $(3, +3)$ critical point and three $(3, +1)$ critical points. This cage is bounded by three ring surfaces. Each of the ring surfaces contains two apical carbon nuclei, as well as the two bridgehead carbons and hence is characterized by substantial curvatures. In Fig. 4.5(c) these three ring surfaces are denoted by their intersections with the σ_h symmetry plane.

The identification of the singularity in Fig. 4.5(b) as being one of the elliptic umbilic type suggests that the structural changes described above can be expressed in terms of the unfolding of the elliptic umbilic catastrophe (eqn (4.3)). In the following, we shall show that this equation does indeed provide a model for the transformation of the equilibrium structure into the cage structure. Moreover, it will be shown to correctly predict other structural changes resulting from those deformations of the system which preserve the σ_h symmetry plane.

The phase portrait of the singularity found in Fig. 4.5(b) indicates that, in a sufficiently small neighbourhood of this singularity, the charge distribution can be represented by a Taylor series expansion of the same form as eqn (4.1), where the behaviour variables (x, y) now refer to a local coordinate system for the σ_h symmetry plane. (In principle, the quadratic term $Q(z)$ must be replaced by a more general function, $\bar{Q}(x, y, z)$ which will enable one to discuss instabilities in $\nabla\rho$ with respect to general deformations in the system's geometry, e.g. a deformation leading to a totally asymmetric configuration. However, for the purpose of the present discussion, we shall continue to use eqn (4.1) with a necessary change in the sign of the term $Q(z)$. This sign change is to ensure that, within the small neighbourhood considered above, the charge distribution increases in directions perpendicular to the x, y symmetry plane.) The structural changes which are observed in the neighbourhood of the bifurcation catastrophe point X_b, and which result from symmetry-preserving deformations, can be described in terms of the perturbations of the function of eqn (4.4), that is, in terms of the unfolding of eqn (4.3).

We have previously noted that the bifurcation set predicted by eqn (4.3) partitions a given control plane $w \neq 0$ into two regions. At any point (u, v) contained in the region which is bounded by the hypocycloid-shaped cross-section of the bifurcation set, the function $f(x, y; \boldsymbol{\mu})$ of eqn (4.3) exhibits two saddle points and another critical point, which is a local maximum if $w < 0$,

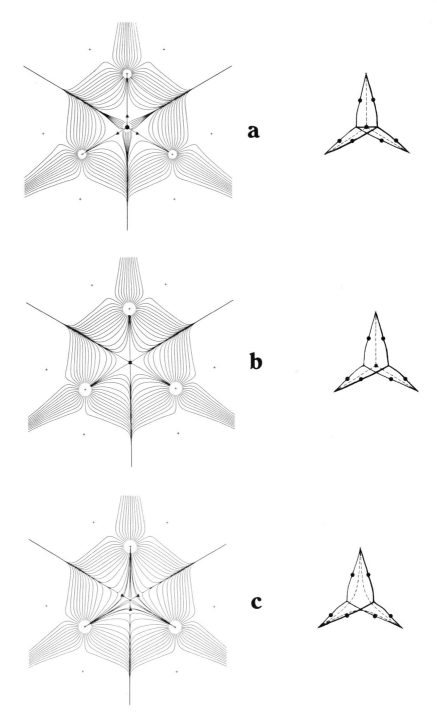

and a local minimum if $w > 0$. In applying eqn (4.3) to the description of the making or opening of a ring structure, we have discarded the case $w < 0$ on the basis that, in this case, the local maximum in $f(x, y; \boldsymbol{\mu})$ would correspond to an extraneous local maximum in the charge distribution at a location where a $(3, +1)$ ring critical point is found. In the present application, this argument does not hold, since, locally, ρ increases in directions perpendicular to the (x, y)-plane. A local maximum in $f(x, y; \boldsymbol{\mu})$ now corresponds to a $(3, -1)$ bond critical point, and a local minimum to a $(3, +3)$ cage critical point. Likewise, a saddle point, i.e. a $(2, 0)$ critical point, in f corresponds to a $(3, +1)$ ring critical point in the charge distribution. It is now apparent that eqn (4.3) correctly describes the transformation of the structure associated with the equilibrium geometry (Fig. 4.5(a)) into the cage structure (Fig. 4.5(c)). It is also seen that the control parameter w appearing in eqn (4.3) is a signed measure of the elongation of the bridgehead internuclear distance with respect to its value at $\mathbf{X}_b$.

The structural model afforded by eqn (4.3) predicts the existence of other structures for the [1.1.1]propellane system. Outside the region enclosed by the bifurcation set in Fig. 4.3(a), two structures, each consisting of two fused rings, are predicted. The first of these structures consists of two four-membered rings with one apical and the two bridgehead carbon atoms in common. This arrangement of the two ring surfaces is stable and must occur over a dense subset of this region of nuclear configuration space. There are three equivalent forms of this structure obtained by a permutation of the apical carbon atom which is common to both rings. The second of these fused-ring structures consists of a three- and a four-membered ring with the bridgehead carbon atoms common to both. In this type of structure, of which there are again three equivalent forms, the $(3, +1)$ ring critical points are arranged in such a manner that the stable manifold (the ring axis) of the four-membered ring intersects non-transversely the unstable manifold (the ring surface) of the three-membered ring. According to the Palis–Smale theorem

---

FIG. 4.5.   Gradient vector fields in the σ_h symmetry plane of C_5H_6 and corresponding molecular graphs (showing just the carbon nuclear framework) for three different values of the internuclear separation between the two bridgehead carbon atoms. The crosses denote the positions of the protons in this plane. The trajectories which terminate at these protons are omitted. The broken lines in the molecular graphs represent profiles of the ring surfaces. These same lines appear as trajectories in the corresponding $\nabla\rho(\mathbf{r})$ plots. (a) The equilibrium geometry. The $(3, -1)$ critical point at the centre of the diagram signifies the existence of a bond path linking the two bridgehead carbon atoms. (b) This diagram is obtained when R_b, the separation between the bridgehead carbon atoms, is increased by 0.4 au from equilibrium value. At this geometry the ring critical points merge with the bond critical point to yield the singularity in $\rho(\mathbf{r})$ denoted by a full square. This is the geometry of the bifurcation catastrophe point and the associated structure is unstable to all motions away from this geometry. If R_b is decreased from its value in (b), one regains the structure shown in (a) with a bond path linking the bridgehead carbons. If R_b is increased, this C–C bond path vanishes and the cage structure shown in (c) is obtained.
(From Bader *et al.* (1981).)

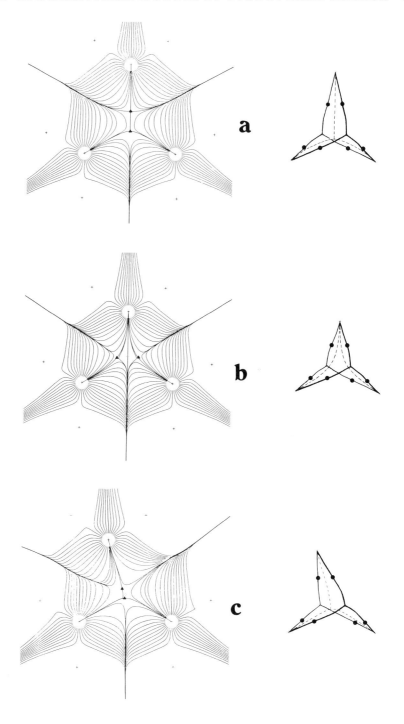

this arrangement is unstable and corresponds to a conflict catastrophe point in control space. A displacement of the system from such a conflict point will result in a structural change to yield one of the three equivalent forms of the stable fused-ring structures.

The above predictions are confirmed by the structural changes observed for a displacement of an apical carbon atom from its position on a C_2 symmetry axis in the catastrophe configuration X_b. A displacement along this axis in a direction away from the axis of the bridgehead carbon atoms yields the conflict structure shown in Fig. 4.6(a). This structure illustrates the non-transversal intersection of the stable and unstable manifolds of two ring critical points. A reversal of the above displacement yields the stable arrangement of two fused four-membered rings with the displaced carbon being common to both rings, Fig. 4.6(b). A similar structure is obtained for a displacement of the apical carbon atom perpendicular to the C_2 axis but still preserving the σ_h symmetry plane (Fig. 4.6(c)).

The above observations can be summarized in the structure diagram shown in Fig. 4.7(b). The catastrophe set in this diagram is similar to that for the H_3^+ system. It consists of three semiaxes of conflict configurations which meet at the bifurcation catastrophe point X_b. These axes partition the control plane ($w = 0$) into three structural regions. Each of these regions is associated with one of the stable fused-ring structures. By continuous extension, structure diagrams for any plane $w \neq 0$ can be constructed. Two such diagrams, one for $w > 0$, the other for $w < 0$ are shown in Fig. 4.7(a) and (c), respectively.

The preceding discussions illustrate the simple way in which the unfolding of the elliptic umbilic singularity (eqn (4.3)) accounts for structural changes which accompany special symmetry-preserving deformations. It is to be realized that the model afforded by this equation works only because the portion of the behaviour space in which these structural changes take place is of dimension two. This is a consequence of the preservation of the σ_h symmetry plane in all the distortions studied so far.

Difficulties are encountered as more general deformations are considered, deformations of the [1.1.1]propellane molecule which require the use of the full three-dimensional behaviour space. At the present time, Thom's classification theorem does not cover situations which involve more than two

FIG. 4.6. These diagrams represent three more structures of C_5H_6 which are obtained for motions away from the configuration of the unstable structure shown in Fig. 4.5(b). The structure in (a) is obtained by an outward displacement of the topmost apical carbon atom. The resulting structure is itself unstable. The structure in (b) is obtained by an inwards displacement of the same carbon atom and the structure in (c) results from the displacement of this carbon from its symmetry axis.
(From Bader *et al.* (1981).)

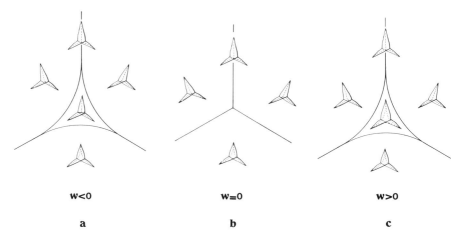

w<0   **w=0**   **w>0**

a   **b**   **c**

FIG. 4.7. Cross-sections of the structure diagram for C_5H_6 for that portion of its control or nuclear configuration space in which the plane of the apical carbon atoms is a symmetry plane. The structure at the origin of the control space is that given in Fig. 4.5(b) for the bifurction catastrophe point. An increase or decrease in w corresponds to an increase or decrease, respectively, in the separation between the bridgehead carbon atoms.
(From Bader *et al.* (1981).)

Table 4.1

Critical points in $\rho(\mathbf{r})$ for H_3^+ predicted by unfolding of elliptic umbilic

Values of control parameters	$(3, -1)$ critical points	
	Number	Location
$w = 0, u = 0 \begin{cases} v < 0 \\ v > 0 \end{cases}$	2	$x = 0, y = \pm\sqrt{-v}$
	2	$x = \pm\sqrt{v}, y = 0$
$w = 0, u \neq 0$	2	$x = \pm[v + (v^2 + u^2)^{1/2}]^{1/2}/\sqrt{2}, y = u/2x$

behaviour variables. Possible ways of overcoming these difficulties are discussed in Bader *et al.* (1981).

References

Bader, R. F. W., Nguyen-Dang, T. T., and Tal, Y. (1979). *J. Chem. Phys.* **70**, 4316.
Bader, R. F. W., Nguyen-Dang, T. T., and Tal, Y. (1981). *Rep. Prog. Phys.* **44**, 893.
Berry, M. V. and Mackley, M. R. (1977). *Phil. Trans. r. Soc., London* **A287**, 1.
Berry, M. V., Nye, J. F., and Wright, F. J. (1979). *Phil. Trans. r. Soc., London* **A291**, 454.

Collard, K. and Hall, G. G. (1977). *Int. J. quantum Chem.* **12**, 623.

Nguyen-Dang, T. T. and Bader, R. F. W. (1982). *Physica* **114A**, 68.

Palis, J. and Smale, S. (1970). *Pure Math.* **14**, 223.

Poston, T. and Stewart, I. (1978). *Catastrophe theory and its applications.* Pitman, London.

Thom, R. (1975). *Structural stability and morphogenesis.* W.A. Benjamin, Reading, Massachusetts.

THE QUANTUM ATOM

There can be no doubt but that in quantum mechanics one has the complete solution to the problems of chemistry.

G. N. Lewis (1933)

5.1 Chemistry and quantum mechanics

5.1.1 From Lewis to quantum mechanical models

Lewis introduced the concept of the electron pair into chemistry in 1916. In this paper (1916) he was able to rationalize the known geometries of carbon compounds by replacing the prevailing notion of a cubical arrangement of an octet of electrons by a tetrahedral arrangement of four pairs of electrons. What particularly served to later convince Lewis (1933) of the potential of the new quantum mechanics, as evidenced by the opening quotation, was its ability to account for observed molecular geometries and so provide a theoretical basis for his electron pair model. The early development of the valence bond theory of directed valence by Slater (1931) and Pauling (1931, 1960) provided the first quantum mechanical rationalization of molecular geometry. The valence bond approximation to the wave function, as expressed in terms of the overlap of hybridized atomic orbitals to describe the pairing of electrons, coupled with the notion of electron spin, was the natural theoretical extension of the pair concept of Lewis. The hybrid orbitals were spatially localized in directions coincident or nearly coincident with observed molecular geometries and this gave theoretical substance to the notion of picturing a bond as a spatially localized pair of electrons. Pauling's emphasis of the use of ionic-covalent resonance structures to account for the differing character of bonded interactions was in turn a restatement of Lewis's idea to represent the bonding between a pair of atoms in terms of tautomeric extremes of electron-sharing possibilities. It was a principal claim of Lewis's first paper (1916) that 'According to the theory which I am now presenting, it is not necessary to consider the two extreme types of chemical combination, corresponding to the very polar and the very nonpolar compounds, as different in kind, but only as different in degree.'

In the description of nature afforded by quantum mechanics, one classifies and characterizes the state of a total system in terms of the eigenvalues of a set of commuting observables acting on an element of the Hilbert space, the state vector. Molecular orbital theory in its canonical representation as originally developed by Mulliken (1928*a–c*, 1932*a*, 1935) and Hund (1933, 1937) was, and is, the theory of *electronic structure* for the prediction and classification of quantum states of many-electron systems. Molecular orbital theory provides a model of the electronic structure of a many-electron system in terms of a set of coupled one-electron states or orbitals which greatly facilitates the prediction and classification of the electronic states of many-electron atoms and molecules. In a molecular system, the one-electron states themselves can be classified as bonding, non-bonding, or antibonding (Mulliken 1932*b*, 1939) and this classification correlates with the observed changes in bond lengths and dissociation energies of the new states generated as a consequence of a change in the occupation of a particular orbital. The delocalized nature of the canonical set of molecular orbitals provided a useful model for understanding the properties of unsaturated systems (Coulson and Longuet-Higgins 1947*a,b*, 1948*a–c*), but was unsatisfactory in providing models of directed bonds. Lennard-Jones (1949*a,b*, 1952) defined equivalent orbitals (which later evolved into localized orbitals) to provide the molecular orbital equivalent of directed valence orbitals and it was only after their introduction that molecular orbital theory found widespread application to structural problems in chemistry. Since the advent of suitable computational methods and machines in the 1960s, molecular orbitals calculated within the self-consistent field approximation and expanded in terms of a finite set of basis functions (Roothaan 1951; Hall 1951) have provided the principal means of obtaining approximate solutions to the wave functions of many-electron systems of chemical interest.

The orbital model has, however, been extended beyond its intended use of predicting and providing an understanding of the electronic structure of a system, by associating the forms of individual orbitals with the assumed spatially localized pairs of bonded or non-bonded electrons, and by attempting to define atomic properties in terms of coefficients of atomic-centred basis functions appearing in the expansion of molecular orbitals. These steps are admittedly arbitrary, as are attempts to define atoms through a partitioning of the Hamiltonian operator. This latter step violates the indistinguishability of the electrons from the outset.

One might imagine that, with the advent of quantum mechanics and its application to chemistry, Dalton's atomic theory would have been reinforced. This has not happened. Quantum mechanics has been shown to account for the properties of isolated atoms and for the total properties of a molecular system. The increased understanding that would result from the discovery of a firm theoretical basis for Dalton's theory has not been obtained because of the lack of a quantum definition of an atom in a molecule. This is not to say that the concepts of atoms and bonds do not appear in the quantum mechanical treatments of chemical systems. They do, but in the reverse manner to that described above. Rather than finding its quantum basis, the atomic and bond concepts are built into an approximate theory to model a real system. Thus, in valence bond theory a molecular wave function is approximated in terms of products of atomic wave functions in appropriate valence states. Moffitt's (1951) 'atoms in molecules approach' to the calculation of molecular binding energies is a further example, as is the success of using atomic functions as a basis for the expansion of molecular orbitals. Properties of an ionic crystal strongly suggest that it is best regarded as a collection of interacting ions. Hence, one may successfully approximate the wave functions for such systems using free-ion wave functions with adjustments for the overlap of non-orthogonal functions as illustrated by the work of Löwdin (1948, 1956) in an early theoretical calculation of the properties of ionic crystals.

In 1929 Dirac worte: 'The underlying physical laws necessary for the mathematical theory of . . . the whole of chemistry are thus completely known, and the difficulty is only that the exact application of these laws leads to equations much too complicated to be soluble.' The stumbling block to obtaining useful approximate solutions to the equations of quantum mechanics for systems of chemical interest has been largely overcome. Overcoming this hurdle, however, has not in itself led to a mathematical theory of the whole of chemistry, at least not to one which is expressible in the language of chemistry. Given a state function, or a good approximation to it, how does one obtain from it a description of a system's properties in terms of its atoms and its bonds which, from a chemist's point of view, summarizes in a concise manner its important chemical properties? The varying successes of the quantum mechanical models of atoms and bonds have served to demonstrate the soundness of the atomic concept but, by building in the idea of atoms, they have not furthered Dalton's theory of atoms in molecules. The increased understanding that would result from the discovery of a firm theoretical basis for Dalton's theory requires a quantum definition of an atom in a molecule.

5.1.2   The role of the charge density in defining structure

Atoms and bonds have meaning in real space and are a reflection of the structure present in real space. This structure is not reflected in the properties of the infinite-dimensional Hilbert space of the state function. Instead, the physical basis for molecular structure should reside in the quantum mechanical function which provides a description of a system as it exists in real space. This function is a system's distribution of charge as defined in eqn (1.3) for a stationary state or as defined now for the general time-dependent case,

$$\rho(\mathbf{r}, \mathbf{X}, t) = N\sum(\text{spins}) \int \{\textstyle\prod_{j \neq i} d\tau_j\} \Psi^*(\mathbf{x}, \mathbf{X}, t)\Psi(\mathbf{x}, \mathbf{X}, t) \qquad (5.1)$$

where Ψ is a properly antisymmetrized solution to the general time-dependent Schrödinger equation,

$$i\hbar\,\partial\Psi/\partial t = \hat{H}\Psi \qquad \text{and} \qquad -i\hbar\,\partial\Psi^*/\partial t = \hat{H}\Psi^*. \qquad (5.2)$$

As in Chapter 1, x denotes the collection of electronic space and spin coordinates, X the nuclear coordinates, and r the space coordinates (x, y, z) of a single electron and $d\tau_j = dx_j dy_j dz_j$. It should be noted that the electronic charge density contains the information needed to determine the distribution of nuclear charge as well, and thus it determines the total distribution of charge. When $r = X_\alpha$, a nuclear position coordinate, ρ exhibits a cusp and through the cusp condition, eqn (E2.2), it determines the nuclear charge Z_α. In addition, the positions of the nuclei are evident in the topology of the electronic charge density, as each nuclear cusp behaves like a local maximum in the charge distribution with $\rho(X_\alpha) \approx Z_\alpha^3/2$.

Molecular orbital theory has played the central role in the definition and understanding of problems of electronic structure. The charge density plays the corresponding role in the definition and understanding of the concepts associated with molecular structure. The previous chapters have shown that atoms, bonds, and structure are indeed consequences of the dominant topological property exhibited by a molecular charge distribution. What remains to be done is to demonstrate that the topological atom and its properties have a basis in quantum mechanics.

5.2   Need for a quantum definition of an atom

5.2.1   Observational basis for a quantum atom

A theory is only justified by its ability to account for observed behaviour. It is important, therefore, to note that the theory of atoms in molecules is a result of observations made on the properties of the charge density. These observations give rise to the realization that a quantum mechanical description of the properties of the topological atom is not only possible but is also necessary, for the observations are explicable only if the virial theorem applies to an atom in a molecule. The original observations are among the most important of the properties exhibited by the atoms of theory (Bader and Beddall 1972). For this reason and for the purpose of emphasizing the observational basis of the theory, these original observations are now summarized. They provide an introduction to the consequences of a quantum mechanical description of an atom in a molecule.

Figure 5.1 displays, in the form of contour plots, the ground-state charge distributions of LiF, LiO, and LiH, each at its equilibrium internuclear separation. Superimposed on each of these plots is the intersection of the

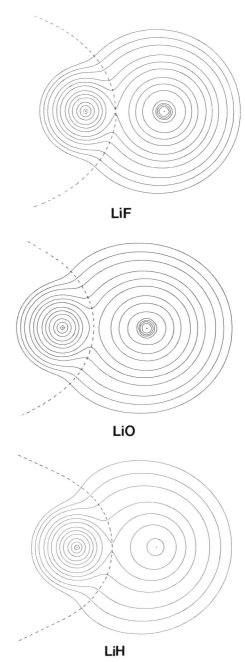

FIG. 5.1.   Contour plots of the ground-state molecular charge distribution of LiF, LiO, and LiH. The intersection of the interatomic surface with the plane shown in the diagram is indicated by a dashed line.

zero-flux interatomic surface, as defined in eqn (2.9), with the plane of the diagram. The distribution of charge within the Li atom so defined is seen to be remarkably similar in all three of these molecules in spite of the very different natures of the neighbouring atom. The net charge of the Li atom (as determined by an integration of $\rho(\mathbf{r})$ over the atomic basin to obtain its average number of electrons followed by its subtraction from the nuclear charge) is very nearly the same for all three molecules (see Table 5.1), the observed variation following the trend anticipated on the basis of decreasing electronegativity of the ligand $F > O > H$. This near constancy in the charge distribution of Li correctly reflects the properties assigned to Li on the basis of the chemistry observed for compounds in which Li is bonded to a more electronegative element—those corresponding to a relatively small, singly-charged positive ion with a tightly bound distribution of electronic charge.

Figure 5.1 illustrates an elementary but important and general observation: the choice of the zero-flux surface for defining an atom maximizes the possibility of assigning an atomic identity to a given mononuclear region of a charge distribution. It is clear that any other choice of a partitioning surface would either include a portion of the neighbouring atom that is very different in all three cases, or omit a portion of the Li atom charge density that changes by only small amounts through the series of molecules. Since the partitioning must exhaust the space of a system, the latter possibility would assign the very similar omitted portions to the very different neighbouring atoms.

Coupled with the observation of near constancy in the charge distributions of the Li atoms in this series of molecules was the further observation that the kinetic energy density distributions exhibited a corresponding degree of constancy. (As demonstrated later, while there is no unique definition of a kinetic energy density, all definitions lead to the same average value for the kinetic energy when integrated over the basin of an atom up to its surface of zero flux. The average kinetic energy of an atom is a well-defined quantity as

Table 5.1

*Some properties of bound Li atoms**

Molecule	Net charge of Li atom	Average electronic kinetic energy of Li atom (au)
$LiF(X^1\Sigma^+)$	+ 0.937	7.354
$LiO(X^2\Pi)$	+ 0.932	7.356
$LiH(X^1\Sigma^+)$	+ 0.913	7.368

*From Bader and Beddall (1972); calculated from state functions close to the Hartree–Fock limit.

a consequence of the zero-flux surface condition.) Thus, it is an observation in this set of molecules and subsequently in others, that a constancy in the distribution of charge for a topologically defined atom leads to a corresponding degree of constancy in the average kinetic energy of the atom, the total spread in values for the Li atoms in the present examples being ~ 32 kJ/mol (see Table 5.1).

According to the virial theorem, the total energy for a system with inverse square forces is equal to minus the average kinetic energy. If one postulates the existence of an atomic statement of the virial theorem, then the above observation predicts that, when the charge distribution of an atom is identical in two different systems, the atom will contribute identical amounts to the total energies in both systems. *Equally important is the more general conclusion that the properties of such a topologically defined atom are directly determined by its charge distribution, the properties changing in direct response to changes in the charge density of the atom.*

The postulation of an atomic virial theorem for the topologically defined atoms leads to a number of important conclusions (Bader and Beddall 1972).

1. The total energy of a molecule is expressible as a sum of atomic energies.

2. The average potential energy of an atom is defined as the average of the virial of the forces exerted on it—as demonstrated later, this is the only non-arbitrary way of partitioning potential energies of interaction between systems.

3. A relationship must exist between the distribution of charge and the virial of the total force exerted on each element of the charge density, the virial field. The form of the atom must be independent of the individual contributions to the forces exerted on it, since these forces change radically between any pair of systems. The charge density of an atom must respond instead only to the sum of the local forces it experiences. The virials of the individual contributions to the forces exerted on the density of the Li ion by the hydride ion differ by thousands of kJ/mol from those exerted by the fluoride ion (the one-electron potential energies differ by 14×10^3 kJ/mol, for example) but the net force in each case corresponds to that emanating from a polarized singly-charged negative ion. The constancy in the charge distribution of Li must be viewed as remarkable when one contemplates these very different contributions to the force exerted on it by its neighbours. One field, however, changes by little—the virial of the total Ehrenfest force exerted on the electrons. This relationship between ρ and the virial field is the basis of the essential observation that, if the distribution of charge for an atom is identical in two different systems, then the atom will contribute identical amounts to the total energy in both systems.

It will be shown that each atom in a system makes an additive contribution to the average value of every system property. This is the principle underlying the cornerstone of chemistry—that atoms and functional groupings of atoms make recognizable contributions to the total properties of a system. In practice, we recognize a group and predict its effect upon the static and reactive properties of a system in terms of a set of properties assigned to the group. In those limiting situations wherein a group is essentially the same in two different systems, one obtains a so-called additivity scheme for the total properties, for in this case the atomic contributions, as well as being additive over each system, are transferable between the systems.

Examples are given later of near perfect transferability of groupings of atoms, cases where the group energies change by less than 4 kJ/mol. It is at this limit that one can determine the properties of atoms in molecules experimentally, as corresponding additive contributions to heats of formation. Such transferability is, of course, the exception rather than the rule but these limiting examples provide the touchstone for obtaining an understanding of the properties of the chemical atom—that these properties change in direct response to the extent of the changes in its form in real space. When this form remains constant, so too do the contributions which the atom makes to the properties of the total system in which it is found.

It is upon the strength of these observations that the theory of atoms in molecules rests, for they form the basis for the identification of the topological atom with the chemical atom of a molecular system. This chapter demonstrates that the postulated existence of the atomic virial theorem is confirmed, as is the complete quantum description of the properties of the topological atom. The development of the quantum mechanics of a subsystem requires careful consideration of the properties of observables and of the derivation of their equations of motion. These fundamental aspects of the quantum description of a total system are considered next, with an emphasis on the changes anticipated in the description of a subsystem.

5.2.2   Observables and their properties for a total system

It is a postulate of quantum mechanics that everything that can be known about a system is contained in the state function Ψ. The value of a physical property is obtained from the state function through the action of a corresponding operator on Ψ. Thus quantum mechanics is concerned with observables, the name given to the linear, Hermitian operators whose action on Ψ yields the values of a system's properties. The Hermitian character of an operator is illustrated for the Hamiltonian $\hat{H}$, an illustration which introduces the quantum mechanical current density and demonstrates its role in the description of the properties of subsystem averages of operators. If

$\Psi^*\Psi d\tau$ is to represent a probability, then the norm, N, of the state function must remain constant in time. Using Schrödinger's equation and its complex conjugate (eqn (5.2)), one has

$$\partial|\Psi|^2/\partial t = \Psi^*\dot\Psi + \dot\Psi^*\Psi = -(i/\hbar)\{\Psi^*(\hat H\Psi) - (\hat H\Psi)^*\Psi\} \tag{5.3}$$

where $\dot\Psi = d\Psi/dt$. Integration of eqn (5.3) over all configuration space yields

$$dN/dt = -(i/\hbar)\int\{\Psi^*(\hat H\Psi) - (\hat H\Psi)^*\Psi\}d\tau. \tag{5.4}$$

In order that the norm remain constant in time it is necessary and sufficient that

$$\int\Psi^*(\hat H\Psi)d\tau = \int(\hat H\Psi)^*\Psi\,d\tau. \tag{5.5}$$

An operator with the property exhibited in eqn (5.5) is said to be Hermitian if it satisfies this equation for all functions Ψ defined in the function space in which the operator is defined. The mathematical requirement for Hermiticity of $\hat H$ expressed in eqn (5.5) places a corresponding physical requirement on the system—that there be a zero flux in the vector current through the surface S bounding the system. To illustrate this and other properties of the total system we shall assume, without loss of generality, a form for $\hat H$ corresponding to a single particle moving under the influence of a scalar potential $\hat V(\mathbf r)$

$$\hat H = -(\hbar^2/2m)\nabla^2 + \hat V(\mathbf r) = \hat p^2/2m + \hat V(\mathbf r). \tag{5.6}$$

It is worthwhile mentioning at this point that all properties of a subsystem defined in real space, including its energy, necessarily require the definition of corresponding three-dimensional density distribution functions. Thus, all the properties of an atom in a molecule are determined by averages over effective single-particle densities or 'dressed operators' and the one-electron picture is an appropriate one.

The potential energy operator in $\hat H$ is a real quantity which does not involve derivatives and, in this case, eqn (5.5) can be rewritten as

$$\int\{\Psi^*(\nabla^2\Psi) - (\nabla^2\Psi)^*\Psi\}d\tau = 0. \tag{5.7}$$

The integral in eqn (5.7), like many to come, can be transformed into a surface integral using the three-dimensional analogue of integration by parts. In one-dimension, one has

$$\int(\phi^*\partial^2\phi/\partial x^2)dx = (\phi^*\partial\phi/\partial x)|_{x_1}^{x_2} - \int(\partial\phi^*/\partial x)(\partial\phi/\partial x)dx \tag{5.8}$$

where x_1 and x_2 represent the limits of integration. In the extension of this result to three dimensions, the contributions arising from these limits combine to yield an integral over the surface S bounding the system. When applied to the complex conjugate term in eqn (5.7) as well, one obtains Green's theorem

$$\int\{\phi^*\nabla^2\phi - \phi\nabla^2\phi^*\}d\tau = \oint dS(\phi^*\nabla\phi - \phi\nabla\phi^*)\cdot\mathbf n \tag{5.9}$$

where $\mathbf{n}$ is the vector normal to the surface S. Use of eqn (5.9) to re-express eqn (5.7) yields

$$\int \{\Psi^* \nabla^2 \Psi - (\nabla^2 \Psi)^* \Psi\} \, d\tau = \oint dS (\Psi^* \nabla \Psi - \Psi \nabla \Psi^*) \cdot \mathbf{n}$$

$$= (2mi/\hbar) \oint dS \, \mathbf{j} \cdot \mathbf{n} \qquad (5.10)$$

where the quantum mechanical vector current density $\mathbf{j}$ is defined as

$$\mathbf{j} = (\hbar/2mi)(\Psi^* \nabla \Psi - \Psi \nabla \Psi^*). \qquad (5.11)$$

Eqn (5.10) can be expressed in terms of the Hamiltonian operator to yield

$$\int \Psi^* (\hat{H}\Psi) d\tau - \int (\hat{H}\Psi)^* \Psi \, d\tau = -i\hbar \oint dS \, \mathbf{j} \cdot \mathbf{n}. \qquad (5.12a)$$

For a bound system, Ψ is square integrable and, hence, Ψ and its derivatives vanish on all elements $d\mathbf{S} = dS\mathbf{n}$ of the surface when the surface is removed to infinity. Thus the right-hand side of eqn (5.10) vanishes for the total system with boundaries at infinity and $\hat{H}$ is Hermitian. However, if the integration is limited to a subsystem Ω bounded by a surface $S(\Omega)$, part or all of which occurs for finite values of the integration variables, then one has

$$\int_\Omega \Psi^* (\hat{H}\Psi) d\tau - \int_\Omega (\hat{H}\Psi)^* \Psi d\tau = -i\hbar \oint dS(\Omega) \mathbf{j} \cdot \mathbf{n} \qquad (5.12b)$$

where integration over a subsystem is indicated by the subscript Ω on the integral sign. In this case the flux in the current density through the surface will not, in general, vanish and one cannot assume that the Hamiltonian integral is equal to its Hermitian conjugate as is true for the total system.

Finally we note that eqn (5.3) can be re-expressed as

$$d\rho/dt + \nabla \cdot \mathbf{j} = 0, \qquad (5.13)$$

which is a statement of the conservation law for a fluid. In its integrated form

$$dN(\Omega)/dt = -\int_\Omega \nabla \cdot \mathbf{j} \, d\tau = -\oint dS(\Omega) \mathbf{j} \cdot \mathbf{n} \qquad (5.14)$$

where $N(\Omega)$, the average number of electrons in Ω, equals the integral of ρ over the region Ω. This law states that the change in the average number of particles in the volume Ω is given by the flux in the vector current density through its bounding surface $S(\Omega)$. The final equality given in eqn (5.14) results from the use of Gauss's theorem which states that the volume integral of the divergence of a vector is equal to the flux in the vector through the surface bounding the system. The divergence of a vector and Gauss's theorem are reviewed in Section E5.3.

Depending on the nature of the state function, some observables yield 'sharp' values—Ψ is an eigenfunction of the observable—while others yield only 'average values'. If two observables $\hat{A}$ and $\hat{B}$ commute, i.e.

$$[\hat{A}, \hat{B}] = (\hat{A}\hat{B} - \hat{B}\hat{A}) = 0, \qquad (5.15)$$

then they possess a common, complete set of orthonormal eigenfunctions. For a system in a stationary state for which the state function is an eigenfunction of the Hamiltonian operator,

$$\hat{H}\psi = E\psi \qquad \text{and} \qquad \hat{H}\psi^* = E\psi^*, \qquad (5.16)$$

all of the observables which form a pairwise commuting set with $\hat{H}$ yield a set of sharp values and this set of numbers is used to classify the state of the system. The ground state of the water molecule, in addition to possessing a sharp value for the energy, is labelled $^1A_1$ as a consequence of the spin and symmetry operators forming such a commuting set with $\hat{H}$. The average value of an observable that is not a member of this commuting set of observables is given by

$$\langle \hat{A} \rangle = \langle \Psi, \hat{A}\Psi \rangle / \langle \Psi, \Psi \rangle = \int \Psi^* \hat{A}\Psi d\tau / \int \Psi^* \Psi d\tau \qquad (5.17)$$

and, of course, for a system in a stationary state, the sharp value for a commuting observable equals its average value.

The equation of motion for the average value of an observable $\hat{A}$ can be obtained directly from Heisenberg's equation for $\hat{A}(t)$ (Messiah 1958, p. 319). The result is

$$d\langle \hat{A} \rangle / dt = (i/\hbar)\langle \Psi, [\hat{H}, \hat{A}]\Psi \rangle + \langle \partial \hat{A}/\partial t \rangle. \qquad (5.18)$$

Equation (5.18) can also be obtained by differentiating the expression for the average value of $\hat{A}$ (eqn (5.17)) assuming Ψ is normalized to unity,

$$d\langle \hat{A} \rangle / dt = \int \{(\partial \Psi^*/\partial t)\hat{A}\Psi + \Psi^* \hat{A}(\partial \Psi/\partial t) + \Psi^*(\partial \hat{A}/\partial t)\Psi\}d\tau,$$

followed by the use of Schrödinger's equations (5.2), and subsequently taking into account the Hermiticity of the Hamiltonian operator,

$$\begin{aligned} d\langle \hat{A} \rangle / dt &= (i/\hbar)\{\langle \hat{H}\Psi, \hat{A}\Psi \rangle - \langle \Psi, \hat{A}\hat{H}\Psi \rangle\} + \langle \partial A/\partial t \rangle \\ &= (i/\hbar)\langle \Psi, (\hat{H}\hat{A} - \hat{A}\hat{H})\Psi \rangle + \langle \partial \hat{A}/\partial t \rangle \\ &= (i/\hbar)\langle \Psi, [\hat{H}, \hat{A}]\Psi \rangle + \langle \partial \hat{A}/\partial t \rangle. \end{aligned} \qquad (5.19)$$

For an observable $\hat{B}$ which commutes with the Hamiltonian and which does not possess an explicit time dependence one has

$$d\langle \hat{B} \rangle / dt = 0 \qquad (5.20)$$

and its value is independent of time. In analogy with classical mechanics the observable $\hat{B}$ is called a constant of the motion. None of the observables considered here have an explicit time dependence and the final term on the right-hand side of eqns (5.18) or (5.19) will be omitted from this point on.

With Schrödinger's equation to describe how the state function changes with time and Heisenberg's equation to determine the corresponding change in the average value of each observable, one has a complete mechanical description of a system. Heisenberg's equation is important from another point of view, as it can be used to obtain relationships of great general importance for particular choices of the observable $\hat{A}$. Examples are the Ehrenfest theorems governing the time rate of change of the average values of an electronic position coordinate $\hat{\mathbf{r}}$ and momentum $\hat{\mathbf{p}} = -i\hbar\nabla$, and the virial theorem which is obtained when $\hat{A} = \hat{\mathbf{r}} \cdot \hat{\mathbf{p}}$. These theorems play important roles in the mechanical description of an atom in a molecule.

With $\hat{A}$ set equal to the product of the mass and the position coordinate for an electron, the commutator is

$$m[\hat{H}, \hat{\mathbf{r}}] = (\hbar/i)(-i\hbar\nabla) = (\hbar/i)\hat{\mathbf{p}} \tag{5.21}$$

and the first of Ehrenfest's relations is obtained,

$$m\,d\langle\hat{\mathbf{r}}\rangle/dt = \langle\hat{\mathbf{p}}\rangle. \tag{5.22}$$

For $\hat{A} = \hat{\mathbf{p}}$, the commutator is

$$[\hat{H}, \hat{\mathbf{p}}] = i\hbar\nabla\hat{V}(\mathbf{r}) \tag{5.23}$$

and the second Ehrenfest relation is obtained

$$d\langle\hat{\mathbf{p}}\rangle/dt = \langle-\nabla\hat{V}(\mathbf{r})\rangle = \langle\hat{\mathbf{F}}(\mathbf{r})\rangle \tag{5.24}$$

where, in analogy with classical definitions, the force operator $\hat{\mathbf{F}}(\mathbf{r})$ is defined as the negative of the gradient of the potential. The operator $\hat{\mathbf{F}}(\mathbf{r})$ will be termed the Ehrenfest force. Equations (5.22) and (5.24) are both quantum analogues of classical relations but, in the quantum case, the relations hold only for the average values of the mechanical properties and not in a pointwise manner for each specific time as they do on a classical trajectory. The two expressions can be combined to yield the analogue of Newton's equation of motion

$$m\,d^2\langle\hat{\mathbf{r}}\rangle/dt^2 = \langle\hat{\mathbf{F}}(\mathbf{r})\rangle. \tag{5.25}$$

The left-hand side of eqns (5.22) and (5.24) may be explicitly evaluated using Schrödinger's equations (5.2) for Ψ and Ψ^*, and expressed in terms of the quantum vector current density, eqn (5.11). For the time derivative of $\langle\hat{\mathbf{r}}\rangle$ one has

$$d\{\Psi^*\hat{\mathbf{r}}\Psi\}/dt = (i/\hbar)\{(\hat{H}\Psi^*)\hat{\mathbf{r}}\Psi - \Psi^*\mathbf{r}(\hat{H}\Psi)\},$$

$$d\{\Psi^*\hat{\mathbf{r}}\Psi\}/dt = (-i\hbar/2m)\{\hat{\mathbf{r}}(\nabla^2\Psi^*)\Psi - \hat{\mathbf{r}}\Psi^*\nabla^2\Psi\}$$

$$= -\hat{\mathbf{r}}\nabla\cdot\mathbf{j}.$$

In general, the notation **AB** will be used to denote a multiplication of two vectors to yield a dyadic. A review of the properties of dyadics is given in

Section E5.2. In particular, the following identity holds

$$\mathbf{V} \cdot (\mathbf{j} \hat{\mathbf{r}}) = \mathbf{j} + \hat{\mathbf{r}} \mathbf{V} \cdot \mathbf{j}$$

and hence

$$d\langle \hat{\mathbf{r}} \rangle / dt = \int \mathbf{j} \, d\tau - \int \mathbf{V} \cdot (\mathbf{j} \hat{\mathbf{r}}) \, d\tau,$$

which, using Gauss's theorem followed by multiplication by m, becomes

$$m \, d\langle \hat{\mathbf{r}} \rangle / dt = m \int \mathbf{j} \, d\tau - m \oint dS (\mathbf{n} \cdot \mathbf{j}) \hat{\mathbf{r}}. \tag{5.26a}$$

The surface term vanishes for a system with boundaries at infinity and one obtains the result that the time derivative of the average position vector is given by the integral of the current density,

$$m \, d\langle \hat{\mathbf{r}} \rangle / dt = m \int \mathbf{j} \, d\tau = \tfrac{1}{2} \{ \langle \hat{\mathbf{p}} \Psi, \Psi \rangle + \langle \Psi, \hat{\mathbf{p}} \Psi \rangle \} = \overline{\langle \hat{\mathbf{p}} \rangle}. \tag{5.26b}$$

The current has the dimensions of a velocity and thus, when multiplied by the mass, it equals a momentum. Similarly, the time derivative of a velocity is an acceleration and hence one should obtain

$$d\overline{\langle \hat{\mathbf{p}} \rangle} / dt = m \int \{ \partial \mathbf{j}(\mathbf{r}) / \partial t \} \, d\tau = \langle \mathbf{F}(\mathbf{r}) \rangle. \tag{5.27}$$

The derivation of eqn (5.27) also entails the vanishing of a surface term, one involving the quantum stress tensor. The stress tensor was first introduced in its relativistic form in 1927 by Schrödinger and its properties were later discussed by Pauli (1933). The derivation of eqn (5.27) follows that given by Pauli (1958).

$$m \, \partial \mathbf{j} / \partial t = \tfrac{1}{2} \{ \hat{H} \Psi^* (\mathbf{V} \Psi) - \Psi^* \mathbf{V}(\hat{H} \Psi) + (\mathbf{V} \Psi^*) \hat{H} \Psi - \mathbf{V}(\hat{H} \Psi^*) \Psi \},$$

$$m \, \partial \mathbf{j} / \partial t = (\hbar^2 / 4m) \{ (-\nabla^2 \Psi^*) \mathbf{V} \Psi + \Psi^* \mathbf{V}(\nabla^2 \Psi)$$
$$- (\mathbf{V} \Psi^*) \nabla^2 \Psi + \mathbf{V}(\nabla^2 \Psi^*) \Psi \} + 1/2 \{ \hat{V}(\mathbf{V} \Psi^*) \Psi$$
$$- \mathbf{V}(\hat{V} \Psi^*) \Psi + \hat{V} \Psi^* \mathbf{V} \Psi - \Psi^* \mathbf{V}(\hat{V} \Psi) \},$$

$$m \, \partial \mathbf{j} / \partial t = - (\hbar^2 / 4m) \{ (\nabla^2 \Psi^*) \mathbf{V} \Psi - \Psi^* \mathbf{V}(\nabla^2 \Psi)$$
$$+ (\mathbf{V} \Psi^*) \nabla^2 \Psi - \mathbf{V}(\nabla^2 \Psi^*) \Psi \} - \Psi \Psi^* \mathbf{V} \hat{V}.$$

Define the stress tensor $\overleftrightarrow{\sigma}$ as

$$\overleftrightarrow{\sigma} = (\hbar^2 / 4m) \{ \Psi^* \mathbf{V}(\mathbf{V} \Psi) + \mathbf{V}(\mathbf{V} \Psi^*) \Psi - \mathbf{V} \Psi^* \mathbf{V} \Psi - \mathbf{V} \Psi \mathbf{V} \Psi^* \} \tag{5.28}$$

and the expression for the time derivative of $\mathbf{j}$ may be expressed as

$$m \, \partial \mathbf{j} / \partial t = + \mathbf{V} \cdot \overleftrightarrow{\sigma} - \Psi^* \Psi \mathbf{V} \hat{V} \tag{5.29}$$

Hence,

$$m \int \{ \partial \mathbf{j}(\mathbf{r}) / \partial t \} \, d\tau = \int \Psi^* (-\mathbf{V} \hat{V}) \Psi d\tau + \oint dS \, \overleftrightarrow{\sigma} \cdot \mathbf{n}$$

where Gauss's theorem has been used to transform the volume integral of $\mathbf{V} \cdot \overleftrightarrow{\sigma}$ into an integral over the surface, an integral which vanishes for a system with infinite boundaries to yield the equations listed under (5.27). While the current density does not appear in the description of the properties of a stationary state in the absence of a magnetic field, it and the quantum stress tensor $\overleftrightarrow{\sigma}$ play important roles in the expressions obtained for the average values of subspace properties and they are introduced here for that reason.

One further theorem of paramount importance obtainable from Heisenberg's equation is the virial theorem as obtained from the operator $\hat{\mathbf{r}} \cdot \hat{\mathbf{p}}$. In this case the commutator is

$$(i/\hbar)\,[\hat{H}, \hat{\mathbf{r}} \cdot \hat{\mathbf{p}}] = (-\hbar^2/m)\nabla^2 - \mathbf{r} \cdot \mathbf{V}\hat{V} = 2\hat{T} - \mathbf{r} \cdot \mathbf{V}\hat{V}$$

where $\hat{T}$ denotes the kinetic energy operator. Hence,

$$d\langle\hat{\mathbf{r}} \cdot \hat{\mathbf{p}}\rangle/dt = 2\langle\hat{T}\rangle + \langle -\hat{\mathbf{r}} \cdot \mathbf{V}\hat{V}\rangle. \tag{5.30}$$

The usual statement of the virial theorem for a stationary state is

$$2\langle\hat{T}\rangle + \langle -\hat{\mathbf{r}} \cdot \mathbf{V}\hat{V}\rangle = 0. \tag{5.31}$$

In the original derivation of the classical virial theorem given by Clausius, an expression corresponding to eqn (5.30) is also obtained. In the classical case one argues that the time average of $d(\hat{\mathbf{r}} \cdot \hat{\mathbf{p}})/dt$ vanishes over a sufficiently long period of time or that the motion is periodic to obtain the equivalent of eqn (5.31).

Alternatively, the virial theorem may be derived directly from eqn (5.29), which expresses the forces in terms of the time derivative of the current density. Dotting the vector $\mathbf{r}$ into eqn (5.29) to obtain the virial of these forces gives

$$m\mathbf{r} \cdot \{\partial\mathbf{j}/\partial t\} = +\mathbf{r} \cdot \mathbf{V} \cdot \overleftrightarrow{\sigma} - \Psi^*\Psi(\mathbf{r} \cdot \mathbf{V}\hat{V}).$$

The term $\mathbf{r} \cdot \mathbf{V} \cdot \overleftrightarrow{\sigma}$ may be re-expressed using the identity

$$\mathbf{V} \cdot (\mathbf{r} \cdot \overleftrightarrow{\sigma}) = \mathrm{Tr}\,\overleftrightarrow{\sigma} + \mathbf{r} \cdot \mathbf{V} \cdot \overleftrightarrow{\sigma}$$

where $\mathrm{Tr}\,\overleftrightarrow{\sigma}$ denotes the trace of the tensor $\overleftrightarrow{\sigma}$,

$$\mathrm{Tr}\,\overleftrightarrow{\sigma} = (\hbar^2/4m)\,\{\Psi^* \nabla^2 \Psi + (\nabla^2 \Psi^*)\Psi$$
$$- \nabla\Psi^* \cdot \nabla\Psi - \nabla\Psi \cdot \nabla\Psi^*\}. \tag{5.32}$$

Each term in eqn (5.32) gives minus one-half of the average kinetic energy when integrated over all space. As demonstrated later in eqn (5.47), the averages of these two ways of expressing the kinetic energy in eqn (5.32) differ by a surface integral and one has

$$- \int \mathrm{Tr}\,\overleftrightarrow{\sigma}\,d\tau = 2\langle\hat{T}\rangle + (\hbar^2/4m)\oint dS\,\mathbf{V}\rho \cdot \mathbf{n}.$$

Use of these results yields

$$m\int \mathbf{r} \cdot \{\partial \mathbf{j}/\partial t\} \, d\tau = 2\langle \hat{T} \rangle + \langle -\mathbf{r} \cdot \nabla \hat{V} \rangle$$
$$+ (\hbar^2/4m)\oint d\mathbf{S} \nabla \rho \cdot \mathbf{n} + \oint d\mathbf{S} \mathbf{r} \cdot \overset{\leftrightarrow}{\sigma} \cdot \mathbf{n}$$

where the final surface term comes from applying Gauss's theorem to the integral of $\nabla \cdot (\mathbf{r} \cdot \overset{\leftrightarrow}{\sigma})$. The two surface integrals vanish for a system with boundaries at infinity. Hence the final result, a statement of the virial theorem for the general time-dependent case,

$$d\langle \hat{\mathbf{r}} \cdot \hat{\mathbf{p}} \rangle/dt = m\int \mathbf{r} \cdot \{\partial \mathbf{j}/\partial t\} \, d\tau = 2\langle \hat{T} \rangle + \langle -\hat{\mathbf{r}} \cdot \nabla \hat{V} \rangle. \qquad (5.33)$$

The virial of a system, denoted by the symbol $\mathscr{V}$, is defined to be the average

$$\mathscr{V} = \langle -\mathbf{r} \cdot \nabla \hat{V}(\mathbf{r}) \rangle = \langle \mathbf{r} \cdot \mathbf{F}(\mathbf{r}) \rangle.$$

It is the average of the virial of the forces acting on the particles which possess the kinetic energy $\langle \hat{T} \rangle$. When $\mathbf{F}(\mathbf{r})$ is the Ehrenfest force acting on the electrons, $\mathscr{V}$ is the potential energy of the electrons. In those cases where $\hat{V}$, the potential energy operator of the system, is a homogeneous function of the coordinates of degree n, one can, by using Euler's theorem, re-express the virial in terms of the average of the potential energy operator

$$\mathscr{V} = -n\langle \hat{V}(\mathbf{r}) \rangle$$

and, consequently, for a stationary state one has

$$2\langle \hat{T} \rangle = n\langle \hat{V} \rangle. \qquad (5.34)$$

For an isolated atom, the potential energy operator for the Coulombic forces is a homogeneous function of degree -1 and in this case one has

$$2\langle \hat{T} \rangle = -\langle \hat{V} \rangle.$$

For a system in a stationary state, the equation of motion for the mean value of $\hat{A}$, eqn (5.18), assumes the form

$$\langle [\hat{H}, \hat{A}] \rangle = 0.$$

Alternatively, this result can be derived directly by making use of the Hermitian property of $\hat{H}$ and Schrödinger's equation for a stationary state (eqn (5.16)),

$$\langle [\hat{H}, \hat{A}] \rangle = \langle \psi, (\hat{H}\hat{A} - \hat{A}\hat{H})\psi \rangle = \langle \hat{H}\psi, \hat{A}\psi \rangle - \langle \psi, \hat{A}\hat{H}\psi \rangle$$
$$= E\{\langle \psi, \hat{A}\psi \rangle - \langle \psi, \hat{A}\psi \rangle\} = 0. \qquad (5.35)$$

If $\hat{H}$ contains a real parameter s such as a nuclear charge or a nuclear coordinate and $\hat{A}$ corresponds to the operator $\partial/\partial s$, then the derivation of eqn (5.35) must be modified to include the term $\partial E/\partial s$, for E contains the same parameters as does $\hat{H}$. The result in this case is

$$\langle [\hat{H}, \hat{A}] \rangle - \langle [E, \hat{A}] \rangle = 0$$

and, when the commutators are evaluated, one obtains the Hellmann–Feynman theorem

$$\langle \psi, (\partial \hat{H}/\partial s)\psi \rangle = \partial E/\partial s. \tag{5.36}$$

When s is a nuclear coordinate $\mathbf{X}_\alpha$, eqn (5.36) gives the Hellmann–Feynman electrostatic theorem

$$\langle \psi, (-\nabla_\alpha \hat{V})\psi \rangle = -\nabla_\alpha E = \mathbf{F}_\alpha. \tag{5.37}$$

Equation (5.35) and the generalized statement of the Hellmann–Feynman theorem can also be derived through a variation of the state function (Epstein 1974*b*). In this case eqn (5.35) is known as the *hypervirial theorem* (Hirschfelder 1960). The variational derivations are important because they demonstrate that the theorems hold not only for the exact state function but also for approximations to these functions when the latter functions, such as Hartree–Fock, unrestricted Hartree–Fock, and multiconfigurational Hartree–Fock, are invariant to the variational parameters. The original such proof of the Hellmann–Feynman theorem was given by Hurley (1954).

It is clear from the use of the Hermiticity of $\hat{H}$ in the derivation of eqn (5.35) that the hypervirial theorem will assume a different form for a subsystem. A derivation of the hypervirial theorem for a subsystem is given at this point to illustrate the consequence of the loss of the Hermitian property of $\hat{H}$. The result is general and applies to any subsystem regardless of its definition. To ensure that the subsystem average of an observable $\hat{A}$ be a real number, one must work with the mean of the average value and its complex conjugate (cc),

$$\frac{1}{2}\left\{ \int_\Omega \psi^*[\hat{H}, \hat{A}]\psi \, d\tau + \int_\Omega ([\hat{H}, \hat{A}]\psi)^* \psi \, d\tau \right\} = \frac{1}{2}\{\langle [\hat{H}, \hat{A}] \rangle_\Omega + \text{cc}\}. \tag{5.38}$$

Working with just the first of the averages, one obtains

$$\langle \psi, [\hat{H}, \hat{A}]\psi \rangle_\Omega = \langle \psi, \hat{H}\hat{A}\psi \rangle_\Omega - \langle \hat{H}\psi, \hat{A}\psi \rangle_\Omega$$
$$+ \langle \hat{H}\psi, \hat{A}\psi \rangle_\Omega - \langle \psi, \hat{A}\hat{H}\psi \rangle_\Omega \tag{5.39}$$

where the term $\langle \hat{H}\psi, \hat{A}\psi \rangle_\Omega$ has been both added and subtracted to maximize the correspondence with eqn (5.35). Making use of Schrödinger's equation as before to get rid of the last two terms on the right-hand side,

$$\langle \psi, [\hat{H}, \hat{A}]\psi \rangle_\Omega = (\hbar^2/2m) \int_\Omega \{(\nabla^2 \psi^*)\hat{A}\psi - \psi^* \nabla^2 (\hat{A}\psi)\} d\tau \tag{5.40}$$

where, as in the derivation of eqn (5.7), the terms involving $\hat{V}$ have cancelled out. Using eqn (5.9) with $\phi^* = \psi^*$ and $\phi = \hat{A}\psi$, gives

$$\langle \psi, [\hat{H}, \hat{A}]\psi \rangle_\Omega = -i\hbar \oint dS(\Omega)\mathbf{j}_A \cdot \mathbf{n} \tag{5.41}$$

where $\mathbf{j}_A$, the current density for an observable $\hat{A}$, is defined in a manner analogous to that for the vector current $\mathbf{j}$ itself,

$$\mathbf{j}_A = (\hbar/2mi)\{\psi^*\mathbf{\nabla}(\hat{A}\psi) - (\mathbf{\nabla}\psi^*)\hat{A}\psi\}. \tag{5.42}$$

The final statement of the subsystem hypervirial theorem is

$$\tfrac{1}{2}\{\langle[\hat{H}, \hat{A}]\rangle_\Omega + \text{cc}\} = -\tfrac{1}{2}\{i\hbar\oint dS(\Omega)\mathbf{j}_A \cdot \mathbf{n} + \text{cc}\}, \tag{5.43}$$

a result first obtained by Epstein (1974); (see also Srebrenik and Bader 1974).

By performing corresponding substitutions on the subsystem analogue of eqn (5.19), the expression for the time derivative of a subsystem expectation value is

$$\tfrac{1}{2}d\{\langle \Psi, \hat{A}\Psi \rangle_\Omega + \langle \hat{A}\Psi, \Psi \rangle_\Omega\}/dt$$
$$= \tfrac{1}{2}\{(i/\hbar)\langle[\hat{H}, \hat{A}]\rangle_\Omega + \text{cc}\} - \tfrac{1}{2}\{\oint dS(\Omega)\mathbf{j}_A \cdot \mathbf{n} + \text{cc}\}$$
$$+ \tfrac{1}{2}\{\oint dS(\Omega)(\partial S/\partial t)\Psi^*\hat{A}\Psi + \text{cc}\}. \tag{5.44}$$

The final term accounts for the dependence of the surface S bounding the region Ω on the time.

The importance of Heisenberg's equation and of the corresponding hypervirial theorem for a stationary state in the description of the properties of the total system is maintained in the description of a subsystem. Indeed, the generalized action principle that is employed to establish the quantum mechanics of a subsystem can be expressed in the form of a variational statement of the Heisenberg equation of motion. For a stationary state, the same principle reduces to a variation of the energy and the resulting theorem is a variational statement of the hypervirial theorem applicable to both the total system and a subsystem. It is to be emphasized, however, that the variational derivations of these statements as obtained through the action principle apply only to a particular class of subsystems, those bounded by a surface of zero flux in the gradient vector of the charge density—the definition of a topological atom. The use of these theorems in the description of a subsystem leads to novel results because of the general non-vanishing of the surface integral in eqns (5.43) and (5.44).

5.3   Need for a subsystem variation principle

An understanding of chemistry requires a regional description of a system. The notion that a molecule can be viewed as a collection of atoms linked by a network of bonds, a notion that has already been shown to be rooted in the topological properties of the charge distribution, is the operational principle underlying our classification and understanding of chemical behaviour. It would appear, therefore, that to find chemistry within the framework of quantum mechanics one must determine the values of observables for pieces of a total system, i.e. for a subsystem. But how is one to choose these pieces? Is there but one, or are there many ways of dividing a system into 'atoms' and its properties into corresponding atomic contributions? If there is an answer to this problem, then the necessary information must be contained in the state function, for Ψ tells us everything we can know about a system.

It is clear from the examples given in the preceding section that one can write down expressions for the expectation values of observables for a subsystem with an arbitrarily defined boundary. One has no guarantee, however, that the results obtained for such an arbitrarily defined region of space will be of physical significance. Indeed, one can show very simply and as one important example that the kinetic energy of an arbitrarily defined region of space is not well-defined.

Consider the expression

$$\nabla^2(\Psi^*\Psi) = \nabla \cdot \{(\nabla\Psi^*)\Psi + \Psi^*(\nabla\Psi)\}$$
$$= (\nabla^2\Psi^*)\Psi + \Psi^*(\nabla^2\Psi) + 2\nabla\Psi^* \cdot \nabla\Psi. \tag{5.45}$$

Thus, in the general many-electron case one has

$$-(\hbar^2/4m)\sum_i\{\Psi^*\nabla_i^2\Psi + \Psi\nabla_i^2\Psi^*\} = (\hbar^2/2m)\sum_i\nabla_i\Psi^* \cdot \nabla_i\Psi$$
$$-(\hbar^2/4m)\sum_i\nabla_i^2(\Psi^*\Psi). \tag{5.46}$$

Since electrons are indistinguishable and Ψ is antisymmetrized, the average of a sum of N one-electron operators can be replaced by N times the average of one of the operators. If the sum of the operators is replaced by N times a single operator in each term in the above equation and is then summed over all the spin coordinates and integrated over the space coordinates of all electrons *but one* (operations denoted by the symbol $\int d\tau'$, see eqn (1.4)), the result is

$$-(\hbar^2/4m)N\int d\tau'\{\Psi^*\nabla^2\Psi + \Psi\nabla^2\Psi^*\} = (\hbar^2/2m)N\int d\tau'\nabla\Psi^* \cdot \nabla\Psi$$
$$-(\hbar^2/4m)N\int d\tau'\nabla^2(\Psi^*\Psi). \tag{5.47}$$

This is readily restated in terms of the one-density matrix, $\Gamma^{(1)}(\mathbf{r}, \mathbf{r}')$ (see eqn (E1.1)), and its trace, the charge density, $\rho(\mathbf{r})$ (eqn (1.4))

$$-(\hbar^2/4m)\{\nabla^2 + \nabla'^2\}\Gamma^{(1)}(\mathbf{r}, \mathbf{r}')|_{r=r'} = (\hbar^2/2m)(\nabla \cdot \nabla')\Gamma^{(1)}(\mathbf{r}, \mathbf{r}')|_{r=r'}$$
$$-(\hbar^2/4m)\nabla^2\rho(\mathbf{r}). \tag{5.48}$$

This equation or, equivalently, eqn (5.47) can be used to define, respectively, the quantities (Bader and Preston 1969)

$$K(\mathbf{r}) = G(\mathbf{r}) + L(\mathbf{r}). \tag{5.49}$$

$K(\mathbf{r})$ and $G(\mathbf{r})$ are kinetic energy densities and $L(\mathbf{r})$, a function of the Laplacian of the charge density. Integration of the final coordinate $\mathbf{r}$ in eqn (5.49) over a region of space Ω yields

$$\int_\Omega K(\mathbf{r})d\tau = \int_\Omega G(\mathbf{r})d\tau - (\hbar^2/4m)\int_\Omega \nabla\cdot\nabla\rho(\mathbf{r})d\tau,$$

$$K(\Omega) = G(\Omega) - (\hbar^2/4m)\oint dS(\Omega)\nabla\rho(\mathbf{r})\cdot\mathbf{n}(\mathbf{r}) = G(\Omega) + L(\Omega) \tag{5.50}$$

where Gauss's theorem has been used to replace the final integral on the right-hand side with an integral over the surface bounding the region Ω. The surface integral is seen to be a measure of the flux in the gradient vectors of the charge density through the surface bounding the region. From the discussion of the gradient vector field of the charge density given in Chapter 3, it is clear that the surface integral will not vanish for an arbitrary surface. There are two cases for which the surface integral will always vanish yielding an equality in the kinetic energy expectation values: (1) when the integral is taken over all space and one has the result first pointed out by Schrödinger, that the average kinetic energy T may be calculated using either of the above forms for the kinetic energy operator,

$$\langle\hat{T}\rangle = T = K = G, \tag{5.51}$$

and (2) when the region Ω is bounded by a surface of zero flux in the gradient vector of the charge density (eqn (2.9) and repeated here as eqn (5.52)),

$$\nabla\rho(\mathbf{r})\cdot\mathbf{n}(\mathbf{r}) = 0 \qquad \text{for all } \mathbf{r} \text{ on the surface } S(\mathbf{r}). \tag{5.52}$$

Thus, the kinetic energy of the topologically defined atom $T(\Omega)$ is a well-defined quantity and one has

$$T(\Omega) = K(\Omega) = G(\Omega) \qquad \text{if } \Omega \text{ is bounded by a zero-flux surface} \tag{5.53}$$

in correspondence with the total system result, eqn (5.51). The expression for the kinetic energy

$$(1/2m)\langle\hat{\mathbf{p}}\psi, \hat{\mathbf{p}}\psi\rangle = (\hbar^2/2m)\langle\nabla\psi\cdot\nabla\psi\rangle$$

is the quantum analogue of the classical expression $|p|^2/2m$. It corresponds to the use of the first term on the right-hand side of eqn (5.46) or the kinetic energy density $G(\mathbf{r})$ in eqn (5.49) in the many-electron case.

Cohen (1979) has shown that all kinetic energy densities, when defined in terms of the generalizations made possible through the use of the Wigner distribution function, yield identical average values when integrated over a region bounded by a surface of zero flux, as in eqn (5.52). It is possible to have a surface through which the net integrated flux is zero, the flux being of opposite sign in different regions of the surface. Such a situation will not, in general, persist for changes in the system's surface, whether the change is real, as caused by nuclear displacements, or virtual, as caused by a mathematical variation of the surface elements. A surface of zero flux in the charge density is maintained at all times through real motions and through virtual changes as brought about by a variation of the charge density. It will be demonstrated later that the virial, as well as the kinetic energy, is not uniquely defined for a subsystem with an arbitrary boundary when either eqn (5.43) or (5.44) is evaluated for the operator $\hat{\mathbf{r}} \cdot \hat{\mathbf{p}}$. Thus, the average total energy of the electrons in a region of space with arbitrary boundaries is not defined. These results have a related and most important consequence: the possible energy functionals, whose variation yields Schrödinger's equation for a total system exhibit different properties with respect to the variations when the functionals are evaluated over regions with arbitrary boundaries. As demonstrated below, a unique variational result is obtained only for a subsystem satisfying eqn (5.52).

It was stated in the introductory chapter that the question 'are there atoms in molecules' is equivalent to asking two equally necessary questions of quantum mechanics: (1) does the state function contain the necessary information to predict a unique partitioning of a molecule into subsystems, and (2) does quantum mechanics provide a complete description of the subsystems so defined? The only method at one's disposal to answer questions of this nature are extrema principles based upon a variational principle. Schrödinger (1926) used such a method in his original derivation of the equation which bears his name. In relatively recent times, two new but equivalent formulations of quantum mechanics, one by Feynman (1948), the other by Schwinger (1951), have appeared, both of them based upon a single principle, the quantum analogue of the principle of least action, or more simply the action principle. The action principle in its generalized form as developed by Schwinger offers an alternative variational approach to the laws of physics, opening the door to new questions and new answers, and herein lies the power of the method. We shall use it to answer the question 'are there atoms in molecules?'

For a system in a stationary state, the variation of the action integral reduces to the variation of an energy integral which is identical to that first constructed by Schrödinger and used by him to derive the wave equation for a stationary state. We shall introduce this subject in the following section by reviewing Schrödinger's derivation of his wave equation and showing how this derivation can be generalized to yield a quantum description of a particular class of quantum subsystems—the topological atom (Srebrenik and Bader 1975; Bader 1988). This demonstration is sufficient to provide an operational understanding of the quantum mechanics of an atom in a molecule.

Chapter 8 gives a full account of Schwinger's principle of stationary action and of how, through its generalization, one obtains a definition of a quantum subsystem and a description of its properties.

5.3.1 Schrödinger's derivation of wave mechanics

Schrödinger concludes his first paper (1926) by stating that 'the function ψ be such as to make the 'Hamilton integral'

$$\int d\tau \{ \hbar^2 \, T(q, \partial\psi/\partial q) + \psi^2 \, V \}$$

stationary while fulfilling the normalizing, accessory condition

$$\int \psi^2 \, d\tau = 1.\text{'}$$

The quantity $T(q, \partial\psi/\partial q)$ is the kinetic energy expressed in terms of the momentum as in the first term on the right-hand side of eqn (5.47) that is, as $|\hat{p}^2|/2m$, and V is the potential energy. Schrödinger considered specifically the problem of the hydrogen atom for which the explicit form of the functional is

$$\mathscr{J}[\psi] = \int d\tau \{ (\hbar^2/2m) \nabla\psi \cdot \nabla\psi + V\psi\psi \} \qquad (5.54)$$

with $V = -e^2/r$. In his first paper Schrödinger did not allow ψ to be complex. The problem is to find the function ψ such that $\mathscr{J}[\psi]$ attains its minimum value subject to the constraint that ψ remain nomalized to unity. In ordinary extrema problems a function is a maximum or a minimum at a single point in the space of its variables and these problems are handled by the methods of differential calculus. In contrast, $\mathscr{J}[\psi]$ is an integral and its stationarity requires that its value be a minimum with respect to the averaging of the integrand over *all* points in configuration space. This is a problem requiring the calculus of variations.

The integral $\mathscr{J}[\psi]$ is extremized in the sense that any arbitrary change in the function $\psi(\mathbf{r})$ from its true value causes no first-order change in the value of the integral. One can think of varying the integral by assuming the existence of the 'correct function $\psi(\mathbf{r})$' and then changing the integral by changing $\psi(\mathbf{r})$ at each point $\mathbf{r}$ into the 'trial function $\phi(\mathbf{r})$' by adding to it the small amount $\eta(\mathbf{r})$

$$\phi(\mathbf{r}) = \psi(\mathbf{r}) + \eta(\mathbf{r}) \qquad (5.55)$$

where $\eta(\mathbf{r})$ is assumed to vanish at the boundaries of the system but is otherwise arbitrary. This varied expression for each value of the coordinate $\mathbf{r}$ is substituted into the integral to yield $\mathscr{J}[\phi]$ and all terms of order η^2 and higher are discarded. The difference between $\mathscr{J}[\phi]$ and $\mathscr{J}[\psi]$ is the first-order difference in $\mathscr{J}$ and is labelled $\delta\mathscr{J}[\psi]$, the variation of $\mathscr{J}$. Since $\psi(\mathbf{r})$ makes $\mathscr{J}[\psi]$ a minimum, $\delta\mathscr{J}[\psi]$ must vanish. Feynman *et al.* (1964, Section 19.1) give a useful and engaging introduction to the calculus of variations in a derivation of Newton's equation by a minimization of the classical action integral.

We shall allow for the possibility of ψ being complex in our review of Schrödinger's work. The functions ψ and ψ^* are treated as independent variables and are varied separately. Proceeding with the variation of ψ as outlined above, one obtains

$$\delta \mathscr{I}[\psi] = \int \{(\hbar^2/2m)\nabla\psi^* \cdot \nabla\eta + V\psi^*\eta\}\,d\tau.$$

To this must be added the effect of the constraint on the normalization of ψ which is written as

$$\int \psi^*\psi\,d\tau - 1 = 0.$$

Substitution of the trial function for ψ yields

$$\int \psi^*\psi\,d\tau + \int \psi^*\eta\,d\tau - 1 = 0$$

or

$$\langle \psi, \eta \rangle = 0. \tag{5.56}$$

Eqn (5.56) is multiplied by a factor λ and then added to the variation of $\mathscr{I}$. Requiring the resultant variation, which is labelled $\delta\mathscr{G}[\psi]$, to be stationary yields

$$\delta\mathscr{G}[\psi] = \delta\mathscr{I}[\psi] + \lambda\langle \psi, \eta \rangle = 0$$

or

$$\delta\mathscr{G}[\psi] = \int \{(\hbar^2/2m)\nabla\psi^* \cdot \nabla\eta + (V + \lambda)\psi^*\eta\}\,d\tau = 0. \tag{5.57}$$

This method of handling the constraint is known as Lagrange's method of undetermined multipliers. One may introduce the constraint in this manner from the very beginning by defining a new functional $\mathscr{G}[\psi]$ as

$$\begin{aligned}\mathscr{G}[\psi] &= \mathscr{I}[\psi] + \lambda\langle \psi, \psi \rangle \\ &= \int \{(\hbar^2/2m)\nabla\psi^* \cdot \nabla\psi + (V + \lambda)\psi^*\psi\}\,d\tau; \end{aligned} \tag{5.58}$$

the variation of $\mathscr{G}[\psi]$ then yields the expression $\delta\mathscr{G}[\psi]$.

We desire an expression for $\delta\mathscr{G}[\psi]$ in which no derivatives of η appear so that the complete integrand is multiplied by $\eta(\mathbf{r})$, the small arbitrary change we make in $\psi(\mathbf{r})$. This is easily done using the equivalent of an integration by parts. One notes that

$$\nabla\cdot(\nabla\psi^*\eta) = \nabla^2\psi^*\eta + \nabla\psi^* \cdot \nabla\eta \quad \text{or} \quad \nabla\psi^* \cdot \nabla\eta = \nabla\cdot(\nabla\psi^*\eta) - \nabla^2\psi^*\eta$$

and, using Gauss's theorem to transform the volume integral of $\nabla\cdot(\nabla\psi^*\eta)$, we obtain

$$\begin{aligned}\delta\mathscr{G}[\psi] = \int \{&-(\hbar^2/2m)\nabla^2\psi^* + (V + \lambda)\psi^*\}\eta\,d\tau \\ &+ (\hbar^2/2m)\oint dS\nabla\psi^* \cdot \mathbf{n}\eta = 0. \end{aligned} \tag{5.59}$$

The surface integral is over the boundary of the system as found for $\mathbf{r} = \infty$

and η is required to vanish on this boundary. This yields

$$\delta\mathcal{G}[\psi] = \int \{ - (\hbar^2/2m)\nabla^2\psi^* + (V - E)\psi^* \}\eta\,d\tau$$
$$= \int \{\hat{H}\psi^* - E\psi^*\}\eta\,d\tau = 0 \qquad (5.60)$$

where the constant λ has been identified with $- E$, the negative of the total energy. This result is to be true for all arbitrary functions $\eta(\mathbf{r})$, including, for example, the case where $\eta(\mathbf{r}) = 0$ for all $\mathbf{r}$ but one arbitrarily chosen narrow range of values. The only way in which the integral in eqn (5.60) can vanish for all such variations η is for the quantity multiplied by η to vanish. This yields

$$\hat{H}\psi^* - E\psi^* = 0. \qquad (5.61a)$$

A similar variation of ψ^* yields the complex conjugate of eqn (5.61)

$$H\psi - E\psi = 0. \qquad (5.61b)$$

Equation (5.61a or b) is called the Euler–Lagrange or simply the Euler equation of the variation and, as first demonstrated by Schrödinger in 1926, variation of $\mathcal{J}[\psi]$ subject to the normalization constraint yields the wave equation. Because of the equivalence of the two kinetic energy densities when integrated over all space (eqn (5.51)) $\mathcal{J}[\psi] = E$ and thus the function ψ which satisfies the wave equation also minimizes the energy of the system. Schrödinger demonstrated that the 'quantum conditions' then postulated for the hydrogen atom could be replaced by this variation principle since the resulting Euler equation yields the quantum numbers n and l in a natural way.

One may obtain eqn (5.61a or b) as the Euler equation in the variation of $\mathcal{G}[\psi]$ without imposing any prescribed boundary conditions on $\eta(\mathbf{r})$, the change or variation of $\psi(\mathbf{r})$. This is accomplished by introducing 'the natural boundary conditions' (Courant and Hilbert 1953). The necessary condition for $\mathcal{G}[\psi]$ to be stationary is that its first variation $\delta\mathcal{G}[\psi]$ as given in eqn (5.59) vanish. If $\mathcal{G}[\psi]$ is stationary with respect to variations which do not have prescribed boundary values, then it is certainly stationary with respect to the smaller class of variations for which $\eta(\mathbf{r})$ vanishes on the boundary which implies Euler's equation. Thus we need consider only that part of the variation which depends on the boundary, a step which yields the 'natural boundary conditions',

$$\mathbf{V}\psi^* \cdot \mathbf{n} = 0 \qquad \text{or} \qquad \mathbf{V}\psi \cdot \mathbf{n} = 0 \text{ for all points on the boundary.} (5.62)$$

This is the necessary boundary condition to obtain Schrödinger's equation as the Euler equation in the variation of $\mathcal{G}[\psi]$. For a bound system, one imposes the further condition that the state functions ψ and ψ^* themselves vanish on the boundary at infinity. This is a further, necessary restriction on the state

function if, as stated in the Born postulate, $\psi^*\psi\,d\tau$ is to represent a probability distribution. This probability must necessarily vanish on the boundaries infinitely far from the attractor which binds the system.

The normalization requirement on ψ can be handled differently by using
the functional,

$$\mathscr{E}[\psi] = \int\{(\hbar^2/2m)\nabla\psi^*\cdot\nabla\psi + V\psi^*\psi\}\,d\tau/\langle\psi,\psi\rangle. \tag{5.63}$$

A variation of $\mathscr{E}[\psi]$ through a variation of ψ as given in eqn (5.55) yields
through first-order in η,

$$\delta\mathscr{E}[\psi] = \int\{(\hbar^2/2m)\nabla\psi^*\cdot\nabla\eta + (V - \mathscr{E}[\psi])\psi^*\eta\}\,d\tau/\langle\psi,\psi\rangle. \tag{5.64}$$

The expansion of $(\langle\psi,\psi\rangle + \langle\psi,\eta\rangle)^{-1}$ followed by that for $(1 + \langle\psi,\eta\rangle/\langle\psi,\psi\rangle)^{-1}$ up to first-order in η have been used to obtain eqn (5.64). Once
again, ridding the expression of $\nabla\eta$ and demanding that η vanish at the
boundaries of the system yields

$$\delta\mathscr{E}[\psi] = \int\{\hat{H}\psi^* - \mathscr{E}[\psi]\psi^*\}\eta\,d\tau/\langle\psi,\psi\rangle = 0. \tag{5.65}$$

Since $\mathscr{E}[\psi] = E$ at the point of variation, one again obtains the wave
equation as the Euler equation.

One can put the variation problem in a form where the usual ideas of
differential calculus can be used to obtain an extremum in some function.
Simply label all possible changes in ψ with a parameter α, such that for some
value of α, say $\alpha = 0$, the varied function will coincide with the function which
extremizes the integral. One possible parametric form, for example, is

$$\psi(\mathbf{r}, \alpha) = \psi(\mathbf{r}, 0) + \alpha\eta(\mathbf{r}) \tag{5.66}$$

where, as before, $\eta(\mathbf{r})$ is any arbitrary function which vanishes at infinity. If we
denote by f the integrand of some integral $I[\psi]$ to be extremized

$$f = f(\psi, \nabla\psi),$$

then the parametrization given in eqn (5.66) gives $I[\psi]$ as a function of α,

$$I[\psi, \alpha] = \int f(\psi(\alpha), \nabla\psi(\alpha))\,d\tau.$$

The condition for an extremum is now given by the usual condition from
differential calculus that $(\partial I/\partial\alpha)_{\alpha=0} = 0$. To obtain the extremum condition
we multiply by $d\alpha$ and evaluate all derivatives at $\alpha = 0$,

$$(\partial I/\partial\alpha)_{\alpha=0}\,d\alpha = \int\{(\partial f/\partial\psi)(\partial\psi/\partial\alpha)_0\,d\alpha$$
$$+ (\partial f/\partial\nabla\psi)(\partial\nabla\psi/\partial\alpha)_0\,d\alpha\}\,d\tau = 0. \tag{5.67}$$

If one makes the following identifications at $\alpha = 0$,

$$(\partial I/\partial\alpha)_0\,d\alpha = \delta I, \qquad (\partial\psi/\partial\alpha)_0\,d\alpha = \delta\psi, \qquad (\partial\nabla\psi/\partial\alpha)_0\,d\alpha = \delta\nabla\psi,$$

then one can write down the general expression for the variation of $I[\psi, \alpha]$ as

$$\delta I = \int \{(\partial f/\partial \psi)\delta \psi + (\partial f/\partial \nabla \psi)\delta \nabla \psi\} \, d\tau. \qquad (5.68)$$

Correspondingly, the expression for a trial function is amended to read

$$\phi(\mathbf{r}) = \psi(\mathbf{r}) + \delta \psi(\mathbf{r}). \qquad (5.69)$$

To proceed beyond eqn (5.68) the procedure is always the same: one rids the expression of $\delta \nabla \psi$ (or $\delta \dot{\psi}$ for a time-dependent system) using an integration by parts to transform the integrand into a quantity multiplied only by $\delta \psi$. Setting this quantity equal to zero yields the Euler equation.

5.3.2   The variational definition of a subsystem and its properties

It will be demonstrated that the generalization of Schrödinger's variation of the functional $\mathscr{G}[\psi]$ to a subsystem has two important consequences: (1) The variation of the subsystem energy functional $\mathscr{G}[\psi, \Omega]$ yields the hypervirial theorem for a subsystem,

$$\delta \mathscr{G}[\psi, \Omega] = (-\varepsilon/2)\{(i/\hbar)\langle[\hat{H}, \hat{G}]\rangle_\Omega + cc\} \qquad (5.70)$$

where ε denotes an infinitesimal and the action of $\varepsilon \hat{G}$ on ψ causes the variation in ψ. Equation (5.70) is the stationary-state analogue of Schwinger's principle of stationary action and it will be shown to be a variational statement of the hypervirial theorem; (2) the generalization represented by eqn (5.70) is possible only if the subsystem is bounded by a surface of zero flux in the gradient vectors of the charge density—the topological atom (Srebrenik and Bader 1975; Bader 1988).

The extension of Schrödinger's energy functional to the many-electron case, including the Lagrange multiplier λ is (in analogy with eqn (5.58))

$$\mathscr{G}[\psi] = \int d\tau \{(\hbar^2/2m)\sum_i \nabla_i \psi^* \cdot \nabla_i \psi + (\hat{V} + \lambda)\psi^* \psi\} \qquad (5.71)$$

where, for brevity, the symbol $\int d\tau$ is used here to denote a sum over the spin coordinates and an integration over the spatial coordinates of all N electrons. The definition of the corresponding functional for a subsystem Ω is

$$\mathscr{G}[\psi, \Omega] = \int_\Omega d\tau_1 \int d\tau' \{(\hbar^2/2m)\sum_i \nabla_i \psi^* \cdot \nabla_i \psi + (\hat{V} + \lambda)\psi^* \psi\} \qquad (5.72)$$

where $\hat{V}$ denotes the full many-electron potential energy operator. The reader is reminded that the symbol $\int d\tau'$ implies a summation over all spins and the integration over the spatial coordinates of all electrons but one. The symbol $\int_\Omega d\tau_1$ implies that the coordinates of electron '1' are integrated over the subsystem Ω. Since ψ is antisymmetrized with respect to the electronic coordinates, all electrons are given equivalent descriptions and it matters not which set of electronic coordinates is integrated over Ω, just as it matters not which set of electronic coordinates is chosen to define the charge density in eqn (5.1). There is in this procedure no partitioning of kinetic or potential energy operators into sets for different regions, a step which does violate the indistinguishability of the electrons. All electrons occupy the total space of the system, as the limits on the electronic coordinates are $\pm \infty$ for each degree of freedom. The symbol $\langle \psi, \psi \rangle_\Omega$ is correspondingly defined as

$$\langle \psi, \psi \rangle_\Omega = \int_\Omega d\tau_1 \int d\tau' \psi^* \psi. \tag{5.73}$$

The trial functions ϕ representing variations in ψ are given by eqn (5.69) and substitution of $\phi(\mathbf{r})$ for $\psi(\mathbf{r})$ into $\mathscr{G}[\psi, \Omega]$ yields $\mathscr{G}[\phi, \Omega]$. At the point of variation, $\phi = \psi$ and $\mathscr{G}[\phi, \Omega]$ equals $\mathscr{G}[\psi, \Omega]$. The variations $\delta\psi$ and $\delta\psi^*$ are not given prescribed values on any of the boundaries, including the boundary of the subsystem. Instead only the natural boundary condition, that $\nabla_i \psi \cdot \mathbf{n}_i$ and $\nabla_i \psi^* \cdot \mathbf{n}$, together with ψ and ψ^*, vanish on all infinite boundaries, will be invoked. The functional $\mathscr{G}[\phi, \Omega]$ is to be varied not only with respect to ϕ, however, but also with respect to the surface defining the subsystem Ω. Only by having the surface itself considered to be a function of ϕ can the definition of the subsystem be determined entirely in a non-arbitrary way by the variational procedure.

The general expression for the variation of an integral including a variation of its surface for variations with respect to ψ is given by

$$\delta_\psi \left\{ \int_\Omega f(\psi, \nabla\psi) \, d\tau \right\} = \int_\Omega \{(\partial f / \partial \psi) \delta\psi + (\partial f / \partial \nabla\psi) \delta\nabla\psi\} \, d\tau$$

$$+ \oint dS(\Omega, \mathbf{r}) f(\psi, \nabla\psi) \delta_\psi S(\Omega, \mathbf{r}). \tag{5.74}$$

where $\delta_\psi S$ denotes a variation of the surface through variations in ψ. The variation of the surface appears in an integral of f over the surface and this term contributes to the first-order change in the functional (Courant and Hilbert 1953).

No particular difficulties arise in passing from the one- to the many-electron case in the variation of $\mathscr{G}[\psi, \Omega]$. Referring to eqns (5.72) and (5.74), one has

$$\delta\mathscr{G}[\psi, \Omega] = \int_\Omega d\tau_1 \int d\tau' \{(\partial f / \partial \psi) \delta\psi + \sum_i (\partial f / \partial \nabla_i \psi) \delta \nabla_i \psi\}$$
$$+ \oint dS(\Omega, \mathbf{r}_1) \int d\tau' f(\psi, \nabla\psi) \delta S(\Omega, \mathbf{r}_1) + \text{cc.} \tag{5.75}$$

In this case $f(\psi, \nabla\psi) \equiv f$ refers to the integrand of eqn (5.72). Only the integral over the surface of the subsystem Ω survives because of the vanishing of ψ and ψ^* on the boundaries at infinity. One again uses an integration by parts based on the identity

$$\mathbf{V}_i \cdot (\mathbf{V}_i \psi^* \delta\psi) = \nabla_i^2 \psi^* \delta\psi + \mathbf{V}_i \psi^* \cdot \delta\mathbf{V}_i \psi$$

to rid the expression of the terms $\delta\mathbf{V}_i\psi = \mathbf{V}_i\delta\psi$. Each of the surface integrals obtained in this manner vanishes except for the integral over the surface of Ω. This step is illustrated in eqn (5.76) which details the variation for one such term,

$$\int_\Omega d\tau_1 \int d\tau' (\partial f / \partial \mathbf{V}_i \psi) \delta\mathbf{V}_i\psi$$

$$= (\hbar^2/2m) \int_\Omega d\tau_1 \int d\tau' \mathbf{V}_i \psi^* \cdot \delta\mathbf{V}_i\psi$$

$$= -(\hbar^2/2m) \int_\Omega d\mathbf{r}_1 \int d\tau' \nabla_i^2 \psi^* \delta\psi$$

$$+ (\hbar^2/2m) \int_\Omega d\mathbf{r}_1 \int d\mathbf{r}_2 \dots \oint dS(\mathbf{r}_i) \dots \int d\mathbf{r}_N \mathbf{V}_i \psi^* \cdot \mathbf{n}_i \delta\psi. \qquad (5.76)$$

The surface integral in eqn (5.76) comes from the application of Gauss's theorem to the term involving $\mathbf{V}_i \cdot (\mathbf{V}_i \psi^* \delta\psi)$. As before, all such surface integrals vanish except for $\mathbf{r}_i = \mathbf{r}_1$ because of the vanishing of $\mathbf{V}_i \psi^*$ on the boundaries at infinity. From this point on, the coordinate $\mathbf{r}_1$ and the volume element $d\tau_1$ will be set equal to $\mathbf{r}$ and $d\tau$, respectively, and $\mathbf{V}_1$ and ∇_1^2 to their corresponding unscripted quantities.

The expression for the variation of $\mathscr{G}[\psi, \Omega]$ including a variation of the surface $S(\Omega, \mathbf{r})$ is, therefore,

$$\delta\mathscr{G}[\psi, \Omega] = \int_\Omega d\tau \int d\tau' \{\hat{H}\psi^* + \lambda\psi^*\} \delta\psi$$

$$+ \oint dS(\Omega, \mathbf{r}) \int d\tau' \{(\hbar^2/2m) \mathbf{V}\psi^* \cdot \mathbf{n}(\mathbf{r}) \delta\psi$$

$$+ \delta_\psi S(\Omega, \mathbf{r}) f(\psi, \mathbf{V}\psi)\} + cc \qquad (5.77)$$

where

$$\hat{H} = -(\hbar^2/2m)\sum_i \nabla_i^2 + \hat{V}. \qquad (5.78)$$

If eqn (5.77) is to be obtained for any variations $\delta\psi$ and $\delta\psi^*$ and any region Ω, it applies to the case where $\Omega = R^3$, i.e. the total system. In this case, all surface terms vanish and $\mathscr{G}[\psi, \Omega]$ and $\delta\mathscr{G}[\psi, \Omega]$ are identical with $\mathscr{G}[\psi]$ and $\delta\mathscr{G}[\psi]$, respectively, and the Euler equations obtained from the variation are Schrödinger's equations

$$\hat{H}\psi = E\psi \qquad \text{and} \qquad \hat{H}\psi^* = E\psi^* \qquad (5.79)$$

with $\hat{H}$ defined as in eqn (5.78) and where $\lambda = -E$. Using eqns (5.79) the expression for $\delta\mathscr{G}[\psi, \Omega]$ reduces at the point of variation to

$$\delta\mathscr{G}[\psi, \Omega] = \oint dS\,(\Omega, \mathbf{r})\int d\tau'\{(\hbar^2/2m)\,\nabla\psi^*\cdot\mathbf{n}(\mathbf{r})\delta\psi$$
$$+ \delta_\psi S(\Omega, \mathbf{r})f(\psi, \nabla\psi)\} + \text{cc.} \qquad (5.80)$$

Further progress towards obtaining a general physical result can be made only by removal of the term involving the variation of the surface of the subsystem. Consider, towards this goal, an alternative expression for the integrand $f(\psi, \nabla\psi)$, one involving the Hamiltonian operator $\hat{H}$. Including the complex conjugate term in eqn (5.80), the integrand $f(\psi, \nabla\psi)$ appears twice and, using eqn (5.46) which relates two forms of the kinetic energy density, one has

$$2f(\psi, \nabla\psi) = \{\psi^*\hat{H}\psi + (\hat{H}\psi)^*\psi\} - 2E\psi^*\psi$$
$$+ 2(\hbar^2/4m)\sum_i\nabla_i^2(\psi^*\psi). \qquad (5.81)$$

Since Schrödinger's equations are assumed to apply, eqn (5.81) simplifies to.

$$2f(\psi, \nabla\psi) = 2(\hbar^2/4m)\sum_i\nabla_i^2(\psi^*\psi). \qquad (5.82)$$

Integration of the right-hand side of eqn (5.82) in the manner indicated in eqn (5.80) transforms it into an integral of the Laplacian of the charge density. A typical term in this integration can be transformed using Gauss's theorem to yield

$$\oint dS(\Omega, \mathbf{r})\int d\tau_2 \ldots \int d\tau_i\nabla_i^2(\psi^*\psi) \ldots \int d\tau_N$$
$$= \oint dS(\Omega, \mathbf{r})\int d\tau_2 \ldots \oint dS(\mathbf{r}_i)\,\nabla_i(\psi^*\psi)\cdot\mathbf{n} \ldots \int d\tau_N, \qquad (5.83)$$

and each such term vanishes for $\mathbf{r}_i \neq \mathbf{r}$ since $\nabla_i\psi^*$ and $\nabla_i\psi$ vanish along with ψ and ψ^* on the infinite boundaries. Thus only the term involving the coordinate $\mathbf{r}$ survives the integration over the sum of operators ∇_i^2. Defining $\rho'(\mathbf{r})$ as the charge density per electron or $\rho(\mathbf{r})/N$,

$$\rho'(\mathbf{r}) = \int d\tau'\psi^*\psi, \qquad (5.84)$$

one obtains for the multiple integration of the sum of terms appearing in eqn (5.82) the result

$$2(\hbar^2/4m)\oint dS(\Omega, \mathbf{r})\nabla^2\rho'(\mathbf{r}).$$

Dividing this contribution equally between the appropriate term appearing explicitly in eqn (5.80) and its complex conjugate yields

$$\delta\mathscr{G}[\psi, \Omega] = \oint dS\,(\Omega, \mathbf{r})\,(\hbar^2/2m)\{\int d\tau'\nabla\psi^*\cdot\mathbf{n}(\mathbf{r})\delta\psi$$
$$+ \tfrac{1}{2}\delta_\psi S(\Omega, \mathbf{r})\nabla^2\rho'(\mathbf{r})\} + \text{cc.} \qquad (5.85)$$

Further progress in obtaining a general principle from this expression for $\delta \mathscr{G}[\psi, \Omega]$ is not possible for an arbitrary region of space. We now demonstrate that this result is transformed into the atomic statement of the hypervirial theorem when the subsystem Ω is restricted to one which satisfies a particular variational constraint. We shall introduce the variational constraint in terms of a trial function ϕ and show how the constraint, which limits the variations to a particular class of subsystems, can be applied from the beginning of the variation process, before the point $\phi = \psi$ is attained. The following conditions are to be fulfilled in the constrained variation:

A region $\Omega(\phi)$ is defined in terms of the trial function ϕ that is bound by a zero-flux surface in $\nabla \rho'_\phi$

$$\nabla \rho'_\phi (\mathbf{r}) \cdot \mathbf{n}(\mathbf{r}) = 0 \qquad \text{for all } \mathbf{r} \text{ in surface } S(\Omega, \mathbf{r}) \qquad (5.86)$$

where the trial density is defined as

$$\rho'_\phi(\mathbf{r}) = \int d\tau' \phi^* (\mathbf{x}, \tau') \phi (\mathbf{x}, \tau'). \qquad (5.87)$$

Recalling that the topological definition of an atom implies the zero-flux surface condition, a region of space bounded by a surface satisfying eqn (5.86) at the point of variation is henceforth called an atom. It is required that, as ϕ tends to ψ, $\Omega(\phi)$ is continuously deformable into the region $\Omega(\psi)$ associated with the atom. The region $\Omega(\phi)$ thus represents the atom in the varied total system, which is described by the trial function ϕ, just as $\Omega(\psi)$ represents the atom when the total system is in the state described by ψ.

Requiring the fulfilment of these conditions amounts to imposing the variational constraint that the divergence of $\nabla \rho_\phi$ integrates to zero at all stages of the variation, i.e. that

$$L(\phi, \Omega) = - (\hbar^2/4m) \int_{\Omega(\phi)} \nabla^2 \rho'_\phi(\mathbf{r}) d\tau = 0 \qquad (5.88a)$$

for all admissible ϕ, which implies

$$\delta L(\psi, \Omega) = \delta \left\{ - (\hbar^2/4m) \int_\Omega \nabla^2 \rho' (\mathbf{r}) d\tau \right\} = 0. \qquad (5.88b)$$

Recalling the expression for the variation of an integral which includes a variation of its surface as given in eqn (5.74) with $f = \nabla^2 \rho$, the variational constraint given in eqn (5.88) leads immediately to a result enabling one to eliminate the term involving the variation of the surface which appears in the general expression for the variation of $\mathscr{G}[\psi, \Omega]$. From eqn (5.74) one has

$$\delta_\psi \left\{ \int_\Omega \nabla^2 \rho'(\mathbf{r}) d\tau \right\} = \int \delta_\psi \{ \nabla^2 \rho'(\mathbf{r}) \} d\tau + \oint dS (\Omega, \mathbf{r}) \delta_\psi S(\Omega, \mathbf{r}) \nabla^2 \rho'(\mathbf{r}),$$

and applying the constraint that $\delta L(\psi, \Omega) = 0$ to this result yields

$$(\hbar^2/4m)\oint dS\,\delta_\psi S(\Omega, \mathbf{r})\nabla^2\rho'(\mathbf{r}) = -(\hbar^2/4m)\int_\Omega \delta_\psi\{\nabla^2\rho'(\mathbf{r})\}\,d\tau. \quad (5.89)$$

The variation of $\nabla^2\rho'$ is easily determined using the recipe given in eqn (5.68). Re-expressing $\nabla^2\rho'$ in terms of ψ and ψ^*, one has

$$\nabla^2\rho' = \int d\tau'\nabla^2(\psi^*\psi) = \int d\tau'\{(\nabla^2\psi^*)\psi + 2\nabla\psi^*\cdot\nabla\psi + \psi^*\nabla^2\psi\} \quad (5.90)$$

Hence the variation is with respect to ψ, $\nabla\psi$, and $\nabla^2\psi$ with the result

$$\int_\Omega \delta_\psi\{\nabla^2\rho'(\mathbf{r})\}\,d\tau = \int_\Omega d\tau\int d\tau'\,\{\nabla^2\psi^*\delta\psi$$
$$+ 2\nabla\psi^*\cdot\delta\nabla\psi + \psi^*\delta\nabla^2\psi\}. \quad (5.91)$$

The Laplacian of ρ is what is termed a divergence expression—its variation yields only surface terms. The integral in eqn (5.91) can be expressed as the divergence of a sum of terms

$$\int_\Omega \delta_\psi\{\nabla^2\rho'(\mathbf{r})\}\,d\tau = \int_\Omega d\tau\int d\tau'\nabla\cdot\{\nabla\psi^*\delta\psi + \psi^*\nabla\delta\psi\},$$

and by Gauss's theorem this can be re-expressed as a surface integral. Thus, the result of imposing the variational constant given in eqn (5.88b) is to replace the integral of the surface variation appearing in the expression for $\delta\mathscr{G}[\psi, \Omega]$ (eqn (5.85)) with the result appearing on the right-hand side of eqn (5.92),

$$(\hbar^2/4m)\oint dS\,\delta_\psi S(\Omega, \mathbf{r})\nabla^2\rho'(\mathbf{r})$$
$$= -(\hbar^2/4m)\oint dS(\Omega, \mathbf{r})\int d\tau'\{(\nabla\psi^*)\delta\psi + \psi^*\delta\nabla\psi\}\cdot\mathbf{n}(\mathbf{r}). \quad (5.92)$$

It is important that no new volume contributions to the variation be obtained by the imposition of the constraint given in eqn (5.88b), or one would no longer obtain Schrödinger's equations as the Euler–Lagrange equation of the variation.

Substitution of the identity in eqn (5.92) into eqn (5.85) and combining terms yields the final result for the variation of the functional $\mathscr{G}[\psi, \Omega]$,

$$\delta\mathscr{G}[\psi, \Omega] = (\hbar^2/4m)\oint dS(\Omega, \mathbf{r})\int d\tau'\,\{\nabla\psi^*\delta\psi$$
$$- \psi^*\delta\nabla\psi\}\cdot\mathbf{n}(\mathbf{r}) + \text{cc}. \quad (5.93)$$

Equation (5.93) is a statement of a physical principle as it can be restated in terms of the flux in the infinitesimal change in vector current density through the surface bounding the atom. In analogy with the definition of the charge density, the single-particle vector current density, $\mathbf{j}(\mathbf{r})$, of a many-particle system is defined as

$$\mathbf{j}(\mathbf{r}) = (\hbar/2mi) \int d\tau' \{\psi^*\nabla\psi - (\nabla\psi^*)\psi\}. \tag{5.94}$$

The infinitesimal change in $\mathbf{j}(\mathbf{r})$ caused by a variation in ψ is

$$\delta_\psi \mathbf{j}(\mathbf{r}) = (\hbar/2mi) \int d\tau' \{\psi^*\delta\nabla\psi - (\nabla\psi^*)\delta\psi\} \tag{5.95}$$

or, equivalently,

$$\delta_\psi \mathbf{j}(\mathbf{r}) = (\hbar/2mi) \int d\tau' \{\psi^*\nabla(\delta\psi) - (\nabla\psi^*)\delta\psi\},$$

and the expression for the variation of the subsystem energy functional constrained to a region bounded by a zero-flux surface becomes

$$\delta\mathscr{G}[\psi, \Omega] = -(i\hbar/2) \oint dS(\Omega, \mathbf{r})\delta_\psi \mathbf{j}(\mathbf{r})\cdot\mathbf{n}(\mathbf{r}) + \text{cc}. \tag{5.96}$$

To demonstrate that the expression for $\delta\mathscr{G}[\psi, \Omega]$ is equivalent to a variational derivation of the hypervirial theorem for an atom in a molecule, it is necessary to take one more very important step which is to consider the variations in ψ to be generated by the action of an operator on ψ. That is, one makes the identifications

$$\delta\psi = -\varepsilon(i/\hbar)\hat{G}\psi \quad \text{and} \quad \delta\psi^* = \varepsilon(i/\hbar)\hat{G}\psi^* \tag{5.97}$$

where $\hat{G}$ is a linear Hermitian operator, an observable, and ε denotes an infinitesimal. At this point we shall use these expressions to substitute for $\delta\psi$ and $\delta\psi^*$ in the expression for $\delta\mathscr{G}[\psi, \Omega]$ and simply state that their use corresponds to identifying the variations in the state function with the action of generators of infinitesimal unitary transformations. Such operators, when acting on a state function, cause changes in the dynamical properties of a system. It is this identification which enabled Schwinger to transform the variation of the action integral into the more powerful principle of stationary action. These ideas are fully developed in Chapter 8 where it will also be shown that the final result to be obtained for $\delta\mathscr{G}[\psi, \Omega]$, using eqn (5.97), is a statement of Schwinger's principle for a stationary state as it applies to an atom in a molecule.

We proceed by defining a current density for the property associated with the operator $\hat{G}$ in analogy with the current $\mathbf{j}_A$ defined in eqn (5.42)

$$\mathbf{j}_G(\mathbf{r}) = (\hbar/2mi) \int d\tau' \{\psi^*\nabla(\hat{G}\psi) - (\nabla\psi^*)(\hat{G}\psi)\}. \tag{5.98}$$

In terms of this vector current the variation in the atomic energy functional is given by

$$\delta\mathscr{G}[\psi, \Omega] = -(\varepsilon/2) \{ \oint dS(\Omega, \mathbf{r})\mathbf{j}_G(\mathbf{r})\cdot\mathbf{n}(\mathbf{r}) + \text{cc}\}. \tag{5.99}$$

Using the expression derived previously for a subsystem statement of the hypervirial theorem, eqn (5.43), we arrive at the atomic statement of the principle of stationary action for a stationary state

$$\delta \mathscr{G}[\psi, \Omega] = -(\varepsilon/2)\{(i/\hbar)\langle\psi, [\hat{H}, \hat{G}]\psi\rangle_\Omega + cc\}. \qquad (5.100)$$

Equation (5.100) is a variational result and it applies only to regions of space bounded by surfaces which satisfy the variational constraint of exhibiting a zero-flux in the gradient vector field of the charge density. It applies, therefore, to an atom, to any linked grouping of atoms, and to the total system. Equation (5.100) is a generalization of Schrödinger's original derivation of his equations to yield the same equations together with the variational statement of the hypervirial theorem. When Ω equals the total space of the system, one has

$$\delta \mathscr{G}[\psi, \Omega] = \delta \mathscr{G}[\psi] = -(\varepsilon/2)\{(i/\hbar)\langle\psi, [\hat{H}, \hat{G}]\psi\rangle + cc\} = 0, \qquad (5.101)$$

a result which yields not only Schrödinger's equations but, in addition, the hypervirial theorem

$$\langle\psi, [H, A]\psi\rangle = 0 \qquad (5.102)$$

and hence the variational development of all the mechanical theorems derivable from it. For an atom, the variation in $\mathscr{G}[\psi, \Omega]$ is proportional to the flux in the vector current density of the generator $\hat{G}$ (eqn (5.99)) and eqn (5.100) is consequently a variational derivation of the subsystem hypervirial theorem. Thus, for an atom, or any functional grouping of atoms, one obtains, in analogy with the total system, Schrödinger's equations and a *variational statement* of the hypervirial theorem, a statement that applies uniquely to systems bounded by zero-flux surfaces, which for any operator $\hat{G}$ is

$$\{(i/\hbar)\langle\psi, [\hat{H}, \hat{G}]\psi\rangle_\Omega + cc\} = \{\int dS(\Omega, \mathbf{r})\mathbf{j}_G(\mathbf{r}) \cdot \mathbf{n}(\mathbf{r}) + cc\}. \qquad (5.103)$$

Through requisite choices for the operator $\hat{G}$, eqn (5.103) determines the force acting on an atom in a molecule and, through the atomic statement of the virial theorem, its energy. It establishes the mechanics of an atom in a molecule, as is demonstrated in the next chapter.

E5.1   The variation of Hamiltonian-based functionals

Schrödinger's use of the expression $(1/2m)\langle\hat{\mathbf{p}}\psi, \hat{\mathbf{p}}\psi\rangle$ to express the kinetic energy in his functional $\mathscr{I}[\psi]$ was based on an analogy with the Hamilton–Jacobi equation of classical mechanics. This is also the form used for the kinetic energy as it appears in the Lagrangian of the action integral, a consequence of the Lagrangian, either quantum or classical, being a function of the field variables and their first derivatives only. Since the Lagrangian degenerates into a functional of the form $\mathscr{G}[\psi]$ for a stationary state, these are termed Lagrangian-based functionals. One may, however, construct an energy functional based upon the Hamiltonian operator which differs from the above by expressing the kinetic energy in terms of the operator $-(\hbar^2/2m)\nabla^2$. We investigate the variational properties of such a functional here. In this section the normalization constant will be handled by division with $\langle \psi, \psi \rangle$ as in eqn (5.63) for $\mathscr{E}[\psi]$.

In analogy with $\mathscr{E}[\psi]$ and its many-electron subsystem counterpart, the atomic form of the energy functional $\mathscr{E}[\psi, \Omega]$ is defined as

$$\mathscr{E}[\psi, \Omega] = \int_\Omega d\tau \int d\tau' \, \{(\hbar^2/2m)\textstyle\sum_i \nabla_i \psi^* \cdot \nabla_i \psi$$
$$+ \hat{V}\psi^*\psi\}/\langle \psi, \psi \rangle_\Omega \tag{E5.1}$$

where

$$\langle \psi, \psi \rangle_\Omega = \int_\Omega d\tau \int d\tau' \, \psi^*\psi. \tag{E5.2}$$

The Hamiltonian-based functional $\mathscr{E}'[\psi, \Omega]$ is defined as

$$\mathscr{E}'[\psi, \Omega] = \tfrac{1}{2}\{\mathscr{H}[\psi, \Omega] + \mathscr{H}^*[\psi, \Omega]\} \tag{E5.3}$$

where

$$\mathscr{H}[\psi, \Omega] = \int_\Omega d\tau \int d\tau' \{ -(\hbar^2/2m)\textstyle\sum_i \psi^* \nabla_i^2 \psi$$
$$+ \hat{V}\psi^*\psi\}/\langle \psi, \psi \rangle_\Omega. \tag{E5.4}$$

The Lagrangian-based functionals employ the form $(1/2m)\langle \hat{\mathbf{p}}\psi, \hat{\mathbf{p}}\psi \rangle$ for the kinetic energy and are, therefore, necessarily real. The functional $\mathscr{H}[\psi, \Omega]$ can, however, be complex, as $\hat{H}$ is not necessarily Hermitian over a subsystem as shown in eqn (5.12). Since one cannot have a complex quantity stationary with respect to a variation, one must vary $\mathscr{H}[\psi, \Omega]$ together with its complex conjugate as defined in eqn (E5.3).

The two types of functionals can be related using eqn (5.46)

$$\mathscr{E}[\psi, \Omega] = \mathscr{E}'[\psi, \Omega] + (\hbar^2/4m)\int_\Omega d\tau \, \nabla^2\rho'(\mathbf{r})/\langle \psi, \psi \rangle_\Omega \tag{E5.5}$$

where only the term involving the coordinate $\mathbf{r}$ survives the integration over the sum of terms $\sum_i \nabla_i^2(\psi^*\psi)$. For a total system, or for a subsystem satisfying the zero-flux surface condition, one has

$$\mathscr{E}[\psi, \Omega] = \mathscr{E}'[\psi, \Omega] = \mathscr{E} \tag{E5.6a}$$

where $\mathscr{E}$ is the energy eigenvalue for a stationary state. Thus the zero-flux surface condition insures that the real part of $\mathscr{H}[\psi, \Omega]$ equals $\mathscr{E}[\psi, \Omega]$. Because of the constraint imposed on the variation of $\mathscr{E}[\psi, \Omega]$ (eqn (5.88b)), one also has

$$\delta\mathscr{E}[\psi, \Omega] = \delta\mathscr{E}'[\psi, \Omega] \tag{E5.6b}$$

and the variations of both functionals yield identical results. Since the function $\sum_i \nabla_i^2(\psi^*\psi)$ is a divergence expression, its addition to the functional $\mathscr{E}[\psi, \Omega]$ to obtain $\mathscr{E}'[\psi, \Omega]$ does not change the Euler equation obtained from the variation. However, as shown below, while the variation of $\mathscr{E}'[\psi, \Omega]$ yields the same results as does $\delta\mathscr{E}[\psi, \Omega]$, it does so without placing restrictions on the surface of Ω.

The variation of $\mathscr{E}[\psi, \Omega]$ with the imposition of the variational constants given in eqn (5.88) at every stage of the variation and including a variation of the surface yields

$$\delta\mathscr{E}[\psi, \Omega] = (\hbar^2/4m) \oint dS(\Omega, \mathbf{r}) \int d\tau' \{\nabla\psi^*\delta\psi$$
$$- \psi^*\delta\nabla\psi\} \cdot \mathbf{n}/\langle\psi, \psi\rangle_\Omega + cc \tag{E5.7}$$

in analogy with the result obtained for the variation of $\mathscr{G}[\psi, \Omega]$ given in eqn (5.93).

The variation of $\mathscr{E}'[\psi, \Omega]$ involves variation with regard to $\delta\psi$ and $\delta\nabla^2\psi$. It is necessary to perform an integration by parts twice in succession to rid the expression of the variations of gradients of ψ. This procedure is illustrated below for a single term in $\mathscr{H}[\psi, \Omega]$

$$- (\hbar^2/2m)\{\partial(\psi^*\nabla_i^2\psi)/\partial\nabla_i^2\psi\}\delta\nabla_i^2\psi$$
$$= - (\hbar^2/2m)\psi^*\delta\nabla_i^2\psi$$
$$= - (\hbar^2/2m)(- \nabla\psi^* \cdot \delta\nabla\psi + \nabla\cdot(\psi^*\delta\nabla\psi))$$
$$= - (\hbar^2/2m)(+ \nabla^2\psi^*\delta\psi + \nabla\cdot(\psi^*\delta\nabla\psi - \nabla\psi^*\delta\psi)). \tag{E5.8}$$

Using this result, the variation of $\mathscr{E}'[\psi, \Omega]$ including a variation of the surface of Ω is found to be

$$\langle\psi, \psi\rangle_\Omega \delta\mathscr{E}'[\psi, \Omega]$$

$$= \frac{1}{2}\int_\Omega d\mathbf{r} \int d\tau' \{\hat{H}\psi^* - \mathscr{H}^*[\psi, \Omega]\psi^*\}\delta\psi$$

$$+ \left(\frac{1}{2}\right)\oint dS(\Omega, \mathbf{r}) \int d\tau' \delta_\psi S(\Omega, \mathbf{r}) \{\psi^*\hat{H}\psi - \mathscr{H}^*[\psi, \Omega]\psi^*\psi\}$$

$$+ (\hbar^2/4m)\oint dS(\Omega, \mathbf{r}) \int d\tau' \{\nabla\psi^*\delta\psi - \psi^*\delta\nabla\psi\} \cdot \mathbf{n} + cc. \tag{E5.9}$$

At the point of variation with $\mathscr{H}[\psi, \Omega] = E$, one obtains Schrödinger's equations as the Euler equations and

$$\delta\mathcal{E}'[\psi, \Omega] = (\hbar^2/4m)\oint dS(\Omega, \mathbf{r})\int d\tau'\{\nabla\psi^*\delta\psi$$
$$- \psi^*\delta\nabla\psi\}\cdot\mathbf{n}/\langle\psi, \psi\rangle_\Omega + \text{cc.} \qquad (E5.10)$$

This result, which is obtained without any restrictions on Ω is identical to that obtained from the variation of $\mathcal{E}[\psi, \Omega]$ by constraining the subsystem to be bounded by a zero-flux surface. It is clear from eqn (E5.9) that, unlike the variation of $\mathcal{G}[\psi, \Omega]$ or $\mathcal{E}[\psi, \Omega]$, one obtains the same result for the variation of $\mathcal{E}'[\psi, \Omega]$ whether or not the surface is varied.

The Lagrangian-based functional $\mathcal{G}[\psi, \Omega]$ or $\mathcal{E}[\psi, \Omega]$ derives directly from the Lagrangian as employed in the quantum action principle. For a total system, both the Lagrangian- and Hamiltonian-based functionals yield identical variational results. This equivalence in variational behaviour is maintained for the corresponding subsystem functionals only if the subsystem is bounded by a zero-flux surface. Only an atomic region ensures an equivalence in both the values and the variational properties of the two types of functionals (eqns (E5.6a,b)) thereby preserving the properties obtained for a total system.

E5.2   Vectors, tensors, and dyadics

A vector in three-dimensional Cartesian space is characterized by three components

$$\mathbf{A} = \mathbf{i}A_x + \mathbf{j}A_y + \mathbf{k}A_z \qquad (E5.11)$$

where $\mathbf{i}$, $\mathbf{j}$, and $\mathbf{k}$ are unit vectors along x, y, and z, respectively. Equation (E5.11) can be expressed as a product of a row and a column matrix as

$$\mathbf{A} = [\mathbf{i} \quad \mathbf{j} \quad \mathbf{k}]\begin{pmatrix} A_x \\ A_y \\ A_z \end{pmatrix} = \mathbf{i}A_x + \mathbf{j}A_y + \mathbf{k}A_z. \qquad (E5.12)$$

Because of the orthogonality of the unit vectors in a scalar or dot product, the scalar product of two vectors $\mathbf{A}$ and $\mathbf{B}$ can be expressed as

$$\mathbf{A}\cdot\mathbf{B} = [A_x \quad A_y \quad A_z]\begin{pmatrix} B_x \\ B_y \\ B_z \end{pmatrix} = A_xB_x + A_yB_y + A_zB_z. \qquad (E5.13)$$

A dot product of two vectors yields a single number, a scalar. It is possible to generate a three-by-three matrix by the multiplication of two vectors. This procedure is denoted by the symbol **AB** which defines the operation

$$\mathbf{AB} = \begin{pmatrix} A_x \\ A_y \\ A_z \end{pmatrix} [B_x \quad B_y \quad B_z] = \begin{pmatrix} A_xB_x & A_xB_y & A_xB_z \\ A_yB_x & A_yB_y & A_yB_z \\ A_zB_x & A_zB_y & A_zB_z \end{pmatrix}. \qquad \text{(E5.14)}$$

The result of this operation is called a dyadic or tensor

$$\overset{\leftrightarrow}{\mathbf{D}} = \mathbf{AB}.$$

A dyadic behaves like a two-headed vector — the scalar product of $\overset{\leftrightarrow}{\mathbf{D}}$ with a vector **A** yields a vector

$$\overset{\leftrightarrow}{\mathbf{D}} \cdot \mathbf{A} = \mathbf{C},$$
$$C_x = D_{xx}A_x + D_{xy}A_y + D_{xz}A_z,$$
$$C_y = D_{yx}A_x + D_{yy}A_y + D_{yz}A_z, \qquad \text{(E5.15)}$$
$$C_z = D_{zx}A_x + D_{zy}A_y + D_{zz}A_z.$$

A scalar is obtained only by taking a scalar product of $\overset{\leftrightarrow}{\mathbf{D}}$ from both sides

$$\mathbf{B} \cdot \overset{\leftrightarrow}{\mathbf{D}} \cdot \mathbf{A} = \mathbf{B} \cdot \mathbf{C}. \qquad \text{(E5.16)}$$

A dyadic is required to describe those directed (vectorial) properties of a system which result from the application of a force or field along directions orthogonal to the observed resultant. The polarizability of a molecule is, for example, described by the polarizability tensor $\boldsymbol{\alpha}$. The dipole $\boldsymbol{\mu}$ induced by an applied field **E** is given by

$$\boldsymbol{\mu} = \boldsymbol{\alpha} \cdot \mathbf{E}. \qquad \text{(E5.17)}$$

Even if **E** is directed along the z-axis, $\mathbf{E} = \mathbf{k}E_z$, the x- and y-components of $\boldsymbol{\mu}$ can differ from zero. Using eqn (E5.15), the three components of $\boldsymbol{\mu}$ are given by

$$\mu_x = \alpha_{xz}E_z,$$
$$\mu_y = \alpha_{yz}E_z,$$
$$\mu_z = \alpha_{zz}E_z.$$

If a dyadic is symmetric, then

$$\mathbf{A} \cdot \overset{\leftrightarrow}{\mathbf{D}} = \overset{\leftrightarrow}{\mathbf{D}} \cdot \mathbf{A},$$

a property possessed by the stress tensor $\overset{\leftrightarrow}{\boldsymbol{\sigma}}$ (eqn (5.28)). An example of an unsymmetrical dyadic is provided by the product **jr** as encountered in the derivation of Ehrenfest's first relationship, eqn (5.26). The identity used in the derivation of this relation

$$\nabla \cdot (\mathbf{jr}) = \mathbf{j} + \mathbf{r}\nabla \cdot \mathbf{j}$$

is obtained from the matrix multiplication

$$\mathbf{V}\cdot(\mathbf{jr}) = \begin{bmatrix} \dfrac{\partial}{\partial x} & \dfrac{\partial}{\partial y} & \dfrac{\partial}{\partial z} \end{bmatrix} \begin{pmatrix} j_x x & j_x y & j_x z \\ j_y x & j_y y & j_y z \\ j_z x & j_z y & j_z z \end{pmatrix}.$$

In tensor notation, an element of the tensor **jr** is

$$T_{kl} = j_k x_l$$

and

$$\frac{\partial T_{kl}}{\partial x_k} = j_{kk} x_l + j_k x_{lk}.$$

Two types of terms appear in the stress tensor $\overset{\leftrightarrow}{\sigma}$ in eqn (5.28), terms of the form $\psi^*\mathbf{V}\mathbf{V}\psi$ and $\mathbf{V}\psi^*\mathbf{V}\psi$. The divergence of this tensor determines the force $-\mathbf{V}\cdot\overset{\leftrightarrow}{\sigma}$. Taking the divergence of $\overset{\leftrightarrow}{\sigma}$ yields terms of the form

$$\mathbf{V}\cdot(\psi^*\mathbf{V}\mathbf{V}\psi) = \mathbf{V}\psi^*\cdot\mathbf{V}\mathbf{V}\psi + \psi^*\mathbf{V}(\nabla^2\psi)$$

and

$$\mathbf{V}\cdot(\mathbf{V}\psi^*\mathbf{V}\psi) = \nabla^2\psi^*\mathbf{V}\psi + \mathbf{V}\psi^*\cdot\mathbf{V}\mathbf{V}\psi.$$

The subtraction of these two contributions as they appear in $\mathbf{V}\cdot\overset{\leftrightarrow}{\sigma}$ yields the terms appearing in the equation for the force preceding eqn (5.28).

E5.3   Divergence of a vector and Gauss's theorem

Consider an element of surface area dS and define **n** to be the outwardly directed unit vector normal to dS. Then $\mathbf{j}\cdot\mathbf{n}$ is the component of the vector current density normal to the surface element dS and the current through this element is $\mathbf{j}\cdot\mathbf{n}\,\mathrm{d}S$. The total current out of some region Ω bounded by a surface $S(\Omega)$ is obtained by summing the contributions from all its surface elements. The resulting surface integral is called 'the flux **j** through the surface $S(\Omega)$', that is

Flux of **j** through the surface $S(\Omega) = \oint \mathrm{d}S(\Omega, \mathbf{r})\mathbf{j}(\mathbf{r})\cdot\mathbf{n}(\mathbf{r}).$ (E5.18)

This expression is applied to the corresponding surface integral of any vector quantity, even if there is no physical flow through the surface. Thus references in the text to a 'flux in the gradient vector of ρ' do not imply any physical flow.

One can imagine the region Ω bounded by $S(\Omega)$ to be divided into an arbitrary number of smaller volumes. The total flux through $S(\Omega)$ is still given by eqn (E5.18) since this total flux is equal to the sum of the fluxes out of all the interior volumes. Consider, for example, a surface S_{ik} between two interior regions Ω_i and Ω_k. The flux out of Ω_i through S_{ik} is $\oint \mathrm{d}S_{ik}\mathbf{j}\cdot\mathbf{n}_i$ which must equal the negative of the flux out of Ω_k, $\oint \mathrm{d}S_{ik}\mathbf{j}\cdot\mathbf{n}_k$, since $\mathbf{n}_i = -\mathbf{n}_k$. Thus the flux out of one region equals the flux into the neighbouring region through their common surface.

Gauss's theorem is proved by considering the flux out of an infinitesimally small cube. Consider a cube with edges of lengths Δx, Δy, and Δz and denote the coordinates of the corner nearest the origin as (x, y, z). The flux of a vector, say $\mathbf{j}$, through the surface of the cube is equal to the fluxes through each of the six faces. The outward flux through the face $\Delta y \Delta z$ which is perpendicular to the x-axis in the negative x direction, called face (1), is $-j_x(x)\Delta y \Delta z$ where $j_x(x)$ is the x-component of $\mathbf{j}$. The flux out of the corresponding face at $x + \Delta x$, face (2), is $j_x(x + \Delta x)\Delta y \Delta z$. The component of j_x at $x + \Delta x$ is slightly different from j_x at x and to first-order one has

$$j_x(x + \Delta x) = j(x) + \frac{\partial j_x}{\partial x} \Delta x.$$

The sum of the fluxes through faces (1) and (2) is

$$\text{Flux out of (1) and (2)} = \frac{\partial j_x}{\partial x} \Delta x \Delta y \Delta z.$$

The total flux through all six faces of the cube is, therefore,

$$\text{Total flux in } \mathbf{j} \text{ out of cube} = \left(\frac{\partial j_x}{\partial x} + \frac{\partial j_y}{\partial y} + \frac{\partial j_z}{\partial z} \right) \Delta x \Delta y \Delta z.$$

This total flux is also given by the surface integral of $\mathbf{j} \cdot \mathbf{n}$ and, hence,

$$\oint dS \mathbf{j} \cdot \mathbf{n} = \nabla \cdot \mathbf{j} \Delta V$$

where ΔV is the volume of the cube. Thus, the divergence of a vector at a given point in space is the net outflow or flux of $\mathbf{j}$ per unit volume in the neighbourhood of the point. From the above demonstration that the total flux from a volume is the sum of the fluxes out of each of its parts, we see that the flux through a surface $S(\Omega)$ bounding a finite volume Ω is given by the sum of the fluxes out of each of its infinitesimal elements dV, or

$$\oint dS(\Omega) \mathbf{j} \cdot \mathbf{n} = \int_\Omega \nabla \cdot \mathbf{j} dV \qquad (E5.19)$$

which is Gauss's theorem.

References

Bader, R. F. W. (1988). *Pure appl. Chem.* **60**, 145.
Bader, R. F. W. and Beddall, P. M. (1972). *J. Chem. Phys.* **56**, 3320.
Bader, R. F. W. and Preston, H. J. T. (1969). *Int. J. quantum Chem.* **3**, 327.
Cohen, L. (1979). *J. Chem. Phys.* **70**, 788.
Coulson, C. A. and Longuet-Higgins, H. C. (1947a). *Proc. r. Soc., London* **A191**, 39.

Coulson, C. A. and Longuet-Higgins, H. C. (1947b). *Proc. r. Soc., London* **A192**, 16.
Coulson, C. A. and Longuet-Higgins, H. C. (1948a). *Proc. r. Soc., London* **A193**, 447.
Coulson, C. A. and Longuet-Higgins, H. C. (1948b). *Proc. r. Soc., London* **A193**, 456.
Coulson, C. A. and Longuet-Higgins, H. C. (1948c). *Proc. r. Soc., London* **A195**, 188.
Courant, R. and Hilbert, D. (1953). *Methods of mathematical physics*, Vol. 1. Wiley Interscience, New York.
Dirac, P. A. M. (1929). *Proc. r. Soc., London* **A123**, 714.
Epstein, S. T. (1974a) *The variation method in quantum chemistry*. Academic Press, New York.
Epstein, S. T. (1974b). *J. Chem. Phys.* **60**, 3351.
Feynman, R. P. (1948). *Rev. mod. Phys.* **20**, 367.
Feynman, R. P., Leighton, R. B., and Sands, M. (1964). *The Feynman lectures on physics*, Vol. II. Addison-Wesley, Reading, Massachusetts.
Hall, G. G. (1951). *Proc. r. Soc., London* **A205**, 511.
Hirschfelder, J. O. (1960). *J. Chem. Phys.* **33**, 1462.
Hund, F. (1933). *Physik* **1**, 163.
Hund, F. (1937). *Physik* **5**, 1.
Hurley, A. C. (1954). *Proc. r. Soc., London* **A226**, 179.
Lennard-Jones, J. E. (1949a). *Proc. r. Soc., London* **A198**, 1.
Lennard-Jones, J. E. (1949b). *Proc. r. Soc., London* **A198**, 14.
Lennard-Jones, J. E. (1952). *J. Chem. Phys.* **20**, 1024.
Lewis, G. N. (1916). *J. Am. chem. Soc.* **38**, 762.
Lewis, G. N. (1933). *J. Chem. Phys.* **1**, 17.
Löwdin, P.-O. (1948). *A theoretical investigation into some properties of ionic crystals*. Almqvist and Wiksell, Stockholm.
Löwdin, P.-O. (1956). *Phil. Mag.* **Suppl. 5**, 1.
Messiah, A. (1958). *Quantum mechanics*, Vol. I. Wiley, New York.
Moffitt, W. (1951). *Proc. r. Soc., London* **A210**, 245.
Mulliken, R. S. (1928a). *Phys. Rev.* **32**, 186.
Mulliken, R. S. (1928b). *Phys. Rev.* **32**, 388.
Mulliken, R. S. (1928c). *Phys. Rev.* **32**, 761.
Mulliken, R. S. (1932a). *Phys. Rev.* **40**, 55.
Mulliken, R. S. (1932b). *Rev. mod. Phys.* **4**, 1.
Mulliken, R. S. (1935). *J. Chem. Phys.* **3**, 375.
Mulliken, R. S. (1939). *Phys. Rev.* **56**, 778.
Pauli, W. (1958) In *Encyclopedia of physics*, Vol. 5, part 1 (ed. S. F. Lügge). Springer, Berlin.
Pauling, L. (1931). *J. Am. chem. Soc.* **53**, 1367.
Pauling, L. (1960). *The nature of the chemical bond* (3rd edn). Cornell University Press, Ithaca, New York.
Roothaan, C. C. J. (1951). *Rev. mod. Phys.* **23**, 69.
Schrödinger, E. (1926). *Ann. d. Phys.* **79**, 361.
Schrödinger, E. (1927). *Ann. d. Phys.* **82**, 265.
Schwinger, J. (1951). *Phys. Rev.* **82**, 914.
Slater, J. C. (1931). *Phys. Rev.* **37**, 481.
Srebrenik, S. and Bader, R. F. W. (1974). *J. Chem. Phys.* **61**, 2536.
Srebrenik, S. and Bader, R. F. W. (1975). *J. Chem. Phys.* **63**, 3945.

THE MECHANICS OF AN ATOM
IN A MOLECULE

I hope and believe that the present statements will prove useful in the elucidation of the magnetic properties of atoms and molecules, and further for explaining the flow of electricity in solid bodies.
— Erwin Schrödinger (1926). The statement following his introduction of the charge and current densities and the quantum equation of continuity in his fourth paper on 'wave mechanics'.

6.1 An atomic view of the properties of matter

6.1.1 The charge and current densities

Schrödinger's belief in the ability of his 'wave mechanics' to provide a description of the properties of matter as demonstrated in the opening quotation was well founded and his mechanics has been used to understand phenomena unknown in 1926. The roles of the charge and current densities in quantum mechanics go beyond their use in the description of the electrical and magnetic properties of matter. The charge density provides a description of the distribution of charge throughout real space and is the bridge between the concept of state functions in Hilbert space and the physical model of matter in real space. Schrödinger subscribed to this same view, pointing out in his fourth paper wherein he defines the electric density at a point in real space, 'that the ψ function itself cannot be and may not be directly interpreted in terms of three-dimensional space—because it is in general a function in configuration space, and not real space.' The zero-flux boundary condition, as stated in terms of the gradient vector field of the charge density, defines a single atom and objects made from collections of atoms. As demonstrated in this chapter, the form and ultimately the properties of an object, microscopic or macroscopic, are determined by its distribution of charge.

A subsystem is an open system, free to exchange charge and momentum with its environment. Thus the current density $\mathbf{j}_G$ for any observable $\hat{G}$ is of particular importance in the mechanics of a subsystem, since a non-vanishing flux in this current implies a fluctuation in the subsystem average value of the property G.

Variation of the energy functional for a subsystem is equal to an infinitesimal times the flux in the current density of the generator $\hat{G}$ through the surface of the subsystem (eqn (5.99) and repeated here as eqn (6.1))

$$\delta\mathcal{G}[\psi, \Omega] = -(\varepsilon/2)\{\oint dS(\Omega, \mathbf{r})\mathbf{j}_G(\mathbf{r})\cdot\mathbf{n}(\mathbf{r}) + cc\}. \tag{6.1}$$

The same surface integral appears in the subsystem statement of the hyper-virial theorem

$$\{(i/\hbar)\langle\psi, [\hat{H}, \hat{G}]\psi\rangle_\Omega + cc\} = \{\oint dS(\Omega, \mathbf{r})\mathbf{j}_G(\mathbf{r})\cdot\mathbf{n}(\mathbf{r}) + cc\} \tag{6.2}$$

and, because of the dependence of both quantities on the current flux, one obtains the atomic statement of the principle of stationary action for a stationary state as given in eqn (5.100). This principle forms the basis for the discussion of the mechanics of an atom in a molecule and is repeated here

$$\delta\mathcal{G}[\psi, \Omega] = -(\varepsilon/2)\{(i/\hbar)\langle\psi, [\hat{H}, \hat{G}]\psi\rangle_\Omega + cc\}. \tag{6.3}$$

The non-vanishing of the flux of a quantum mechanical current in the absence of a magnetic field is what distinguishes the mechanics of a subsystem from that of the total system in a stationary state. The flux in the current density will vanish through any surface on which ψ satisfies the natural boundary condition, $\nabla\psi\cdot\mathbf{n} = 0$ (eqn (5.62)), a condition which is satisfied by a system with boundaries at infinity. Thus, for a total system the energy is stationary in the usual sense, $\delta\mathcal{G}[\psi] = 0$, and the usual form of the hyper-virial theorem is obtained with the vanishing of the commutator average,

$$\langle\psi, [\hat{H}, \hat{G}]\psi\rangle = 0. \tag{6.4}$$

As noted in the previous chapter, eqn (6.4) is a consequence of the Hermitian property of $\hat{H}$, a property not enjoyed in general, by a subsystem.

To further investigate the relationship between the variation of an energy functional and the current consider the time derivative of the density of a property G. Assume the corresponding operator $\hat{G}$ to have no explicit time dependence. Let

$$\rho_G = \Psi^*\hat{G}\Psi$$

where Ψ is a solution to Schrödinger's time-dependent equations (eqn (5.2)). Then

$$d\rho_G/dt + \nabla\cdot\mathbf{j}_G = (i/\hbar)\Psi^*[\hat{H}, \hat{G}]\Psi.$$

This result is obtained through the use of eqn (5.2) and by the addition and subtraction of the term $\Psi^*\hat{H}\hat{G}\Psi$, the same procedure followed in the derivation of the subsystem hypervirial theorem, eqn (5.40). The usual quantum equation of continuity for the density $\rho = \Psi^*\Psi$ as derived in eqn (5.19),

$$d\rho/dt + \nabla\cdot\mathbf{j} = 0,$$

is so-called because of its direct correspondence with the equation for a classical fluid. The presence of the commutator in the expression for $\dot{\rho}_G (= d\rho_G/dt)$ has no classical analogue and is quantum mechanical in origin. The expression for $\dot{\rho}_G$ is analogous to that for $\dot{\rho}$ only when the commutator vanishes.

The magnitude of the commutator average is related to the product of the uncertainties in the energy and in the observable $\hat{G}$ (Messiah 1958, pp. 299–301),

$$\Delta E \Delta G \geqslant \tfrac{1}{2}|\langle[\hat{H}, \hat{G}]\rangle| \tag{6.5}$$

where ΔG, the uncertainty in the value of G, is given by its root-mean-square deviation (the square root of the fluctuation in the average value $\langle\hat{G}\rangle$)

$$\Delta\langle\hat{G}\rangle = \{\langle\hat{G}^2\rangle - (\langle\hat{G}\rangle)^2\}^{1/2}. \tag{6.6}$$

The non-vanishing of the subsystem average of the commutator implies a fluctuation in the value of the observable $\hat{G}$ over the subsystem as measured by the flux of its vector current density through the surface of the subsystem. Thus one anticipates and finds non-vanishing fluctuations in subsystem expectation values for observables which do not commute with $\hat{H}$.

Even in a stationary state, that is, an energy eigenstate, where both $\dot{\rho}$ and $\dot{\rho}_G$ vanish, one finds that a non-vanishing div $\mathbf{j}_G$ throughout some volume Ω is associated with an uncertainty in the value of G. When G is a constant of the motion and the values E and G are simultaneously well defined, the net flux in the current through the boundary of Ω vanishes, as it does when Ω refers to the total system. Thus, when the generator $\hat{G}$ of an infinitesimal unitary transformation possesses a sharp value the variation $\delta\mathscr{G}[\hat{G}\psi, \Omega]$ vanishes and the energy functional is stationary.

When $\hat{G}$ does not possess a sharp value in a stationary state, there is a non-vanishing current whose net outflow from any infinitesimal region is determined by the corresponding commutator,

$$\mathbf{V}\cdot\mathbf{j}_G = (\mathrm{i}/\hbar)\psi^*[\hat{H}, \hat{G}]\psi.$$

The energy is not stationary over a volume Ω in such a situation, its change being determined by the flux of the current of G through the surface (eqn (6.1)) or, equivalently, by the average of the commutator (eqn (6.3)). It is clear from this discussion that $\hat{H}$ retains the property of Hermiticity over a subsystem, i.e. that

$$\int_\Omega \psi^*\hat{H}(\hat{G}\psi)\mathrm{d}\tau = \int_\Omega \hat{H}\psi^*(\hat{G}\psi)\mathrm{d}\tau,$$

only when $\hat{H}$ and $\hat{G}$ commute.

Because of the presence of the surface term in eqn (6.2), the hypervirial theorem for a subsystem leads to important physical results which have no counterpart for the total system. It is our immediate goal to use eqn (6.3) to develop the mechanics of an atom in a molecule through the derivation of the atomic statements of the Ehrenfest force law and virial theorem for which the surface terms describe, respectively, the force acting on the atom in terms of the pressure exerted over its surface and the contribution of this surface force to the atomic virial.

6.1.2   Variational derivation of the atomic force law

The atomic statement of the principle of stationary action, eqn (6.3), yields a variational derivation of the hypervirial theorem for any observable $\hat{G}$, a derivation which applies only to a region of space Ω bounded by a surface satisfying the condition of zero flux in the gradient vector field of the charge density,

$$\nabla \rho(\mathbf{r}) \cdot \mathbf{n}(\mathbf{r}) = 0 \qquad \text{for all } \mathbf{r} \text{ on the surface of } \Omega. \qquad (6.7)$$

One may explicitly evaluate the variation in the functional $\mathscr{G}[\psi, \Omega]$ in eqn (6.3) for a particular observable by applying the corresponding infinitesimal unitary transformation to the state function in the functional. This procedure is illustrated in the general time-dependent case in Chapter 8. According to eqn (6.1), the result of this variation is given by the surface integral of the flux in the current $\mathbf{j}_G$. Therefore, an equivalent derivation of the variational result is obtained by using the subsystem statement of the hypervirial theorem (eqn (6.2)) and this is the procedure to be followed here. It should be borne in mind that the use of the hypervirial theorem (eqn (6.2)), which is of general applicability, in lieu of the atomic statement of the principle of stationary action (eqn (6.3)), is restricted to regions of space satisfying the zero-flux boundary condition, since the principle is restricted to such regions.

We shall use the principle of stationary action to obtain a variational definition of the force acting on an atom in a molecule. This derivation will illustrate the important point that the definition of an atomic property follows directly from the atomic statement of stationary action. To obtain Ehrenfest's second relationship as given in eqn (5.24) for the general time-dependent case, the operator $\hat{G}$ in eqn (6.3) and hence in eqn (6.2) is set equal to $\hat{\mathbf{p}}_1$, the momentum operator of the electron whose coordinates are integrated over the basin of the subsystem Ω. The Hamiltonian in the commutator is taken to be the many-electron, fixed-nucleus Hamiltonian

$$\hat{H} = -(\hbar^2/2m)\sum_i \nabla_i^2 - \sum_i\sum_\alpha Z_\alpha e^2 (|\mathbf{r}_i - \mathbf{X}_\alpha|)^{-1} + \sum_{i<j}\sum e^2 (|\mathbf{r}_i - \mathbf{r}_j|)^{-1}$$

$$+ \sum_{\alpha<\beta}\sum Z_\alpha Z_\beta e^2 (|\mathbf{X}_\alpha - \mathbf{X}_\beta|)^{-1},$$

$$\hat{H} = \hat{T} + \hat{V}. \qquad (6.8)$$

The symbol $\hat{V}$ will be used to denote the complete potential energy operator, the sum of the electron–nuclear $\hat{V}_{en}$, electron–electron $\hat{V}_{ee}$, and nuclear–nuclear $\hat{V}_{nn}$ potential energy operators,

$$\hat{V} = \hat{V}_{en} + \hat{V}_{ee} + \hat{V}_{nn}. \tag{6.9}$$

The commutator of this Hamiltonian and the momentum operator of a single electron is, as in eqn (5.23), equal to $i\hbar\nabla\hat{V}$.

The method of obtaining the subsystem average of the commutator and hence of the force acting on the atom Ω, is determined by the definition of the functional $\mathscr{G}[\psi, \Omega]$ via eqn (6.3). It is demonstrated in Section 6.2 that the mode of integration used in the definition of the subsystem functional $\mathscr{G}[\psi, \Omega]$, (see eqn (5.72) and discussion following) is the only one which leads to a physically realizable boundary condition. Because of eqn (6.3), this same mode of integration defines the atomic average of the commutator and thus of the atomic force, $\mathbf{F}(\Omega)$,

$$(N/2)\{(i/\hbar)\langle\psi, [\hat{H}, \hat{\mathbf{p}}_1]\psi\rangle_\Omega + cc\} = N\int_\Omega d\tau_1 \int d\tau'\{\psi^*(-\nabla_1\hat{V})\psi$$

$$= \mathbf{F}(\Omega). \tag{6.10}$$

The result is multiplied by N, the total number of electrons, in the definition of an atomic property. The reader is reminded that the mode of integration indicated by $N\int d\tau'\psi^*\psi$ as used in this definition of an atomic average is the same as that employed in the definition of the electronic charge density, $\rho(\mathbf{r})$ (eqns (1.3) and (1.4)). From this point on the subscript '1' will be dropped from the coordinates of the electron whose coordinates are integrated only over Ω and all single-particle, unlabelled coordinates and operators will refer to this electron.

The corresponding variation of $\mathscr{G}[\psi, \Omega]$, subject to the constraint which gives rise to the zero-flux boundary condition (eqn (6.7)), may be determined by evaluating the surface integral of the current flux appearing in eqn (6.1) for the operator $\hat{G} = \hat{\mathbf{p}} = -i\hbar\nabla$. Multiplication of the surface integral plus its complex conjugate by the factor N to match the atomic average defined in eqn (6.10) gives

$$(N/2)\{\oint dS(\Omega, \mathbf{r})\mathbf{j}_\mathbf{p}(\mathbf{r})\cdot\mathbf{n}(\mathbf{r}) + cc\} = -\oint dS(\Omega, \mathbf{r})\overleftrightarrow{\sigma}(\mathbf{r})\cdot\mathbf{n}(\mathbf{r}) \tag{6.11}$$

where $\overleftrightarrow{\sigma}(\mathbf{r})$ is the quantum mechanical stress tensor introduced for the one-electron case in eqn (5.28). Its many-electron analogue is defined as

$$\overleftrightarrow{\sigma}(\mathbf{r}) = (N\hbar^2/4m)\int d\tau'\{\nabla(\nabla\psi^*)\psi + \psi^*\nabla\nabla\psi - \nabla\psi^*\nabla\psi - \nabla\psi\nabla\psi^*\}, \tag{6.12}$$

a result which may be expressed in terms of the one-electron density matrix $\Gamma^{(1)}$ (eqn (E1.1)) as

$$\overleftrightarrow{\sigma}(\mathbf{r}) = (\hbar^2/4m)\{(\nabla\nabla + \nabla'\nabla') - (\nabla\nabla' + \nabla'\nabla)\}\Gamma^{(1)}(\mathbf{r}, \mathbf{r}')|_{\mathbf{r}=\mathbf{r}'}. \tag{6.13}$$

The stress tensor is a symmetric dyadic and its mathematical properties were reviewed at the end of Chapter 5. It has the dimensions of pressure, force/unit area, or, equivalently, of an energy density. The quantum stress tensor plays a dominant role in the description of the mechanical properties of an atom in a molecule and in the local mechanics of the charge density.

Combining eqns (6.10) and (6.11) yields eqn (6.14), the atomic force law for a stationary state (Bader 1980; Bader and Nguyen-Dang 1981),

$$\mathbf{F}(\Omega) = - \oint dS(\Omega, \mathbf{r}) \overset{\leftrightarrow}{\sigma}(\mathbf{r}) \cdot \mathbf{n}(\mathbf{r}). \tag{6.14}$$

The force may be equivalently expressed using Gauss's theorem as an integration of the force density $-\nabla \cdot \overset{\leftrightarrow}{\sigma}(\mathbf{r})$ over the basin of the atom,

$$\mathbf{F}(\Omega) = N \int_{\Omega} d\tau \int d\tau' \psi^*(-\nabla \hat{V}) \psi = - \int_{\Omega} d\tau \nabla \cdot \overset{\leftrightarrow}{\sigma}(\mathbf{r}). \tag{6.15}$$

Equation (6.14) has a classical analogue which states that the force exerted on the matter contained in a region Ω is equal to the negative of the pressure acting on each element of the surface bounding the region. A local form of the force law is readily obtained in the same manner as used in the derivation of the time derivative of the current density in eqn (5.29). No problems arise in extending the expression to the many-electron case and for a stationary state the result is

$$\mathbf{F}(\mathbf{r}) = N \int d\tau' \psi^*(-\nabla \hat{V}) \psi = - \nabla \cdot \overset{\leftrightarrow}{\sigma}(\mathbf{r}) \tag{6.16}$$

which is clearly the differential form of the integrated force law in eqn (6.15). The integrated and differential force laws have a number of important consequences which are now explored.

The potential energy operator $\hat{V}$ averaged in eqns (6.10) and (6.16) is the many-particle operator defined in eqns (6.8) and (6.9). The operator $-\nabla \hat{V}$ (eqn (6.17)) is the force exerted at the position $\mathbf{r}$ of electron 1 by all of the other electrons and the nuclei in the system, each of the other particles being held fixed in some arbitrary configuration ($\nabla \equiv \nabla_1$ and $\mathbf{r} \equiv \mathbf{r}_1$),

$$-\nabla_1 \hat{V} = \sum_\alpha Z_\alpha e^2 \nabla_1 (|\mathbf{r}_1 - \mathbf{X}_\alpha|)^{-1} - \sum_{j>1} e^2 \nabla_1 (|\mathbf{r}_1 - \mathbf{r}_j|)^{-1}$$

$$= \mathbf{F}_1 = - e^2 \sum_\alpha Z_\alpha \frac{(\mathbf{r}_1 - \mathbf{X}_\alpha)}{|\mathbf{r}_1 - \mathbf{X}_\alpha|^3} + e^2 \sum_{j>1} \frac{(\mathbf{r}_1 - \mathbf{r}_j)}{|\mathbf{r}_1 - \mathbf{r}_j|^3}. \tag{6.17}$$

The integration implied by $d\tau'$ in eqn (6.16) averages this force on the electron at $\mathbf{r}$ over the motions (i.e. positions) of all of the remaining particles in the system and the result is *the force density* $\mathbf{F}(\mathbf{r})$, *the force exerted on the electron at* $\mathbf{r}$ *by the average distribution of the remaining particles in the total system.* Integration of this force density over the basin of the atom Ω then yields the average electronic or *Ehrenfest force* exerted on the atom in the system. Even though the force operator $-\nabla \hat{V}$ involves the coordinates of all the particles in the system, and includes their mutual interaction, the mode of integration employed in eqn (6.16) yields a corresponding density in real space whose integration over an atom with a boundary defined in real space yields the force acting on the atom.

The direct evaluation of the average value of this operator requires the information contained in the two-electron density matrix, yet, according to equations (6.14) and (6.16), this force, in both its differential and integrated forms is determined by the stress tensor which requires only the one-electron density matrix for its evaluation. One can view eqn (6.16) as a statement that the forces acting on a particle arising from the electrostatic interactions between the particles and describable in terms of the gradient of a potential energy operator are balanced by a force $-\nabla \cdot \overleftrightarrow{\sigma}$, which is purely quantum in origin. The virial of the Ehrenfest force, which determines the potential energy of the electrons, is also describable in terms of the stress tensor $\overleftrightarrow{\sigma}$; thus, the mechanics of a quantum system is determined by the information contained in the one-electron density matrix.

An atomic surface for an atom Ω is the union of some number of interatomic surfaces denoted by $S(\Omega|\Omega', \mathbf{r})$, there being one such surface for each bonded neighbour Ω'. Thus, the force acting on an atom as given in eqn (6.14) can be expressed as a sum of surface terms

$$\mathbf{F}(\Omega) = -\sum_{\Omega' \neq \Omega} \oint dS(\Omega|\Omega', \mathbf{r})\overleftrightarrow{\sigma}(\mathbf{r}) \cdot \mathbf{n}(\mathbf{r}). \tag{6.18}$$

The sum in this equation runs over the surfaces shared with atoms bonded to Ω, the atoms linked to Ω by atomic interaction lines. This expression for the force acting on an atom provides the physical basis for the model in which a molecule is viewed as a set of interacting atoms. It isolates, through the definition of structure, the set of atomic interactions which determines the force acting on each atom in a molecule for any configuration of the nuclei.

6.1.3   The atomic virial theorem

Following the order of the theorems derivable from the Heisenberg equation given in Chapter 5, we now consider the use of the virial operator $\mathbf{r} \cdot \mathbf{p}$ (see eqn (5.30)) in the statement of the principle of stationary action (eqn (6.3)) to obtain the atomic statement of the virial theorem. The virial theorem may be obtained by a scaling of the electronic coordinates (Löwdin 1959) and, as shown in Chapter 8, the use of the virial operator as the generator of an infinitesimal unitary transformation is indeed equivalent to a scaling of the electronic coordinate $\mathbf{r}$ (Srebrenik and Bader 1975; Bader and Nguyen-Dang 1981).

Multiplication of the commutator average appearing in eqn (6.2) by $N/2$ for $\hat{G} = \hat{\mathbf{r}} \cdot \hat{\mathbf{p}}$ yields the result

$$(N/2)\{(i/\hbar)\langle\psi, [\hat{H}, \hat{\mathbf{r}} \cdot \hat{\mathbf{p}}]\psi\rangle_\Omega + cc\}$$

$$= 2N \int_\Omega d\tau \int d\tau'(- \hbar^2/4m)\{\psi^*\nabla^2\psi + (\nabla^2\psi^*)\psi\}$$

$$+ N \int_\Omega d\tau \int d\tau'\psi^*(- \mathbf{r} \cdot \nabla\hat{V})\psi = 2T(\Omega) + \mathscr{V}_b(\Omega). \qquad (6.19)$$

The first term is twice the average electronic kinetic energy of the atom $T(\Omega)$ expressed in terms of the usual Laplacian operator and equal to the kinetic energy $K(\Omega)$ defined in eqns (5.47) and (5.50). Since the atom satisfies the zero-flux boundary condition, $T(\Omega)$ also equals the kinetic energy $G(\Omega)$ defined in the same equations as $\langle\mathbf{p} \cdot \mathbf{p}\rangle_\Omega/2m$. The two expressions for the kinetic energy differ by $L(\Omega)$, the average of the Laplacian of the charge density, which by Green's theorem reduces to a surface integral of the flux in $\nabla\rho$ through the surface of the atom. Consequently, as previously stressed, the average electronic kinetic energy of an atom is a well-defined quantity. The second term arising from the commutator and labelled $\mathscr{V}_b(\Omega)$, is the integrated average of the virial of the Ehrenfest force acting on an electron in the basin of the atom

$$\mathscr{V}_b(\Omega) = N \int_\Omega d\tau \int d\tau'\psi^*(-\mathbf{r} \cdot \nabla\hat{V})\psi.$$

Substitution of the operator $\hat{\mathbf{r}} \cdot \hat{\mathbf{p}}$ for $\hat{G}$ in the expression for the current density and multiplication of the surface term in eqn (6.2) by $N/2$ yields

$$(N/2)\{\oint dS(\Omega, \mathbf{r})\mathbf{j}_{\mathbf{r} \cdot \mathbf{p}} \cdot \mathbf{n}(\mathbf{r}) + cc\} = - (N\hbar^2/4m)\{\oint dS(\Omega, \mathbf{r})\int d\tau'[\psi^*\nabla(\mathbf{r} \cdot \nabla\psi)$$

$$- \nabla\psi^*(\mathbf{r} \cdot \nabla\psi) + \psi\nabla(\mathbf{r} \cdot \nabla\psi^*)$$

$$- \nabla\psi(\mathbf{r} \cdot \nabla\psi^*)] \cdot \mathbf{n}(\mathbf{r})\}$$

$$= - \oint dS(\Omega, \mathbf{r})\mathbf{r} \cdot \overleftrightarrow{\sigma}(\mathbf{r}) \cdot \mathbf{n}(\mathbf{r})$$

$$- (\hbar^2/4m)\oint dS(\Omega, \mathbf{r})\nabla\rho(\mathbf{r}) \cdot \mathbf{n}(\mathbf{r}) \qquad (6.20)$$

where the final line is obtained through the use of the identity $\nabla(\mathbf{r} \cdot \nabla\psi) = \nabla\psi + \mathbf{r} \cdot \nabla\nabla\psi$. The negative of the first term on the left-hand side of eqn (6.20) is labelled $\mathscr{V}_s(\Omega)$, and is the virial of the Ehrenfest forces exerted on the surface of the atom. The quantity $\overleftrightarrow{\sigma} \cdot \mathbf{n}$ in eqn (6.21) is the outwardly directed force per unit area of surface and $\mathbf{r} \cdot \overleftrightarrow{\sigma} \cdot \mathbf{n}$ is the virial of this force,

$$\mathscr{V}_s(\Omega) = \oint dS(\Omega, \mathbf{r})\mathbf{r} \cdot \overleftrightarrow{\sigma}(\mathbf{r}) \cdot \mathbf{n}(\mathbf{r}). \qquad (6.21)$$

The second term in eqn (6.20) is $L(\Omega)$, defined in eqn (5.50), expressed in its equivalent surface form. Equating the commutator and surface results followed by some rearranging of terms yields

$$- 2T(\Omega) = \mathscr{V}_{b}(\Omega) + \mathscr{V}_{s}(\Omega) + L(\Omega). \tag{6.22}$$

Since the atom Ω is bounded by a surface of zero flux, $L(\Omega) = 0$ and one obtains the atomic statement of the virial theorem,

$$- 2T(\Omega) = \mathscr{V}_{b}(\Omega) + \mathscr{V}_{s}(\Omega) \tag{6.23}$$

or

$$- 2T(\Omega) = \mathscr{V}(\Omega). \tag{6.24}$$

where $\mathscr{V}(\Omega)$, the sum of the surface and basin terms, is the total virial for the atom. While the partitioning of the virial into basin and surface contributions is dependent upon the choice of origin (an origin can always be found which causes the surface virial to vanish), the value of the total virial $\mathscr{V}(\Omega)$ is, as evident from its equality with twice the kinetic energy, independent of this choice.

Equation (6.24) is identical in form to the virial theorem for a total system—the negative of twice the average kinetic energy of the electrons equals the virial of the forces exerted on them. It is worthwhile here to summarize the ways in which this result is dependent upon the zero-flux boundary condition, eqn (6.7). (1) The use of the principle of stationary action to obtain a variational derivation of this theorem is restricted to a region satisfying eqn (6.7). (2) Satisfaction of eqn (6.7) ensures the vanishing of the term $L(\Omega)$ which arises from the surface flux of the current density $\mathbf{j}_{\mathbf{r} \cdot \mathbf{p}}$ (eqns (6.20) and (6.22)). (3) The vanishing of $L(\Omega)$ is also necessary for the kinetic energy $T(\Omega)$ to be well defined. There is no statement corresponding to eqn (6.24), variational or otherwise, for a subsystem with arbitrary boundaries.

For a stationary state, a local statement of the virial theorem can be obtained using the identity

$$\mathbf{V} \cdot (\mathbf{r} \cdot \overleftrightarrow{\sigma}) = \mathrm{Tr}\,\overleftrightarrow{\sigma} + \mathbf{r} \cdot \mathbf{V} \cdot \overleftrightarrow{\sigma}. \tag{6.25}$$

As previously noted in eqn (5.32), the trace of the stress tensor is given in terms of the kinetic energy densities defined in eqn (5.50) by

$$\mathrm{Tr}\,\overleftrightarrow{\sigma}(\mathbf{r}) = - K(\mathbf{r}) - G(\mathbf{r}) \tag{6.26}$$

or, equivalently, as

$$\mathrm{Tr}\,\overleftrightarrow{\sigma}(\mathbf{r}) = - 2G(\mathbf{r}) - L(\mathbf{r}). \tag{6.27}$$

Substituting this result into eqn (6.25) and rearranging yields

$$- 2G(\mathbf{r}) = - \mathbf{r} \cdot \mathbf{V} \cdot \overleftrightarrow{\sigma} + \mathbf{V} \cdot (\mathbf{r} \cdot \overleftrightarrow{\sigma}) - (\hbar^{2}/4m)\mathbf{V}^{2}\rho(\mathbf{r}). \tag{6.28}$$

For a stationary state the local virial $- \mathbf{r} \cdot \mathbf{V} \cdot \overleftrightarrow{\sigma}$ equals the virial of the

Ehrenfest force density $\mathbf{F}(\mathbf{r})$ as can be seen by taking the virial of eqn (6.16)

$$\mathbf{r} \cdot \mathbf{F}(\mathbf{r}) = N \int d\tau' \psi^* (- \mathbf{r} \cdot \nabla \hat{V}) \psi = - \mathbf{r} \cdot \nabla \cdot \overleftrightarrow{\sigma}. \tag{6.29}$$

Thus, the local statement of the virial theorem is, term for term, the differential form of the integrated theorem in eqn (6.22). Because of this correspondence, one can define the density corresponding to the total virial $\mathscr{V}(\Omega)$ as

$$\mathscr{V}(\mathbf{r}) = - \mathbf{r} \cdot \nabla \cdot \overleftrightarrow{\sigma} + \nabla \cdot (\mathbf{r} \cdot \overleftrightarrow{\sigma}) \tag{6.30}$$

and the local form of the virial theorem can be written as

$$(\hbar^2/4m)\nabla^2 \rho(\mathbf{r}) = 2G(\mathbf{r}) + \mathscr{V}(\mathbf{r}). \tag{6.31}$$

The kinetic energy density $G(\mathbf{r})$ is necessarily positive and eqn (6.31) demonstrates that, in those regions where electronic charge is locally concentrated, i.e. where the Laplacian of the charge density is negative, the electronic potential energy density $\mathscr{V}(\mathbf{r})$ is in local excess over the ratio of 2:1 for the average value of T to $\mathscr{V}$ in the virial theorem. Equation (6.31) is unique in relating a property of the electronic charge density to the local components of the total energy. It will be used extensively in the characterization of bonding and in the prediction of the mechanisms of generalized Lewis acid–base reactions.

From eqn (6.30) it is clear that the virial of the electronic forces, which is the electronic potential energy, is totally determined by the stress tensor $\overleftrightarrow{\sigma}$ and hence by the one-electron density matrix. The atomic statement of the virial theorem provides the basis for the definition of the energy of an atom in a molecule, as is discussed in the sections following Section 6.2.2.

6.1.4   Correspondence between local and subsystem mechanics

It has been shown (Bader 1980; Bader and Nguyen-Dang 1981; Section 8.4.2) that all of the local expressions obtained here for a stationary state, can be obtained in their general time-dependent form from relations satisfied by the energy–momentum tensor, a quantity which summarizes the principal properties of the quantum field, $\psi^*\psi$ (Morse and Feshbach 1953). A subsystem is an open system, free to exchange charge and momentum with its neighbours. These local expressions, as shown here for a stationary state, contain the terms which yield the surface integrals of current flux associated with an open system. The surface terms are obtained as a direct result of the variation of the subsystem energy or action integrals, terms which do not appear in the variation of the corresponding expressions for a total system with boundaries at infinity. Thus the mechanics of a subsystem reflects local properties of the quantum field $\psi^*\psi$ which are absent for a system with boundaries at infinity and thus closed to the exchange of charge and momentum.

6.2 Atomic properties

6.2.1 Single-particle basis for atomic properties

The atomic statements of the Ehrenfest force law and of the virial theorem establish the mechanics of an atom in a molecule. As was stressed in the derivations of these statements, the mode of integration used to obtain an atomic average of an observable is determined by the definition of the subsystem energy functional $\mathcal{G}[\psi, \Omega]$. It is important to demonstrate that the definition of this functional is not arbitrary, but is determined by the requirement that the definition of an open system, as obtained from the principle of stationary action, be stated in terms of a physical property of the total system. This requirement imposes a single-particle basis on the definition of an atom, as expressed in the boundary condition of zero flux in the gradient vector field of the charge density, and on the definition of its average properties.

It has been shown that the principle of stationary action for a stationary state applies to a system bounded at infinity and to one bounded by a surface of zero flux in $\nabla\rho(\mathbf{r})$. It is demonstrated in Chapter 8, through a variation of the action integral, that the same boundary conditions are obtained in the general time-dependent case. One may seek the most general solution to the problem of defining an open system by asking for the set of all possible subsystems to which the principle of stationary action is applicable. Thus, one must consider the variation of the energy functional $\mathcal{G}[\psi, \{\Omega_i\}]$ defined as

$$\mathcal{G}[\psi, \{\Omega_i\}] = \int_{\Omega_1} d\tau_1 \int_{\Omega_2} d\tau_2 \dots \int_{\Omega_N} d\tau_N \{(\hbar^2/2m)\sum_i \nabla_i\psi^* \cdot \nabla_i\psi$$
$$+ (\hat{V} + \lambda)\psi^*\psi\} \tag{6.32}$$

where $\{\Omega_i\}$ denotes a set of subspaces, the integration of the coordinates of electron i being restricted to the region Ω_i. (In the time-dependent case, one performs a corresponding integration of the many-particle Lagrangian density over a set of separate regions.) Carrying through the same variational procedure as followed in the variation of $\mathcal{G}[\psi, \Omega]$, one finds (Srebrenik and Bader 1975) that the condition for the satisfaction of the principle of stationary action is that each subsystem Ω_i be bounded by a surface S_i satisfying a zero-flux boundary condition of the form

$$\nabla_i\rho_i(\mathbf{r}_i) \cdot \mathbf{n}(\mathbf{r}_i) = 0 \qquad \text{for all } \mathbf{r}_i \in S_i \tag{6.33}$$

where

$$\rho_i(\mathbf{r}_i) = \int_{\Omega_1} \dots \int_{\Omega_j} \dots \int_{\Omega_N} (\prod d\tau_{j\neq i})\psi^*\psi.$$

The quantity $\rho_i(\mathbf{r}_i)$ is the probability density that one electron is at $\mathbf{r}_i$ when each of the remaining electrons is in one of the subsystem volumes, Ω_j. Thus $\rho_i(\mathbf{r}_i)$ does not, in general, describe a physically realizable distribution of charge and it requires the diagonal element of the full N-particle density matrix for its evaluation. In only one instance does ρ_i assume physical meaning. This occurs when all the surfaces S_i but one are taken at infinity, in which case ρ_i reduces to $(1/N)\rho$, where ρ is the measurable charge density. The defining condition of the subspaces Ω_i, eqn (6.33), then reduces to the zero-flux surface condition on $\nabla\rho$ (eqn (6.7)). Thus, out of the complete set of mathematical solutions to the problem of determining subsystems that preserve the variational properties of a total system, only one solution is physically realizable.

6.2.2   Definition of atomic properties

From the preceding discussion, the mode of integration used in the definition of an atomic property is determined by the atomic variation principle and is the same as that used in the definition of the charge density itself. The atomic average of an observable $\hat{A}$ is given by

$$A(\Omega) \equiv \langle \hat{A} \rangle_\Omega = \int_\Omega d\tau \int d\tau' (N/2) \{\psi^* \hat{A}\psi + (\hat{A}\psi)^* \psi\}. \qquad (6.34)$$

An atomic property is therefore, determined by the integration of a corresponding property density $\rho_A(\mathbf{r})$ over the basin of the atom where

$$\rho_A(\mathbf{r}) = (N/2) \int d\tau' \{\psi^* \hat{A}\psi + (\hat{A}\psi)^* \psi\} \qquad (6.35)$$

and

$$A(\Omega) = \int_\Omega d\tau \rho_A(\mathbf{r}). \qquad (6.36)$$

One notes that, if $\hat{A} = 1$, the property density in eqn (6.35) reduces to the electronic charge density. The operator $\hat{A}$ is either a one-electron operator, a function of the electronic coordinate $\mathbf{r}$, $\hat{A}(\mathbf{r})$, or else it is a sum of operators each of which contains the coordinate $\mathbf{r}$, as illustrated in the averaging of the commutator $[\hat{H}, \hat{\mathbf{p}}]$ in eqn (6.10). The operator $\hat{A}$ in this case corresponds to $-\nabla\hat{V}$, eqn (6.17), where the gradient is taken with respect to the coordinate $\mathbf{r}$, and every term in the sums over the nuclear and electronic coordinates involves the distance of another particle from the electron at $\mathbf{r}$. Thus the averaging of this operator, as indicated in eqn (6.35), yields a density representing (N times) the force exerted on the density of charge of the electron at $\mathbf{r}$ by the remaining particles averaged over all possible configurations, just as the charge density is (N times) the density of charge of one electron at **r** as determined by averaging the motion of the remaining particles over all possible configurations. Because each contribution to the operator $\hat{A}$ involves the coordinate **r**, all atomic properties can be expressed using either the one-density matrix $\Gamma^{(1)}(\mathbf{r}, \mathbf{r}')$ (eqn (E1.1)) for a one-electron operator, expressed as $\hat{A}(\mathbf{r}, \mathbf{r}')$ in the most general case

$$A(\Omega) = \int_{\Omega} d\tau \{\hat{A}(\mathbf{r}, \mathbf{r}')\Gamma^{(1)}(\mathbf{r}, \mathbf{r}')|_{\mathbf{r} = \mathbf{r}'}\}, \tag{6.37}$$

or the diagonal elements $\Gamma^{(2)}(\mathbf{r}_1, \mathbf{r}_2)$ of the two-density matrix (eqn (E1.3)) for a two-electron operator $\hat{A}(\mathbf{r}_1, \mathbf{r}_2)$

$$A(\Omega) = \int_{\Omega} d\tau_1 \int d\tau_2 \hat{A}(\mathbf{r}_1, \mathbf{r}_2)\Gamma^{(2)}(\mathbf{r}_1, \mathbf{r}_2) \tag{6.38}$$

where $\Gamma^{(2)}$ is normalized to $N(N-1)/2$ pairs. The use of the operator $\hat{\mathbf{r}} \cdot \hat{\mathbf{p}}$ as the generator yields the definition of the electronic potential energy and appearing in the commutator $[\hat{H}, \hat{\mathbf{r}} \cdot \hat{\mathbf{p}}]$ is a sum of two-electron operators of the form $e^2/(|\mathbf{r}_1 - \mathbf{r}_j|) = e^2/r_{1j}$ with $j > 1$. They define the electron–electron repulsion energy of Ω as

$$V_{ee}(\Omega) = \int_{\Omega} d\tau_1 \int d\tau_2 (e^2/r_{12})\Gamma^{(2)}(\mathbf{r}_1, \mathbf{r}_2). \tag{6.39}$$

The quantity $V_{ee}(\Omega)$ equals the repulsion energy of the electrons in Ω, $V_{ee}(\Omega, \Omega)$ and one-half the repulsion of the electrons in Ω with that of those in the remainder of the system Ω', $V_{ee}(\Omega, \Omega')$. Since the average value of the operator for the total system is

$$\langle \hat{V}_{ee} \rangle = V_{ee}(\Omega, \Omega) + V_{ee}(\Omega, \Omega') + V_{ee}(\Omega', \Omega) + V_{ee}(\Omega', \Omega'), \tag{6.40}$$

one necessarily has $V_{ee}(\Omega, \Omega') = V_{ee}(\Omega', \Omega)$.

The most important consequence of the definition of an atomic property as given in eqn (6.34) or (6.36) is that the average value of an observable for the total system $\langle \hat{A} \rangle$, is given by the sum of its atomic contributions $A(\Omega)$,

$$\langle \hat{A} \rangle = \sum_{\Omega} A(\Omega). \tag{6.41}$$

Equation (6.41) is true for both one- and two-particle operators. Equation (6.41) states that each atom makes an additive contribution to the value of every property for a total system. *This is the principle underlying the cornerstone of chemistry—that atoms and functional groupings of atoms make recognizable contributions to the total properties of a system.* In practice, we recognize a group and predict its effect upon the static and reactive properties of a system in terms of a set of properties assigned to the group. In the limiting case of a group being essentially the same in two different systems, one obtains a so-called additivity scheme for the total properties, for in this case the atomic contributions, as well as being additive in the sense of eqn (6.41), are transferable between molecules.

Even a property not represented by a linear Hermitian operator can be expressed as a sum of atomic contributions, as in eqn (6.41). The polarizability of a molecule, for example, which is determined by the first-order response of the charge density to an electric field, is not directly expressible as an average over a corresponding operator. This is not to say, however, that the polarizability cannot be expressed as an additive atomic property, as is indeed done empirically. The atomic contributions to the molecular polarizability and magnetic susceptibility are defined and discussed in Chapter 8. Equation (6.41) is not usually applied directly to vector or tensor moments, such as the dipole or quadrupole moments of a charge distribution, which involve the definition of a single origin. Molecular moments are more usefully related to a sum of atomic contributions with a local origin and this is done by using moments in addition to the one in question. This is illustrated for the dipole moment which, along with a number of other properties determined directly by the charge density, is introduced below.

The electron population of an atom in a molecule, its average number of electrons $N(\Omega)$, is obtained by setting the operator $\hat{A}$ equal to 1 in which case $\rho_A(\mathbf{r})$ reduces to the electronic charge density $\rho(\mathbf{r})$,

$$N(\Omega) = \int_\Omega \rho(\mathbf{r})\,d\tau. \tag{6.42}$$

The net charge on an atom $q(\Omega)$, is given by the sum of its nuclear charge $Z_\Omega e$ and its average electronic charge $- N(\Omega)e$,

$$q(\Omega) = (Z_\Omega - N(\Omega))e. \tag{6.43}$$

Setting $\hat{A}$ equal to r_Ω, the radial distance of an electron from the nucleus of the atom, or some power n of this distance yields the corresponding average over the charge density of the atom,

$$r^n(\Omega) = \int_\Omega r_\Omega^n \rho(\mathbf{r})\,d\tau. \tag{6.44}$$

The atomic volume $v(\Omega)$ is defined as a measure of the region of space enclosed by the intersection of its interatomic surfaces and an envelope of the charge density of some chosen value. An atomic surface is the union of some number of interatomic surfaces, there being one such surface for each bonded neighbour and, if the atom is not an interior atom, some portions which may be infinitely distant from the attractor. It is these latter, open portions of the atomic surface which are replaced with an envelope of the charge density, a surface on which $\rho(\mathbf{r})$ has a constant value. An envelope of the charge density can be used to define the 'van der Waals' shape or size of a molecule in relation to its non-bonded interactions with other molecules in the gaseous and crystalline phases (Bader *et al.* 1967; Bader and Preston 1970; Bader *et al.* 1987*a*). The 0.001-au density envelope has been shown to yield values of the molecular diameters of methane and other inert monatomic and polyatomic gases in good agreement with the equilibrium diameters of these molecules as determined by the second virial coefficient or viscosity data fitted with a Lennard-Jones 6–12 potential. The 0.001-au density envelope encloses over 98 per cent of the electronic charge of a hydrogen atom in a hydrocarbon molecule and over 99 per cent of the electronic charge of the atoms carbon to neon. The 0.002-au contour has proved to provide a useful measure of molecular size in crystals.

The first moment of an atom's charge distribution $M(\Omega)$, is obtained by averaging the vector r_Ω, with origin at the nucleus, over the charge density of the atom,

$$M(\Omega) = -e \int_\Omega r_\Omega \rho(r) d\tau. \tag{6.45}$$

The first moment provides a measure of the extent and direction of the dipolar polarization of the atom's charge density by determining the displacement of the atom's centroid of negative charge from the position of its nucleus. The dipole moment of a neutral molecule is expressed as

$$\mu = e \sum_\Omega Z_\Omega X_\Omega - e \int r \rho(r) d\tau \tag{6.46}$$

where the electronic and nuclear position vectors r and X_Ω, respectively, are measured from a common, arbitrary origin. They are related to the vector r_Ω with nucleus Ω as origin by $r = r_\Omega + X_\Omega$ and use of this relationship in eqn (6.46) yields

$$\mu = \sum_\Omega q(\Omega) X_\Omega + \sum_\Omega M(\Omega) = \mu_c + \mu_a. \tag{6.47}$$

Thus the total molecular dipole moment can be equated to a sum of atomic charges and first moments. The first term in eqn (6.47), μ_c, is the contribution from the interatomic charge transfer, while the second term, μ_a, arises from the polarizations of the individual atomic distributions. In general, both terms are important in determining μ (Bader *et al.* 1987*b*).

Another important atomic polarization is measured by the axial components of the quadrupole moment. The quadrupolar polarization of an atomic density measured with respect to the z-axis is given by

$$Q_{zz}(\Omega) = -e \int_\Omega (3z_\Omega^2 - r_\Omega^2) \rho(r) d\tau \tag{6.48}$$

with corresponding definitions for the x- and y-axes. Defined in this manner, the quadrupole moment tensor is traceless—the sum of the three diagonal elements is zero. Since it corresponds to a real symmetric matrix, it can be diagonalized to obtain three principal axes and their corresponding moments of polarization as defined above. Each moment equals zero for a spherically symmetric charge distribution while, for a sphere flattened at the poles of the z-axis, an oblate spheroid, $0 < Q_{zz}(\Omega) = -\frac{1}{2}Q_{xx}(\Omega) = -\frac{1}{2}Q_{yy}(\Omega)$. For an atom with electronic charge depleted in a plane and concentrated along the perpendicular (z) axis, $Q_{zz}(\Omega) < 0$. These three situations are illustrated by the corresponding distortions of a sphere in Fig. 6.1. The quadrupole moment is the charge density analogue of a π population of the orbital model. The physical moment has the advantage, however, of being definable even when no plane or axis of symmetry exists for the definition of a π system.

The definition of the energy of an atom in a molecule requires detailed consideration from a number of points of view and the following section is devoted to that task. The definition is shown to follow directly from the atomic statement of the virial theorem and, once having established this fact, the underlying equations are readily put down. We give the final equations here. The energy of an atom in a molecule, $E_e(\Omega)$, is purely electronic in origin and is defined as

$$E_e(\Omega) = T(\Omega) + \mathcal{V}(\Omega) \tag{6.72}$$

where the electronic kinetic energy $T(\Omega)$ and the electronic virial $\mathcal{V}(\Omega)$ are defined in the atomic statement of the virial theorem, eqn (6.24). Because of this theorem, $E_e(\Omega)$ satisfies the following alternative statements of the virial theorem,

$$E_e(\Omega) = -T(\Omega) = +\tfrac{1}{2}\mathcal{V}(\Omega). \tag{6.73}$$

The potential energy of an atom in a molecule is thus defined to be the

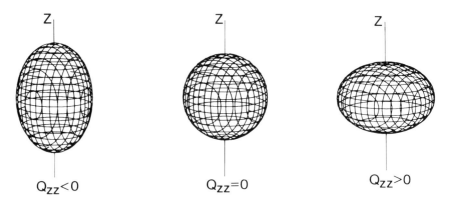

FIG. 6.1. Pictorial demonstration of quadrupolar polarizations of an electronic charge distribution.

average of the virial of the forces exerted on its electrons. Why this step is essential to the partitioning of the total energy and what are its consequences, are the topics of the following section.

6.3 Energy of an atom in a molecule

6.3.1 Physical constraints on partitioning the energy

A system which is isolated and characterized by its own Hamiltonian is separated from other systems to a degree sufficient to exclude, to any measurable extent, the exchange of identical particles between it and any other system. The Hamiltonian is unique to a given system and through Schrödinger's equation, eqn (5.2) for a time-dependent state or eqn (5.16) for a stationary state, it determines the state function. The state function in turn, determines all properties of the system, including its decomposition into, and the properties of, its subsystems. There is no Hamiltonian or state function for a subsystem of a total system. One cannot partition a many-particle Hamiltonian operator into distinct sets of operators without violating the indistinguishability of identical particles. However, all electrons contribute equally to the charge density at any point in real space in a manner determined by Schrödinger's equation and a partitioning of space treats all electrons equivalently. Thus, the definition of an atomic energy proceeds not through a *spatial* partitioning of the Hamiltonian or of the elements of the abstract Hilbert space on which it acts, but rather through a partitioning of the Hamiltonian into a sum of effective one-electron contributions. This is accomplished through the use of the virial operator for a single electron. It is shown that this operator projects out from the total potential energy operator *which includes all interparticle interactions*, that part belonging to a single electron. This projection, since it is the virial of a force acting at a point in space, yields an energy density which may be integrated over individual spatial regions to yield corresponding average potential energies.

6.3.2 The virial and the partitioning of an energy of interaction

The discussion will be confined to the energy of a stationary state with the state function expressed for a rigid nuclear framework within the Born–Oppenheimer approximation. The Hamiltonian is given in eqn (6.8) and yields the average energy

$$E = \langle \psi, \hat{H}\psi \rangle. \tag{6.49}$$

It is the partitioning of the potential energy of interaction that is the stumbling block in partitioning a total energy. It is shown that the virial operator can be used to obtain a spatial partitioning of the average potential energy. Since the potential energy operator $\hat{V}$ is a homogeneous function of degree -1 with respect to the electronic *and* nuclear coordinates, it may be alternatively expressed in terms of the virials of the corresponding forces exerted on the electrons and the nuclei using Euler's theorem as

$$\sum_i(-\mathbf{r}_i\cdot\mathbf{V}_i\hat{V}) + \sum_\alpha(-\mathbf{X}_\alpha\cdot\mathbf{V}_\alpha\hat{V}) = \hat{V} \tag{6.50}$$

or as

$$\sum_i\mathbf{r}_i\cdot\hat{\mathbf{F}}_i + \sum_\alpha\mathbf{X}_\alpha\cdot\hat{\mathbf{F}}_\alpha = \hat{V}. \tag{6.51}$$

A force operator $\hat{\mathbf{F}}_\kappa$ for particle κ, $-\mathbf{V}_\kappa\hat{V}$, is the classical force exerted on this particle by all the particles in the system. Correspondingly, one may consider the virial of this force, $\mathbf{r}_\kappa\cdot\hat{\mathbf{F}}_\kappa$, to be the potential energy operator for the particle since, according to eqn (6.51), the sum of such operators for all the particles in the system yields the potential energy operator $\hat{V}$.

This idea is further illustrated in terms of the simplest possible interaction, that between just two particles which interact by an inverse square force. Their energy of interaction is described by $\hat{v}_{12}$. How much of this energy of interaction belongs to particle 1 and how much to particle 2? The answer to this question is provided by expressing the potential energy in terms of the virials of the forces, $\hat{\mathbf{F}}_1 = -\mathbf{V}_1\hat{v}_{12}$ and $\mathbf{F}_2 = -\hat{\mathbf{V}}_2\hat{v}_{12}$, which the particles exert on one another,

$$\mathbf{r}_1\cdot\hat{\mathbf{F}}_1 + \mathbf{r}_2\cdot\hat{\mathbf{F}}_2 = \hat{v}_{12}. \tag{6.52}$$

Equation (6.52) expresses the simple but important result that each particle's share of the potential energy is given by the virial of the force exerted on it by the other. The virial operator $\mathbf{r}_\kappa\cdot\hat{\mathbf{F}}_\kappa$ is like a projection operator in that it projects from $\hat{V}$, that part of the potential energy operator belonging to particle κ. In this elementary case, each share of the potential energy is dependent upon the choice of origin used in the definition of the vectors $\mathbf{r}_1$ and $\mathbf{r}_2$. This does not turn out to be the case when this idea is used to spatially partition the potential energy of a many-electron system. If one denotes by $\sum_\kappa-\dot{\mathbf{r}}_\kappa\cdot\mathbf{V}_\kappa$ the complete set of virial operators in eqn (6.50), one has

$$\left(\sum_\kappa-\mathbf{r}_\kappa\cdot\mathbf{V}_\kappa\right)^n\hat{V} = \hat{V}$$

and the complete set of virial operators is idempotent as required for projection operators.

We shall first demonstrate that the identification of a particle's virial with its share of the potential energy of interaction leads to the correct physics for the total system and then show that it forms the basis for the definition of the energy of an atom in a molecule.

The first summation in eqn (6.51) which defines the electronic share of the potential energy operator should, together with kinetic energy operator $\hat{T}$ of eqn (6.8), define the electronic Hamiltonian of the system and this, in turn, the electronic energy E_e. Thus,

$$E_e = \langle \hat{H}_e \rangle = \langle (\hat{T} + \sum_i \mathbf{r}_i \cdot \hat{\mathbf{F}}_i) \rangle. \tag{6.53}$$

This Hamiltonian defines the average electronic potential energy V_e to be the average virial of the forces exerted on the electrons, the electronic virial,

$$V_e = \mathscr{V} = \langle \sum_i \mathbf{r}_i \cdot \hat{\mathbf{F}}_i \rangle. \tag{6.54}$$

Correspondingly, the nuclear energy of the system is purely potential in origin and is given by averaging the nuclear share of $\hat{V}$ given in eqn (6.51).

$$E_n = \langle (\sum_\alpha \mathbf{X}_\alpha \cdot \hat{\mathbf{F}}_\alpha) \rangle = - \sum_\alpha \mathbf{X}_\alpha \cdot \nabla_\alpha E. \tag{6.55}$$

The force $-\nabla_\alpha E$ is the Hellmann–Feynman electrostatic force exerted on nucleus α as defined in eqn (5.37). The total energy E is thus given by

$$E = E_e + E_n. \tag{6.56}$$

When a molecule is in an equilibrium configuration, the external forces exerted on the nuclei, the forces $-\nabla_\alpha E$, vanish as does E_n and, in this case, $E = E_e$.

The identification of the potential energy of the electrons with the virial of the forces exerted on them as in eqn (6.54) is more than a question of semantics as it yields a deeper physical insight into the energy of a molecule with a rigid nuclear framework. It is because the nuclei are clamped that the interpretation of the energy is not straightforward. The molecule can be in mechanical equilibrium in a given arbitrary configuration $\mathbf{X}$ only if each of the nuclear forces $-\nabla_\alpha E$ is balanced by an applied external force. There is an extended form of Hamilton's principle in classical mechanics which is used to handle just this sort of problem—to obtain the equations of motion for a system subject to forces of constraint and which, in addition, may be non-conservative. In this situation the classical action integral is replaced by the variational integral

$$I = \int_{t_1}^{t_2} dt \, (\sum_i p_i^2 / 2m + \sum_i \mathbf{r}_i \cdot \mathbf{F}_i), \tag{6.57}$$

which is to be compared with the expression for E_e given in eqn (6.53). In analogy with this classical result, it has been shown (Srebrenik *et al.* 1978) that the electronic energy of a molecule, E_e, can be obtained from a variation principle if the variation is made subject to the constraints imposed on the system via the external forces that must be applied to the nuclei to maintain the configuration $\mathbf{X}$.

Thus, it is established that the energy E_e is the minimum energy of a set of electrons moving in the field of a rigid nuclear framework. The electronic virial $\mathscr{V}$ (eqn (6.54)) that defines the electronic potential energy can be expressed as

$$\mathscr{V} = \langle \hat{V} \rangle + \sum_\alpha \mathbf{X}_\alpha \cdot \nabla_\alpha E$$

$$= \langle \hat{V}_{ee} \rangle + \langle \hat{V}_{en} \rangle + \langle \hat{V}_{nn} \rangle + \sum_\alpha \mathbf{X}_\alpha \cdot \nabla_\alpha E. \tag{6.58}$$

This equation shows that when the external forces on the nuclei vanish, the electronic energy E_e equals the usual total energy E of eqn (6.49). It is more informative however, to explicitly evaluate the average of the virial of the forces appearing in eqn (6.54). This yields

$$\mathscr{V} = \langle \hat{V}_{ee} \rangle + \langle \hat{V}_{en} \rangle - \sum_\alpha \mathbf{X}_\alpha \cdot \mathbf{F}_{\alpha e} \tag{6.59}$$

where $\mathbf{F}_{\alpha e}$ is the electronic contribution to $\mathbf{F}_\alpha$, the force on nucleus α. That is,

$$\mathbf{F}_\alpha = -\nabla_\alpha E = \mathbf{F}_{\alpha n} + \mathbf{F}_{\alpha e}$$

$$= -Z_\alpha e^2 \sum_{\beta \neq \alpha} Z_\beta \frac{(\mathbf{X}_\beta - \mathbf{X}_\alpha)}{|\mathbf{X}_\beta - \mathbf{X}_\alpha|^3} + Z_\alpha e^2 \int \rho(\mathbf{r}) \frac{(\mathbf{r}_1 - \mathbf{X}_\alpha)}{|\mathbf{r}_1 - \mathbf{X}_\alpha|^3} \, d\tau \tag{6.60}$$

and

$$-\sum_\alpha \mathbf{X}_\alpha \cdot \mathbf{F}_{\alpha e} = \sum_\alpha \mathbf{X}_\alpha \cdot \nabla_\alpha E + \sum_\alpha \mathbf{X}_\alpha \cdot \mathbf{F}_{\alpha n}.$$

The virial of the nuclear–nuclear repulsive forces equals the nuclear–nuclear repulsive energy. That is,

$$\sum_\alpha \mathbf{X}_\alpha \cdot \mathbf{F}_{\alpha n} = -e^2 \sum_\alpha \sum_{\beta \neq \alpha} Z_\alpha Z_\beta \mathbf{X}_\alpha \cdot \frac{(\mathbf{X}_\beta - \mathbf{X}_\alpha)}{|\mathbf{X}_\beta - \mathbf{X}_\alpha|^3}$$

$$= +e^2 \left(\frac{1}{2}\right) \sum_\alpha \sum_{\beta \neq \alpha} Z_\alpha Z_\beta (|\mathbf{X}_\beta - \mathbf{X}_\alpha|)^{-1} = \langle \hat{V}_{nn} \rangle.$$

In an equilibrium configuration, $\mathbf{F}_{\alpha e}$ is balanced by the nuclear contribution $\mathbf{F}_{\alpha n}$ and the virial of the electronic contributions $\mathbf{F}_{\alpha e}$ appearing in eqn (6.59) reduces to the nuclear–nuclear repulsive energy V_{nn} in agreement with eqn (6.58) for the case where $\nabla_\alpha E = 0$. Thus, out of the total potential energy operator $\hat{V}$, $\hat{V}_{ee}$ is purely electronic, as $\hat{V}_{nn}$ is purely nuclear, but $\hat{V}_{en}$ contains contributions form both sets of particles. The electronic share of this potential energy of interaction is given by the projection of the virial operators for the electrons,

$$\langle (-\sum_i \mathbf{r}_i \cdot \nabla_i \hat{V}_{en}) \rangle = \langle \hat{V}_{en} \rangle - \sum_\alpha \mathbf{X}_\alpha \cdot \mathbf{F}_{\alpha e} = \langle \hat{V}_{en} \rangle + \sum_\alpha \mathbf{X}_\alpha \cdot \mathbf{F}_{e\alpha} \tag{6.61}$$

where $\mathbf{F}_{e\alpha} = -\mathbf{F}_{\alpha e}$ is the force which nucleus α exerts on the electrons and $\sum_\alpha \mathbf{X}_\alpha \cdot \mathbf{F}_{e\alpha}$ is the virial of the forces which the nuclei exert on the electrons. The corresponding nuclear projection is

$$\langle (-\sum_\alpha \mathbf{X}_\alpha \cdot \nabla_\alpha \hat{V}_{en}) \rangle = \sum_\alpha \mathbf{X}_\alpha \cdot \mathbf{F}_{\alpha e}, \tag{6.62}$$

which is the virial of the forces which the electrons exert on the nuclei. The addition of this result to V_{nn} gives the nuclear share of the potential energy as given in eqn (6.55).

Thus the potential energy of the electrons moving in a fixed-nucleus framework consists of the electron–electron repulsion energy, the electron–nuclear attractive energy, and the virial of the forces which the nuclei exert on the electrons (eqn (6.59)). For configurations not far removed from an equilibrium configuration, this latter term approaches V_{nn} in value and, for a system at equilibrium, it equals V_{nn}. Since $\mathscr{V}$, the average potential energy of the electrons, is the virial of the forces exerted on the electrons (eqn (6.54)), it satisfies the virial theorem derived previously in eqn (5.31),

$$- 2T = \mathscr{V} = V_e \tag{6.63}$$

and the electronic energy E_e, therefore, satisfies the corresponding relationships,

$$E_e = - T = \tfrac{1}{2}\mathscr{V}. \tag{6.64}$$

The equations represented in (6.63) and (6.64) are true for any configuration of the nuclei. The molecular virial theorem may also be expressed as

$$- 2T = \langle \hat{V} \rangle + \sum_\alpha \mathbf{X}_\alpha \cdot \mathbf{V}_\alpha E. \tag{6.65}$$

the form originally given by Slater (1933).

6.3.3   The electronic energy of an atom

The importance of defining an electronic energy and of being able to express it in the form of eqn (6.53) is that it enables one to express $\hat{H}_e$ and hence E_e as a sum of single-particle contributions,

$$\hat{H}_e = \sum_i \hat{H}_i = \sum_i \{ - (\hbar^2/2m)\nabla_i^2 - \mathbf{r}_i \cdot \mathbf{V}_i \hat{V} \} \tag{6.66}$$

and

$$E_e = N \langle \hat{H}_i \rangle = N \int d\tau \int d\tau' \{ \psi^* (- \hbar^2/2m) \nabla_i^2 - \mathbf{r}_i \cdot \mathbf{V}_i \hat{V} \} \psi. \tag{6.67}$$

It should not escape the readers' attention that the mode of integration used in eqn (6.67) is identical to that used in the definition of an atomic average. Thus *the electronic energy is* N *times the average energy of a single electron*, and E_e may be expressed as an integral over an effective single-particle energy density

$$E_e = \int d\tau \{ K(\mathbf{r}) + \mathscr{V}_b(\mathbf{r}) \} \tag{6.68}$$

where $K(\mathbf{r})$ is the kinetic energy density defined in eqn (5.50) and $\mathscr{V}_b(\mathbf{r})$ is the virial field,

$$\mathscr{V}_b(\mathbf{r}) = N \int d\tau' \{ \psi^* (- \mathbf{r} \cdot \mathbf{V} \hat{V}) \psi \} = - \mathbf{r} \cdot \mathbf{V} \cdot \overleftrightarrow{\sigma}, \tag{6.69}$$

the field which determines the atomic virial $\mathscr{V}_b(\Omega)$ (eqns (6.19) and (6.29)). The virial field $\mathscr{V}_b(\mathbf{r})$ is N times the potential energy density of one electron at $\mathbf{r}$ as determined by its average energy of interaction with all of the other particles in the system. It is an exact prescription of the 'average field' experienced by a single electron in a many-electron system.

Thus the virial operator for a single electron defines a single-particle energy density required for the partitioning of the potential energy of an interacting system and for the definition of the energy of an atom in a molecule. Once the average energy is expressed as an integral over an energy density, the energy of any region is obtained simply by integrating the density over that region, assuming the region has a physical significance. It is here that quantum mechanics through the atomic statement of stationary action plays a crucial role. The virial theorem for a subsystem is restricted to an atomic region, one bounded by a zero-flux surface. In addition, as noted before, the kinetic energy is not well defined for a region with arbitrary boundaries. Thus, only the energy of an atomic region satisfies the quantum mechanical virial theorem, in addition to being well defined.

The above discussion used the total system to illustrate the principle of using the virial operator to define a potential energy density but the virial density appearing in the integral for the total system cannot be taken directly over to the atomic case. The virial for a quantum subsystem consists of a surface as well as a basin contribution, eqn (6.23), and the virial density thus contains in addition to $\mathscr{V}_{b}(\mathbf{r})$, the term $\mathbf{V} \cdot (\mathbf{r} \cdot \boldsymbol{\sigma})$. This definition of a virial density is in agreement with that derived from the local statement of the virial theorem, eqn (6.30). The term $\mathbf{V} \cdot (\mathbf{r} \cdot \boldsymbol{\sigma})$ does not appear in the integrated virial for a total system since, in this case, the surface term vanishes. Its presence in the local expression for $\mathscr{V}(\mathbf{r})$ ensures that this quantity and its integrated value for an atom $\mathscr{V}(\Omega)$ are independent of the choice of origin.

The subtraction of the kinetic energy density $G(\mathbf{r})$ from each side of the statement of the local virial theorem given in eqn (6.31) yields a definition of an energy density $E_{e}(\mathbf{r})$ as

$$E_{e}(\mathbf{r}) = G(\mathbf{r}) + \mathscr{V}(\mathbf{r}) = -K(\mathbf{r}). \qquad (6.70)$$

The second equality given in eqn (6.70) follows from the definition of $K(\mathbf{r})$ in eqn (5.49). The integration of this energy density over a region of space bounded by a surface of zero flux in $\nabla \rho$ yields an energy $E_{e}(\Omega)$ which will satisfy the various statements of the atomic virial theorem,

$$\int_{\Omega} E_{e}(\mathbf{r}) \mathrm{d}\tau = \int_{\Omega} \{ G(\mathbf{r}) + \mathscr{V}(\mathbf{r}) \} \, \mathrm{d}\tau \qquad (6.71)$$

or

$$E_{e}(\Omega) = T(\Omega) + \mathscr{V}(\Omega). \qquad (6.72)$$

Because of the atomic virial theorem, eqn (6.24), the atomic energy $E_{e}(\Omega)$

satisfies the following relationships, which are the direct analogues of the all space results given in eqn (6.64) for E_e,

$$E_e(\Omega) = - T(\Omega) = \tfrac{1}{2}\mathscr{V}(\Omega). \qquad (6.73)$$

Because of eqn (6.73) and the vanishing of the Laplacian of the charge density over an atomic basin, the following identities hold,

$$E_e(\Omega) = - K(\Omega) = - G(\Omega) = \tfrac{1}{2}\int \mathrm{Tr}\,\sigma(\mathbf{r})\,\mathrm{d}\tau. \qquad (6.74)$$

It is to be emphasized that all of the above relationships, together with the atomic statements of the virial theorem (eqns (6.72) and (6.74)), remain true when Ω refers to the total system. It is in this sense that an atom is a quantum subsystem.

From its definition in eqn (6.71) in terms of an integral of an energy density, it is clear that, like other atomic properties (eqn (6.41)), the sum of the energies of the atoms in a system equals the total electronic energy

$$E_e = \sum_\Omega E_e(\Omega) \qquad (6.75)$$

and, when there are no forces acting on any of the nuclei in the system, this sum equals the total molecular energy E (eqn (6.49)).

One may also view the possibility of defining an electronic potential energy density as the result of $\mathscr{V}(\mathbf{r})$ being expressible in terms of the quantum stress tensor $\overset{\leftrightarrow}{\sigma}(\mathbf{r})$, eqn (6.30). The stress tensor is a function of just the coordinate $\mathbf{r}$ because it is a functional of the one-electron density matrix, one whose form implicitly includes the two-electron interactions. The electronic virial also contains the nuclear–nuclear energy of repulsion (eqn (6.58)). Thus, the electronic energy $E_e(\Omega)$ includes a partitioning of this contribution to the energy, through the virial of the forces which the nuclei exert on the electrons (eqn (6.61)) even though $\langle \hat{V}_{nn} \rangle$ does not directly involve the electronic coordinate $\mathbf{r}$. The result is a non-arbitrary partitioning of the total energy of a molecule. Section 5.2.1 demonstrated the observational basis for this definition of the energy of an atom in a molecule by showing that the energy, like all properties of an atom in a molecule, changes in direct response to changes in the atom's distribution of charge. Thus, when the form of an atom is the same in two different molecules, it makes the same contribution to the total energy in both molecules. The transferability of atoms and their properties between different molecules is illustrated in Section 6.4.2.

6.3.4   Potential energy contributions to an atomic energy

The individual contributions to the electronic potential energy of an atom are obtained by determining the action of the operator $-\mathbf{r}\cdot\nabla$ on $\hat{V}$ in the expression for the basin virial, eqn (6.19). The basin virial is the virial of the Ehrenfest force and the force density $\mathbf{F}(\mathbf{r})$ of eqn (6.16) is evaluated as the first step in the determination of $\mathscr{V}(\Omega)$. To keep proper track of the subscripts, the coordinates which are integrated over the atomic basin are again subscripted with '1' and so $\mathbf{r} \cdot \mathbf{V} \equiv \mathbf{r}_1 \cdot \mathbf{V}_1$. The expression for $-\mathbf{V}_1 \hat{V}$ is given in eqn (6.17) and, with the vectors $\mathbf{r}_{1j}$, from electron j to electron 1, and $\mathbf{r}_{1\alpha}$, from nucleus α to electron 1, defined as

$$\mathbf{r}_{1j} = \mathbf{r}_1 - \mathbf{r}_j \quad \text{and} \quad \mathbf{r}_{1\alpha} = \mathbf{r}_1 - \mathbf{X}_\alpha,$$

the virial of this force with respect to $\mathbf{r}_1$ may be expressed as

$$-\mathbf{r}_1 \cdot \mathbf{V}_1 \hat{V} = -\sum_\alpha Z_\alpha e^2 r_{1\alpha}^{-1} - \sum_\alpha \mathbf{X}_\alpha \cdot \mathbf{V}_1 (Z_\alpha e^2 r_{1\alpha}^{-1})$$
$$+ (N-1)\{e^2 r_{12}^{-1} + \mathbf{r}_2 \cdot \mathbf{V}_2 (e^2 r_{12}^{-1})\} \tag{6.76}$$

where the sum over j has been replaced by $(N-1)$ times the interaction between electrons 1 and 2. The two-electron terms can be put into a more useful form by applying the identity (eqn (6.52))

$$(\mathbf{r}_1 \cdot \mathbf{V}_1 + \mathbf{r}_2 \cdot \mathbf{V}_2) r_{12}^{-1} = -r_{12}^{-1}$$

to $(N-1)/2$ of the final terms in eqn (6.76) which yields

$$\{(N-1)/2\}\{e^2 r_{12}^{-1} + (\mathbf{r}_2 \cdot \mathbf{V}_2 - \mathbf{r}_1 \cdot \mathbf{V}_1)(e^2 r_{12}^{-1})\}. \tag{6.77}$$

Averaging of this final operator expression for the virial of $\hat{V}$ in the manner indicated in eqn (6.69) for the potential energy density $\mathscr{V}_b(\mathbf{r})$, which is the virial of the Ehrenfest force eqn (6.29), yields

$$\mathscr{V}_b(\mathbf{r}) = N \int d\tau' \psi^*(-\mathbf{r}_1 \cdot \mathbf{V}_1 \hat{V})\psi = -\mathbf{r}_1 \cdot \mathbf{V}_1 \sigma(\mathbf{r})$$
$$= [-\sum_\alpha Z_\alpha e^2 r_{1\alpha}^{-1} - \sum_\alpha \mathbf{X}_\alpha \cdot \mathbf{V}_1 (Z_\alpha e^2 r_{1\alpha}^{-1})]\rho(\mathbf{r})$$
$$+ \int d\tau_2 e^2 r_{12}^{-1} \Gamma^{(2)}(\mathbf{r}_1, \mathbf{r}_2)$$
$$+ \int d\tau_2 (\mathbf{r}_2 \cdot \mathbf{V}_2 - \mathbf{r}_1 \cdot \mathbf{V}_1)(e^2 r_{12}^{-1})\Gamma^{(2)}(\mathbf{r}_1, \mathbf{r}_2). \tag{6.78}$$

The nuclear–electron terms in the operator on the right-hand side refer to the electronic coordinate $\mathbf{r}_1$, which is excluded from the integration in eqn (6.78). The subscript '1' is not indicated on the coordinate appearing in ρ. The two-electron terms involve the coordinates $\mathbf{r}_1$ and $\mathbf{r}_2$ and integration of $\psi^*\psi$ over the coordinates of the remaining electronic coordinates and multiplication by the factor $N(N-1)/2$, as indicated, yields the two-electron density matrix $\Gamma^{(2)}(\mathbf{r}_1, \mathbf{r}_2)$ (eqn (E1.3)). Integration of the density in eqn (6.78) for the final electronic coordinate $\mathbf{r}_1 \equiv \mathbf{r}$ over the basin of the atom yields the basin virial $\mathscr{V}_b(\Omega)$ as indicated in eqn (6.79)

$$\mathscr{V}_b(\Omega) = V_{en}(\Omega) + \sum_\alpha \mathbf{X}_\alpha \cdot \mathbf{F}_\alpha(\Omega) + V_{ee}(\Omega) + V(\Omega, \Omega'). \tag{6.79}$$

The term $V_{en}(\Omega)$ is the attractive energy of interaction between the electronic charge density in the basin of atom Ω and all of the nuclei in the system,

$$V_{en}(\Omega) = - \sum_{\alpha} Z_{\alpha} e^2 \int_{\Omega} r_{1\alpha}^{-1} \rho(\mathbf{r}) \, d\tau. \tag{6.80}$$

The primary contribution to this energy is from the interaction of the density in the basin of atom Ω with its own nucleus, a quantity labelled $V_{en}^0(\Omega)$ where

$$V_{en}^0(\Omega) = - Z_{\Omega} e^2 \int_{\Omega} r_{\Omega}^{-1} \rho(\mathbf{r}) \, d\tau. \tag{6.81}$$

The term $V_{ee}(\Omega)$ is the contribution of the electron–electron repulsion energy to the energy of atom Ω as previously defined in eqn (6.39). It equals the energy of repulsion of the electrons in Ω, $V_{ee}(\Omega, \Omega)$ and one-half of the energy of repulsion between the electrons in Ω with those in the remainder of the system, $V_{ee}(\Omega, \Omega')$,

$$V_{ee}(\Omega) = V_{ee}(\Omega, \Omega) + V_{ee}(\Omega, \Omega'). \tag{6.82}$$

The second region Ω or Ω' indicated in $V_{ee}(\Omega, \Omega')$ refers to the region of integration for electron '2'.

The quantity $\mathbf{F}_{\alpha}(\Omega)$ is the force exerted on the electronic charge density in the basin of atom Ω by nucleus α,

$$\mathbf{F}_{\alpha}(\Omega) = - Z_{\alpha} e^2 \int_{\Omega} (\mathbf{r}_{1\alpha}/r_{1\alpha}^3) \rho(\mathbf{r}) \, d\tau. \tag{6.83}$$

This force is opposite in sign to the force which the charge density exerts on nucleus α and, thus, the sum of $\mathbf{F}_{\alpha}(\Omega)$ over all the atoms in a molecule, i.e. over the total charge distribution, will equal the negative of $\mathbf{F}_{\alpha e}$, the electronic contribution to the Hellmann–Feynman force exerted on nucleus α, eqn (6.60). Thus, the sum of the virials $\mathbf{X}_{\alpha} \cdot \mathbf{F}_{\alpha}(\Omega)$ over all the atoms and all the nuclei in the system will equal a sum of the virials $\mathbf{X}_{\alpha} \cdot \mathbf{F}_{\alpha e}$ over the nuclei. By using eqns (6.59) and (6.58), one obtains

$$\sum_{\alpha} \sum_{\Omega} \mathbf{X}_{\alpha} \cdot \mathbf{F}_{\alpha}(\Omega) = - \sum_{\alpha} \mathbf{X}_{\alpha} \cdot \mathbf{F}_{\alpha e} = \sum_{\alpha} \mathbf{X}_{\alpha} \cdot \nabla_{\alpha} E + \langle \hat{V}_{nn} \rangle. \tag{6.84}$$

When there are no net forces exerted on the nuclei and $E_e = E$, the sum of the virials of the forces which the nuclei exert on the electrons equals the nuclear–nuclear repulsion energy, V_{nn}. Thus, the sum $\sum_{\alpha} \mathbf{X}_{\alpha} \cdot \mathbf{F}_{\alpha}(\Omega)$ may be interpreted as the share of the nuclear–nuclear repulsion energy belonging to atom Ω. This spatial partitioning of an energy between fixed point charges is accomplished by relating it to the virial of a force which is determined by an integration of the charge density over the space of a molecule, an atom at a time.

The final term in eqn (6.79) arises from the projection of the electron–electron repulsive potential energy from $\hat{V}$ by the single-particle operator

$-\mathbf{r}\cdot\mathbf{V}$. In the integration of the final term in the virial density given in eqn (6.78) to obtain $V(\Omega, \Omega')$, there is a cancellation of contributions when the coordinates of both electrons are in Ω and the final expression is

$$V(\Omega, \Omega') = \int_{\Omega} d\tau_1 \int_{\Omega'} d\tau_2 (\mathbf{r}_2\cdot\mathbf{V}_2 - \mathbf{r}_1\cdot\mathbf{V}_1) e^2 r_{12}^{-1} \Gamma^{(2)}(\mathbf{r}_1, \mathbf{r}_2) \quad (6.85)$$

where, as before, Ω' refers to the space occupied by the remainder of the system. This term vanishes for the total system and its vanishing for a partitioning into atomic contributions implies that the energy of repulsion between the electrons in Ω with those in Ω' is equally shared between the two regions. This required symmetry in the interactions between indistinguishable particles is contained in the expression for $V_{ee}(\Omega)$, as detailed in eqn (6.40) and is thus automatically obtained when $V(\Omega, \Omega')$ vanishes.

While the individual contributions to the virial $\mathscr{V}_b(\Omega)$ can be given physical interpretations, care must be exercised in this regard. The value of the total virial $\mathscr{V}(\Omega)$ is independent of the choice of origin used in the definition of the virial operator, but this is not the case for the basin and surface terms treated separately. The origin-dependent terms appearing in the total virial for an atom are

$$\sum_{\alpha} \mathbf{X}_{\alpha}\cdot\mathbf{F}_{\alpha}(\Omega) + V(\Omega, \Omega') + \mathscr{V}_s(\Omega)$$

where $\mathscr{V}_s(\Omega)$ is the surface virial defined in eqn (6.21). Since the value of the total virial is independent of origin, as are the values of the electron–nuclear and electron–electron contributions, $V_{en}(\Omega)$ and $V_{ee}(\Omega)$, respectively, the sum of the three origin-dependent terms is also origin-independent. It has been shown using numerical examples (Srebrenik and Bader 1975) that the same choice of origin which causes the surface integral to vanish (thereby making the functional $\mathscr{G}[\psi, \Omega]$ stationary with respect to the virial as generator) causes $V(\Omega, \Omega')$ to vanish as well. In this case the atom's virial reduces to

$$\mathscr{V}(\Omega) = V_{en}(\Omega) + V_{ee}(\Omega) + V_{nn}(\Omega), \quad (6.86)$$

the virial of the nuclear forces equalling $V_{nn}(\Omega)$, the atom's share of the nuclear–nuclear repulsion energy for a system in an equilibrium configuration. The contributions $V_{en}(\Omega)$ and $V_{ee}(\Omega)$ are well-defined as is $\mathscr{V}(\Omega)$, in terms of either $-2T(\Omega)$ or the stress tensor, and one may use eqn (6.86) to determine $V_{nn}(\Omega)$. For a system in an equilibrium configuration, the sum over the atomic contributions for each term in eqn (6.86) will yield the value of the corresponding quantity for the total system, that is,

$$\sum_{\Omega} \mathscr{V}(\Omega) = \langle \hat{V} \rangle, \quad \sum_{\Omega} V_{en}(\Omega) = \langle \hat{V}_{en} \rangle, \quad \sum_{\Omega} V_{ee}(\Omega) = \langle \hat{V}_{ee} \rangle,$$

$$\sum_{\Omega} V_{nn}(\Omega) = \langle \hat{V}_{nn} \rangle. \quad (6.87)$$

The breakdown of the average potential energy of an atom into its contributions as given in eqn (6.86) is illustrated and discussed in the following section.

6.4 Properties of atoms in molecules

It is the operational essence of the atomic hypothesis that one can assign properties to atoms and groupings of atoms in molecules and on this basis identify them in a given system or use their properties to predict the behaviour of the system in which they are found. The primary purpose of this section is to demonstrate that the quantum atoms transform this atomic hypothesis into an atomic theory of matter by identifying the atoms of chemistry and defining their properties. This section is not a review of applications, but is rather intended to introduce and illustrate the uses of various atomic properties.

The properties of atoms in molecules are calculated using the program with the corresponding acronym **PROAIM**, which was developed over a period of years in the author's laboratory and described in the literature (Biegler-Konig *et al.* 1982). The first step in this series of programs is performed by **EXTREME** which locates and classifies the critical points in a charge distribution. The information from **EXTREME** defines the structure of a molecule and enables one to calculate the trajectories which terminate at the $(3, -1)$ critical points and thence define each of the interatomic surfaces comprising the surface of a given atom. An atomic property, eqn (6.36), is then determined by numerical integration of the corresponding property density, eqn (6.35), over the basin of the atom as defined by its stored surface coordinates. A related program enables one to perform corresponding property integrations over an interatomic surface to evaluate the surface integrals appearing in and derived from the atomic hypervirial theorem, eqn (6.2).

A check on the accuracy of the numerical integrations of the atomic properties is provided by the evaluation of $L(\Omega)$. This quantity, defined in eqn (5.50), is proportional to the integral of the Laplacian of the charge density over the basin of the atom or, equivalently, to the flux in the gradient vector field of ρ through the surface of the atom,

$$L(\Omega) = -(\hbar^2/4m) \int_\Omega \nabla^2 \rho(\mathbf{r}) \, d\tau = -(\hbar^2/4m) \oint dS(\Omega, \mathbf{r}) \nabla \rho(\mathbf{r}) \cdot \mathbf{n}(\mathbf{r}). \qquad (6.88)$$

Because of the quantum boundary condition of zero flux, the value of $L(\Omega)$ should equal zero and the extent to which this is true in any given case is a direct test of how well the atomic surface has been approximated in the integration procedure. The value of $L(\Omega)$ is a measure of the difference in the integrated values of the two equivalent expressions, $K(\Omega)$ and $G(\Omega)$, for $T(\Omega)$, the average kinetic energy of the atom. In favourable cases the values of $L(\Omega)$ fall in the range 1×10^{-4} to 1×10^{-5} au, which represent errors in the estimates of the kinetic energy of less than 0.4 kJ/mol and it is generally possible, using reasonable amounts of computer time, to attain values for $L(\Omega)$ less than 3×10^{-3} au.

There is a practical problem encountered in the application of the theory. This concerns the use of the atomic virial theorem to define the energy of an atom in a molecule. Charge densities are obtained from calculations using basis sets of finite size and, consequently, they do not properly describe the Hellmann–Feynman forces acting on the nuclei. Thus, one finds that the forces do not vanish for the configuration of the nuclei which yields the lowest energy and the requirement of the virial theorem for an equilibrium geometry, that the ratio $\gamma = \langle \hat{V} \rangle / \langle \hat{T} \rangle = -2$, is not satisfied. Under these conditions, the average values of $\langle \hat{T} \rangle$ and $\langle \hat{V} \rangle$ must be multiplied by the factors $(1 + \gamma)$ and $(1 + 1/\gamma)$, respectively, in order to obtain the correct results that $E = -\langle \hat{T} \rangle = \frac{1}{2} \langle \hat{V} \rangle$. Therefore, each atomic kinetic energy $T(\Omega)$ calculated from the wave function, is multiplied by the factor $(1 + \gamma)$ in order that their sum, using eqn (6.75), will equal the negative of the total energy E. The factor $(1 + \gamma)$ is, in general, little different from -1, its correct value for an equilibrium geometry.

6.4.1   Properties determined by the electronic charge density

Pauling (1960) defined electronegativity to be 'the power of an atom in a molecule to attract electrons to itself'. This concept has proved to be extremely useful and it is reflected in the net charges on the atoms found in diatomic molecules. The atomic charges in a diatomic molecule are a direct measure of the relative abilities of the two atoms to attract and bind electronic charge within their basins. The variation in the charge on atom A in AB where A and B vary across the second row of the periodic table, Li → F and including hydrogen are displayed in Fig. 6.2. The densities used in the determination of these atomic charges are obtained from ground-state wave functions that are expressed in terms of a large set of Slater-type basis functions and they are close to the Hartree–Fock limit (see references to Cade and Huo (1973, 1975) and Cade and Wahl (1974) at the end of Chapter 2). Other properties of the atoms and bonds in these molecules are given in the Appendix, as are their internuclear separations and state symbols. Each atom withdraws charge from elements to the left of it and donates charge to those on its right, with H appearing between C and N. The orderings are as anticipated with C and H possessing almost equal electronegativities. As illustrated later, the electronegativity of C relative to H increases with the degree of unsaturation and with the extent of geometric strain. This result is anticipated on the basis of the orbital model which predicts the electronegativity of C to increase as the s character of its hybrid bonds to H increases. Most of the secondary variation in charges across the table are explicable in terms of the extent of charge transfer being limited by either the

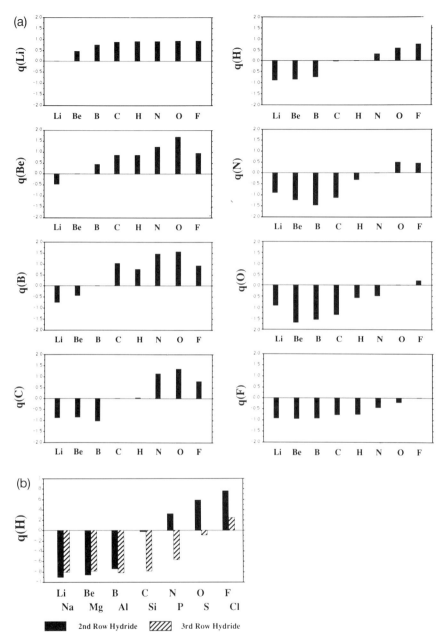

FIG. 6.2. (a) Bar graphs of the charges on the atoms in the ground states of the diatomic molecules AB where both A and B = Li, Be, B, C, H, N, O, F. This is the ordering of increasing electronegativity as determined by theory—all charges to the left of the position of the reference atom are negative, all those to its right are positive. (b) Bar graphs of the charge on hydrogen $q(H)$, for the second- and third-row diatomic hydrides. The reader is referred to the Appendix, Tables A2 and A3, for characterizations of the critical points in these molecules and for a listing of their atomic properties. The results are obtained from near Hartree–Fock wave functions.

number of valence electrons on the donor or vacancies on the acceptor. The charges on the third-row elements Na → Cl are also given relative to H in their hydrides and, as anticipated, H advances towards the electronegative end of the scale in this row relative to its position *vis-à-vis* the second-row elements. Unexpectedly, sodium and magnesium are slightly less electropositive than their second-row congeners.

The charge distributions of the second- and third-row hydrides are illustrated in Fig. 6.3 in the form of contour maps. The extent and direction of charge transfer and its effect on the charge distribution are reflected in the behaviour of the interatomic surface of zero flux which is indicated for each molecule. In LiH the surface envelops what is essentially a Li ion while in HF the total charge distribution is dominated by the forces exerted by the F nucleus (Bader and Beddall 1973; Bader and Messer 1974). The charges $q(H)$ of AH are also characteristic of the stable polyatomic species AH_n, the two values usually differing by less than 0.05 e, and they reflect the chemical behaviour of the hydrides. The hydrides of Li, Be, and B, for example, are all hydridic, expelling molecular hydrogen from water and, for all of these, $q(H) < 0$. There is a sharp break in the value of $q(H)$ for methane for which $q(H) \approx 0$ and this is a faithful reflection of the non-polar nature of this molecule. It has a low solubility in water and does not dissociate. The remaining hydrides, NH_3, H_2O, and HF, are all increasingly polar with $q(H) > 0$ and the ordering of the charges accounts for the aqueous solution of ammonia being basic and that of HF being acidic.

The ability to determine the charge on an atom in a molecule removes the necessity of defining a numerical electronegativity scale. The concept, however, remains useful and one may use the atomic populations to demonstrate that they recover the basic idea underlying electronegativity—to predict the degree of charge transfer between two atoms. Since hydrogen can either donate or accept but a single electron, the electron population of hydrogen in AH may be used to define an electronegativity per electron of A relative to hydrogen. This electronegativity is given by $X(A) = 1 - N(H)$, where $N(H)$ is the population of H in AH. A positive or negative value for $X(A)$ implies that A has a greater or lesser bonding electron affinity than does hydrogen, respectively. If the $X(A)$ are meaningful, then the difference $|X(A) - X(B)|$ should determine the charge transfer per valence electron in AB. Using this concept, the population of A in AB is predicted to be

$$N(A)_{AB} = [N(A)_a - \{X(B) - X(A)\}v]$$

where v is the number of valence electrons on the donor A or the number of vacancies on the acceptor B, whichever value is limiting. $N(A)_a$ is the electron population of the isolated A atom. Examples of predicted and actual atomic populations for A are: NF, 6.56 (6.56); NO, 6.46 (6.50); CF, 5.21 (5.22); CO, 4.78 (4.65); CN, 4.94 (4.88); LiC, 2.12 (2.12). With the added stipulation that the charge transferred per vacancy, $|X(A) - X(B)|$, cannot exceed v, then all diatomic fluoride populations are predicted to within a maximum error of 0.08 electrons. This method of predicting the charge transfer between atoms yields significantly larger errors only for some compounds of the elements Be and B.

The charge distributions of the diatomic hydrides illustrate a general phenomenon—that a significant degree of interatomic charge transfer is accompanied by a polarization of the valence densities of the atoms in a direction counter to that of the charge transfer. The polarizations are in response to the electric field created by the charge transfer, the acceptor atom polarizing towards the positively charged donor atom which is itself polarized away from the negatively charged acceptor. This polarization of the donor atom is particularly pronounced when it possesses a greater number of valence electrons than there are vacancies on the acceptor atom, as illustrated by the first moments (eqn (6.45)) for the atoms in the diatomic hydrides given in Table 6.1. Also listed are the molecular dipole moments and their charge transfer contributions, eqn (6.47). In general, the magnitude of the molecular dipole is less than that of μ_c the charge transfer contribution, because of the opposing atomic polarizations. In some instances, the atomic polarizations determine the direction of the molecular moment. The polarizations of Li and Na in their hydrides are quite small as they correspond to tightly bound core densities. For the second row, the atomic polarizations are largest for the diffuse valence density on Be and B. They are larger still for their third-row congeners and Si where, because of the larger, 10-electron K–L core, the valence density is less tightly bound and more polarizable.

Attempts to assign atomic charges on the basis of measured dipole moments are unrealistic as such a procedure ignores the polarizations of the atomic densities. Such an attempt corresponds to assuming the molecular charge distribution to be composed of a set of spherically symmetric atomic densities, each centred on its own nucleus, a physically unacceptable model even in the limit of an ionic system. It should be evident from a comparison of the charge distribution in the non-bonded region of the A atoms that the reduction in magnitude or reversal in sign of the dipole moment, which occurs after LiH in the second-row and after Na in the third-row, is a consequence of an atomic polarization and is not indicative of a sudden increase in the electronegativity of the A atom. The extent of the physical distortion of those atoms for which the atomic polarizations are greatest is reflected in the values of their non-bonded radii. A *non-bonded radius*, $r_n(\Omega)$, is defined as the axial distance from a nucleus to an outer contour of the charge density on its non-bonded side. The 0.001- or 0.002-au contour is chosen since, as discussed in Section 6.4.1, the corresponding density envelopes provide good approximations to the experimentally determined van der Waals sizes and corresponding radii for molecules in the gas or solid phase,

(a)

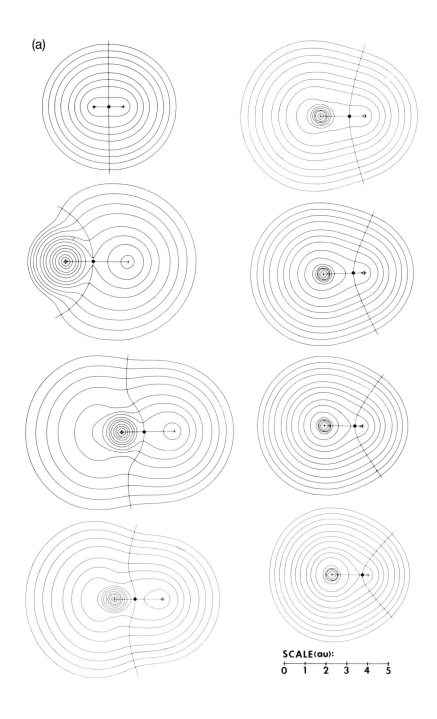

SCALE(au):

0   1   2   3   4   5

(b)

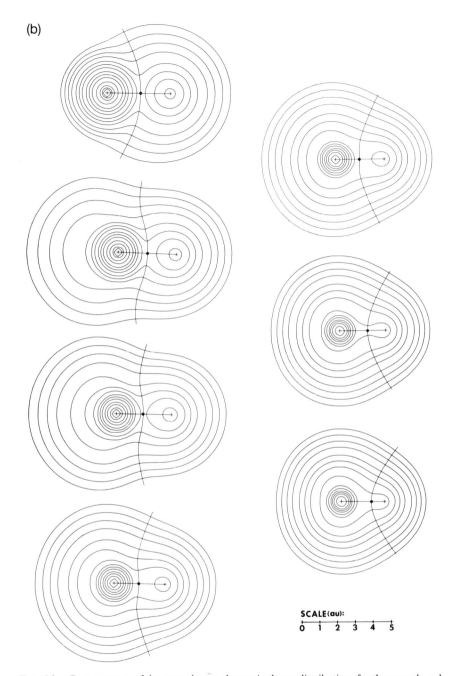

SCALE(au):

0　1　2　3　4　5

FIG. 6.3. Contour maps of the ground-state electronic charge distributions for the second- and thrid-row diatomic hydrides showing the positions of the interatomic surfaces. The first set of diagrams (a) also includes a plot for the ground state of the H_2 molecule. The outer density contour in these plots is 0.001 au. The remaining contours increase in value according to the scale given in the Appendix (Table A2). (a) The left-hand side H_2 $^1\Sigma_g^+$, LiH $^1\Sigma^+$, BeH $^2\Sigma^+$, BH $^1\Sigma^+$; right-hand side CH $^2\Pi$, NH $^3\Sigma^-$, OH $^2\Pi$, HF $^1\Sigma^+$. (b) The left-hand side: NaH $^1\Sigma^+$, MgH $^2\Sigma^+$, AlH $^1\Sigma^+$, SiH $^2\Pi$; right-hand side: PH $^3\Sigma^-$, SH $^2\Pi$, HCl $^1\Sigma^+$.

Table 6.1

*Dipole moments of diatomic hydrides**

AH(X)	$q(A) =$ $Z_A - N(A)$	$\mathbf{M}(A)$	$\mathbf{M}(H)$	$\mu(CT)$	$\mu(AH)$	$r_n(A)$
LiH	+ 0.911	− 0.001	+ 0.387	− 2.747	− 2.361	1.88
BeH	+ 0.868	+ 1.520	+ 0.571	− 2.203	− 0.112	4.68
BH	+ 0.754	+ 1.950	+ 0.493	− 1.761	+ 0.682	4.34
CH	+ 0.032	+ 0.807	− 0.121	− 0.068	+ 0.618	3.86
NH	− 0.323	+ 0.183	− 0.176	+ 0.633	+ 0.640	3.51
OH	− 0.585	− 0.224	− 0.148	+ 1.073	+ 0.701	3.23
FH	− 0.760	− 0.449	− 0.104	+ 1.317	+ 0.764	2.97
NaH	+ 0.810	+ 0.017	+ 0.133	− 2.890	− 2.739	2.64
MgH	+ 0.796	+ 1.701	+ 0.302	− 2.602	− 0.599	5.18
AlH	+ 0.825	+ 2.274	+ 0.360	− 2.568	+ 0.066	4.97
SiH	+ 0.795	+ 1.976	+ 0.428	− 2.285	+ 0.119	4.49
PH	+ 0.579	+ 1.464	+ 0.317	− 1.569	+ 0.212	4.16
SH	+ 0.094	+ 0.587	− 0.009	− 0.239	+ 0.339	3.86
ClH	− 0.241	− 0.006	− 0.103	+ 0.581	+ 0.471	3.68

*A negative value for μ implies the direction A^+H^- for the dipole. The units are atomic units; 1 au = 2.542 D. The nonbonded radius $r_n(A)$ is to the 0.001 au contour of ρ.

respectively. A *bonded radius*, $r_b(\Omega)$, of an atom is correspondingly defined as the distance from its nucleus to the associated $(3, -1)$ or bond critical point. The non-bonded radii for Li and Na are close to the values for the corresponding singly-charged ions while those for the strongly back-polarized atoms are all considerably greater than are their values in the free atomic state (Table 6.1). (The 0.001- and 0.002-au radii of the free atoms are given in the Appendix (Table A5).)

The presence of such a large and diffuse (weakly bound) charge distribution has important chemical consequences, imparting to the molecule the characteristics of a strong Lewis base. A classic example of this behaviour is the carbon atom in the CO molecule (Fig. 6.4). This molecule has a near-zero dipole moment because of very pronounced polarizations of the atomic densities, particularly that of carbon, which oppose the considerable charge transfer moment. The charge on oxygen is $- 1.33\,e$ and the magnitudes of the opposing atomic dipoles are $|\mathbf{M}(O)| = 0.98$ au and $|\mathbf{M}(C)| = 1.72$ au with the non-bonded radius on carbon exceeding its free atomic value by 0.15 au. (Properties of the atoms and the bonds in the molecules shown in Fig. 6.4 are given in Table 7.8.) The physical importance of the atomic polarization of carbon is reflected in the ability of CO to act as a Lewis base, particularly in the formation of metal carbonyls. The considerable difference in the electro-negativities of C and O is reflected in the relatively large dipole moment, $|\mu| = 1.11$ au, of the formaldehyde molecule, $H_2C=O$ (Fig. 6.4). The charge transfer from C to O in formaldehyde where $q(O) = -1.24\ e$, is only slightly less than it is in CO. Unlike CO, however, the charge transfer contribution dominates the final moment in formaldehyde because of the close to halving of the atomic dipole on carbon which results from the use of its non-bonded density in the formation of non-polar bonds to the hydrogen atoms (Table 7.8).

The loss of the non-bonded charge on the carbon atom in CO that occurs in the formation of H_2CO and the accompanying dramatic reduction in its atomic moment is illustrated by the charge distributions of Fig. 6.4. Also illustrated in Fig. 6.4 is the progression of changes in the atomic charge distributions which occur through the series $OC \rightarrow CO_2 \rightarrow SCO \rightarrow CS_2 \rightarrow CS$. The charge on the oxygen atom in CO, CO_2, and OCS varies by only $\pm 0.03\ e$ about the mean value of $-1.30\ e$. When carbon is bonded only to oxygen, its charge loss per oxygen is correspondingly constant. Note the similarity in the distribution of charge density over the oxygen atom and over the bonded portion of the carbon atom (up to a planar surface through the C nucleus, perpendicular to the CO axis) in all of the bonds between C and O shown in Fig. 6.4, even in OCS where the net charge on C is considerably reduced because of a transfer of charge from sulphur. As a consequence, the C–O interatomic surface exhibits the same characteristic shape in all of the molecules and the bonded radii change by only small amounts (Table 7.8). A C–O bond in these molecules is relatively insensitive in its gross features to the neighbouring bond, or lack of it, on carbon.

The electronegativity difference between C and S is less than between O and C and the distribution of charge in a C–S bond shows a wider variation in its properties. The net charge on C in CS_2 is not twice the magnitude found in CS, as it is in the oxygen analogues.

The contribution to an atom's population from orbitals of π symmetry or pseudo π symmetry found in systems with axial or planar symmetry, respectively, is easily determined by the separate integration of their corresponding densities over an atomic basin (Wiberg and Wendolowski 1981). The resulting σ and π populations are frequently found to undergo opposing polarizations. Examples of this in a familiar context are provided by the Hückel π populations predicted for a system of conjugated double bonds as compared to the total atomic populations as determined by theory. The π populations of the carbon atoms in the allyl and pentadienyl cations, starting from a terminal carbon atom, are 0.48 and 0.97 in allyl and 0.63, 1.00, and 0.61 for pentadienyl. These values are very similar to the predicted Hückel populations of $\frac{1}{2}$ and 1 for allyl and 2/3, 1, and 2/3 for pentadienyl. However, the atoms with the smallest π populations bear the smallest net positive

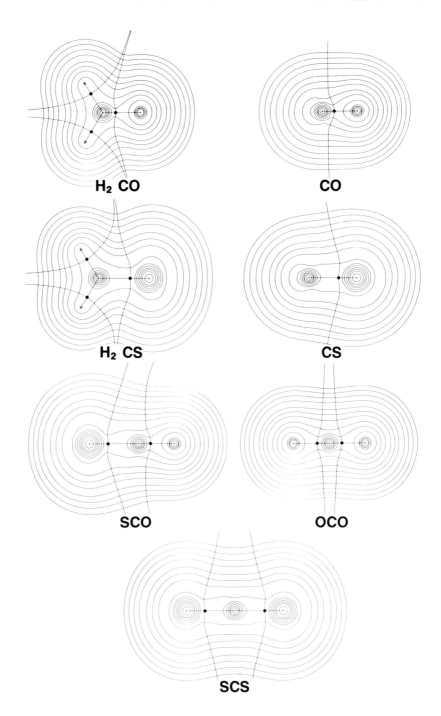

charges because of an opposing polarization of the σ density. The net charges on the atoms in the same order as the π populations given above are $+ 0.09$ and $+ 0.22$ for allyl and $+ 0.09$, $+ 0.16$, and $+ 0.01$ for pentadienyl (see Fig. 6.8). Thus atomic net charges cannot be assigned on the basis of Hückel π populations as is done in models based on an assumed relationship between the π density at a carbon nucleus and the ^{13}C chemical shift observed in nuclear magnetic resonance spectra.

Polarization of the π density is of particular importance when an unsaturated system is bonded to a π electron donor or abstracter group (Libit and Hoffmann 1974; Slee *et al.* 1988). This is illustrated by the polarization of the π density of the carbonyl group as caused by a substituent X bonded to carbon of the carbonyl group in the substituted formyl compounds HXC=O. A π donating group causes the π population of the oxygen atom, rather than that of the carbon atom, to increase while a π withdrawing group causes the π population of oxygen to decrease relative to that for X = H. This is illustrated in Fig. 6.5 which plots the π population on oxygen versus the resonance parameter σ_R^0 of Taft (1956, 1960), which provides an empirical measure of the relative π-donating ($\sigma_R^0 < 0$), π-abstracting ($\sigma_R^0 > 0$) ability of a group.

Examples of opposing polarizations of the σ and π density distributions which result in an alternation in the corresponding atomic populations are provided by the substituted benzenes (Bader and Chang 1989a). This is illustrated in Fig. 6.6 which gives the changes in the π and σ charges on the carbon atoms of the phenyl group ϕ relative to their values in benzene. The amino group is π-donating and, as in the substituted carbonyls, the polarization of the π density in the bonds to the carbon atom bearing the substituent causes its π populations to decrease and that of the atoms attached to it, the ortho carbon atoms, to increase, i.e. the π charge $q_\pi(C)$ becomes more negative. This effect alternates around the ring producing the pattern of charge increase at ortho and para carbon atoms and its decrease at the meta atoms characteristic of ortho–para directing groups. The σ populations change in just the opposite way with the σ charge becoming more negative at the meta carbon atoms. The nitro group is π electron withdrawing and, thus, the π population of the carbon bearing the substituent is increased and the accompanying alternation causes corresponding increases at the meta position and decreases at the ortho and para positions, as is characteristic of meta directing groups. The amino group increases the π population of the phenyl group by $0.084\,e$ and activates the ring towards electrophilic aromatic substitution relative to benzene, while the nitro group decreases the π population by $0.099\,e$ and deactivates the ring.

FIG. 6.4.   Contour maps of the ground-state electronic charge distributions for CO, H_2CO, CO_2, SCO and CS_2, CS, and H_2CS showing the positions of the interatomic surfaces. The outer density contour in these plots is 0.001 au. The remaining contours increase in value according to the scale given in the Appendix (Table A2).

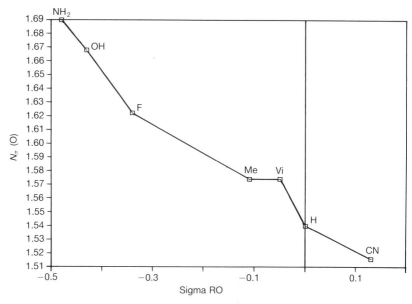

FIG. 6.5. The contribution to the atomic population on oxygen from the orbitals of π symmetry, the population $N_\pi(O)$, plotted versus the Taft's σ_{R_0} parameter for a series of substituted formyl derivatives, XHC=O; Me $\equiv$ methyl, Vi $\equiv$ vinyl.

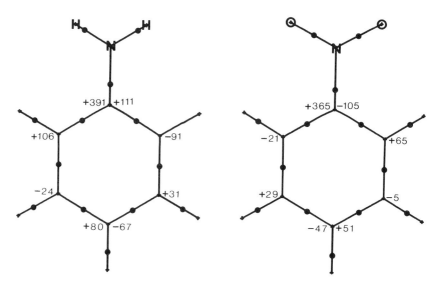

FIG. 6.6. Molecular graphs showing the bond critical points, calculated from theoretically determined charge densities for aniline and nitrobenzene. The numbers to the right of each structure give the changes in the π contribution to the charge on carbon relative to their values in benzene, $\Delta q_\pi(C)$, while those to the left of each structure give the changes in the σ contribution to the charge on carbon relative to their values in benzene, $\Delta q_\sigma(C)$.

The π-donating/withdrawing ability of a substituent X in the substituted phenyls ϕ-X is found to be the same as that observed in the substituted formyl derivatives, XHC=O. The effect of most of the same group of substituents on the charge distribution and moment of the ethyl group has also been studied (Slee *et al.* 1988). The ordering of the total charge withdrawal by X is found to be the same in all three series of molecules.

Information corresponding to the π populations of the orbital model is recovered in the quadrupole polarization of the atomic charge densities, a property of the total charge density. A quadrupolar polarization of the *density* along the z-axis, eqn (6.48), has the form of a d_{z^2} *orbital*, a removal of charge from a plane and its concentration in an axial direction perpendicular to the plane (Fig. 6.1) for $Q_{zz} < 0$. In benzene and ethylene, with the z-axis perpendicular to the plane containing the nuclei, $Q_{zz}(C) = -3.34$ and $= -3.38$ au, respectively, corresponding to the presence of a single π electron. With z taken as the internuclear axis in acetylene, $Q_{zz}(C)$ is large and positive, equal to $+4.14$ au corresponding to a torus-like concentration of π density about the z-axis as reflected in the negative values for $Q_{xx}(C) = Q_{yy}(C) = -2.07$ au. In the planar methyl cation molecule with a nearly vacant p_π orbital, $Q_{zz}(C) = +1.22$ au and the carbon atom density appears as an oblate spheroid (Fig. 6.1) with $Q_{zz} > 0$. This orbital vacancy is partly filled in the tertiary-butyl cation by inductive and hyperconjugative electron release from the methyl groups and the moment $Q_{zz}(C)$ is correspondingly reduced to $+0.43$ au in this molecule.

Figure 6.7 illustrates the close correspondence between the polarization of an atomic density as measured by $Q_{zz}(\Omega)$ and its π population by plotting the

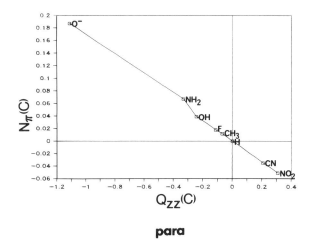

para

FIG. 6.7.   A plot of the π population of the para carbon atom in substituted benzenes, $N_\pi(C)$, versus its z component of the quadrupole moment, $Q_{zz}(C)$.

π population of the para carbon atom in a series of substituted benzenes versus its quadrupolar polarization along the z-axis. Such a near linear relationship is obtained for both the ortho and para positions, but the values for the meta position are not so well correlated (Bader and Chang 1989). A fragment of the benzenium ion (Fig. 6.8), the protonated σ-complex of benzene and an example of the arenium ion intermediate occurring in electrophilic aromatic substitution, is thought to resemble the pentadienyl cation. The benzenium ion intermediate is not planar and does not possess

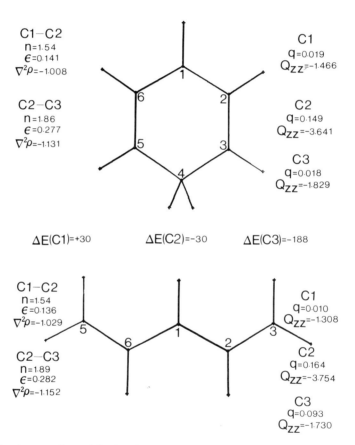

FIG. 6.8. A comparison of the C–C bond and atomic properties of the carbon atoms in the pentadienyl cation with the corresponding fragment (as indicated by the numbering of the atoms) in the benzenium ion, protonated benzene. The bond properties compared are; bond order n, bond ellipticity ε, and the Laplacian at the bond critical point, $\nabla^2\rho_b$. The atomic properties compared are; the net charges on the carbons $q(C)$ and their quadrupole moments $Q_{zz}(C)$. Also given are the differences in energy of the carbon atoms, $\Delta E(C) = E(C)[C_6H_7^+] - E(C)[C_5H_7^+]$ in kJ/mol.

orbitals of 'π' symmetry. However, one may still compare the quadrupolar polarizations of corresponding carbon atoms in the two molecules. A comparison of properties of corresponding atoms and bonds in these two cations is summarized in Fig. 6.8 to illustrate how the theory enables one to obtain a quantitative determination of the degree of similarity in a grouping of atoms as it occurs in two different molecules and, in doing so, to provide the necessary base for understanding the observed differences.

6.4.2   Transferability of atomic properties

The discussion accompanying Fig. 5.1 emphasized that the use of the zero-flux surface for the definition of an atom or functional grouping of atoms maximizes the extent of the transferability of its properties between systems, a characteristic essential to the role of the atomic concept in chemistry. The idea illustrated above, that the benzenium ion can be considered to be the union of a pentadienyl and a methylene group by basing the group recognition on the comparison of its properties in different systems, is an expression of this concept. By defining a group and its properties and enabling one to determine the effect of its presence on the properties of another group, the theory of atoms in molecules parallels the most important of all chemical codifiers, the substituent effect. Studies referred to above have illustrated the ability of the theory to quantify and make understandable the effects of a range of substituents on the properties of the ethyl, formyl, and phenyl groups. It is the purpose of the present discussion to explore the limiting case where a group is transferable with little or no change in its properties and, correspondingly, its perturbation of the remainder of the system is minimized.

It is possible to experimentally measure the energy of an atom in a molecule as an additive contribution to the molecule's heat of formation in those instances where a class of molecules exhibits an additivity scheme for the energy. Essential to the theoretical prediction and understanding of this experimental observation is the property of the atoms expressed in eqn (6.75), that their properties, including their energies, be additive to yield the total property value for a molecule. It is demonstrated here that the energies of the methyl and methylene groups, as defined by theory, predict the *additivity and transferability* of the group energies as is observed experimentally in normal hydrocarbon molecules. Their properties also predict and account for the deviations in this additivity scheme that are observed for small cyclic molecules, deviations which serve to define the strain energy. The ultimate test of any theory is its ability to predict what can be experimentally measured. By appealing to the limiting case of near transferability of atomic properties, one can demonstrate that the atoms of theory are the atoms of chemistry (Bader 1986; Bader *et al.* 1987*b*; Wiberg *et al.* 1987).

The study of the molar volumes of the normal hydrocarbons provided the earliest example of the additivity of group properties (Kopp 1855). The experimentally determined heats of formation for the same homologous series of molecules, $CH_3(CH_2)_mCH_3$, also obey a group additivity scheme (Franklin 1949; Prosen 1946; Benson 1968; Pittam and Pilcher 1972). It is possible to fit the experimental heats of formation for this series, beginning with $m = 0$, with the expression,

$$\Delta H_f^0(298) = 2A + mB, \tag{6.89}$$

where A is the contribution from the methyl group and B that from the methylene group. The generally accepted values for A and B at $25°C$ are -42.34 and -20.63 kJ/mol, respectively. The group enthalpy corrections from 298 to 0 K are additive for the n-alkanes (Wiberg 1984), as are the group zero-point energy corrections (Schulman and Disch 1985). Thus, the calculated energies of the vibrationless molecules in their equilibrium geometries should exhibit the same additivity of the energy as represented by eqn (6.89) and the additivity is indeed mirrored by the single determinantal SCF energies at both the 6–31G*/6–31G* and 6–31G**/6–31G* levels of approximation. (The notation implies that the SCF calculations were performed using the basis set listed first, at the optimized geometry for the basis set listed second; Hariharan and Pople 1972; Krishnan et al. 1980.) The calculated molecular energies E for the n-alkanes satisfy the relationship,

$$E = 2E(CH_3) + mE(CH_2). \tag{6.90}$$

The quantity $E(CH_3)$ is one-half the energy of ethane, equal to -39.61912 au, and $E(CH_2)$ is the energy increment per methylene group equal to -39.03779 au, using the 6–31G**/6–31G* calculated results. These group values fit the calculated energies to within ± 0.00014 au, an average deviation smaller than the experimental one. The calculated results indicate that the corrections to the energy arising from the correlation of the electronic motions, a contribution neglected in a single determinantal calculation, should also obey a group additivity scheme. This indeed appears to be the case as is demonstrated and discussed later.

The distributions of charge for molecules in this series are illustrated in Fig. 1.1 in terms of an outer envelope of the charge density ρ and, specifically for the five- and six-carbon members, in Fig. 6.9 in the form of contour maps of ρ. The latter maps show the bond paths linking the nuclei and indicate the intersection of the interatomic surfaces with the plane of the diagram. The intersection of these same surfaces with the density envelopes are shown in Fig. 1.1 and they define the methyl and methylene groups as envisaged by chemists and as defined by theory. The diagrams show qualitatively what the atomic properties will demonstrate quantitatively, that the methyl and

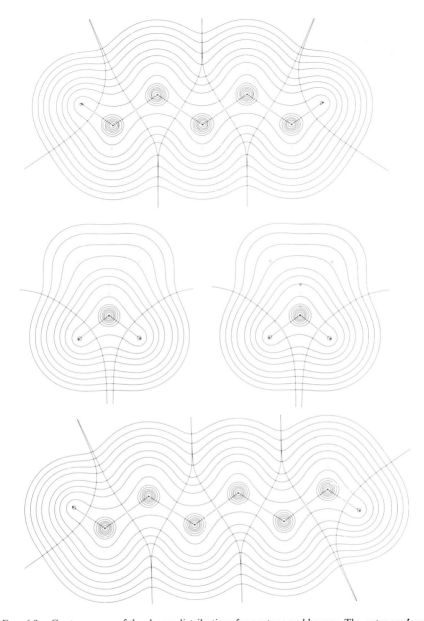

FIG. 6.9. Contour maps of the charge distributions for pentane and hexane. The outer contour in these plots is 0.001 au. The intersections of the interatomic surfaces with the plane of the diagram are indicated, as are the bond paths. The upper and lower diagrams are in the plane containing the carbon nuclei and a terminal proton of each methyl group. The middle diagrams are for the central methylene group in pentane (on the left) and for one of the two equivalent central methylene groups in hexane (on the right) in the plane containing the carbon nucleus and the two protons. The positions of out-of-plane nuclei are indicated by open crosses. The methyl groups, the methylene groups bonded to a methyl group, and the central methylene groups yield superimposable pairs of charge distributions.

Table 6.2

Relative properties of methyl and methylene groups in normal hydrocarbons (*Bader et al. 1987*)

Molecule	Methyl group*		Methylene group†			
	$\Delta N(CH_3)$	$\Delta E(CH_3)$ (kJ/mol)	$\Delta N(CH_2)$		$\Delta E(CH_2)$ (kJ/mol)	
Ethane	0.000	0.0				
Propane	+ 0.017	− 45.6	− 0.034		+ 90.8	
Butane	+ 0.018	− 45.2	− 0.018		+ 45.2	
Pentane	+ 0.017	− 41.4	− 0.018	+ 0.001	+ 45.6	+ 0.0
Hexane	+ 0.018	− 41.4	− 0.018	+ 0.000	+ 44.4	− 4.6

*Differences relative to methyl group in ethane.
†Differences relative to standard methylene group. The second set of entries beginning with pentane are for methylene groups bonded only to other methylenes.

methylene groups in this series of molecules are transferable with little change in their form and, hence, with little change in their properties.

The physical properties of the *n*-alkanes indicate that the molecules are non-polar and this is reflected in the small magnitudes of the net charges of the carbon and hydrogen atoms and of the molecular moments. Hydrogen is slightly more electronegative than carbon in saturated hydrocarbons, and the order of group electron-withdrawing ability in hydrocarbons without geometric strain is $H > CH_3 > CH_2 > CH > C$. In ethane, methyl is bonded to methyl, while in the other molecules of the *n*-alkanes it is bonded to methylene from which it withdraws charge. Table 6.2 lists the populations and energies of the methyl groups for $m > 0$ relative to their values in ethane. To within the accuracy of the numerical integrations of the atomic properties (which in general are $\pm 0.001\ e$ and $\pm 4.0\ kJ/mol$), the energy and population of the methyl group are seen to be constant when it is bonded to a methylene group. Thus the methyl group is essentially the same in all the members of the homologous series past ethane. This transferable methyl group is more stable relative to methyl in ethane by an amount $\Delta E = -43.4 \pm 2.0\ kJ/mol$, and its electron population is greater by an amount $\Delta N = 0.0175\ e$.

The charge and energy gained by the methyl group is taken from the methylene group. What is remarkable, and what accounts for the additivity observed in this series of molecules, is that the energy gained by methyl is equal to the energy lost by methylene. In propane, where the methylene group transfers charge to two methyl groups, its energy relative to the increment in eqn (6.90) is $E(CH_2) - 2\Delta E$, and its net charge is (necessarily) $+ 2\Delta N$ where ΔE and ΔN are the quantities defined above for the methyl group. Thus the *energy as well as the charge is conserved* relative to the group energies defined in eqn (6.90). In butane, a methylene group is bonded to a single methyl group and, correspondingly, its energy is $E(CH_2) - \Delta E$ and its net charge is $+ \Delta N$. The corresponding methylene groups in pentane and hexane, those bonded to a single methyl grup, have the same energies and net charges as a methylene group in butane. Thus, the charge transer to methyl is damped by a single methylene group, and the central methylene group in pentane and the two such groups in hexane (see Fig. 6.9) should have a zero net charge and an energy equal to the increment $\Delta E(CH_2)$. This is what is found to within the uncertainties in the integrated values, their calculated net charges being $0.0005\ e \pm 0.0002\ e$ and the maximum deviation in the energy from the standard value being within the integration error of ≈ 4 kJ/mol or 1 kcal/mol. (It should be borne in mind that the total energy of a methylene group is 10.25×10^4 kJ/mol.) Therefore, methylene groups bonded only to other methylenes, as found in pentane, hexane, and all succeeding members of the series, possess a zero net charge and contribute the standard increment $E(CH_2)$ to the total energy of the molecule. The underlying reason for the observation of additivity in this series of molecules is the fact that the change in energy for a change in population, the quantity $\Delta E/\Delta N$, is the same for both the methyl and methylene groups. The small amount of charge shifted from methylene to methyl makes the same contribution to the total energy.

It is to be emphasized that the energies assigned to the methyl and methylene groups are independently determined by the theory of atoms in molecules (eqn (6.72)). The fact that this assignment leads to an energy for the transferable methylene group equal to the value $E(CH_2)$ in eqn (6.90), an equation which mirrors the experimental additivity of the energy eqn (6.89), confirms that the theoretically defined atoms are responsible for the experimentally measured increments to the heat of formation, and that quantum mechanics predicts the properties of atoms in molecules just as it does the properties of the total molecule. It is a straightforward matter to use quantum mechanics to relate a *spectroscopically* determined energy to the *theoretically* defined difference in energy between two states of a system. In a less direct, but no less rigorous manner, quantum mechanics also relates the difference in the *experimentally* determined heats of formation of butane and pentane to the corresponding *theoretically* defined energy of the methylene group.

The additivity of the energy in the *n*-alkanes is obtained in spite of small differences in group properties, differences which necessarily result from a change in the nature of the bonded neighbour. Thus, there are two kinds of methyl groups: the one unique to ethane and the transferable methyl group which is bonded to a methylene group. There are three kinds of methylene groups: the one unique to propane and two transferable forms, one bonded to a methyl, the other bonded only to other methylene groups. Other properties of these groups exhibit the same pattern of transferable values as do their energies and populations. This is illustrated by the properties listed in Table 6.3. The first moments for the methyl and methylene groups are calculated as for an atom, eqn (6.45), with the integration extending over the basins of all the atoms in the group and with the carbon nucleus as origin. The magnitude of the moment for the methyl group is unique for ethane but nearly constant for the remaining molecules. The methylene groups exhibit two transferable values, one for the methylene attached to a single methyl and another, slightly smaller, for a methylene attached only to other methylenes. In all cases the hydrogen atoms form the negative end of the group moment. The directions of these moments are also transferable. The moment of the methyl group is parallel to the bond path linking it to the neighbouring carbon nucleus to within $0.5°$ through the series. The direction of the methylene moment, in those cases where it is not dictated by symmetry, lies $0.05°$ off the axis bisecting the HCH angle in pentane and $1.1°$ and $1.2°$ off this axis in butane and hexane, respectively. These group moments along with the group charges can be used, as are the atomic values in eqn (6.46), to determine the total dipole moment of a hydrocarbon molecule.

The volumes of the methyl and methylene groups reflect the same pattern of transferable values. The volume of the methyl group in ethane is slightly larger than that of the methyl groups in the succeeding members of the series whose volumes vary by only ± 0.02 au from their average value. The volume of the methylene group bonded to two methyls in propane is slightly greater than the transferable volume of a methylene bonded to one methyl. The methylene group bonded only to other methylenes also possesses a characteristic volume.

Also listed in Table 6.3 are the group values for the electronic correlation energy calculated from the density functional expression of Langreth et al. (1983; Hu and Langreth 1985). It is a complicated functional of the charge density $\rho(\mathbf{r})$, involving powers and gradients of $\rho(\mathbf{r})$. The values of this quantity for the methyl and methylene groups listed under the heading ε exhibit the same pattern of values as do the other group properties. The values of ε for the transferable methyl and methylene groups exhibit variations which are remarkably small, particularly when one bears in mind the complicated form of the functional. While no functional for the correlation energy which yields values of chemical accuracy yet exists, these results indicate that each of the transferable methyl and methylene groups should make a characteristic and essentially constant contribution to the total correlation energy of a normal hydrocarbon molecule, a result anticipated on the basis of the ability of SCF calculations to recover the experimental additivity of the energy.

Table 6.3
Properties of methyl and methylene groups in normal hydrocarbons (Bader et al. 1987)

Molecule	Methyl group			Methylene group*		
	$|M(CH_3)|^†$ (au)	$\varepsilon^‡$ (au)	Molar volume (cm³)	$|M(CH_2)|^†$ (au)	$\varepsilon^‡$ (au)	Molar volume (cm³)
Ethane	0.253	− 0.3456	19.77			
Propene	0.244	− 0.3484	19.73	0.307	− 0.3180	14.18
Butane	0.243	− 0.3486	19.70	0.305	− 0.3208	14.12
Pentane	0.249	− 0.3486	19.71	0.310 0.304	− 0.3210 − 0.3238	14.08 13.98
Hexane	0.245	− 0.3485	19.74	0.306 0.300	− 0.3208 − 0.3237	14.09 14.03

*The second set of entries beginning with pentane are for methylene groups bonded only to other methylenes.
†Calculated with carbon nucleus as origin.
‡Correlation energy from a density expression.

Table 6.4 lists the individual contributions to the potential energies, as defined and discussed in eqn (6.86), of the carbon and hydrogen atoms of the methyl groups in propane to hexane, all relative to their values in hexane. These data illustrate a very important property of an atom—that it responds only to the total force exerted on it and not to the individual contributions to this force. If it were not for this property of the atom and its energy being determined by some net field (the virial of the Ehrenfest force, eqns (6.30) and (6.78)) rather than by the individual contributions to this field, contributions which as illustrated here differ markedly even for members of a homologous series, there would be no chemically recognizable atoms or functional groups.

The remarkable constancy in the total properties of these atoms in the different molecules is reflected in the values of their populations $N(\Omega)$, of their first and second moments $r(\Omega)$ and $r^2(\Omega)$ as defined in eqn (6.44), and their electronic kinetic energies $T(\Omega)$. If the distribution of charge is the same for each type of atom, then the quantity $V_{ne}^0(\Omega)$, the potential energy of interaction of the nucleus of atom Ω with its charge distribution (eqn (6.81)), should also be the same. This is indeed the case, the largest difference being 1.7 kJ/mol for a hydrogen atom and 12.1 kJ/mol for a carbon atom. The interaction of all the nuclei in the molecule with the charge density in atom Ω, the quantity $V_{ne}(\Omega)$ defined in eqn (6.80), increases by very large amounts with the removal of each CH_2 group, changing by 6.3×10^3 kJ/mol for H and by 38.5×10^3 kJ/mol for C between the first and final members of the series. The electron–electron and nuclear–nuclear repulsion energies, $V_{ee}(\Omega)$ and $V_{nn}(\Omega)$, respectively, decrease by large amounts with the removal of each methylene group. The sum of the three contributions to give $\mathscr{V}(\Omega)$, the total potential energy of the atom, is, however, the same for each kind of atom through the series. This must be, since $\mathscr{V}(\Omega)$ is equal to twice the total energy of the atom by the virial theorem, eqn (6.73). It is the virial of all the forces acting on atom Ω. This quantity, along with the kinetic energy of the atom, is conserved when the atom's distribution of charge remains unchanged. It must be considered remarkable that a methyl group with a total energy in excess of 100 000 kJ/mol, can be transferred between molecules—in reality and in theory—with changes in its energy of approximately 4 kJ/mol. It is still more remarkable when it is recalled that the individual contributions to the energy of a carbon atom change by 8000 to 20 000 kJ/mol between members of the series.

It has been demonstrated that atoms are indeed pieces of matter in real space with the properties anticipated in Section 1.2. A consequence of the necessity of two identical pieces of matter possessing the same properties as stated there, is that one may use symmetry to define the energy of groupings of atoms. The energy of an ethyl group is for example, one-half the energy of n-butane as demonstrated above. Similarly, the energy of the 2-methyl propyl group will be given by one-half the energy of 2,5-dimethyl hexane. Using the

Table 6.4

Properties of H and C atoms in CH_3 relative to CH_3 in hexane* (Wiberg et al. 1987)

Atom	Molecule	$N(\Omega)$, e	$r(\Omega)$	$r^2(\Omega)$	$T(\Omega) = -E(\Omega)$	$V^0_{en}(\Omega)$	$V_{en}(\Omega)$	$V_{ee}(\Omega)$	$V_{nn}(\Omega)$	$V(\Omega) = 2E(\Omega)$
H_1	Hexane	1.025	1.135	1.649	0.6266	−1.3102	−9.3780	4.3442	4.3503	−1.2532
	Pentane	0.000	0.000	−0.001	+1.3	−1.7	1491.1	−747.7	−746.8	−2.5
	Butane	0.000	0.000	+0.001	+1.7	−1.7	3313.3	−1659.0	−1659.4	−3.3
	Propane	0.000	0.000	+0.001	−0.4	+0.0	5619.1	−2810.4	−2808.3	+0.8
H_2	Hexane	1.028	1.139	1.653	0.6278	−1.3125	−10.5354	4.6370	4.6429	−1.255
	Pentane	0.000	−0.001	−0.003	−0.4	−0.4	1595.8	−798.3	−798.3	+0.8
	Butane	0.000	−0.001	−0.003	+0.4	−1.3	3740.9	−1864.8	−1877.8	−0.8
	Propane	0.000	−0.001	−0.003	−1.3	+1.3	6310.7	−3155.2	−3153.1	+2.5
C	Hexane	5.936	5.984	9.029	37.7494	−89.5288	−138.1137	37.6551	24.9598	−75.4988
	Pentane	−0.001	−0.006	−0.022	−5.4	−11.3	9763.8	−4880.2	−4872.3	+11.3
	Butane	−0.001	−0.002	−0.004	+0.4	−0.4	22380.2	−11182.6	−11198.1	−0.8
	Propane	+0.001	+0.004	+0.009	+1.7	−12.1	38314.1	−19164.0	−19153.5	−3.3

* The energies for hexane are given in au, and for the other compounds the differences with respect to hexane are given in kJ/mole. The values of $r(\Omega)$ and $r^2(\Omega)$ are in au.

demonstrated additivity of the zero point energies and contributions to the heats of formation from 0 to 298 K, together with the observation that minor perturbations in saturated hydrocarbons are damped by a single methylene group, one can determine the heats of formation of these groups in the same manner and predict that their sum should equal the heat of formation of 2-methyl pentane. The value of -173.6 kJ/mol predicted in this manner equals the experimental value (Keith 1990).

Theory predicts what can be observed and performed in the laboratory. Therefore, it should be possible to use theory to construct a molecule in the same manner that this is done in the laboratory. Chemists do not begin a synthesis starting with separate beakers of nuclei and electrons, corresponding to the theoretical description of a molecule starting from the Hamiltonian, but rather from pieces of other molecules. The first requirement in developing a theoretical parallel to the experimental synthesis is the identification of the atoms and functional groups of chemistry, coupled with the definition of their properties, for it is these groups and their properties that are manipulated and linked together in a synthesis. The second requirement is that one be able to predict the changes in the properties of a group when it is transferred from one molecule to another. The first requirement is fulfilled by the theory of atoms in molecules, the initial step in the development of the quantum mechanics of an open system. The second requirement will be met with the development of the requisite perturbation theory for the quantum mechanics of an open system.

The synthesis of a molecule and a prediction of its properties from theoretically defined atoms in real space is already possible in those cases where, as exemplified above for the normal hydrocarbons, groups can be transferred between molecules with negligible changes in their distributions of charge and hence in their properties. One can take advantage of the observation that functional groups frequently appear to undergo only minor perturbations when transferred between corresponding environments, as occurs in the repeating units of a peptide, for example. It should be possible in such cases to construct a molecule from the atoms of theory and predict its properties with acceptably small errors. This has been demonstrated by the synthesis of the simplest of dipeptides glycylglycine, from the appropriate fragments of formylglycine and glycylamine (Chang and Bader 1990). The glycyl groups are defined by cutting glycylamine and formylglycine along their C–N interatomic surfaces as indicated by the vertical bars in the formulae $H_2NCH_2C(=O)|NH_2$ and $HC(=O)|NHCH_2CO_2H$, and the dipeptide is synthesized by joining the free sides of the interatomic surfaces of the C and the N atoms to yield $H_2NCH_2C(=O)|NHCH_2CO_2H$. The surfaces are joined and the geometry of the dipeptide is determined by the anti-parallel linking of the bond path vectors from the C and the N nuclei to the $(3, -1)$ critical points of the C–N bonds in the reactant groups. This results in a near perfect matching of the two reactant C–N interatomic surfaces and in a predicted C–N bond length which is in error by only 0.001Å. This construction yields a molecular graph that is superimposable on the graph for the dipeptide, correctly predicting that the geometrical parameters and the charge distributions of the fragments remain almost unchanged when combined to form the dipeptide. As a consequence, the atomic properties, including the net charges, moments and energies of the atoms, can be transferred with only small error in this synthesis. The total energy of the dipeptide predicted by summing the energies of the atoms in both fragments is in error by only 12 kJ/mol. In addition, the nonbonded shape or van der Waals envelope of the dipeptide as determined by the 0.001 au envelope of the charge density is also accurately reproduced, as is its "reactive surface", a quantity determined by the Laplacian of the electronic charge distribution and discussed in Section 7.3.2.

The functional groupings which serve as the building blocks in such a theoretical synthesis can be determined using relatively large sets of basis functions, with or without correlative corrections. Access to a bank of these theoretically defined building blocks would enable one to synthesize a macromolecule and obtain operationally useful predictions of its properties. Equally if not more important, is the ability to use the same method of synthesis to predict the properties of just that portion of a macromolecule that is of interest by terminating the synthesis at the interatomic surfaces which define the relevant fragment. With further development of the quantum mechanics of an open system, these building blocks would serve to provide the zeroth-order approximation to the system of interest, theory enabling one to take into account the perturbations arising from their coupling in a new environment. It is the goal of the theory of atoms in molecules to do with atoms of theory everything that one does with atoms in the laboratory.

6.4.3   The origin of strain energy in cyclic hydrocarbon molecules

The hybridization model predicts that the smaller bond angles found in a molecule with angular strain should result in an increase in the p character of the strained C–C bonds and hence in an increase in the s character of the associated C–H bonds (Walsh 1947). Orbital models relate an increase in electronegativity of a carbon atom relative to that of a bonded hydrogen to an increase in the s character of its bonding hybrid orbital. Thus, it follows that the presence of geometric strain in a hydrocarbon molecule should result in an increase in the electronegativity of carbon relative to hydrogen. In their classic study of strain in the cyclopropane molecule, Coulson and Moffitt (1949) emphasized this point by showing that the bond lengths and bond angles of the methylene group in cyclopropane resemble those for ethylene.

The argument for an increase in electronegativity with increasing s charcter is based on energy, an s electron being more tightly bound than a p electron. The theory of atoms in molecules shows that the electronegativity of a carbon atom does indeed increase and its energy decrease as the extent of geometric strain increases.

Relative to its population in the standard methylene group, each hydrogen in cyclopropane transfers 0.045 e to carbon, reducing the net charge on the carbon atom from $+ 0.196$ to $+ 0.106$ e (Bader et $al.$ 1987a). While this charge transfer leads to an increase of 65.3 kJ/mol in the stability of the carbon atom, it results in a decrease of 52.3 kJ/mol in the stability of each hydrogen atom. Thus, the methylene group in cyclopropane is calculated to be 39.3 kJ/mol less stable than the standard transferable methylene group. This yields a total strain energy for the molecule three times this, or 118.0 kJ/mol, in good agreement with the generally accepted value based on the experimental heats of formation of 115.1 kJ/mol. The methylene group in cyclopropane is more stable than the same group in ethylene by only 8.8 kJ/mol. In terms of the charge transfer within the group and its energy, it resembles more closely the ethylene fragment (where $q(C) = + 0.080$ e) than it does the standard methylene group.

In the less strained cyclobutane, the transfer of charge from hydrogen to carbon relative to the populations in the standard methylene group is reduced to 0.014 e for axial H and 0.012 e for the other, and the charge on carbon is $+ 0.170$ e. The hydrogens are destabilized by 37.7 kJ/mol and the carbon stabilized by only 10.5 kJ/mol, to give an energy increase of 27.2 kJ/mol for each methylene group. This yields a predicted strain energy of 108.8 kJ/mol, a value which again is in agreement with the experimental value of 110.9 kJ/mol. Experimentally, the heat of formation of cyclohexane is found to be six times the heat of formation of the standard methylene group and to possess no strain energy. An axial hydrogen in this molecule is calculated to possess 0.007 more electrons, ($N(H) = 1.099$ e) and be more stable by 6.7 kJ/mol than an equatorial hydrogen. The atomic populations and energies of a methylene group in cyclohexane differ little from their values in the standard group and the energy of the group differs by only 0.0001 au or 0.25 kJ/mol from the standard value. Thus, in agreement with experiment, cyclohexane is predicted to be strain-free when its energy is compared with six times the energy of the standard methylene group.

When a carbon atom is subjected to an increasing amount of geometric strain, as measured by the departure of the C–C–C bond path angles from the tetrahedral value, an increasing amount of charge is transferred from H to C in each CH_2 group relative to the populations in the standard methylene group. The same effect occurs to an even greater extent for the bridgehead carbon atoms in the bicyclic and propellane molecules. The bridgehead carbon atom in [1.1.0]bicyclobutane, the most strained of the bicyclic molecules, withdraws charge from the bridging methylene groups and from its bonded hydrogen and possesses a net negative charge of -0.121 e. This carbon atom is only ≈ 42 kJ/mol less stable than a carbon atom in acetylene which has a net charge of -0.177 e. The bicyclics and propellane molecules are among the most geometrically strained of the hydrocarbon molecules. As the number of bridging methylene groups in these molecules is decreased, the bridgehead carbon atoms increase in both electronegativity and stability. There is a physical flow of charge from the peripheral methylene groups to the bridgehead carbons in the bicyclics and in the propellanes as the geometrical strain in these molecules is increased. These redistributions of charge within a methylene group or between CH_2 and a C–H or a C group, lead to a stabilization of the carbon atom which, in the classical sense, is most strained and to an even greater destabilization of the hydrogen or neighbouring methylene groups. The overall result is a strain energy equal to the increase in the energy of the group relative to that of the standard. Thus, theory provides an atomic explanation of the origin of strain energy in cyclic hydrocarbons. The reader is referred to the paper by Wiberg *et al.* (1987) for a full discussion of this topic. The few examples discussed here are introduced to emphasize that the strain energies calculated for cyclopropane and cyclobutane and the predicted absence of strain in cyclohexane, all of which are in agreement with experiment, are all predicted by the theory of atoms in molecules. The energy, as defined in eqn (6.72), of the standard transferable methylene group, as defined by the zero-flux boundary condition (eqn (6.7)) and as found in the pentane and hexane molecules, serves as the basis for the determination of these results. Not only does theory predict the transferability of atoms and groupings of atoms without change, it also correctly predicts the measured changes in their energies when these groups are perturbed.

While the excellence of the agreement of the relative energies of the methylene group in the cyclic molecules with the measured strain energies may be to some extent due to the fortuitous cancellation of errors in the contributions not specifically considered, namely the correlation energy, the zero-point energy, and $\Delta(\Delta H_f^0)$ between 0 and 298 K, the nature of the results leaves no doubt as to the correctness of the interpretation that has been given, that the atoms of theory recover the experimentally measured properties of atoms in molecules.

6.4.4   Origin of rotation and inversion barriers

In addition to predicting a total energy, quantum mechanics enables one to calculate the average values of its kinetic and potential energy contributions, the latter consisting of the nuclear–electron attractive potential energy (V_{ne}), and the electron–electron and the nuclear–nuclear repulsive energies (V_{ee} and V_{nn}, respectively). These energies, along with the contribution to the potential energy of the electrons arising from the virials of the external forces exerted on the nuclei (eqn (6.58)), are the only energy quantities defined in terms of the usual fixed-nucleus Hamiltonian in the absence of external fields. Model-independent predictions and interpretations of energy changes are, therefore, restricted to these quantities and the atomic contributions to these quantities. Since the theory of atoms in molecules enables one to calculate all well-defined mechanical properties at the atomic level, new emphasis is placed on the study of these energies.

The electron–nuclear interaction energy V_{ne} is the only attractive inter-action in a molecular system and it is the decrease in the potential energy resulting from this interaction that is responsible for the formation of a bound molecular state from the separated atoms. Because of the virial theorem and in the absence of external forces acting on the nuclei, the total energy E equals $\frac{1}{2}V$ (eqn (6.64)) where $V = V_{ne} + V_{ee} + V_{nn}$ is the total potential energy. Similarly, the change in energy ΔE between two states free of external forces equals $\frac{1}{2}\Delta V$. Energy changes associated with relatively large reductions in internuclear separations, such as those encountered in the formation of molecules from atoms, lead to a decrease in the attractive potential energy and to increases in the electron–electron and nuclear–nuclear energies of repulsion. The attractive potential energy, however, does not necessarily decrease when the total energy decreases for a rearrangement of atoms within a molecule. Thus, it is possible for the total energy to decrease and for a system to become more stable in spite of an increase in the attractive energy, because of an even larger reduction in the energies of repulsion. Correspond-ingly, it is possible for the energy to increase and for a system to become less stable in spite of a decrease in the attractive interactions.

One anticipates that the changes in the electron–electron and nuclear–nuclear repulsions will in general, parallel one another. One fre-quently finds, as anticipated on the basis of simple arguments, that the changes in the repulsive energies V_{ee} and V_{nn} are nearly equal and that each is approximately one-half the magnitude of the change in the attractive poten-tial energy. Consider the formation of AB from neutral atoms A and B with internuclear separation R. The new interactions without charge transfer are: nuclear repulsion $= Z_a Z_b / R = N_a N_b / R$, which is also the approximate increase in the electron repulsion energy, where each of these is one-half the magnitude of the approximate nuclear–electron energy change of $-(Z_a N_b / R + Z_b N_a / R)$. The same near cancellation occurs when there is a transfer of charge A → B, but the change in the internal energies to the same order, $-I_A + A_B$ (where I and A are the ionization potential electron affinity, respectively) should be included. Thus, whether a change in state is accompanied by an increase or decrease in the total energy is determined by the difference in two larger, competing energy changes of approximately equal magnitude.

Denote the electron–nuclear attractive potential energy as V_a and the sum of the electron–electron and nuclear–nuclear repulsive energies as V_r. We shall term Case I the instance where a decrease in total energy E is a result of a decrease in V_a and a smaller increase in V_r (see Fig. 6.10). As noted above, this is generally the case for a process accompanied by significant decreases in internuclear separations such as forming a molecule from separated atoms. The equivalent of Case I for the reverse process, is an increase in energy accompanying an increase in internuclear separations. Table 6.5 lists values of ΔV_a and ΔV_r for the dissociation of N_2, BF (which is isoelectronic with N_2), and LiF obtained from CI calculations which give dissociation energies close to the observed values. The magnitudes of ΔV_{nn} and ΔV_{ee} are nearly equal and each is approximately one-half of ΔV_{ne}, but overall the increase in V_a exceeds the decrease in V_r and $\Delta E > 0$. The value of V_{nn} is larger for N_2 than for BF since $Z_a Z_b$ is a maximum when $Z_a = Z_b$.

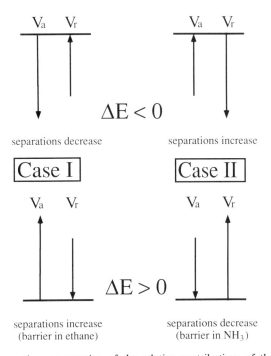

FIG. 6.10. Diagramatic representation of the relative contributions of the changes in the attractive (V_a) and repulsive (V_r) energies to barriers characterized by an increase or decrease in internuclear separations.

Table 6.5

Dissociation, rotation, and inversion barriers

Molecule	E (au)	$-V/T$	ΔE (kJ/mol)	ΔV_a (au)	ΔV_r (au)	ΔV_{nn} (au)	ΔV_{ee} (au)	ΔE(expt) (kJ/mol)
Dissociation								
N_2	−109.29881	1.99874	771	48.4820	−47.6421	−24.3126	−23.3295	957
BF	−124.43859	2.00073	666	38.0934	−37.4888	−19.0998	−18.3890	762
LiF	−107.22269	2.00088	529	20.6937	−20.2494	−9.1426	−11.1068	576
Internal rotation								
C_2H_6	−79.25733	2.00026	12.6	0.3307	−0.3217	−0.1684	−0.1533	12.3
CH_3NH_2	−95.25167	2.00034	8.4	0.1613	−0.1557	−0.0838	−0.0719	8.3
CH_3OH	−115.08644	2.00054	4.2	0.0926	−0.0901	−0.0490	−0.0411	4.5
Inversion								
NH_3	−56.43695	1.99950	22.6	−0.2047	0.2229	0.1187	0.1042	24.3
PH_3	−342.64992	1.99997	143.1	−0.4946	0.5979	0.2549	0.3431	131.8
H_3O^+	−76.56923	1.99989	8.8	−0.0995	0.1086	0.0753	0.0333	
H_2O	−76.29326	2.00008	138.1	−0.2245	0.3375	0.2143	0.1232	

Following similar arguments, one would expect Case II, where the decrease in energy results from a decrease in the repulsive contributions to ΔE, to be characteristic of processes dominated by increases in internuclear separations of bonded atoms or, equivalently, for the reverse process, an energy increase resulting from an increase in repulsive interactions because of dominant decreases in internuclear separations. Considered from the point of view of $\Delta E > 0$, Case I is exemplified by barriers to internal rotation and Case II by barriers to inversion. Payne and Allen (1977) have discussed various combinations of energy components in the classification of barriers as 'repulsive' or 'attractive' dominant.

We shall discuss the barrier to internal rotation in ethane as an example of Case I and the inversion barrier in NH_3 and the barrier to bending in H_2O, as examples of Case II (Bader *et al.* 1990). Calculations at the single-determinantal level using the $6-311++G(2d, 2p)$ basis set for both geometry optimization and energy determination were performed for all three molecules. Such a calculation is known to recover the barrier to rotation in ethane, as evidenced by a comparison of the calculated and experimental results reported in Table 6.5. The results reported for ammonia and water are obtained from configuration interaction calculations using all single and double configurations generated from the orbital set obtained using the large set of basis functions.

In general, the dominant geometry change encountered in the formation of a rotationally eclipsed conformer from a staggered one is an increase in the length of the bond about which rotation occurs. In ethane the C–C separation increases by 0.0140 Å in forming the eclipsed conformer and this increase is more than 10 times larger than the accompanying decreases in the C–H separations of 0.0010 Å. The CCH bond angles open by 0.4° and there is an overall decrease of 0.0069 Å in the separations of the hydrogens on a given carbon. Reference to Table 6.5 shows that, as anticipated on the basis of the general arguments, the rotational barrier is a result of a decrease in the repulsive potential energy and an even larger increase in the attractive potential energy (a decrease in its magnitude).

The increase in the C–C separation in ethane dominates the changes in V_{nn} and V_{ee}, both these quantities undergoing decreases of corresponding magnitude when the staggered (S) is rotated into the eclipsed (E) conformer (Table 6.5 and Fig. 6.11). The same geometrical increase results in a lessening of the attractive interactions within the molecule, with the increase in V_a exceeding the decrease in the repulsive interactions. Thus, the barrier to internal rotation in ethane results from a reduction in the attractive interactions between the nuclei and the electrons and in spite of an accompanying reduction in the electron–electron and nuclear–nuclear repulsions. The energy changes which give rise to the rotation barrier are characteristic of Case I and represent a small reversal of the process of C–C bond formation.

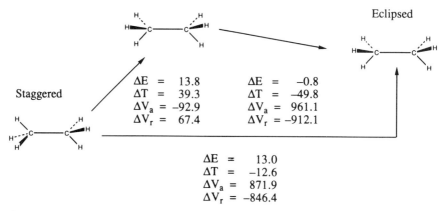

Eclipsed

$$\Delta E = 13.8 \qquad \Delta E = -0.8$$
$$\Delta T = 39.3 \qquad \Delta T = -49.8$$
$$\Delta V_a = -92.9 \qquad \Delta V_a = 961.1$$
$$\Delta V_r = 67.4 \qquad \Delta V_r = -912.1$$

Staggered

$$\Delta E = 13.0$$
$$\Delta T = -12.6$$
$$\Delta V_a = 871.9$$
$$\Delta V_r = -846.4$$

FIG. 6.11.   Energy changes in kJ/mol for the rotation of the staggered conformer of ethane (S) into the eclipsed conformer (E) and into the frozen eclipsed conformer (FE).

The changes in the charge distribution of ethane caused by the internal rotation are relatively small (Fig. 6.12). The principal change is the replacement of the curved C–C interatomic surface in the S conformer with the planar one in the E conformer, a consequence of the replacement of a threefold alternating axis of symmetry with a simple threefold axis. With reference to Fig. 6.12, one sees that the upper arm of the atomic surface of an eclipsed hydrogen atom is moved inwards relative to its form in the S conformer and its volume, which equals 38.44 au in the S geometry, is slightly decreased, by 0.24 au. The remainder of the basin of a hydrogen atom is nearly unperturbed by the rotation. The changes in ρ at the bond critical points are small, being largest for the C–C bond for which the value of ρ_b decreases by 0.0071 au from a value of 0.2495 au in the S conformer. Its value for the C–H bond increases by 0.0008 au from its value of 0.2908 au. The C–H bond critical point shifts by 0.0017 au towards the proton in the eclipsed geometry, a harbinger of the small transfer of charge from H to C which accompanies the rotation. The C–H bonds in ethane exhibit a small ellipticity with major axes directed so as to be tangent to the cone obtained by rotation of the C–H bonds of a methyl group about the C–C axis. This ellipticity approximately doubles in value to 0.014 when the group is rotated into the eclipsed geometry. Thus, the extent to which electronic charge is accumulated in the tangent plane containing the C–H bond as opposed to the one perpendicular to it, while small, is increased when the molecule attains the staggered geometry.

    The atomic contributions to the energy changes are given in Table 6.6. The changes in the energies of the individual atoms are small, a reflection of the

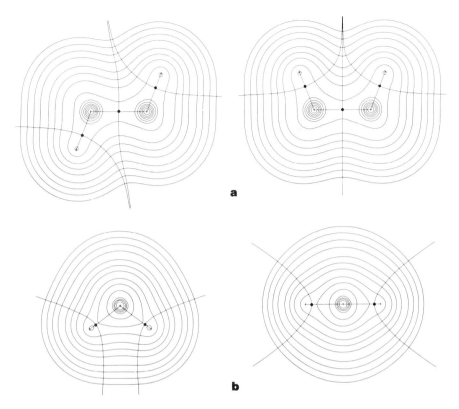

Fig. 6.12. Contour maps of the electronic charge density for the staggered and eclipsed conformers of ethane (a) and for the bent and linear conformers of water (b). The outermost contour has the value 0.001 au. The remaining contours increase in value according to the scale given in the Appendix (Table A2). The bond paths, the interatomic surfaces, and the bond critical points are also indicated.

small perturbations of the atomic charge distributions which accompany the internal rotation. The net charge on a H atom in the staggered form is $q(H) = -0.061\ e$. On changing to the eclipsed geometry, there is a transfer of 0.003 e from each hydrogen to its bonded carbon atom. Each hydrogen atom is stabilized by a small amount in the eclipsed geometry in spite of the small decrease in its population. Each carbon atom is destabilized by an amount approximately three times the magnitude of the stabilization of the hydrogens bonded to it and this is the source of the barrier at the atomic level. The transfer of charge to carbon leads to a decrease in $V_a^0(C)$, the potential energy of interaction of the carbon nucleus with its own charge distribution eqn (6.81), with $\Delta V_a^0(C) = -35.6$ kJ/mol. The energy change for a carbon atom, however, is dominated by the increase in the C–C separation. This increase,

Table 6.6

Atomic properties for dissociation, rotation, and inversion barriers

Molecule	Ω	$E(\Omega)$ (au)	$q(\Omega)$ (au)	$\Delta E(\Omega)$ (kJ/mol)	Barrier* (kJ/mol)	$\Delta N(\Omega)$ (au)	$\Delta V_a(\Omega)$ (au)	$\Delta V_r(\Omega)$ (au)	$\Delta V_a^0(\Omega)$ (au)	$r_b(H)^\dagger$ (au)	$\Delta r_b(H)$ (au)
N_2	N	-54.64940	0.000	385.8	771.1	0.000	24.0774	-23.7836	1.4239		
BF	B	-24.21290	0.902	-1001.2	669.9	0.902	10.8928	-11.6554	-2.2924		
	F	-100.22582	-0.902	1671.1		-0.902	27.1239	-25.8508	5.7786		
LiF	Li	-7.36270	0.932	-182.0	529.3	0.932	5.4975	-5.6362	-0.7821		
	F	-99.85995	-0.932	711.3		-0.932	15.1559	-14.6141	5.1026		
C_2H_6	C	-37.67215	0.184	10.9	14.2	0.010	0.0878	-0.0802	-0.0135		
	H	-0.65224	-0.061	-1.3		-0.003	0.0261	-0.0270	0.0012	0.7766	-0.0017
NH_3	N	-54.91746	-1.005	-152.7	23.1	0.145	-0.8874	0.7711	-0.5634		
	H	-0.50649	0.335	58.6		-0.048	0.2274	-0.1829	0.0468	0.5112	-0.0226
H_2O	O	-75.53993	-1.172	-423.0	138.1	0.310	-1.8776	1.5552	-1.3985		
	H	-0.37697	0.586	280.7		-0.155	0.8223	-0.6085	0.2066	0.3447	-0.0604

*The small difference between the calculated barrier and the sum of the $\Delta E(\Omega)$ values is the result of errors in the integrated atomic values.
$\dagger r_b(H)$ is the bonded radius of hydrogen.

while leading to a reduction in the repulsive interactions between the two carbon atoms, gives rise to an even greater reduction in the attractive interaction of each carbon nucleus with the charge density on the other carbon atom. ($\Delta V_a(C)$ consists of $\Delta V_a^0(C)$, which is less than zero, and the interaction of each of the other nuclei with the charge density of the carbon atom in question. Since the C–H bond lengths have decreased in the eclipsed geometry, and only the C–C length has increased, the increase in $V_a(C)$ is the result of the latter effect.) Thus, the carbon atoms are destabilized in the eclipsed geometry as a result of a decrease in the attractive interactions of each carbon nucleus with the charge density of the other. These energy changes of the carbon atom dominate the change in energy for the total system and hence the barrier in ethane is an example of Case I.

Orbital arguments concerning the origin of the barrier to rotation in ethane have been reviewed and discussed by Pitzer (1983). The explanation favoured by him is based upon a model in which the molecular orbitals for the staggered geometry are first frozen in form and the nuclear framework of the staggered geometry is rotated without change into the eclipsed geometry. This leads to a small negative barrier, a result which is interpreted to mean that simple electrostatic interactions between the ends of the molecule are not important in the ethane barrier. The freezing of the orbitals destroys their orthogonality, as required by the Pauli principle, and the change in ρ obtained from their re-orthogonalization as carried out in the second step, results in so-called overlap or exchange repulsion contributions to the energy, that is, to 'Pauli repulsions' between the C–H bond orbitals.

While, as shown in Fig. 6.12, the charge distribution of ethane changes so that the basins of the eclipsed hydrogen atoms do not overlap in forming the E conformer, these changes in ρ do not correspond to a compression of the hydrogen atoms. A pair of eclipsed hydrogens do not share a common interatomic surface—they do not touch, as occurs in a closed-shell repulsion—and, as reflected in the decrease in the repulsive contributions to their energy (Table 6.6), they are not compressed. Instead, the spatial extent of each is diminished by a transfer of a small portion of its space and its contained charge to the carbon atoms, each of whose volumes increases by 0.47 au. To demonstrate that the increased C–C bond length in the E conformer and hence the barrier are not a result of repulsions between eclipsed hydrogens, one can inquire into the origin of the barrier resulting from an internal rotation of the rigid nuclear framework of ethane from its staggered to an eclipsed geometry without recourse to a model state. Such a configuration of the nuclei, while physically realizable, does not lie on the reaction path and it possesses an energy slightly in excess of the eclipsed transition-state geometry. Forces act on the nuclei in this constrained geometry such as to move them into the positions they occupy in the eclipsed transition state. Since the major geometrical change between the staggered and eclipsed equilibrium geometries is a lengthening of the C–C bond, one anticipates that the principal forces operative in the frozen eclipsed geometry are forces of repulsion acting on the two carbon nuclei. The changes in the potential and kinetic energies for the nuclear motions: staggered to frozen eclipsed, S → FE, frozen eclipsed to eclipsed, FE → E, and S → E are summarized in Fig. 6.11.

Because forces act on the nuclei in the FE conformation, the general statements of the virial theorem, which include the contributions of the virials of the nuclear forces to the total electronic energy, must be used (eqn (6.65)). The contribution of nucleus α to the virial of the forces acting on the electrons is equal to $-\mathbf{X}_\alpha \cdot \mathbf{F}_\alpha$, where $\mathbf{X}_\alpha$ is the position vector of nucleus α and $\mathbf{F}_\alpha = -\nabla_\alpha E$, the net force acting on it. When the forces on the nuclei vanish, one obtains the corresponding equilibrium statements of the virial theorem, $T = -E$ and $E = \frac{1}{2}V$. The general statement corresponding to $T = -E$ is

$$T = -E + \sum_\alpha \mathbf{X}_\alpha \cdot \mathbf{F}_\alpha. \qquad (6.91)$$

When the nuclear virial is positive, T must exceed $|E|$ to balance the contribution to the virial arising from the repulsive forces acting on the nuclei. Since we are taking differences between a state where the forces do not vanish and one where they do, the total nuclear virial for the non-equilibrium state appears in the difference. Thus, for the reaction S → FE, the change in kinetic energy equals minus the change in total energy plus the nuclear virial (eqn (6.92)),

$$\Delta T = -\Delta E + \sum_\alpha \mathbf{X}_\alpha \cdot \mathbf{F}_\alpha. \qquad (6.92)$$

Since ΔT is positive for the reaction S → FE (Fig. 6.11) and exceeds the magnitude of ΔE, the nuclear virial is positive, showing that this contribution to the potential energy of the electrons is dominated by repulsive forces acting on the nuclei. The corresponding expression for the change in potential energy is

$$\Delta V = \Delta V_a + \Delta V_r = 2\Delta E - \sum_\alpha \mathbf{X}_\alpha \cdot \mathbf{F}_\alpha. \qquad (6.93)$$

In this case V_a decreases and V_r increases, both quantities changing in the opposite direction to that for the overall reaction S → E, but the increase in repulsive energy arising from the nuclear–nuclear and electron–electron forces is less than the increase in the magnitude of the attractive energy and overall the change in the potential energy is less than zero, i.e. the attractive contributions dominate. Thus, the increase in energy encountered in the formation of the frozen eclipsed conformer comes not from an increase in the usual potential energy contributions, but from the virials of the repulsive forces acting on the carbon nuclei in the non-equilibrium geometry. This is also evident in the expression for the total energy change expressed as $\Delta E = \Delta T + \Delta V$. The fact that $\Delta T > \Delta E$ shows that ΔT includes the contribution of a repulsive nuclear virial. The atomic contributions to the change

S → FE, like those for S → E, come primarily from the carbon atoms, $\Delta T(C)$ = 16.7 kJ/mol and $\Delta V_a(C) = -149.8$ kJ/mol, while $\Delta T(H) = 0.8$ kJ/mol.

The geometry changes encountered in the reaction FE → E result in a decrease of only 0.4 kJ/mol in the total energy. However, the relaxation in the geometry, particularly the lengthening of the C–C internuclear separation, causes sizeable changes in the potential energy contributions, primarily those of the carbon atoms. The contribution from the nuclear virial vanishes in reaching this metastable geometry and the attractive and repulsive potential energy contributions change in the direction dictated by Case I for the partial unmaking of the C–C bond, with $\Delta V_a > 0$ and $\Delta V_r < 0$. The kinetic energy undergoes a decrease which, aside from the small ΔE value, equals the loss of the nuclear virial. The value of $\Delta T(H)$ is only 0.4 kJ/mol for this change showing that essentially the whole of the decrease in T comes from the carbon atom contributions. Thus, the energy increase associated with the rigid rotation of one methyl with respect to another to give the non-equilibrium eclipsed geometry arises from the contribution of the virials of the repulsive forces generated on the carbon nuclei and this change in energy overrides the decrease in the potential energy V.

One can investigate the nature of the changes in the charge density over the basins of the carbon atoms to better understand the accompanying changes in energy. In the S conformation of ethane, each carbon atom is polarized towards the other with $|\mathbf{M}(C)| = 0.030$ au. This value decreases to 0.025 au in the FE conformation. Thus, electronic charge is shifted from the binding to the antibinding region of the carbon nuclei (Berlin 1951) when the molecule is rotated without the possibility of relaxing the C–C separation. These counter-polarizations contribute to the forces of repulsion acting on the carbon nuclei in the FE conformation. They are destroyed and replaced by an even larger polarization of each carbon atom into its binding region, $|\mathbf{M}(C)| = 0.039$ au, when the FE conformation is allowed to relax to the force-free E conformation. The value of the quadrupole moment $Q_{zz}(\Omega)$, eqn (6.48), for a carbon atom in the S conformation is $+0.031$ au, a polarization which corresponds to the transfer of charge from along the C–C axis to a torus-like distribution about this axis. (For comparison, the same moment for a carbon atom in acetylene has the value of 4 au.) This moment is increased to 0.074 au in the FE conformation and undergoes a further increase to 0.092 au in the E conformation. The effect is a progressive transfer of charge density from along the C–C internuclear axis to a torus of density about this axis, a polarization which contributes to the lengthening of the C–C bond and to the shortening of the C–H bonds as occurs in the E conformation.

The principal changes in the charge distribution of ethane upon internal rotation occur along the C–C bond path and within the basins of the carbon atoms, with only relatively small changes occurring along the C–H bond paths and within the basins of the hydrogen atoms. It is difficult to rationalize these observed changes in the charge density with those anticipated on the basis of the requirements of orthogonality of the C–H bond orbitals (Pitzer 1983). One might argue that the greater C–C separation found in the E conformation is a consequence of the exchange repulsions between the C–H bond orbitals. However, the changes in density and energy for the reaction S → FE do not correspond to those anticipated on the basis of the overlap of occupied bond orbitals, as even here the primary changes in density are along the C–C axis and within the basins of the carbon atoms and the change in potential energy is attractive rather than repulsive. The barrier in this case, which is of almost the same size as the equilibrium barrier, results from the creation of repulsive forces on the carbon nuclei whose presence increases the total energy of the system by increasing the electronic kinetic energy of the carbon atoms. Relaxing the geometry by lengthening the C–C separation to its value in the E geometry, replaces these forces of nuclear repulsion with a barrier arising from an increase in the attractive potential energy of the carbon atoms.

To understand the origin of the barrier in ethane requires an understanding of why the C–C bond is lengthened in the eclipsed geometry. The change S → E transforms an alternating axis of symmetry into a simple threefold axis. The alternating field of the protons exerted the length of the C–C axis in the Hamiltonian of the S conformation is replaced by a more pronounced field with a threefold symmetry in the Hamiltonian for the E conformation. This change in the Hamiltonian has the effect of increasing the quadrupolar polarization of the electronic charge density along the C–C axis, removing charge from this axis and concentrating it in a torus-like distribution about the axis. As a consequence, the electronic charge in the C–C binding region is less effective in binding these nuclei and the C–C separation increases. This same quadrupolar polarization slightly increases the binding of the protons.

The observations that a rotational barrier is characterized by a lengthening of the bond about which the rotation occurs and by an increase in the attractive and a decrease in the repulsive potential energies as found for ethane appear to be general. That is, rotational barriers belong to Case I energy changes. The same observations are found for the barriers to internal rotation in methylamine and methanol (Table 6.5), in carboxylic acids and esters for rotation about the C–O bond, and for rotation about the C–N bond in formamide (Wiberg and Laidig 1987).

The inversion barrier is found to be an example of a Case II energy change for the molecules NH_3, PH_3, H_3O^+, and H_2O (Table 6.5). Only the first and last of these molecules is discussed in detail. The A–H internuclear separations in ammonia and water are found to decrease when the pyramidal or bent geometry is transformed into its respective planar or linear form, by 0.0130 Å in NH_3 and by 0.0180 Å in H_2O. The approach of the protons towards the A nucleus, while resulting in a lowering of the electron–nuclear attractive potential energy (Table 6.5) upon attainment of the planar or linear geometry, leads to an even greater increase in the repulsive interactions. The inversion barrier in these molecules is thus a consequence of an increase in the repulsive interactions that outweighs an accompanying decrease in the attractive potential energy, an example of a Case II energy change. Rauk *et al.* (1970), in the first successful calculation of the barrier in ammonia, give the same explanation for its origin. The decreases in internuclear separations in these molecules result in an increase, rather than a decrease, in the total energy of the system, a behaviour opposite to that found for Case I.

The hybridization model predicts an increase in the electronegativity of the A atom relative to hydrogen in the attainment of the planar or linear geometries. As discussed in Section 6.4.3, this is the predicted consequence of the increase in the s-character of its bonds to hydrogen, from sp^3 to sp^2 or from p to sp, coupled with the fact that s electrons are more tightly bound than are p electrons. On this basis one predicts a transfer of charge from H to A to accompany the shortening of the A–H bonds in the formation of the planar or linear geometries. This prediction and the associated energetic consequences are found to be correct as indicated by the data in Table 6.6.

The molecular charge distributions of pyramidal and bent molecules undergo larger perturbations during inversion or linearization, as a consequence of the accompanying change in hybridization, than does a charge distribution for a molecule undergoing internal rotation. This greater reorganization of the charge is reflected in the larger changes in the atomic energies, changes which are a direct measure of the extent of change in the distribution of charge over the basin of each atom. The sizeable redistribution of charge caused by the change in hybridization of the N atom in attaining the planar form of ammonia is reflected in the value of $Q_{zz}(N)$ changing from -2.92 au in the pyramidal molecule to -4.15 au in the planar geometry. This moment is characteristic of a doubly occupied $p\pi$ orbital on the N atom. Table 6.6 also lists the bonded radius of hydrogen in the equilibrium geometry and the change in this radius when the barrier is attained. The shifts in the A–H interatomic surface towards the proton are much larger for the inversion and linearization barriers than for the rotation barrier in ethane (Fig. 6.12).

The change in the relative electronegativities of N and H or of O and H and the associated changes in their stabilities resulting from the change in hybridization upon inversion or linearization have a direct parallel with the changes in the properties of the carbon and hydrogen atoms of the methylene group of a normal hydrocarbon when it is transferred to a cyclic system with geometric strain. The electronegativity of the carbon, nitrogen, and oxygen atoms is increased relative to hydrogen in each case. The resulting transfer of charge from H to A destabilizes the hydrogens to a greater extent than the A atom is stabilized and the energy is increased. In terms of the mechanical properties of atoms, the strain energy of a methylene group in a cyclic hydrocarbon, the inversion barrier in ammonia, and linearization barrier in water have a common origin.

The transfer of charge from H to A is larger in the linearization of water than in the pyramidalization of ammonia and the energy barrier and changes in atomic energies are correspondingly greater for H_2O than for NH_3. The pattern of energy changes, however, is the same for the barrier in water as it is for the barrier in ammonia, with the hydrogens being destabilized to a greater extent than the oxygen atom is stabilized. The repulsive contributions to the energy changes of the hydrogen atoms decrease in both molecules and the increase in the repulsive interactions which dominates the barriers in these molecules is localized within the A atoms.

The change in $V_a^0(\Omega)$—the change in the attractive interaction of the nucleus of atom Ω with its own charge density, eqn (6.81)—is stabilizing for the A atom and destabilizing for the hydrogens in each of the molecules. The observation that the magnitude of the decrease in $V_a^0(\Omega)$ is considerably greater than is its increase for all the hydrogens in a given molecule shows that the binding of the charge transferred to the A atom is significantly increased and is one of the principal contributors to the attractive energy contributions to the barriers in these molecules.

The origin of the barrier for the rotation of planar formamide by 90° or 270° investigated by Wiberg and Laidig (1987) is an interesting case, as here the rotation about the C–N bond results in the pyramidalization of the N atom and its formal hybridization changes from sp^2 to sp^3. Thus, as for the inversion barriers discussed above, the N atom is found to gain electronic charge from its bonded neighbours and to increase in stability when the amide group is rotated from a non-planar into its planar form. In this molecule, however, the destabilization of the atoms which donate charge to the N atom in the planar geometry is less than the stabilization acheived by the N atom and the equilibrium geometry of this molecule is planar. Here, as is typical of a Case I energy change and of an internal rotation, the principal geometrical change is a lengthening of the bond about which rotation occurs (as in the inversion of ammonia, the bonds to N are shortest when its s character is greatest, i.e. in its planar form) and the predicted barrier of 66.9 kJ/mol is a result of a decrease in the repulsive energies and of a larger decrease in the stabilizing nuclear–electron attractive interactions. Because of the transfer of charge from N to C which accompanies the loss of planarity, all of the lengthening of the C–N bond, which equals 0.15 au, is taken up by an increase in the bonded radius of the carbon atom, from 0.83 to 1.04 au. The resonance model accounts for the relative stability of the planar geometry by invoking a resonance structure wherein the N atom donates charge to the carbonyl oxygen atom, a resonance interaction which is lost upon rotation. The resonance model is, therefore, in direct contradiction with the hybridization model for these systems which predicts the N to be most electronegative in the planar structure. The theory of atoms in molecules shows the resonance model to be wrong in this instance (Wiberg and Laidig 1987). Not only is the predicted direction of the charge transfer for nitrogen incorrect, the properties of the oxygen atom, including its geometrical parameters, are found to change by only small amounts compared to the changes undergone by the C and N atoms, the magnitude of its energy change being 20 to 30 times smaller.

6.4.5   Perfect transferability—an unattainable limit

The examples of the group properties in hydrocarbon molecules given above demonstrate that the properties of an atom change in direct response and proportion to changes in its distribution of charge. Whether the form of an atom changes by a little or by a lot, its energy and other properties change by corresponding amounts. It was the observation of a paralleling constancy in the kinetic energy and the charge distribution of an atom, as discussed in Section 5.2, which led to the development of the theory of the quantum atom. This relationship between the spatial form of an atom and its properties is most important, for it is because of this relationship that we are able to identify an atom in different systems.

On the basis of the demonstrated relationship between the transferability of the charge density of an atom and the corresponding constancy in its contribution to the energy, one might anticipate that the theorem of Hohenberg and Kohn (1964) would apply to regions of space bounded by the zero-flux surface condition. Their theorem, which forms the basis of density functional theory, states that the ground-state wave function and energy are unique functionals of the charge density, $\rho(\mathbf{r})$. The general problem of extending this theorem to a subsystem has been considered by Riess and Münch (1981). They have demonstrated that the theorem is applicable to any finite, but otherwise arbitrary subsystem of some total, bounded system. As a corollary to their proof, it is shown that perfect transferability of an atom between systems is an unattainable limit (Bader and Becker 1988).

Riess and Münch are able to show that the density $\rho(\mathbf{r})$ associated with the total system is a unique functional (i.e. totally determined by) of $\rho_\Omega(\mathbf{r})$, where the latter is the restriction of the density to an arbitrary subdomain Ω of the total system. The only assumption that must be made in obtaining this result is that the total system, while arbitrarily large, must be bounded. Having shown that $\rho(\mathbf{r})$ is a unique functional of $\rho_\Omega(\mathbf{r})$, it follows, together with the original theorem of Hohenberg and Kohn, that the ground-state energy and the ground-state wave function are unique functionals of the ground-state density of an arbitrary subdomain of a total, bounded system. The sub-domain density $\rho_\Omega(\mathbf{r})$ thus uniquely determines the expectation value of any spin-free observable associated with the subdomain Ω, or with any other subdomain, or with the total domain of the system. It is a corollary of this proof that, if the density over a given atom Ω, or any portion thereof, is identical in two molecules A and B, i.e. $\rho_{\Omega A}(\mathbf{r}) = \rho_{\Omega B}(\mathbf{r})$, then $\rho_A(\mathbf{r}) = \rho_B(\mathbf{r})$ and the two molecules are identical. Therefore, the limit of perfect transferability of atomic properties is unattainable. No restriction, however, is placed on how closely this limit can be approached. For the examples of the hydrogen atoms in the transferable methyl groups of the n-alkanes given above and taking into account as well the limits on the experimental accuracy of measured heats of formation, the existence of the limit may not be of practical importance.

The work of Riess and Münch demonstrates that the energy of an atom in a molecule, as well as its other properties, are indeed determined by its distribution of charge. Thus, this important physical characteristic of an atom has a basis in theory. The atoms and their properties provide visual proof of the theorem of Hohenberg and Kohn by showing that similar distributions of charge possess similar energies. There is nothing in the original derivation of the theorem concerning the dependence of local contributions to the total energy on the local properties of the charge density. The actual functional which relates the total energy to the charge density is unknown and, in terms of quantum mechanics, a knowledge of the one-density matrix, and not just its diagonal elements which define $\rho(\mathbf{r})$, is required for the determination of the energy of a system. The quantum stress tensor $\overleftrightarrow{\sigma}(\mathbf{r})$ determines the potential energy density of the electrons, $\mathscr{V}(\mathbf{r})$ (eqn (6.30)), and their kinetic energy density (eqn (6.27)), while it itself is a functional of the one-density matrix (eqn (6.13)). Thus, the integration of the sum of the two energy densities or of either one of them separately, together with the use of the atomic virial theorem (eqn (6.73)) will yield the electronic energy of an atom, or of the total system. If the system is not in an equilibrium configuration, the subtraction of the virial of the nuclear forces (as determined by $\rho(\mathbf{r})$ in the electrostatic theorem) from E_e for the total system will yield the total energy E (eqns (6.55) and (6.56)). Thus, the energies E_e and E are known functionals of the one-density matrix, $\Gamma^{(1)}(\mathbf{r}, \mathbf{r}')$. A reader interested in a unified account of density-functional theory and of its application to atoms, molecules, and chemical concepts is referred to Parr and Yang, 1989.

While known quantum mechanical relationships require more information than is contained in just $\rho(\mathbf{r})$ for the determination of the energy, only a portion of the information contained in the one-density matrix is required for this purpose (Bader 1980). The one-density matrix is a function of six variables. The physical information of $\Gamma^{(1)}(\mathbf{r}, \mathbf{r}')$ is however, contained in the neighbourhood of its diagonal elements $\mathbf{r} = \mathbf{r}'$. To establish this, one expresses $\Gamma^{(1)}(\mathbf{r}, \mathbf{r}')$ in a new system of coordinates defined by

$$\mathbf{X} = \tfrac{1}{2}(\mathbf{r} + \mathbf{r}') \quad \text{and} \quad \mathbf{x} = \tfrac{1}{2}(\mathbf{r} - \mathbf{r}') \tag{6.94}$$

and considers the properties of $\Gamma^{(1)}$ in the neighbourhood of a point along the diagonal $(\mathbf{X}, 0)$. Towards this end, we perform a Taylor series expansion of $\Gamma^{(1)}$ in the neighbourhood of this point. Including terms up to second-order in $\delta\mathbf{X}$ and $\delta\mathbf{x}$, this yields

$$\Gamma^{(1)}(\mathbf{X} + \delta\mathbf{X}, \delta\mathbf{x}) = \Gamma^{(1)}(\mathbf{X}, 0) + [\partial\Gamma^{(1)}(\mathbf{X}, 0)/\partial\mathbf{X}] \cdot \delta\mathbf{X}$$
$$+ \tfrac{1}{2}\delta\mathbf{X} \cdot [\partial^2\Gamma^{(1)}(\mathbf{X}, 0)/\partial\mathbf{X}^2] \cdot \delta\mathbf{X}$$
$$+ [\partial\Gamma^{(1)}(\mathbf{X}, 0)/\partial\mathbf{x}] \cdot \delta\mathbf{x}$$
$$+ \tfrac{1}{2}\delta\mathbf{x} \cdot [\partial^2\Gamma^{(1)}(\mathbf{X}, 0)/\partial\mathbf{x}^2] \cdot \delta\mathbf{x}$$
$$+ \partial\mathbf{X} \cdot [\partial^2\Gamma^{(1)}(\mathbf{X}, 0)/\partial\mathbf{X}\partial\mathbf{x}] \cdot \delta\mathbf{x}. \qquad (6.95)$$

Each term in eqn (6.95) has a direct physical interpretation when the derivatives are evaluated after expressing $\Gamma^{(1)}$ in terms of its natural orbital expansion. Retaining the same ordering of terms as appears in eqn (6.95), this yields

$$\Gamma^{(1)}(\mathbf{X} + \delta\mathbf{X}, \delta\mathbf{x}) = \rho(\mathbf{X}) + \nabla\rho(\mathbf{X}) \cdot \delta\mathbf{X} + \tfrac{1}{2}\delta\mathbf{X} \cdot \nabla\nabla\rho(\mathbf{X}) \cdot \delta\mathbf{X}$$
$$+ (2mi/\hbar)\mathbf{j}(\mathbf{X}) \cdot \delta\mathbf{x} + (8m/\hbar^2)\delta\mathbf{x} \cdot \overleftrightarrow{\sigma}(\mathbf{X}) \cdot \delta\mathbf{x}$$
$$+ (2mi/\hbar)\delta\mathbf{X} \cdot \nabla\mathbf{j} \cdot \delta\mathbf{x}. \qquad (6.96)$$

The terms in the expansion derived from derivatives with respect to $\mathbf{X}$ are identical to those obtained by taking the corresponding derivatives of the charge density $\rho(\mathbf{X})$ itself. The dyadic $\nabla\nabla\rho$ is the Hessian matrix of ρ, whose eigenvectors and eigenvalues determine the properties of the critical points in the charge distribution. The trace of this term is the Laplacian of the charge density, $\nabla^2\rho$.

The derivatives with respect to $\mathbf{x}$ sample the off-diagonal behaviour of $\Gamma^{(1)}$ and generate terms related to the current density $\mathbf{j}$ and the quantum stress tensor $\overleftrightarrow{\sigma}$. The first-order term is proportional to the current density, and this vector field is the $\mathbf{x}$ complement of the gradient vector field $\nabla\rho$. The second-order term is proportional to the stress tensor. Considered as a real symmetric matrix, its eigenvalues and eigenvectors will characterize the critical points in the vector field $\mathbf{j}$ and its trace determines the kinetic energy densities $K(\mathbf{r})$ and $G(\mathbf{r})$. The cross-term in the expansion is a dyadic whose trace is the divergence of the current density.

The second-order expansion given in eqn (6.96) recovers all of the physical quantities needed to describe a quantum system and determine its properties: the charge density ρ and its gradient vector field $\nabla\rho$ define atoms and determine many of their properties in a stationary state; the current density determines the system's magnetic properties and the change in ρ in a time-dependent system; and, finally, the stress tensor $\overleftrightarrow{\sigma}$ determines the local and average mechanical properties of the system. Thus, one does not need all the information in the state function ψ to describe a quantum mechanical system and determine its properties. The expansion in eqn (6.96) demonstrates that, even with the existing quantum mechanics, we do not require much more information than that contained in ρ itself, the other extreme to the over-complete knowledge afforded by $\psi^*\psi$.

Further investigation into the properties of $\Gamma^{(1)}$ in the vicinity of its diagonal elements would seem to be a worthy endeavour. In a stationary state, in the absence of a magnetic field, $\mathbf{j}$ and $\mathbf{V j}$ vanish and, in this situation, eqn (6.96) reduces to

$$\Gamma^{(1)}(\mathbf{X} + \delta\mathbf{X}, \delta\mathbf{x}) = \rho(\mathbf{X}) + \mathbf{V}\rho(\mathbf{X})\cdot\delta\mathbf{X} + \tfrac{1}{2}\delta\mathbf{X}\cdot\mathbf{V V}\rho(\mathbf{X})\cdot\delta\mathbf{X}$$

$$+ (8m/\hbar^2)\delta\mathbf{x}\cdot\overleftrightarrow{\sigma}(\mathbf{r})\cdot\delta\mathbf{x}. \tag{6.97}$$

A critical point in the full field $\Gamma^{(1)}(\mathbf{X}, \mathbf{x})$ is one for which its first derivatives with respect to both $\mathbf{X}$ and $\mathbf{x}$ vanish. Thus, any critical point in $\rho(\mathbf{r})$ is also a critical point in $\Gamma^{(1)}$ because of the vanishing of the vector current density in a stationary state. The gradient vector field of the charge density is thus the gradient vector field of $\Gamma^{(1)}$ in a stationary state and the topology of $\Gamma^{(1)}$ is susceptible to study. A critical point in $\Gamma^{(1)}$ is characterized by the eigenvectors and eigenvalues of two Hessian matrices, $\mathbf{V V}\rho(\mathbf{r})$ and $\overleftrightarrow{\sigma}(\mathbf{r})$. The formation of a degenerate critical point in the Hessian of ρ signals an instability in the gradient vector field of ρ and a change in molecular structure. The stress tensor $\overleftrightarrow{\sigma}$ determines the local mechanics of a charge distribution and it is possible that the eigenvectors and eigenvalues of this second Hessian at the degenerate critical point could summarize the mechanical consequences of the change in structure. The quantum stress tensor provides a link between the mechanical and structural properties of a molecular system.

6.5   The hypervirial theorem and the definition of bond properties

The opening section of this chapter stressed the importance of the presence of the surface integral in the hypervirial theorem for an open system, eqn (6.2). Unlike the theorem for a total system, eqn (6.4), in which case the average of the commutator of any observable $\hat{G}$ with the Hamiltonian $\hat{H}$ vanishes, the corresponding result for an atom in a molecule is proportional to the flux in the effective single-particle vector current density of the property $\hat{G}$ through the atomic surface. As a result, the hypervirial theorem plays an important role in determining the properties of an atom in a molecule. It also enables one to relate an atomic property to a sum of bond contributions, as is now demonstrated.

6.5.1   An atomic property expressed as a sum of bond contributions

We shall be interested in observables that have non-vanishing commutators with the Hamiltonian $\hat{H}$. Suppose the commutator of $\hat{H}$ and $\hat{G}$ is related to the observable $\hat{A}$ by the relation

$$(i/\hbar)[\hat{H}, \hat{G}] = \hat{A}. \tag{6.98}$$

In this case $\hat{A}$ and $\hat{G}$ shall be referred to as associated operators. The averaging of this commutator and its complex conjugate over the atom Ω, as indicated in eqn (6.2), and multiplication by $N/2$ then yield the average value of $\hat{A}$ for an atom in a molecule, the quantity $A(\Omega)$. In such a case, the commutator average equals the change in the value of the property A when a free atom Ω combines to form a molecule, since this average for the isolated atom vanishes (eqn (6.4)). If one denotes the change in the property A by $\Delta A(\Omega)$, then from eqn (6.2) one obtains

$$\Delta A(\Omega) = (N/2)\{\oint dS(\Omega, \mathbf{r})\, \mathbf{j}_G(\mathbf{r}) \cdot \mathbf{n}(\mathbf{r}) + cc\}. \tag{6.99}$$

Thus the change in the value of the property A that occurs as a result of an atom entering into chemical combination is given by the flux in the vector current density of its associated generator $\hat{G}$ through the atomic surface.

An atomic surface $S(\Omega, \mathbf{r})$ is, in general, composed of a number of interatomic surfaces, there being one such surface for each atom Ω' linked to Ω by an atomic interaction line or, in the case of a bound system, by a bond path. That is,

$$S(\Omega, \mathbf{r}) = \sum_{\Omega' \neq \Omega} S(\Omega|\Omega', \mathbf{r}) \tag{6.100}$$

where $S(\Omega|\Omega', \mathbf{r})$ denotes the interatomic surface between atom Ω and Ω'. Thus, the surface integral in eqn (6.99) can be broken up into a sum of integrals over the interatomic surfaces bounding the atom Ω and the value of $\Delta A(\Omega)$ is thereby equated to sum of terms, there being one term for each atom linked to atom Ω by a bond path. If each such surface integral, including the factor $(N/2)$, is denoted by $A(\Omega|\Omega')$, eqn (6.99) becomes

$$\Delta A(\Omega) = \sum_{\Omega' \neq \Omega} A(\Omega|\Omega') \tag{6.101}$$

where

$$A(\Omega|\Omega') = (N/2)\{\oint dS(\Omega|\Omega', \mathbf{r})\, \mathbf{j}_G(\mathbf{r}) \cdot \mathbf{n}(\mathbf{r}) + cc\}. \tag{6.102}$$

In this manner the atomic average of the observable satisfying eqn (6.98) for the associated generator $\hat{G}$ is expressible as a sum of contributions where each contribution is determined by an integral over the surface which the atom shares with each of its bonded neighbours. This atomic average is necessarily the change in the value of the property relative to its value in the isolated atom. An example is provided by eqn (6.18), which shows that the

Ehrenfest force acting on an atom in a molecule, $A(\Omega) \equiv \mathbf{F}(\Omega)$, is the vector sum of the forces exerted on it by each of its bonded neighbours, forces which vanish when the atom is free. The force exerted on atom Ω by its bonded neighbour Ω', the quantity $\mathbf{F}(\Omega|\Omega')$, is determined by the pressure $\overleftrightarrow{\sigma} \cdot \mathbf{n}$ acting on each of the surface elements $\mathrm{d}S(\Omega|\Omega', \mathbf{r})$.

6.5.2   Surface integrals proportional to the energy

The dimensions of the property A are those of its associated generator divided by the time. Thus, when $\hat{G}$ equals the virial operator $\mathbf{r} \cdot \mathbf{p}$, which has the dimensions of action, the property A has the dimensions of energy and equals twice the electronic kinetic energy plus the virial of the Ehrenfest forces acting over the basin of the atom (eqn (6.19)). For a free atom, the sum $2T(\Omega) + \mathscr{V}_\mathrm{b}(\Omega)$ vanishes while for a bound atom it equals the negative of the virial of the forces acting on the surface of the atom, the term $\mathscr{V}_\mathrm{s}(\Omega)$ (eqns (6.20) and (6.21)). Thus, $-\mathscr{V}_\mathrm{s}(\Omega)$ equals twice the change in the kinetic energy of the atom plus the change in its basin virial (the change in the potential energy of the free atom) when the atom becomes bonded to other atoms

$$2\Delta T(\Omega) + \Delta \mathscr{V}_\mathrm{b}(\Omega) = -\mathscr{V}_\mathrm{s}(\Omega). \tag{6.103}$$

According to eqn (6.103), the surface virial measures the disparity in the changes in the values of the electronic kinetic and potential energies over the basin of the atom required for the satisfaction of the virial theorem, as occasioned by the formation of chemical bonds.

The classical virial theorem, when applied to a system of particles confined to a volume V, equates the negative of twice the kinetic energy of the particles to the virials of the internal and external forces acting upon them. The virial of the external forces is equal and opposite to the virial of the forces which the molecules exert on an element of surface area of the container, the quantity $p\mathbf{r} \cdot \mathbf{n}\, \mathrm{d}S$ where p is the pressure. Integration over the surface of the container is transformed using Green's theorem into the integral of $3p\, \mathrm{d}V$ over its volume and the virial of the external forces is $3pV$. For a system of N point particles with kinetic energy $(3/2)NkT$, this yields the familiar virial equation of state, $pV = NkT -$ (a term determined by the ensemble average of the virial of the intermolecular forces or of the forces between the nuclei and electrons of the molecules in the system).

The analogous result is obtained using quantum mechanics with the pV operator being defined as $\widehat{(PV)} = (2/3)\{\hat{T} - \frac{1}{2}\hat{\mathscr{V}}\}$ (Hirschfelder *et al.* 1967; these authors define the virial as $-\frac{1}{2}\mathscr{V}$). The analogy is complete, with the pressure exerted on an element of surface area in the quantum case given by $-\overleftrightarrow{\sigma} \cdot \mathbf{n}\, \mathrm{d}S$ and its virial by $-\mathbf{r}\overleftrightarrow{\sigma} \cdot \mathbf{n}\, \mathrm{d}S$. The pressure in this case is not uniform throughout the system or over the atomic surface and, instead, each interatomic surface makes its own contribution to the surface virial, a quantity denoted by $\mathscr{V}_s(\Omega|\Omega')$ in eqn (6.104),

$$\mathscr{V}_s(\Omega) = \sum_{\Omega' \neq \Omega} \mathscr{V}_s(\Omega|\Omega') \tag{6.104}$$

where

$$\mathscr{V}_s(\Omega|\Omega') = \oint dS(\Omega|\Omega', \mathbf{r})\mathbf{r}\cdot\vec{\sigma}\cdot\mathbf{n}.$$

For a molecule in an equilibrium configuration, $\Delta T(\Omega) = -\Delta E(\Omega)$ and one has

$$\Delta E(\Omega) - \tfrac{1}{2}\Delta\mathscr{V}_b(\Omega) = \tfrac{1}{2}\sum_{\Omega' \neq \Omega} \mathscr{V}_s(\Omega|\Omega'), \tag{6.105}$$

showing that each $\mathscr{V}_s(\Omega|\Omega')$ measures the potential energy contribution to $\Delta E(\Omega)$ arising from the pressure exerted on the interatomic surface with atom Ω'. The virial of the surface force has the dimensions of force per unit length, or surface tension and $\mathscr{V}_s(\Omega|\Omega')$ can be interpreted as providing a measure of the surface energy associated with the bond to atom Ω'.

The separation of the virial into basin and surface contributions requires a choice of origin and a natural and physically useful choice is to place the origin of the position vector for atom A, the vector $\mathbf{r}_a$, at the nucleus of the atom in question. If the vector from the nucleus of atom A to that of atom B is denoted by $\mathbf{R}_{ab}$, then $\mathbf{R}_{ab} = \mathbf{r}_a - \mathbf{r}_b$ and, recalling that the surface normals are related by $\mathbf{n}_a = -\mathbf{n}_b$, one can write an expression for the sum of the surface virials for the interatomic surface between atoms A and B as

$$\mathscr{V}_s(A|B) + \mathscr{V}_s(B|A) = \mathbf{R}_{ab} \cdot \oint dS(A|B, \mathbf{r})\vec{\sigma}\cdot\mathbf{n}_a. \tag{6.106}$$

For a diatomic molecule, the integral on the right-hand side of eqn (6.106) is a measure of the contribution of the interatomic surface energy to the total energy change ΔE.

For real $\hat{G}$, the statement of the atomic hypervirial theorem yielding the atomic average of the commutator is (see eqn (5.43))

$$(N/2)\{\langle[\hat{H}, \hat{G}]\rangle_\Omega + cc\} = \{(N\hbar^2/4m)\oint dS\{\nabla\psi^*\hat{G}\psi + \hat{G}\psi^*\nabla\psi$$
$$- \psi^*\nabla(\hat{G}\psi) - \nabla(\hat{G}\psi^*)\psi\}\cdot\mathbf{n}. \tag{6.107}$$

An operator, which provides a measure of the distortion of the gradient vector field of an atom from its radial form characteristic of the free state to that found for a bound atom, is obtained from the commutator of $\hat{H}$ and r^2, r being the radial distance of the electron from the nucleus. Substitution of the result

$$[\hat{H}, r^2] = -(\hbar^2/m)(3 + 2\mathbf{r}\cdot\nabla) \tag{6.108}$$

into eqn (6.107) for $\hat{G} = r^2$ yields

$$3N(\Omega) + \int_\Omega \mathbf{r}\cdot\nabla\rho\,d\tau = \oint dS\rho(\mathbf{r})\mathbf{r}\cdot\mathbf{n}(\mathbf{r}). \tag{6.109}$$

The first term of the commutator yields three times the average number of electrons in an atom. The second integral represents another way of counting electrons over the basin of an atom, a way which reflects the distortion of the gradient vector field of the charge density that is caused by the formation of chemical bonds. Because the nucleus acts as an attractor in the gradient vector field, the vector $\nabla\rho$ and the radial vector $\mathbf{r}$ are parallel and oppositely directed for the undistorted gradient vector field of a free atom (Fig. 6.13). Since the surface integral vanishes in this case, the integral of the product $\mathbf{r}\cdot\nabla\rho$ over the basin of the free atom yields $-3N(\Omega)$. For the distorted field of a bound atom the two vectors are no longer antiparallel and their dot product integrates to a number which is less than $3N(\Omega)$. The stronger the interaction with a neighbouring atom, the greater the distortion of the gradient vector field, the smaller the overlap of $\mathbf{r}$ and $\nabla\rho$, and the larger the difference between the two ways of counting electrons in eqn (6.109), a difference which is measured by the integral over the atomic surface.

The surface integral may, of course, be expressed as a sum, there being an integral over the interatomic surface for each atom bonded to the one in question. Reference to Fig. 6.13 makes clear that the distortion of the gradient vector field away from radial symmetry is a result of the formation of the interatomic surfaces, that is, as a result of bonds to neighbouring atoms. The trajectories of $-\nabla\rho$, which are radially directed in the neighbourhood of the nucleus, bend to avoid a bond critical point and then run parallel to the associated interatomic surface. Each surface integral relates the distortion of the gradient vector field which results from the formation of the associated bond path to a corresponding average number of electrons. A value for this number of electrons may be determined for each bond in a molecule by combining the surface integrals for each pair of atoms linked by a bond path in the manner indicated in eqn (6.110)

$$\oint dS(A\,|\,B,\mathbf{r})\rho(\mathbf{r})\mathbf{r}_a\cdot\mathbf{n}_a(\mathbf{r}) + \oint dS(B\,|\,A,\mathbf{r})\rho(\mathbf{r})\mathbf{r}_b\cdot\mathbf{n}_b(\mathbf{r})$$

$$= \mathbf{R}_{ab}\cdot\oint dS(A\,|\,B,\mathbf{r})\rho(\mathbf{r})\mathbf{n}_a(\mathbf{r}) \quad (6.110)$$

where, as in eqn (6.106), $\mathbf{R}_{ab}$ is the vector linking nucleus A to nucleus B. In a polyatomic molecule there is one such surface integral for every bond path. The values of this integral for some diatomic molecules in their ground states are (Bader and Wiberg 1987): H_2, 1.080; B_2, 3.097; C_2, 4.176; N_2, 4.902; N_2^+, 4.474; CH, 1.919; O_2, 3.930; O_2^+, 4.219.

If one reinstates the units of the commutator, $\hbar^2/m$, then the surface integrals in eqns (6.109) and (6.110) have the dimensions of energy $\times$ (length)^2. Thus, division of the integral for the pair of atoms linked by a bond path in eqn (6.110) by R_{ab}^2 yields a quantity with the dimensions of energy. Since the value of the surface integral is determined by the changes in the gradient vector fields of the atoms resulting from the formation of the bond path and hence of the bond which links them, the energy it determines should be

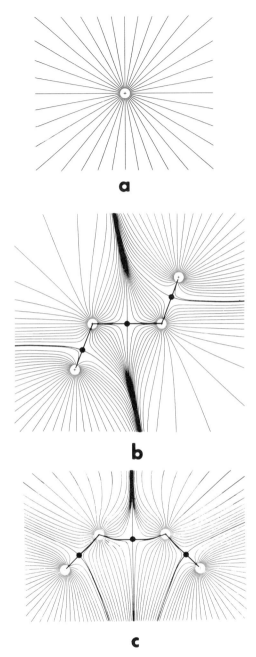

Fɪɢ. 6.13. (a) The trajectories of $\mathbf{V}\rho$ representing the gradient vector field of a free atom. (b), (c) The distorted gradient vector fields of bound atoms. The plane in (b) contains H–C–C–H nuclei of staggered ethane. The plane shown in (c) contains the H–C–C–H nuclei representing the bridging C–C bond in bicyclo-butane (see Table 6.7). The C–C bridging bond is very curved (strained) and, correspondingly, its energy is significantly less than that of a C–C bond in an acyclic hydrocarbon.

proportional to the corresponding change in the energy of the atoms, the energy of the bond. The sum of the bond energies determined in this manner for a given molecule should, therefore, equal the heat of atomization of the molecule in its vibrationless state at 0 K. This procedure works well for systems with only small or no interatomic charge transfer (Bader and Wiberg 1987). Table 6.7 lists the values of the bond energies determined in this manner for a number of acyclic, cyclic, and bicyclic hydrocarbon molecules, including molecules with strained C–C bonds. The charge densities were obtained from SCF calculations and the proportionality constants for the C–H and C–C bond energies were determined by fitting the experimental energies of atomization of methane and ethane at 0 K. The agreement between the calculated atomization energy ΔE_a^c and the experimental value ΔE_a^0 is satisfactory and the individual bond energies exhibit the anticipated trends, with the C–C bond energy increasing with the degree of branching (compare propane, iso-butane, and neo-pentane) and decreasing with a decrease in ring size in cyclic systems. It is found that the constant of proportionality linking the value of the integral and the energy approaches the limiting value of $1/\pi$ as the charge density approaches that obtained at the Hartree–Fock limit.

The purpose of these examples is to illustrate how one can relate a molecular property to a sum of bond contributions in a rigorous manner. The method has the further advantage of yielding directly the change in the value of the property from that observed when the atoms of the molecule are in their free state. The method is applicable to any property associated with a generator $\hat{G}$ via its commutator with the Hamiltonian of the system.

A quantity of some physical interest, whose commutator average with $\hat{H}$ yields the surface integral of the flux in the gradient vector of the charge density, is given by the operator $- \ln \rho'$, where $\rho' = \rho/N$, the density normalized to one. In the density formulation of quantum mechanics the entropy of a system is defined as $S[\hat{\rho}] = - \text{Tr}(\hat{\rho} \ln \hat{\rho})$ where $\hat{\rho}$ is the density operator. The quantity $- \ln \rho'$ has been termed the 'surprisal' (Alhassid and Levine 1977) and it, along with $S[\hat{\rho}]$, have been shown to be constants of the motion. The commutator average of $\hat{H}$ and $- \ln \rho'$ for an atom in a molecule also vanishes identically as a result of the zero-flux surface condition. Thus

$$\tfrac{1}{2}\{\langle [\hat{H}, -\ln \rho'] \rangle_\Omega + \text{cc}\} = (\hbar^2/2m) \int_\Omega \nabla^2 \rho'(\mathbf{r}) d\tau$$

$$= (\hbar^2/2m) \oint dS(\mathbf{r}) \nabla \rho'(\mathbf{r}) \cdot \mathbf{n}(\mathbf{r}) = 0. \quad (6.111)$$

Since this result is obtained for both stationary and non-stationary states, the energy and Lagrangian functionals are stationary with respect to a variation with $- \ln \rho'$ as generator. The averaging of this generator over the basin of

Table 6.7

Theoretical bond energies in hydrocarbons

Molecule	Bond energy (kJ/mol)		Energy at atomization (kJ/mol)	
	CH	CC	ΔE_a^c	ΔE_a^o
Methane	438.52		1754.1	1754.1
Ethane	440.32	331.00	2972.9	2972.9
Propane	441.83(1) 440.53(2) 439.15(3)	339.28	4199.9	4201.0
n-Butane	440.62(1) 441.58(2) 439.86(3)	347.52(4) 339.95(5)	5432.5	5431.5
n-Pentane	438.52(1) 440.03(2) 441.50(3) 439.32(4)	349.99(5) 341.00(6)	6659.4	6661.7
iso-Butane	443.55(1) 437.98(2) 439.82(3)	348.44	5441.8	5437.9
neo-Pentane	438.99	357.40	6697.6	6677.4
Cyclopropane	444.42	297.23	3558.2	3556.0
Cyclohexane	442.50(e) 440.03(a)	335.60	7310.7	7367.2
Bicyclo[1.1.0]butane	438.94(1) 442.08(2) 436.73(3)	263.38(4) 303.30(5)	4112.0	4106.2

the atom yields

$$I(\Omega) = -\int_\Omega \rho' \ln\rho' \, d\tau,$$

a quantity analogous to the missing information function. This quantity has the property of decreasing in value when a free atom becomes bound (Bader 1980).

References

Alhassid, Y. and Levine, R. D. (1977). *J. Chem. Phys.* **67**, 4321.

Bader, R. F. W. (1980). *J. chem. Phys.* **73**, 2871.

Bader, R. F. W. (1986). *Can. J. Chem.* **64**, 1036.

Bader, R. F. W. and Becker, P. (1988). *Chem. Phys. Lett.* **148**, 452.

Bader, R. F. W. and Beddall, P. M. (1973). *J. Am. chem. Soc.* **95**, 305.

Bader, R. F. W. and Chang, C. (1988a). *J. Phys. Chem.* **93**, 2946.

Bader, R. F. W. and Chang, C. (1989b). *J. Phys. Chem.* **93**, 5095.

Bader, R. F. W. and Messer, R. R. (1974). *Can. J. Chem.* **52**, 2268.

Bader, R. F. W. and Nguyen-Dang, T. T. (1981). *Adv. Quantum Chem.* **14**, 63.

Bader, R. F. W. and Preston, H. J. T. (1970). *Theor. Chim. Acta* **17**, 384.

Bader, R. F. W. and Wiberg, K. B. (1987). *Density matrices and density functionals* (ed. R. Erdahl and V. H. Smith), 677, D. Reidel Pub. Co.

Bader, R. F. W., Henneker, W. H. and Cade, P. E. (1967). *J. Chem. Phys.* **46**, 3341.

Bader, R. F. W., Carroll, M. T., Cheeseman, J. R., and Chang, C. (1987a). *J. Am. chem. Soc.* **109**, 7968.

Bader, R. F. W., Larouche, A., Gatti, C., Carroll, M. T., MacDougall, P. J., and Wiberg, K. B. (1987b). *J. Chem. Phys.* **87**, 1142.

Bader, R. F. W., Cheeseman, J. R., Laidig, K. E., Wiberg, K. B. and Breneman, C. (1990). *J. Am. chem. Soc.* **112**, 6530.

Benson, S. W. (1968). *Thermochemical kinetics.* Wiley, New York.

Berlin, T. (1951). *J. Chem. Phys.* **19**, 208.

Biegler-König, F. W., Bader, R. F. W., Tang, T. H. (1982). *J. comput. Chem.* **3**, 317.

Chang, C. and Bader, R. F. W. (1990). Unpublished results.

Coulson, C. A. and Moffitt, W. E. (1949). *Phil. Mag.* **40**, 1.

Franklin, J. L. (1949). *Ind. Engng Chem.* **41**, 1070.

Hariharan, P. C. and Pople, J. A. (1972). *Chem. Phys. Lett.* **16**, 217.

Hirschfelder, J. O., Curtiss, C. F., and Bird, R. B. (1967). *Molecular theory of gases and liquids.* Wiley, New York.

Hohenberg, P. and Kohn, W. (1964). *Phys. Rev.* **B136**, 864.

Hu, C. D. and Langreth, D. D. (1985). *Phys. Scr.* **32**, 391.

Keith, T. (1990). Personal communication.

Kopp, H. (1855). As reported in Glasstone, S. (1946). *Textbook of physical chemistry* (2nd edn), p. 525. Van Nostrand, New York.

Krishnan, R., Frisch, M. J., and Pople, J. A. (1980). *J. Chem. Phys.* **72**, 4244.

Langreth, D. D. and Mehl, M. J. (1983). *Phys. Rev.* **B28**, 1809.

Libit, L. and Hoffmann, R. (1974). *J. Am. chem. Soc.* **96**, 1370.

Löwdin, P.-O. (1959). *J. Mol. Spectrosc.* **3**, 46.

Messiah, A. (1958). *Quantum mechanics*, Vol. 1. North-Holland, Amsterdam.

Morse, P. M. and Feshbach, H. (1953). *Methods of theoretical physics*, Vol. I. McGraw-Hill, New York.

Parr, R. G. and Yang, W. (1989). *Density-functional theory of atoms and molecules*, Oxford University Press, New York.

Pauling, L. (1960). *The nature of the chemical bond* (3rd edn). Cornell University Press, Ithaca, New York.

Payne, P. W. and Allen, L. C. (1977). *Applications of electronic structure theory.* (*Modern theoretical chemistry*), Vol. IV (ed. H. F. Schaefer), 29, Plenum Press, New York.

Pittam, D. A. and Pilcher, G. (1972). *J. chem. Soc., Faraday Trans. 1* **68**, 2224.

Pitzer, R. M. (1983). *Acc. chem. Res.* **16**, 207.

Prosen, E. J., Johnson, W. H., and Rossini, F. D. (1946). *J. Res. nat. Bur. Stand.* **37**, 51.

Rauk, A., Allen, L. C., and Clementi, E. (1970). *J. Chem. Phys.* **52**, 4133.

Riess, I. and Münch, W. (1981). *Theor. Chim. Acta* **58**, 295.

Schrödinger, E. (1926). *Ann. d. Physik* **81**, 109.

Schulman, J. M. and Disch, R. L. (1985). *Chem. Phys. Lett.* **113**, 291.

Slater, J. C. (1933). *J. Chem. Phys.* **1**, 687.

Slee, T., Larouche, A., and Bader, R. F. W. (1988). *J. Phys. Chem.* **92**, 6219.

Srebrenik, S. and Bader, R. F. W. (1975). *J. Chem. Phys.* **63**, 3945.

Srebrenik, S., Bader, R. F. W., and Nguyen-Dang, T. T. (1978). *J. Chem. Phys.* **68**, 3667.

Taft, R. W. (1956). In *Steric effects organic chemistry* (ed. M. S. Newman), Chapter 13. Wiley, New York.

Taft, R. W. (1960). *J. Phys. Chem.* **64**, 1805.

Walsh, A. D. (1947). *Nature, London* **159**, 167, 712.

Wiberg, K. B. (1984). *J. comput. Chem.* **5**, 197.

Wiberg, K. B. and Laidig, K. E. (1987). *J. Am. chem. Soc.* **109**, 5935.

Wiberg, K. B. and Wendolowski, J. J. (1981). *Proc. nat. Acad. Sci. USA* **78**, 6561.

Wiberg, K. B., Bader, R. F. W., and Lau, C. D. H. (1987). *J. Am. Chem. Soc.* **109**, 1001.

CHEMICAL MODELS AND THE LAPLACIAN
OF THE CHARGE DENSITY

Indeed it seems hardly likely that much progress can be made in the solution of the difficult problems relating to chemical combination by assigning in advance definite laws of force between the positive and negative constituents of an atom, and then on the basis of these laws building up mechanical models of the atom.

G. N. Lewis (1916)

7.1 The physical basis of the Lewis electron pair model

7.1.1 The electron pair at any cost

The 1916 paper of Lewis, introducing the concept of the electron pair, is one of the most important papers to be published in chemistry in this century. It is a revelation to read this paper and realize the extent to which his ideas continue to dominate chemical thinking. One must marvel at the intuitive leap that was involved in advancing the notion of the importance of the electron pair, coming as it did before the development of quantum mechanics or the still later introduction of the concept of electron spin. Indeed the opening quotation from the paper by Lewis was penned in response to Bohr's then recently advanced mechanical model of the atom. So strong was his conviction regarding the existence of a closely coupled pair of electrons (we would say a localized pair) as being necessary for the understanding of chemical bonding, he felt forced to postulate the failure of Coulomb's law for small interelectronic separations, and to thus necessarily reject any model which made use of this law. It is clear from the quotation given at the beginning of Chapter 5 that Lewis later (1933) abandoned this extreme position.

The Lewis model is second only to the molecular structure hypothesis itself in providing a conceptual basis for much of present-day chemical thinking, particularly with regard to models of molecular geometry and reactivity. It is essential that the physical basis of this model be established if one is to place all of the vital concepts of chemistry on a firm theoretical footing.

It is shown in this chapter that, while the Lewis model of the electron pair does not correspond to the existence of spatially localized pairs of electrons, it does find a more abstract but no less real physical expression in the topological properties of the Laplacian of the charge density, the quantity $\nabla^2 \rho(\mathbf{r})$ as defined in eqn (2.3) (Bader and Essén 1984; Bader et al. 1984). The

Laplacian of the charge density $\nabla^2\rho(\mathbf{r})$, along with $\rho(\mathbf{r})$ and $\nabla\rho(\mathbf{r})$, serve to define the conceptual models of chemistry and they provide the necessary basis for the theoretical description of these models. The Laplacian of the charge density $\nabla^2\rho(\mathbf{r})$, as demonstrated in the two preceding chapters, plays a dominant role throughout the theory of atoms in molecules. It is shown here that the Laplacian provides a link between theory and the chemical models of geometry and reactivity that are based upon the Lewis model.

What Lewis (1916) considered to be one of the major accomplishments of his model, was the unified description it provided of chemical bonding. 'However, according to the theory which I am now presenting, it is not necessary to consider the two extreme types of chemical combination, corresponding to the polar and the very nonpolar compounds, as different in kind, but only as different in degree.' The different degrees of polarity corresponded to the equal or unequal sharing of the electron pair between the 'kernels' of the two atoms, and the two resulting extremes of bonding are to-day referred to as the covalent and ionic limits. It is shown in Section 7.4 that the Laplacian of the charge density provides a physical basis for the classification of the continuous spectrum of atomic interactions and their mechanical consequences, which lie between these two extremes of bonding.

7.1.2   Are electrons localized in pairs?

The concept of the localized electron pair has been a central theme in the development of bonding theories since Lewis first postulated that a chemical bond was a consequence of a shared electron pair. However, the total electron density distribution in a molecule shows no indication of discrete bonding or non-bonding electron pairs. The principal topological property of the charge density of a many-electron system is that, in general, it exhibits local maxima only at the positions of nuclei. When one sums the individual orbital densities in the determination of the total charge density, all suggestions of spatially localized patterns of charge and the associated nodes disappear to yield the relatively simple topology exhibited by the *total* charge density ρ. While this topology, as demonstrated in preceding chapters, provides a faithful mapping of the concepts of atoms, bonds, and structure, it does not provide any indication of the maxima in a charge distribution which would correspond to the spatially localized pairs of bonded and non-bonded electrons as anticipated on the basis of the Lewis model. The possibility of obtaining a set of localized orbitals does not imply any physical localization of electrons. A choice of a particular localized set is arbitrary, because the summation of the orbital densities of any set, obtained by a unitary transformation of another, yields the same *total* density. Thus, using the forms of individual orbitals to describe the charge distribution is both incorrect and ambiguous. Does a keto oxygen for example, possess an 's' and a 'p' lone pair or two equivalent lone pairs as pictured on the basis of the canonical and localized sets of orbitals, respectively? The forms exhibited by individual orbitals are not evident in the topology of the total charge distribution. These comments are, of course, not criticisms of the orbital model, which is *the* model for the prediction and understanding of the *electronic structure* of a many-electron system, but rather of the misuse of this model to describe the *spatial structure* of the electron distribution through illustrations of individual orbitals or their corresponding densities.

The concept of a localized electron pair implies that there exists a region of real space in which there is a high probability of finding two electrons of opposite spin and for which there is a correspondingly small probability of exchange of these electrons with the electrons in other regions. Such physical localization of charge is a direct result of the operation of the Pauli exclusion principle as it affects the pair distribution function for electrons (Bader and Stephens 1975). The charge density $\rho(\mathbf{r}_1)$, the density of electrons at $\mathbf{r}_1$, is given by N times the probability density of finding one electron at $\mathbf{r}_1$. Its integral over the coordinates of one electron gives N, the total number of electrons. Similarly, the pair density $\rho(\mathbf{r}_1, \mathbf{r}_2)$, defined in eqns (E7.7) or (E1.3), the density of pairs of electrons with coordinates $\mathbf{r}_1$ and $\mathbf{r}_2$, is given by $N(N-1)/2$ times the probability density that one electron is at $\mathbf{r}_1$ when the other is at $\mathbf{r}_2$. Its integral over the coordinates of both electrons gives $N(N-1)/2$, the total number of distinct pairs of electrons. The antisymmetry requirement of the Pauli principle is imposed on the pair density through the introduction of the so-called Fermi hole. The theoretical development of this most important effect which the resulting Fermi correlation has on the pair distribution function is discussed in Section E7.1. We give here a qualitative discussion of how the operation of the Pauli exclusion principle can indeed lead to the localization of pairs of electrons, but follow this with a demonstration that the requirements for achieving such localization are, in general, not met.

As an electron moves through space, it carries with it a *doppelgänger*, its Fermi hole. The Fermi hole is a distribution function defined relative to an uncorrelated pair density, which determines the *decrease* in the probability of finding an electron with the same spin as some reference electron, relative to a given position of the reference electron. The value of the Fermi hole at the position of an α (or β) reference electron equals the total density of α (or β) electrons at that point, thereby ensuring that there is a zero probability of finding density of another α (or β) electron at the position of the reference electron. The Fermi hole, when integrated over all space for a fixed position of the reference electron, corresponds to the removal of one electronic charge of identical spin, and one may equally interpret the Fermi hole as a distribution function which describes the spatial delocalization of the charge of the reference electron. An electron can go only where its hole goes and, if the Fermi hole is localized, then so is the electron (Bader and Stephens 1975).

Consider an electron of α spin in the proximity of a nucleus. Because the electron is tightly bound in a deep potential well, its Fermi hole is strongly localized in its immediate vicinity. If the Fermi hole is localized so as to equal its maximum value ρ^α—the total density of α electrons—at every point over this region of space, then all other electrons of α spin are excluded from the potential well. The same result will be obtained for an electron of β spin and, as a consequence, a pair of electrons is confined to a region of space from which all other electrons of both α and β spin are excluded. Thus, Fermi correlation does not act directly to 'pair up' electrons. Rather, since there is no Fermi correlation between electrons of opposite spin, an α, β pair is obtained as a result of all other electrons of both α and β spin being excluded from a given region of space in which they are both bound by some attractive force. Repulsions between the electrons, of course, act in opposition to this effect and one finds that the long-range nature of the Coulomb force disrupts and limits the localization of charge (Sinanoğlu 1962).

In general, one finds such extreme localization of the Fermi hole for an electron pair in the core of an atom (Bader and Stephens 1975). In these cases it is possible to find a spherical boundary such that the contained Fermi correlation is maximized, a result which, as shown in Section E7.1, is equivalent to minimizing the fluctuation in the average population of the contained charge. It is in this sense of being able to extremize the degree of isolation of the electrons to a given region of space through a variation of the boundary that the electrons are referred to as being *physically localized*. It corresponds to finding a region of space such that the exchange of the electrons within the region with the rest of the system is minimized. Any change in its boundary increases the exchange and decreases the contained Fermi correlation and the degree of localization. The same spatial localization of the Fermi hole is found in a few simple hydrides such as LiH, BeH_2, and BH_3 in which, in addition to the core on Li, Be, or B, there are, respectively, one, two, and three pairs of electrons, each localized on a proton. However, in molecules where the bonding is predominantly covalent, and particularly when there are four or more electron pairs in a valence shell such as in CH_4, bounded regions do not exist which *maximize* the contained Fermi correlation and the electrons are not physically localized. Instead, the Fermi hole of each electron, and hence each electron itself, is delocalized to a significant extent throughout the valence shell of the central atom.

It is shown in Section E7.1 that the orbital requirement for physical localization of pairs of electrons is not that a set of orbitals be localized, but that each orbital be localized to its own separate region of space. Such separate localization is apparent even in the canonical set of orbitals for the same molecules, LiH, BeH_2, and BH_3, etc., which, as noted above, exhibit localization of electron pairs. Localized orbitals are in general, however, not that localized and they are not localized to separate regions of space (Daudel *et al.* 1976).

A physically localized pair of electrons is maximally isolated from the remainder of the system in that its exchange and all of its correlative interactions with the remainder of the system are minimized. Hence, the degree of physical localization as determined by the properties of the Fermi correlation parallels the relative importance of the intra- and interpair correlation energies as determined by various 'correlated pair' theories. In LiH, BeH_2, and BH_3 (Ebbing and Henderson 1965; Kutzelnigg 1973), for example, intrapair correlation accounts, respectively, for over 90, for 86, and for 77 per cent of the total correlation energy, in systems for which the corresponding Fermi localizations are 96, 93, and 82 per cent. As pointed out by Davidson (1972), by Sinanoğlu and Skutnik (1968), and by Kutzelnigg (1973), this is not the general result and, in molecules such as CH_4 and NH, the intrapair correlation contribution to the total correlation energy drops to 56 and 20 per cent, respectively.

The study of the localizability of the Fermi correlation and the related fluctuation in the average electron population, of the relative importance of the intra- and interpair contributions to the total correlation energy and the study of the degree of the separate localizability of orbitals all provide similar information and all indicate that, in general, the physics of a many-electron system is not dominated by the behaviour and properties of individual electron pairs. In addition, localized pairs of electrons, bonded or non-bonded, are not evident in the topological properties of the charge density and, moreover, the pair density does not, in general, define regions of space beyond an atomic core in which pairs of electrons are localized.

7.1.3   Properties of the Laplacian of the charge density

Where then to look for the Lewis model, a model which in the light of its ubiquitous and constant use throughout chemistry must most certainly be rooted in the physics governing a molecular system? If one reads the introductory chapter on fields in Morse and Feshbach's book *Methods of theoretical physics* (1953), one finds a statement to the effect that the Laplacian of a scalar field is a very important property, for it determines where the field is locally concentrated and depleted. The Laplacian of the charge density at a point **r** in space, the quantity $\nabla^2\rho(\mathbf{r})$, is defined in eqn (2.3). This property of the Laplacian of determining where electronic charge is locally concentrated and depleted follows from its definition as the limiting difference between the two first derivatives which bracket the point in question as defined in eqn (2.2) and illustrated in Fig. 2.2.

To make clear what is meant by a 'local charge concentration' as defined by a second derivative, consider the function $f(x)$ displayed in Fig. 7.1 along with its first and second derivative. Since $f(x)$ is a steeply rising function in the region of x_1, the magnitude of the slope at $x_1 - \Delta x$ is greater than that at $x_1 + \Delta x$ and, since both slopes are negative, the curvature of $f(x)$ at x_1 is positive. The function exhibits a shoulder in the region of the point x_2 and the slope at $x_2 + \Delta x$ is of greater magnitude than at $x_2 - \Delta x$. Thus, the curvature of $f(x)$ is negative in the region where $f(x)$ exhibits a shoulder. The function exhibits a point of inflexion between x_1 and x_2 where the curvature is zero. From eqn (2.2) for the curvature, one sees that, where the curvature of $f(x)$ is negative, the value of $f(x)$ at x is greater than the average of its values at the neighbouring points, $x + dx$ and $x - dx$, with the reverse being true where the curvature is positive. It is in this sense that $f(x)$ is said to be concentrated in regions where $d^2f(x)/dx^2 < 0$ and depleted where $d^2f(x)/dx^2 > 0$. To graphically emphasize the fact that $f(x)$ is concentrated where its curvature is negative, the negative of the curvature is plotted. Note that the curvature exhibits its largest negative value at x_2 and f(x) *is said to be maximally concentrated at that point, the point where the negative of the curvature exhibits a maximum.*

The slope of $f(x)$ (Fig. 7.1) is always negative. Thus, $f(x)$ itself does not exhibit either maxima or minima, i.e. points where $f'(x) = 0$, aside from the maximum at $x = 0$. We see that the existence of regions where a function is locally concentrated or depleted as determined by the sign of its second derivative does not imply the corresponding existence of maxima or minima in the function itself.

All of the above considerations carry over into three dimensions and, in particular, it follows from eqn (2.2) that the value of $\rho(\mathbf{r})$ is greater than the average of its values over an infinitesimal sphere centred on $\mathbf{r}$ when the sum of the three curvatures of ρ is negative, that is, when $\nabla^2\rho(\mathbf{r}) < 0$, and $\rho(\mathbf{r})$ is less than this average when $\nabla^2\rho(\mathbf{r}) > 0$. The charge density decays exponentially from a nucleus and, in general, the curvature of $\rho(\mathbf{r})$ along a radial line from the nucleus is positive, as for $f(x)$ at x_1 in Fig. 7.1. Its two curvatures perpendicular to a radial line are, however, negative. Thus, $\nabla^2\rho(\mathbf{r})$ may change sign even in the absence of a shoulder in the charge distribution, the presence of which in any event is of relatively rare occurrence. It is important to distinguish between local maxima and minima or shoulders in the charge density and corresponding extrema in its Laplacian distribution, and this is best done by first comparing these two distributions for an atom with a spherically symmetric distribution of charge.

Figure 7.2 displays the value of the charge density and of its associated Laplacian distribution as a function of position in the y, z-plane for an argon atom. Since the charge distribution is spherically symmetric, the same display is obtained for any plane containing the nucleus. The charge density exhibits

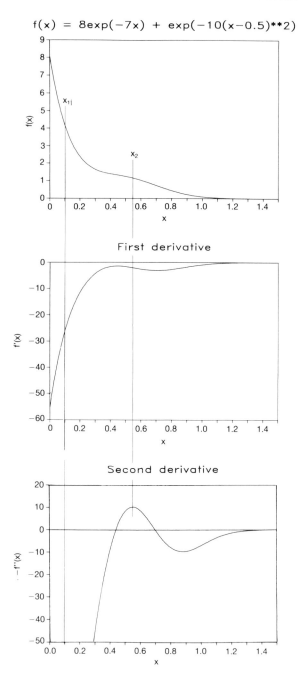

FIG. 7.1.   Plots of a monotonically decreasing function $f(x)$ and its first and second derivatives. The negative of the second derivative is shown to emphasize that a function is concentrated in a region where its second derivative is negative.

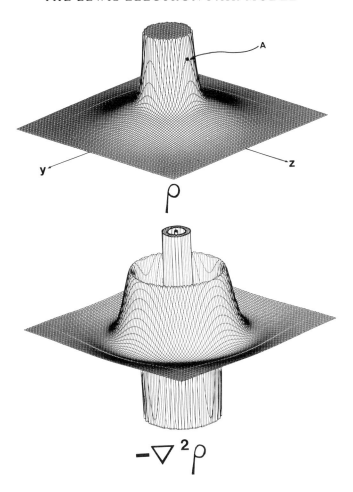

FIG. 7.2. Relief maps of the charge density and the negative of its Laplacian for a plane containing the nucleus of an argon atom. Function values above an arbitrary magnitude are not shown. At the point A, the curvature of ρ along z (the radial curvature) is positive while the curvature along the y-axis (a tangential curvature) is negative. In addition to the spike-like maximum in $-\nabla^2\rho$ at the nucleus, there are two shells of charge concentration and three shells of charge depletion corresponding to the three quantum shells in this atom.

a single maximum at the position of the nucleus (the maximum value is not shown) and decays monotonically for motion away from the nucleus along any radial line. The shell structure of an atom is not evident in the distribution of charge throughout three-dimensional space. Shell structure is made evident in the radial distribution function, a one-dimensional distribution obtained by expressing the charge density in terms of spherical polar coordinates followed by its integration over the angular coordinates. The radial distribution function is the probability of finding electronic charge in the infinitesimal volume lying between two concentric spheres of radii r and $r + dr$. Thus a maximum in a radial distribution function determines a value of r at which one is most likely to find electronic charge averaged over all angles, but the actual distribution of charge in three-dimensional space does not exhibit a maximum at the corresponding value of r.

To understand the very different topology exhibited by the Laplacian of the charge distribution, we consider a spherical surface of constant density. An axis through the origin, say the z-axis, is normal to this surface and intersects it at the point $(0, 0, z)$. The charge density increases with decreasing z and, in general, the curvature along such a radial line is found to be positive. Thus $\partial \rho / \partial z < 0$ and $\partial^2 \rho / \partial z^2 > 0$. Since the curvature of ρ along z is positive, the value of $\rho(0, 0, z)$ is less than the average of its values at the points $(0, 0, z \pm dz)$. The charge density is however, a maximum at $(0, 0, z)$ in the tangent plane, $\partial \rho / \partial x = \partial \rho / \partial y = 0$, and, hence, the corresponding curvatures are negative (see Fig. 7.2). Thus $\rho(0, 0, z)$ is greater than the average of its values at ($\pm$ dx, 0, z) and $(0, \pm dy, z)$. If the sum of the magnitudes of the two negative curvatures, those perpendicular to a radial line, exceeds the value of the positive or parallel curvature, then $\nabla^2 \rho(0, 0, z) < 0$ and the value of ρ at this point is greater than that obtained by averaging ρ over all points in the neighbourhood of the point $(0, 0, z)$. By determining a radius in an atom at which $\nabla^2 \rho(\mathbf{r})$ is a minimum (or a maximum), one determines a radius of a sphere on which electronic charge is maximally concentrated (or depleted).

The observation that ρ exhibits a maximum at a nuclear position does not imply that the contraction of ρ towards the nucleus is the dominant one at all distances from a nucleus. Indeed this cannot be so, for the integral of $\nabla^2 \rho(\mathbf{r})$ over an atom free or bound must vanish, as a consequence of the zero-flux surface condition. Because of this, in addition to the regions where the contraction of ρ towards the nucleus along a radial line is dominant and $\nabla^2 \rho(\mathbf{r}) > 0$, there must exist a region or regions where the magnitudes of the curvatures of ρ perpendicular to a radial line are dominant, yielding negative values for the Laplacian of ρ. Thus $\nabla^2 \rho(\mathbf{r})$ exhibits spherical nodes in an atom (values of the radius for which $\nabla^2 \rho = 0$) and their number is found to be related to its shell structure (Fig. 7.2). There are pairs of regions (one negative, one positive) for each principal quantum shell, with the innermost region being a spike-like maximum in charge concentration (Bader and Beddall 1972; Bader and Essén 1984).

The number of shells defined in this manner by the Laplacian of the charge density or in terms of the number of maxima in the radial distribution function can be less than the principal quantum number for elements past Ca. The Laplacian distribution, while exhibiting the required number of outer maxima and minima for elements with $Z > 40$, does not undergo the corresponding changes in sign (Sagar et al. 1988; Shi and Boyd 1988). The addition of relativistic corrections to a wave function causes the outer charge distribution to contract inwards for atoms with large Z. Savin (personal communication) has observed that even with this contraction, the missing shell structure is not present in the Laplacian distribution.

The presence of a shell of charge concentration and one of charge depletion in the Laplacian distribution for each quantum shell is demonstrated in Fig. 7.3 for the four quantum shells of the Kr atom. While the diagram displays the behaviour of $-\nabla^2\rho$ along a given radial line, it should be borne in mind that this is a representation of the structure present in three-dimensional

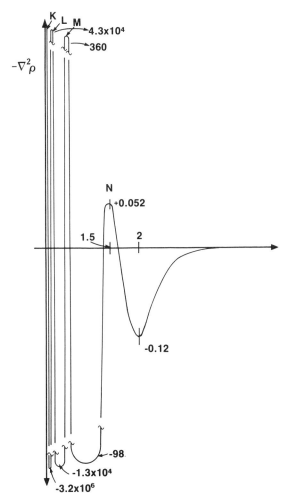

Fig. 7.3.   Schematic representation of the profile of $-\nabla^2\rho$ for the krypton atom along a radial line. Unlike a radial distribution plot, which is a one-dimensional function, the Laplacian distribution displays shell structure in three-dimensional space.

space. The Laplacian distribution, unlike the radial distribution function, exhibits shell structure in real space for a free atom and also for an atom in a molecule. In this and other displays of the Laplacian, the function plotted is $-\nabla^2\rho$, such that a maximum in the function plotted corresponds to a maximum in the concentration of charge. The first spherical node in $\nabla^2\rho$ for a one-electron atom in its ground state occurs at a radius $r = 1/Z$, and this is very nearly true for a many-electron atom as well, becoming increasingly so as the nuclear charge Z increases (Bader and Essén 1984). Thus, the innermost region of the argon atom, where $\nabla^2\rho$ is negative, in Fig. 7.2 appears in the display as a spike-like region of charge concentration. There are two more regions where $\nabla^2\rho < 0$, separated and bounded by regions where $\nabla^2\rho > 0$, corresponding to the presence of a total of three pairs of spherical shells of alternating charge concentration and charge depletion in agreement with the presence of three quantum shells in this atom.

The value of the radius r for which $-\nabla^2\rho$ attains its maximum value in each succeeding shell defines the radius of a sphere on which electronic charge is maximally concentrated in each shell. These values are listed in Table 7.1 and, like the corresponding maxima in the radial distribution function whose most probable r values are very similar, they exhibit periodic variations. The radii decrease across a row of the periodic table for a given shell and the values for succeeding shells increase down a family of elements. Most importantly, the Laplacian of the charge density defines the chemically important valence-shell charge concentration, the shell of charge concentration which, upon chemical combination, is distorted to yield maxima corresponding in number and relative position to the electron pairs anticipated by the Lewis and related models, such as the VSEPR model of molecular geometry (Gillespie 1972). The description of these changes requires a study of the topology of the Laplacian distribution.

7.1.4   The valence shell charge concentration (VSCC)

Extrema in the Laplacian of ρ are classified by rank and signature in the same way as are critical points in the charge density. The critical points in the Laplacian occur where $\nabla(\nabla^2\rho) = 0$ and the eigenvalues of the Hessian of $\nabla^2\rho$ are the principal curvatures of $\nabla^2\rho$ at the critical point. The topological discussions will always refer to the negative of the Laplacian, the quantity $-\nabla^2\rho$. Since charge is concentrated where $\nabla^2\rho < 0$, a local maximum in $-\nabla^2\rho$ is synonymous with a maximum in the concentration of electronic charge. Thus a local maximum in $-\nabla^2\rho$, a $(3, -3)$ critical point with $\nabla^2\rho < 0$, will denote a local concentration in electronic charge and a local minimum in $-\nabla^2\rho$, a $(3, +3)$ critical point, and, with $\nabla^2\rho > 0$, will denote a local depletion in electronic charge.

Table 7.1

(a) Radii* r_n of spheres of maximum charge concentration for quantum shells $n \geqslant 2$

	Li	Be	B	C	N	O	F	Ne
r_2	2.49	1.59	1.19	0.94	0.78	0.66	0.57	0.50

	Na	Mg	Al	Si	P	S	Cl	Ar
r_2	0.44	0.40	0.36	0.33	0.30	0.28	0.26	0.24
r_3	3.44	2.55	2.08	1.76	1.52	1.34	1.20	1.08

	K	Ca	Ga	Ge	As	Se	Br	Kr
r_2	0.23	0.21	0.13	0.12	0.12	0.12	0.11	0.11
r_3	0.98	0.90	0.49	0.47	0.45	0.43	0.41	0.39
r_4	4.94	3.77			2.18	1.83	1.65	1.50

(b) Values* of $-\nabla^2\rho$ at r_n

n	Li	Be	B	C	N	O	F	Ne
2	0.002	0.027	0.141	0.506	1.447	3.480	7.592	15.29

	Na	Mg	Al	Si	P	S	Cl	Ar
2	29.99	55.61	96.15	157.4	247.4	375.3	553.3	794.7
3	0.0002	0.002	0.009	0.038	0.115	0.273	0.592	1.196

	K	Ca	Ga	Ge	As	Se	Br	Kr
2	1115.0	1544.0	18494	22126	26303	31124	36615	42961
3	2.219	3.818	109.0	140.2	179.7	226.9	288.3	361.4
4	-3.1E-5	-1.2E-4			-0.026	-0.025	-0.010	0.052

*Radii and values of $-\nabla^2\rho$ are in atomic units, where 1 au of length = a_0 = 0.52918 Å. Units of Laplacian are e/a_0^5 = 24.100 $e/\text{Å}^5$.

The outer quantum shell of an atom is divided into an inner region over which $\nabla^2\rho < 0$ and an outer one over which $\nabla^2\rho > 0$. The portion of the shell over which $\nabla^2\rho < 0$, is called the *valence-shell charge concentration or VSCC*. Within this shell is the sphere over whose surface the valence electronic charge is maximally and uniformly concentrated.

Each point on the surface of the sphere of maximum concentration in the valence shell of charge concentration is a $(1, -1)$ critical point: the two curvatures tangent to the surface are zero reducing the rank to one and the curvature perpendicular to the surface is negative since $-\nabla^2\rho$ is a maximum on the surface of the sphere. In general this surface persists when the atom is in chemical combination (the derivative of $-\nabla^2\rho$ normal to the surface remains equal to zero and its corresponding curvature is negative), but the sphere is distorted and the surface is no longer one of uniform concentration, as the two tangential curvatures assume either positive or negative values. The topology of the Laplacian of ρ on this surface is equivalent to the description of the hills and valleys of a surface terrain. If a local maximum is formed on the surface, then the two tangential curvatures are negative and the result is a $(3, -3)$ critical point in $-\nabla^2\rho$. If the two tangential curvatures assume positive values, a local minimum is formed on the surface and the result is a $(3, +1)$ critical point in $-\nabla^2\rho$. If the two curvatures assume values of opposite sign, there is a saddle point in the surface and the result is a $(3, -1)$ critical point in $-\nabla^2\rho$. The local maxima that are created within the valence shell of charge concentration provide a one-to-one mapping of the electron pairs of the Lewis model, both bonded and non-bonded.

Just as the energy changes of interest to chemistry are but small fractions of the total energy of a system, so the changes induced in the charge density by the interactions between atoms are only small 'ripples' in the total density. These small changes are magnified and made evident by the Laplacian of the charge density. What is remarkable is that the description of the changes so obtained corresponds to the Lewis model of spatially localized electron pairs.

7.1.5 · Bonded and non-bonded charge concentrations

The structure exhibited by the Laplacian through its collection of valence-shell critical points is characteristic of a given atom in a given classical valence state. In methane and methyl fluoride there is a local maximum in the VSCC of the carbon along the C–F and each of the C–H bond paths (Fig. 7.4a). This figure displays the zero envelope of the Laplacian distribution which separates the VSCC from regions of charge depletion. This surface encompasses the four local charge concentrations or $(3, -3)$ critical point in $-\nabla^2\rho$ in the VSCC of carbon and is contiguous with the VSCC's on the protons in methane. The greatly reduced charge concentration on the C–F

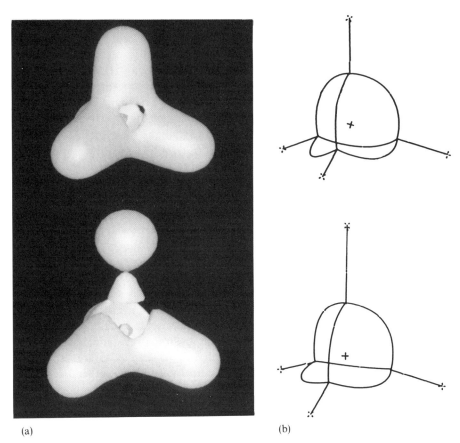

(a)   (b)

FIG. 7.4. Representations of the Laplacian distributions of methane and methylfluoride. The figures in (a) are displays of the zero envelope of $\nabla^2\rho(\mathbf{r})$, those in (b) of the atomic graphs. The envelope encompassing the inner shell charge concentration on carbon appears as a small sphere. The envelopes of the bonded maxima in the VSCC of carbon also encompass the protons in CH_4 and CH_3F. There is a transfer of charge from C to F in CH_3F and the bonded maximum along the C–F axis is reduced to the small region lying between the C nucleus and the envelope on F. An atomic graph displays the connectivity of the critical points in a VSCC. The carbon nucleus is denoted by a solid cross, the positions of the remaining nuclei by open crosses. There is a bonded maximum, a $(3, -3)$ critical point in $-\nabla^2\rho$, at each of the four vertices.

axis in methyl fluoride is separate from the remainder of the VSCC on carbon. Each of the bonded maxima in the VSCC of carbon is linked to the other three by unique pairs of trajectories of the gradient of $-\nabla^2\rho$, which originate at intervening saddle points on the surface of charge concentration (Fig. 7.4b). They are $(3, -1)$ critical points in $-\nabla^2\rho$. These lines are the analogues of the bond paths defined by the corresponding trajectories of the gradient of the charge density. This network of lines, in the case of the

Laplacian distribution called *the atomic graph*, partitions the surface of charge concentration into four segments and the basic structure is that of a tetrahedron with curved faces. In the centre of each face there is a local minimum in the surface of charge concentration, a $(3, + 1)$ critical point in $- \nabla^2 \rho$. Thus, the spherical surface of uniform charge concentration of a free carbon atom is transformed, upon the formation of methane, into one with four tetrahedrally arranged concentrations of charge, together with centres of local charge depletion in each of the resulting triangular faces. The surface of charge concentration within the VSCC of the carbon atom in the molecule is no longer spherical, the different critical points lying at slightly different distances from the nucleus, but all within a range of 0.010 Å about the average value of 0.545 Å. This value represents a slight increase over the free-atom value of 0.500 Å. The replacement of a hydrogen in CH_4 by a fluorine leaves the basic structure of the atomic graph for carbon unchanged, but it does cause major changes in the extents of charge concentration and charge depletion (Fig. 7.4a). Of chemical significance is the increase in the extent of charge depletion in the triangular face opposite the fluorine atom, a point illustrated and discussed in Section 7.3.2.

The manner in which the uniform sphere of charge concentration for an oxygen atom is distorted upon formation of the water molecule to yield local charge concentrations in its valence shell is illustrated in some detail. Relief maps of the charge density and the negative of its Laplacian are shown in Fig. 7.5 for two planes of the water molecule. In the plane perpendicular to the one containing the nuclei, ρ exhibits a single maximum at the position of the oxygen nucleus as anticipated on the basis of its general topological behaviour. Thus, as previously noted, the topology of ρ itself offers no direct evidence to support a model that assumes the existence of two localized, non-bonded electron pairs. There are, however, small departures from circular symmetry in ρ giving rise to variations in its local curvatures and the corresponding display of the Laplacian of ρ shows, in a rather dramatic fashion, the existence of two local concentrations of electronic charge in the valence shell of the oxygen atom in this non-bonded plane. In such a two-dimensional display, the uniform sphere of maximum charge concentration appears as a circular ring in the free atom (see Fig. 7.2), while here it appears with two maxima and two minima. The third curvature at each of the maxima is negative, showing that the two-dimensional maxima are indeed $(3, - 3)$ critical points in $- \nabla^2 \rho$ which indicate the presence of two local concentrations of non-bonded electronic charge.

In the plane of the nuclei, ρ again exhibits maxima only at the positions of the nuclei and ridges linking the oxygen nucleus to the protons, indicating the presence of two bond paths. The negative of the Laplacian distribution, on the other hand, in addition to the maxima at the positions of the nuclei, exhibits two equivalent bonded maxima in the valence-shell charge concen-

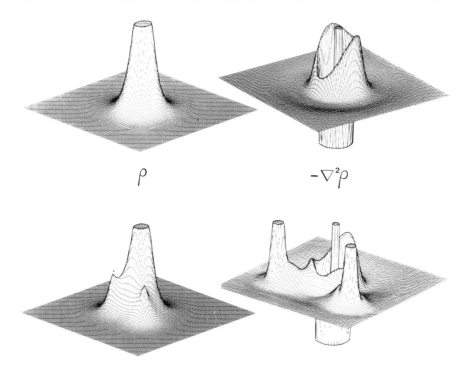

ρ $-\nabla^2\rho$

FIG. 7.5. Relief maps for ρ and $-\nabla^2\rho$ for the two symmetry planes of the water molecule: lower maps are for the bonded plane (plane containing the nuclei) and upper ones for the non-bonded plane. There is a spike-like concentration of charge at the position of the oxygen nucleus (which in this and remaining diagrams is terminated at an arbitrary value), surrounded by the inner shell of charge depletion. The valence shell of charge concentration, the VSCC, of oxygen is not uniform but exhibits maxima and saddles in these two-dimensional relief maps. In the non-bonded plane ρ exhibits a single local maximum at the position of the oxygen nucleus. In this same plane the VSCC of oxygen exhibits two local maxima ($(3, -3)$ critical points in $-\nabla^2\rho$) and two saddle points in $-\nabla^2\rho$. In the bonded plane, ρ exhibits a local maximum at each of the nuclei, as does $-\nabla^2\rho$. In addition, the VSCC of oxygen also exhibits two local maxima, one along each bond path to a proton. What appears as a third maximum in the VSCC of oxygen is not a local maximum but *another* view of the $(3, -1)$ saddle point between the two non-bonded maxima appearing in the upper diagram. The value of $-\nabla^2\rho$ at this saddle point is less than its value at a non-bonded maximum. Thus, the lower diagram makes clear the general observation that bonded charge concentrations are smaller than non-bonded concentrations. The two diagrams show that the uniform sphere of charge concentration of an oxygen atom is distorted to yield four local maxima or local charge concentrations and four intervening saddle points.

tration of the oxygen atom which prove to be $(3, -3)$ critical points. The third, even larger maximum in the two-dimensional display of the VSCC of oxygen in the bonded plane is actually another view of the $(3, -1)$ critical point linking the two non-bonded maxima. This same critical point appears in the non-bonded plane as a saddle point indicating that it does indeed possess one positive curvature. One notes that the magnitude of $\nabla^2\rho$ at this saddle point is less than that at the positions of the non-bonded maxima but greater than that at the positions of the bonded maxima. Clearly, the non-bonded charge concentrations are of considerably greater magnitude than are the bonded concentrations. The second saddle point in $-\nabla^2\rho$, illustrated in these two diagrams and lying between the two bonded maxima, is a $(3, +1)$ critical point. The relatively large angle formed between the two non-bonded charge concentrations causes a change in the atomic graph from that found for carbon in methane. The saddle point between the bonded maxima and two of the local minima in neighbouring faces coalesce to yield a single $(3, +1)$ critical point between the bonded maxima and one edge of the tetrahedron is missing in the atomic graph for oxygen.

Thus the VSCC of the oxygen atom in the water molecule exhibits four local maxima in $-\nabla^2\rho$, i.e. four local concentrations of electronic charge—two bonded concentrations (each lies on a bond path linking the oxygen to a proton) and two larger and equivalent non-bonded concentrations. Orbital models of non-bonded electron pairs offer alternative descriptions for an oxygen atom in water or in a ketone: two distinct pairs, one s-like and the other p-like when two orbitals of the canonical set are invoked, or two equivalent pairs when described within some transformed, localized set of orbitals. The total charge density has a unique set of properties for each quantum state, properties that are independent of the choice of orbital set and of the consequences observed on excitation or ionization.

The four local charge concentrations form the corners of a distorted tetrahedron, with the angle between the bonded maxima of $103°$ being less than the angle of $138°$ found between the non-bonded maxima. Lewis, of course, assumed a tetrahedral arrangement for four valence electron pairs and one has here the beginnings of a theory of molecular geometry, a subject pursued in the following section.

There is no suggestion in this demonstration of the correspondence between the number and relative positions of local charge concentrations and the electron pairs assumed in the Lewis model that the former proves the existence of the latter. As discussed above, there is, in general, no evidence for the existence of localized pairs of electrons beyond the core regions of atoms, free or bound. Thus the presence of a $(3, -3)$ critical point in $-\nabla^2\rho$ does not imply the existence of a localized pair of electrons. It does show that electronic charge is locally concentrated at that point. However, the faithfulness of the mapping between the assumed pairs of the model and the local maxima in $-\nabla^2\rho$, as will be further illustrated, leaves no doubt that the number of local charge concentrations present in the valence shell of an atom is related to an electron count and that the Lewis model has a validity which extends beyond its statement in terms of the model of localized electron pairs.

7.2   A physical basis for the VSEPR model of molecular geometry

7.2.1   Résumé of the VSEPR model

The valence-shell electron pair repulsion or VSEPR model of Gillespie (1972) is a natural extension of the localized electron pair model of Lewis. It has become the most successful and widely used model for the prediction of geometries of closed-shell molecules. We give here only a brief review of the basic tenets of this model as regards the number, relative orientation, and size of the localized electron pairs assumed by the model for differing numbers of valence electrons. Sufficient examples are given to demonstrate that these extensions to the assumed properties of the Lewis pair are recovered in the corresponding properties of the local charge concentrations as defined by the Laplacian of the electronic charge density. More details of the comparison of the VSEPR model and the properties of the VSCC of the central atom are given by Bader *et al.* (1984, 1988).

The VSEPR model has two basic assumptions: (1) the valence charge density is spatially localized into pairs of electrons; and (2) the geometrical arrangement of the ligands about an atom is that which maximizes the interpair separation, for both the bonded and non-bonded pairs. Initially, one does not distinguish between bonded and non-bonded pairs but simply determines the most probable arrangement of all the pairs on the surface of a sphere by maximizing the smallest distance between any two pairs. If all the pairs are bonded and equivalent, this most probable arrangement determines the geometry. In general, however, not all the pairs in the valence shell are equivalent and in terms of the electron-pair sphere model this means that the electron pair domains are not all the same size. This inequivalence leads to the following subsidiary postulates.

1. Non-bonding or lone pairs have larger domains than bonding pairs in the same valence shell.

2. Bonding pair domains in the valence shell of a central atom decrease in size with increasing electronegativity of the ligand, and increase in size with increasing electronegativity of the central atom.

3. Double and triple bond domains which are composed of two and three electron pairs, respectively, are larger than single-bond electron pair domains.

The spherical surface, on which the electron pairs are assumed to be localized in Gillespie's model, is identified with the sphere of maximum charge concentration in the VSCC of the central A atom, and the localized pairs of electrons are identified with the local concentrations of electronic charge present on the sphere of maximum charge concentration.

7.2.2   Valence-shell charge concentrations with four maxima

The hydrides AH_n, A = C, N, and O, are assumed to possess four pairs of electrons in the valence shell of A. The properties of the bonded and non-bonded charge concentrations in the VSCC of the A atom are listed in Table 7.2. In each case, as illustrated above for C and O, the surface of the sphere of maximum charge concentration exhibits four local maxima in $-\nabla^2\rho$, but it is no longer completely spherical, as it was in the free atom. The radial distance of a bonded charge concentration is increased by 0.02 to 0.04 Å over its value in the free atom, while the non-bonded charge concentrations are drawn slightly closer to the nucleus (Table 7.2). The magnitude of $\nabla^2\rho$ is considerably greater for a non-bonded maximum than it is for a bonded maximum and electronic charge is concentrated to a greater extent in the former than in the latter. The radial curvature of $-\nabla^2\rho$, the curvature perpendicular to the surface of the sphere of charge concentration listed as μ_3 in Table 7.2, is always greater for the non-bonded than for the bonded charge concentrations and the former concentrations are 'thinner' in radial extent than are the bonded ones.

Also of importance to the VSEPR model are the relative sizes of the bonding and non-bonding electron pairs. The estimate of the size of the local maximum in $-\nabla^2\rho$ is given in terms of its area on the sphere of charge concentration. This area is given by the expression $2\pi r_A^2 (1 - \cos \alpha/2)$ where the angle α is the average of the angles subtended at the nucleus by the lines linking the nucleus to the edges of the charge concentration in question (Bader *et al.* 1984). The boundary of a charge concentration is defined by the position of the neighbouring saddle points in $\nabla^2\rho$, or points where $\nabla^2\rho$ changes sign if $\nabla^2\rho > 0$ at the saddle point. In agreement with the VSEPR model, the area of a non-bonded charge concentration is considerably greater than that of a bonded concentration in the same molecule. These observations regarding the properties of the bonded and non-bonded charge concentrations are found to be of general applicability, regardless of the number of electron pairs present on the A atom. They are understandable in terms of the VSEPR postulate that a non-bonded charge concentration is attracted by a single nucleus, while a bonded one is attracted by, and shared between, two nuclei.

The transformation of a bonded into a non-bonded charge concentration on passage from CH_4 to NH_3 results in a decrease in the angle between the bonded charge concentrations, the angle bAb, and, hence, to an angle greater than tetrahedral between the non-bonded and bonded charge concentrations, the angle nAb. Correspondingly, the non-bonded charge concentration in NH_3 occupies a larger fraction of the surface of maximum charge concentration than does a bonded one. Thus a non-bonded charge concentration, while radially more contracted than a bonded one, is laterally more diffuse and

Table 7.2

Bonded and non-bonded charge concentration on A in AX_n*

AX_n	R(AX) (calc.) (Å)	Angle XAX (calc.) (°)	Bonded charge concentrations					Angle nAb (°)	Non-bonded charge concentrations					Net Charge on X
			r_A (Å)	$-\nabla^2\rho$ (au)	Angle bAb (°)	Area (Å²)	μ_3 (au)		r_A (Å)	$-\nabla^2\rho$ (au)	Angle nAn (°)	Area (Å²)	μ_3 (au)	
CH_4	1.087	109.5	0.612	1.158	109.5	0.99	6.2							−0.04
SiH_4	1.477	109.5	Charge transfer to hydrogen											−0.74
NH_3	1.007	105.7	0.510	1.685	104.4	0.71	19.5	114.2	0.388	2.299		1.12	100.0	+0.34
PH_3	1.408	95.4	0.945	0.638	96.1	1.61	2.5	120.8	0.775	0.325		2.49	7.8	−0.63
OH_2	0.947	103.8	0.441	2.057	101.0	0.38	51.4	104.4	0.339	4.138	134.0	0.50	210.1	+0.56
SH_2	1.330	94.3	0.841	0.806	95.4	1.17	4.4	105.4	0.696	0.554	133.7	2.20	17.3	−0.34
NF_3	1.337	102.5	0.473	0.990	97.6	0.45	28.2	119.6	0.358	4.450	119.6	0.70	294.7	−0.36
PF_3	1.548	97.5	Charge transfer to fluorine						0.750	0.502	119.8	2.15	12.1	−0.90
$ClCl_2^+$	2.002	106.3	0.653	0.717	104.3	1.48	25.3	101.9	0.615	1.148	140.9	1.88	48.9	
ClF_2^+	1.576	100.2	0.658	0.671	99.0	0.66	27.9	102.0	0.604	1.541	142.8	1.84	62.2	−0.18
S			Charge transfer to oxygen						0.678	0.785		1.56	24.4	
SO_2	1.423	118.7	0.426	1.769	124.0	1.82	54.1	118.0						
O									0.346	3.697		182.0		−1.26

*All results calculated (calc.) using the 6–21G** basis set.

possesses a larger surface area. In H_2O where there are two non-bonded charge concentrations, the angle between the bonded concentrations is further reduced to $103°$. This angle is to be compared with an angle of $138°$ between the non-bonded concentrations. The effects on the local charge concentrations in the VSCC of A of replacing A by its neighbouring element in the same family of the periodic table and of increasing the electronegativity of the ligands are also illustrated by the data in Table 7.2. Subsiduary postulate (3) can be illustrated using the sizes of the C–C bonded maxima in ethane, ethylene, and acetylene which, expressed in terms of surface areas, are 0.62, 1.24, and 1.97 $Å^2$, respectively. By including the size of a non-bonded charge concentration in the VSCC of carbon in singlet methylene, which is 0.94 $Å^2$, one establishes that, for carbon, a doubly bonded charge concentration occupies a larger fraction of the VSCC than does a non-bonded one. The same is true for sulphur in SO_2 where the size of the non-bonded charge concentration is 1.56 $Å^2$, while the size of the doubly bonded concentration is 1.82 $Å^2$.

All of these observations regarding the properties of the local charge concentrations in the VSCC of the A atom are in complete accord with the tenets of the VSEPR model. The primary observation in the 10-electron hydrides is that of the presence of four local concentrations of charge in the valence shell of A. A non-bonded charge concentration is larger than a bonded one and occupies a larger fraction of the surface of charge concentration than does a bonded one. The angle subtended by two non-bonded concentrations is greater than that subtended by two bonded concentrations in an equilibrium geometry.

7.2.3   Valence-shell charge concentrations with five and six maxima

The molecule ClF_3 possesses five pairs of electrons in the valence shell of the Cl atom. The most probable arrangement of five electron pairs in the VSEPR model is a trigonal bipyramid. The five sites are not equivalent and, hence, the model predicts that the two non-bonded pairs of electrons in ClF_3 should occupy two of the three spatially less restricted equatorial sites, thereby predicting the 'T-shaped' geometry observed for this molecule. In addition to studying the properties of the VSCC of the Cl atom in the equilibrium geometry, two other constrained higher-energy geometries were studied (Bader *et al.* 1984). They were obtained by assigning all three fluorine atoms to equivalent equatorial sites to yield a planar D_{3h} geometry 192 kJ/mol higher in energy or to one axial and two equatorial sites to yield a tripod-like geometry, 263 kJ/mol less stable than the 'T-shaped' geometry. The calculated equilibrium geometry is in good agreement with experiment. The calculated values, followed by the experimental ones in parentheses, of the equatorial and axial bond lengths and FCF angle are 1.68(1.70) Å, 1.61(1.60) Å and 85.4°(87.5°), respectively. Relief maps of the negative of the Laplacian of ρ for two planes of the equilibrium geometry are displayed in Fig. 7.6. The properties of the local charge concentrations in the VSCC of the A atom are given in Table 7.3.

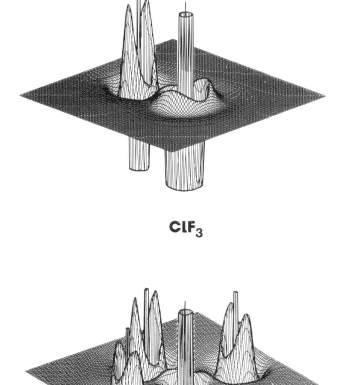

ClF$_3$

FIG. 7.6.   Relief maps of $-\nabla^2\rho$ for the equatorial (upper) and axial (lower) planes of ClF$_3$. The chlorine atom exhibits three shells of charge concentration. The equatorial plane shows the presence of two non-bonded and one smaller bonded charge concentration. In the axial plane there are three bonded charge concentrations and a fourth apparent maximum which is actually another view of the $(3, -1)$ critical point between the two non-bonded maxima. Thus the VSCC of Cl possesses two non-bonded and three bonded concentrations of charge.

Table 7.3

Bonded and non-bonded charge concentrations in ClF_3, SF_4, SF_4O, and ClF_5

Molecule and geometry	Bond	Bonded charge concentrations						Non-bonded charge concentrations					
		r_A (Å)	$-\nabla^2\rho$ (au)	Area (Å²)	μ_3 (au)	Angle b_aAb_e (°)	Angle nAb (°)	r_A (Å)	$-\nabla^2\rho$ (au)	Area (Å²)	μ_3 (au)	Angle nAn (°)	Net charge on F or O
ClF_3 (F–Cl–F, n n)	Equatorial(e)	0.670	0.595	0.66	23.1	83.6		0.605	1.58	1.52	61.9	147.8	− 0.30
	Axial(a)	0.687	0.319	0.41	14.6		96.5						− 0.52
ClF_3 (F–Cl–F, n F)	Equatorial	0.661	0.374		21.1	120(b_eAb_e)	90.0						
	Axial							0.601	1.787		68.7	180.0	− 0.49
ClF_3 (n–Cl–F, F F)	Equatorial	0.663	0.428		20.9	73.1	139.3	0.607	1.529		59.5	134.2	
	Axial	0.672	0.526		19.7		76.5	0.605	1.635		62.5		

SF$_4$ 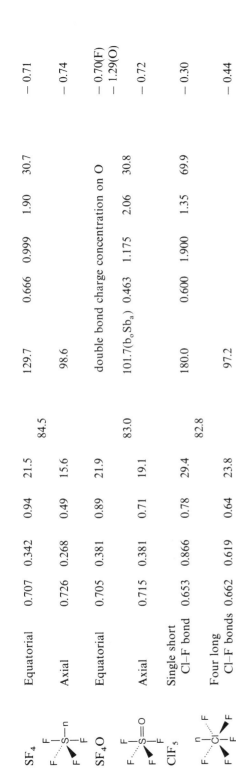	Equatorial	0.707	0.342	0.94	21.5		129.7	0.666	0.999	1.90	30.7	−0.71
	Axial	0.726	0.268	0.49	15.6	84.5	98.6					−0.74
SF$_4$O	Equatorial	0.705	0.381	0.89	21.9		double bond charge concentration on O					−0.70(F) −1.29(O)
	Axial	0.715	0.381	0.71	19.1	83.0	101.7(b$_o$,Sb$_a$)	0.463	1.175	2.06	30.8	−0.72
ClF$_5$	Single short Cl–F bond	0.653	0.866	0.78	29.4		180.0	0.600	1.900	1.35	69.9	−0.30
	Four long Cl–F bonds	0.662	0.619	0.64	23.8	82.8	97.2					−0.44

There are five local charge concentrations present in the valence shell of the chlorine atom in all three geometries. The upper diagram in Fig. 7.6 is for the equatorial plane of the equilibrium geometry and shows two relatively large non-bonded concentrations and one bonded concentration opposite the single F present in this plane. The second diagram contains all four nuclei, and three bonded charge concentrations are present. The fourth maximum in this two-dimensional display is not a $(3, -3)$ critical point, i.e. not a charge concentration, but another view of the $(3, -1)$ critical point located between the two non-bonded charge concentrations. This and other examples demonstrate that the local charge concentrations defined by the Laplacian of the charge density are able to recover violations of the octet rule that the Lewis model must assume to occur for atoms in the third and later rows of the periodic table.

In the equilibrium geometry of ClF_3 the size and magnitude of the local charge concentrations in the VSCC of Cl decrease in the order: non-bonded > equatorial bonded > axial bonded (Table 7.3). The angle between the two equatorial non-bonded maxima is relatively large and equal to 148°, while the axial bonded–non-bonded angle is opened to 97°. All of these observations are in accord with the VSEPR model of molecular geometry. In the two geometries of higher energy, the nClb angle is decreased to 90° in the D_{3h} geometry while in the least stable tripod-like geometry the corresponding angle is further reduced to 77°. The angle between the axial bonded and equatorial non-bonded charge concentrations in this latter geometry is also less than 90°, and equals 86°. Except for planar ClF_3, the smallest inter-maxima angle is a bClb angle.

The molecule SF_4 also exhibits an expanded valency shell, with five pairs of valence electrons on sulphur. The observed bisphenoid geometry of this molecule is correctly predicted by assuming that the non-bonded pair occupies an equatorial site in the most probable trigonal bipyramidal geometry. Five local charge concentrations are indeed found in the VSCC of the sulphur atom. The non-bonded maximum is the largest of the charge concentrations with respect to magnitude and size (Table 7.3). The relative properties of the equatorial and axial bonded maxima in SF_4 are the same as those discussed above for the equilibrium geometry of ClF_3, including the observation that the equatorial maxima are of greater magnitude and more tightly bound than are the axial maxima.

The square-based pyramid geometry observed for ClF_5 is rationalized on the basis of the most probable octahedral arrangement of six electron pairs. The unique ClF bond lying along the fourfold axis and opposed to the non-bonded pair has a length of 1.62 Å compared to the remaining ClF bonds which have a length of 1.72 Å. The FClF angle formed with the unique F is 87.5°. Figure 7.7 shows the unique bonded and the opposed non-bonded charge concentrations and two of the four equivalent bonded concentrations

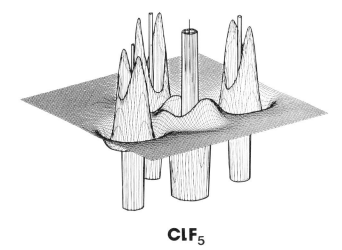

ClF$_5$

FIG. 7.7.   A relief map of $-\nabla^2\rho$ in a symmetry plane of ClF$_5$ containing the chlorine nucleus, the axial fluorine nucleus, and two of the four fluorine nuclei in the base of the square-base pyramid. The figure clearly illustrates the relative sizes of the charge concentrations; non-bonded > axial bonded > basal bonded.

in this molecule. The diagram illustrates that, in accordance with the model, there are a total of six local concentrations of charge present in the VSCC of Cl in this molecule, five bonded and one non-bonded.

The single non-bonded maximum in ClF$_5$ is of greater magnitude than those of the two such maxima in ClF$_3$ (Table 7.3). In ClF$_3$ and SF$_4$ the axial fluorines are nearest to the non-bonded maxima and they possess the longest bond lengths. Correspondingly, their bonded maxima are of smaller magnitude and are less tightly bound than are the equatorial maxima. In ClF$_5$ the four equivalent fluorines are nearest to the non-bonded maximum. They possess the longest bond lengths and their associated bonded maxima on chlorine are of smaller magnitude and are less tightly bound than that for the single fluorine with the shortest bond length. The bond angle formed by a long and a short bond is calculated to be very similar for all three molecules, $85.5 \pm 1.1°$, as are the corresponding angles between their bonded maxima, $83.6 \pm 0.9°$. In all three molecules the angle between the non-bonded maximum on chlorine or sulphur and the bonded maximum of the longest ligand bond is greater than 90°, ranging from 96.5° in ClF$_3$ to 98.6° in SF$_4$. It is also found that the fluorine ligands with the longest bond lengths possess the largest net negative charges (Table 7.3). Thus the behaviour of the local charge concentrations, as defined by the Laplacian of the charge density, mimic all of the consequences of the repulsions assumed to exist between the localized pairs of electrons in the VSEPR model.

7.2.4   Effects of charge transfer on the properties of the VSCC

The degree of charge transfer accompanying bond formation affects the topology of an atom's VSCC. The interatomic surface separating the basins of the two atoms linked by a bond path is defined by the topology of the charge density as made evident in its gradient vector field. While the zero-flux property of such surfaces ensures that the integral of the Laplacian over the basin of an atom equals zero for either a free or a bound atom (eqn (6.88)), there is no direct constraint on the topology of the Laplacian within the atomic basin of a bound atom or even a requirement that the VSCC must persist on bonding, or remain within the atomic surface of the given atom. Thus, an atom which loses most of its valence charge density on bonding, as Li does in many of its compounds, does not possess a VSCC in its bonded form. In the other examples the VSCC initially associated with the free atom is shared with more electronegative bonded neighbours. For a bond between identical atoms and for polar bonds between dissimilar atoms, there are, in general, two bonded maxima, one associated with each atom. Thus, a single bonded pair of the Lewis model is represented by two bonded maxima—one in each VSCC. The same is true for multiple bonds. The bonded pair or pairs in the Lewis model of the C–C bonds in ethane, ethylene, and acetylene, for example, are all respresented by a bonded maximum of increasing size on each carbon atom through the series.

From the many examples of Laplacian distributions that have now been studied it appears that fluorine is unique in its behaviour. The charge density of a fluorine atom is tightly bound and very localized in all of its compounds. In other than ionic compounds, the VSCC of fluorine exhibits only a torus of non-bonded charge encircling the bond axis and no bonded charge concentration, not even in F_2 (Bader and Essén 1984). Rather than three negative curvatures in $-\nabla^2\rho$ (which denote a local maximum in $-\nabla^2\rho$), the two curvatures perpendicular to the bond axis are positive for the bonded and non-bonded axial critical points in the VSCC of F. Such a $(3, +1)$ critical point denotes the presence of a local minimum in the surface of the sphere of maximum charge concentration. The radial curvature at these points is still negative—the sphere of charge concentration persists—and, because of the magnitude of this curvature, its sign dominates the value of the Laplacian. Only in HF has the VSCC of fluorine been found to exhibit a bonded maximum.

The uniform torus of charge concentration of F can be perturbed by the presence of low-lying unoccupied orbitals centred on a neighbouring atom (MacDougall and Bader 1986). In these cases there is a polarization of the π density on F towards its bonded neighbour, counter to the direction of the much greater transfer of σ density.

Finally, in the case of a Lewis acid with an incomplete valence shell, such as BH_3, the surface of the sphere of maximum charge concentration is punctured. Thus, the VSCC of B in BH_3 or of C in CH_3^+ is reduced to a belt of charge concentration in the plane containing the protons. There is, in these molecules, direct access to the region of charge depletion associated with the core from either side of this plane, and they are very electrophilic in nature.

7.2.5   Summary

The examples and accompanying discussion that have been given above demonstrate a remarkable mapping of the number and properties of the localized electron pairs assumed in the VSEPR model of molecular geometry on to the number and properties of the local maxima found in the VSCC of an atom, as determined by the Laplacian of the electronic charge density. So faithful is the mapping in recovering all of the nuances of the VSEPR model that one can conclude that the properties of the Laplacian of the charge density as displayed in the VSCC of the central atom provide the physical basis for this model and for the Lewis model of electron pairs. It has also been demonstrated here and in other examples (Bader *et al.* 1988) that the total energy of a molecule is minimized for the geometry which maximizes the separations between the charge concentrations, but at this point no theoretical reason has been advanced as to why this should be so.

   A discussion of why the Laplacian exhibits local concentrations of electronic charge where the VSEPR model assumes localized electron pairs to be is given in Section E7.2, which deals with the localization of the Fermi hole. It is shown there that the number and relative positions of the local maxima in the VSCC of the A atom in a molecule AX_n are a consequence of the partial localization of electron pairs in its valence region which results from the ligand field operating in concert with the Pauli principle. With this final demonstration, all of the postulates of the VSEPR model, including the effect of the Pauli principle on the most likely arrangement of electron pairs, are recovered in the properties of the VSCC of the central atom. A knowledge of these properties may be used to predict molecular geometries in a modified VSEPR model that requires only a single postulate, namely, that the most stable geometry of a molecule is the one which maximizes the separations between local maxima in the VSCC of the central atom.

7.3   The Laplacian of ρ and chemical reactivity

7.3.1   The Laplacian and the local energy of ρ

The Laplacian of the charge density plays a central role in the theory of atoms in molecules where it appears as an energy density, that is,

$$L(\mathbf{r}) = -(\hbar^2/4m)\nabla^2\rho(\mathbf{r}). \tag{7.1}$$

The integral of $L(\mathbf{r})$ over an atom Ω to yield $L(\Omega)$,

$$L(\Omega) = \int_\Omega L(\mathbf{r})d\tau = (-\hbar^2/4m)\int_\Omega \nabla^2\rho(\mathbf{r})d\tau$$

$$= (-\hbar^2/4m)\oint dS(\Omega, \mathbf{r})\nabla\rho(\mathbf{r})\cdot\mathbf{n}(\mathbf{r}) = 0, \tag{7.2}$$

vanishes because of the zero-flux boundary condition, eqn (6.7), which defines an atom in a molecule. The demonstration that an atom is an open quantum subsystem is obtained by a variation of Schrödinger's energy functional $\mathscr{G}[\psi, \Omega]$, eqn (5.58), for a stationary state and, as shown in Chapter 8, by a variation of the action integral, the time integral of the Lagrangian $\mathscr{L}[\Psi, \Omega]$, for a time-dependent system. In each case the zero-flux boundary condition is introduced by imposing the variational constraint that

$$\delta\left\{\int_\Omega \nabla^2\rho(\mathbf{r})d\tau\right\} = 0$$

at every stage of the variation. The possibility of introducing the constraint in this manner is a consequence of the property of the functionals $\mathscr{G}[\psi, \Omega]$ and $\mathscr{L}[\Psi, \Omega]$ that, at the point of variation where the appropriate Schrödinger equation is satisfied, they both reduce to an integral of the density $L(\mathbf{r})$. The property given in eqn (7.2) is common for an atom and for the total system and it is this property which endows them with similar variational properties, thereby making possible the generalization of the principle of stationary action to an atom in a molecule.

The two kinetic energy densities $K(\mathbf{r})$ and $G(\mathbf{r})$ differ by $L(\mathbf{r})$ (eqn (5.48)) and it is because $L(\Omega)$ vanishes for an atom that $T(\Omega)$, the electronic kinetic energy of an atom, enjoys the satisfaction of the same identity as does T for the total system, that is,

$$T(\Omega) = K(\Omega) = G(\Omega). \tag{7.3}$$

It was demonstrated in Chapter 6 that the density $L(\mathbf{r})$ appears in the local expression for the virial theorem, eqn (6.31). This is an important result, since it relates a property of the charge density to the local contributions to the energy; it is repeated here

$$(\hbar^2/4m)\nabla^2\rho(\mathbf{r}) = 2G(\mathbf{r}) + \mathscr{V}(\mathbf{r}). \tag{7.4}$$

The electronic potential energy density $\mathscr{V}(\mathbf{r})$, the virial of the forces exerted on the electrons (eqn (6.30)), and the electronic kinetic energy density, $G(\mathbf{r})$ (eqn (5.49)), define the electronic energy density,

$$E_e(\mathbf{r}) = G(\mathbf{r}) + \mathscr{V}(\mathbf{r}). \tag{7.5}$$

Because $L(\Omega)$ vanishes for an atom, integration of eqn (7.4) over the basin of an atom yields the atomic virial theorem

$$2T(\Omega) = -\mathscr{V}(\Omega), \qquad (7.6)$$

and, as a consequence, the electronic energy of an atom in a molecule satisfies the identities,

$$E_e(\Omega) = \int_\Omega E_e(\mathbf{r})\,d\tau = -T(\Omega) = \tfrac{1}{2}\mathscr{V}(\Omega). \qquad (7.7)$$

The potential energy density $\mathscr{V}(\mathbf{r})$ is everywhere negative, while the kinetic energy density is everywhere positive. Thus, the sign of the Laplacian of the charge density determines, via eqn (7.4), which of these two contributions to the total energy is in excess over their average virial ratio of 2:1. *In regions of space where the Laplacian is negative and electronic charge is concentrated, the potential energy dominates both the local total electronic energy* $E_e(\mathbf{r})$ *and the local virial relationship. Where the Laplacian is positive and electronic charge is locally depleted, the kinetic energy is in local excess.*

An energy density is dimensionally equivalent to a force per unit area or a pressure. Thus, the Laplacian may alternatively be viewed as a measure of the pressure exerted on the electronic charge density relative to the value of zero required to satisfy a local statement of the virial theorem, i.e. $\mathscr{V}(\mathbf{r}) + 2G(\mathbf{r}) = 0$. In regions where the Laplacian is negative, the charge density is tightly bound and compressed above its average distribution. In regions where the Laplacian is positive, the charge density is expanded relative to its average distribution, the pressure is positive, and the kinetic energy of the electrons is dominant.

7.3.2   Lewis acid–base reactions

The discussion so far has focused on the properties of the local charge concentrations of the Laplacian distribution and on how they recover the Lewis and VSEPR models of electron pairs. The Lewis model, however, encompasses chemical reactivity as well through the concept of a generalized acid–base reaction. Complementary to the local maxima in the VSCC of an atom for the discussion of reactivity are its local minima. A local charge concentration is a Lewis base or a nucleophile, while a local charge depletion is a Lewis acid or an electrophile. A chemical reaction corresponds to the combination of a 'lump' in the VSCC of the base combining with the 'hole' in the VSCC of the acid. The recovery of this fundamental chemical property is graphically illustrated in Fig. 7.8 which illustrates the facility for the non-bonded charge concentration on the carbon of carbon monoxide to react with the hole present on the boron atom in BH_3. The zero envelope of $\nabla^2\rho$

shown in this figure, since it displays the sites of the local charge concentrations and charge depletions, defines a system's reactive surface. In terms of the local virial theorem, eqn (7.4), the reaction of a nucleophile with an electrophile is a reaction of a region with excess potential energy on the base atom with a region of excess kinetic energy on the acid atom. The accompanying rearrangement of the charge is such that, at every stage of the reaction, $L(\Omega)$ remains equal to zero for each atom. Thus, reductions in the magnitudes of the local concentrations or depletions of charge require opposing changes in other parts of the atom to satisfy the constraint on its charge distribution given in eqn (7.2).

The effect of the replacement of H by F in methane on the properties of the VSCC of the carbon atom (see Fig. 7.9) illustrates the internal compensations which maintain the satisfaction of eqn (7.2). While, as pointed out previously (Fig. 7.4), the structure of the atomic graph for carbon is left unchanged by the substantial charge transfer from carbon to fluorine which accompanies the substitution of H by F (q(C) increases from $+0.24$ in methane to $+0.87$ in fluoromethane, q(H) changes from -0.06 to -0.04, and q(F) $= -0.74$), it does cause major changes in the extent of charge concentration and depletion. As a result of the charge transfer, the bond critical point for the C–F bond path is found only 0.43 Å from the C nucleus and, as discussed in Section 7.2.4, the associated bonded charge concentration now lies within the boundary of the F atom. If the loss of this charge concentration and its associated excess of potential energy were not compensated for, the energy of the carbon atom would increase significantly and, in addition, the zero-flux surface condition would no longer be satisfied. Consequently, the three remaining bonded maxima, those linking the protons, increase in magnitude, the values of $-\nabla^2\rho$ increasing from 0.72 au in methane to 1.42 au in methyl fluoride, as do its values at the intervening saddle points (Fig. 7.9). As a result of this increase in the concentration of charge about the perimeter of the face of the VSCC opposite the F, the extent of charge depletion in this face is increased. Thus the value of $\nabla^2\rho$ at the $(3, +1)$ critical point in the centre of each face increases from $+0.12$ au in methane to $+0.30$ au in the face opposite F and to $+0.20$ au in the faces opposed to the H atoms. These $(3, +1)$ critical points define the minima in the sphere of charge concentration in the VSCC of carbon, the sites of maximum charge depletion. They are also the sites of nucleophilic attack and the carbon of CH_3F is clearly more susceptible to such attack than is the carbon of methane. The attack is predicted to occur at the ring critical point opposite the F atom, as indicated by the arrow in Fig. 7.9. The carbon of methanol exhibits similar behaviour, the value of $\nabla^2\rho$ at the face critical point opposite the oxygen being only slightly less positive than the corresponding value in methyl fluoride.

The positions of the local charge concentration and depletion, together with their magnitudes, are determined by the positions of the corresponding

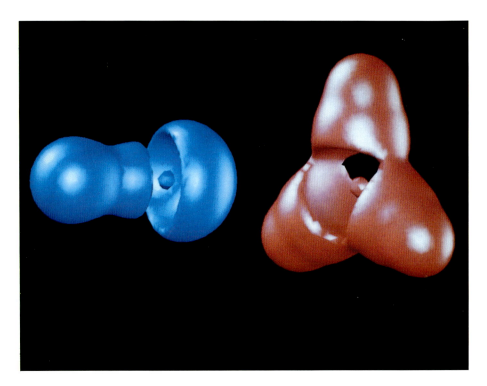

Fig. 7.8. Isovalue surfaces of $-\nabla^2\rho$ for carbon monoxide (blue) and borane (red). The surface drawn is the $\nabla^2\rho = 0$ surface, the reactive surface. The molecules are oriented so that the 'lump' in the VSCC of carbon is aligned with the hole in the VSCC of boron. The VSCC of boron is reduced to a belt-like distribution lying in the plane of the protons, giving the base atom direct access to the core of boron. The zero-valued surfaces encompassing the inner shell charge concentration on carbon and boron are also evident. Note the torus of charge depletion encircling the carbon nucleus. This feature corresponds to the localization of the π^* orbitals on carbon.

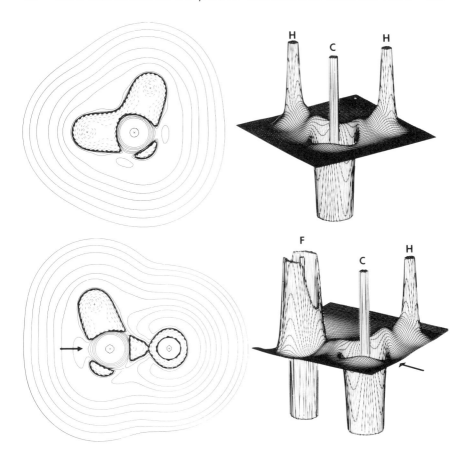

FIG. 7.9.   Contour and relief maps of the Laplacian distribution for CH_4 and CH_3F, in a plane containing HCH nuclei in the former and HCF nuclei in the latter. Solid contours denote positive values, dashed contours negative values of $\nabla^2\rho$. The contour values in these and succeeding diagrams of $\nabla^2\rho$ are given in the Appendix (Table A2); the outermost contour in each plot is $+\,0.002$ au. The inner contour encircling each nucleus in the contour plots of $\nabla^2\rho$ encompasses the negative spike-like charge concentration present on each nucleus. The relief maps display the negative of the Laplacian distributions. The relief displays are mirror images of the nuclear arrangements shown in the contour plots. There is a spike-like charge concentration at each nucleus which is somewhat broader for a proton. The arrow points to the $(3,\ +\ 1)$ critical point which is most susceptible to nucleophilic attack.

critical points in the VSCCs of the respective base and acid atoms. This information enables one to predict positions of attack within a molecule and, hence, the geometries of approach of the reactants. For example, a keto oxygen in the formamide molecule has two large non-bonded charge concentrations in the plane of the nuclei ($\nabla^2\rho = \ -\ 6.25$ and $\ -\ 6.30$ au), while the nitrogen atom exhibits two such maxima of lesser magnitude ($\nabla^2\rho =$ -2.14 au) above and below this plane (Fig. 7.10). On the basis of this information one correctly predicts that the formamide molecule will preferentially protonate at the keto oxygen (Birchall and Gillespie 1963), specifically at the position of the largest of the two charge concentrations and in the

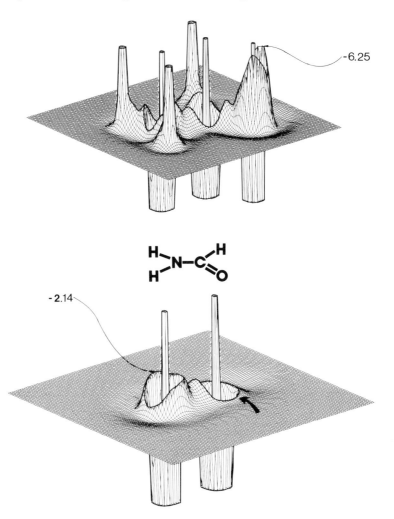

FIG. 7.10. Displays of $-\nabla^2\rho$ for the formamide molecule in the plane containing the nuclei and in a perpendicular plane containing the C–N axis. The values refer to the non-bonded maxima in the VSCCs of the oxygen and nitrogen atoms. The VSCC of carbon exhibits three bonded maxima in the upper figure. The Laplacian is positive over much of the VSCC of carbon in the plane illustrated in the lower diagram. The two points in this plane where $\nabla^2\rho$ is most positive are the 'holes' in the VSCC of carbon, the points of nucleophilic attack (as indicated by an arrow).

plane of the nuclei. There are holes in the VSCC of a carbonyl carbon and they determine the position of nucleophilic attack at this atom. These holes are above and below the plane of the nuclei of the keto grouping (see Fig. 7.10) and the corresponding critical points for a number of ketones are positioned to form angles of $110 \pm 1°$ with respect to the C=O bond axis (see Fig. 7.20, as well). This is the angle of attack predicted for the approach of a nucleophile to a carbonyl carbon (Burgi and Dunitz 1983).

Similar predictions have been made for the Michael addition reaction, specifically for the nucleophilic attack of an unsaturated carbon in acrylic acid, CH_2=CH–CO_2H, and methyl acrylic acid (Carroll *et al.* 1989). The properties of the Laplacian distribution correctly predict that the attack occurs at the terminal carbon of the methylene group, the carbon of the unsubstituted acid being most reactive, and that the approach of the nucleophile will be from above or below the plane of the nuclei along a line forming an angle of $115°$ with the C=C bond axis, the latter prediction being in agreement with calculations of the potential energy surface for this reaction. The reader is referred to Bader and Chang (1989) for a discussion of the use of the Laplacian distribution in the prediction of the sites of electrophilic attack in a series of substituted benzenes.

Electrostatic potential maps have been used to make predictions similar to these (Scrocco and Tomasi 1978). Such maps, however, do not in general reveal the location of the sites of nucleophilic attack (Politzer *et al.* 1982), as the maps are determined by only the classical part of the potential. The local virial theorem, eqn (7.4), determines the sign of the Laplacian of the charge density. The potential energy density $\mathscr{V}(\mathbf{r})$ (eqn (6.30)) appearing in eqn (7.4) involves the full quantum potential. It contains the virial of the Ehrenfest force (eqn (6.29)), the force exerted on the electronic charge at a point in space (eqns (6.16) and (6.17)). The classical electrostatic force is one component of this total force.

The Laplacian distribution has been used to predict the structures and geometries of a large number of hydrogen-bonded complexes by aligning the $(3, +3)$ critical point, a local charge depletion on the non-bonded side of the proton in the acid HF, with the $(3, -3)$ critical point of the base, a local concentration of charge, for which $-\nabla^2\rho$ attains its largest value (Carroll *et al.* 1988). The molecular graphs for the non-linear complexes in this study are shown in Fig. 7.11. With only a few exceptions, the geometries of the complexes predicted in the SCF calculations (which agree with experiment where comparison is possible), are those predicted by the properties of the Laplacian as outlined above. Figure 7.12 illustrates the Laplacian distribution for three of the bases involved in this study. They form an interesting set, as the Lewis model of localized pairs fails for two of these molecules. The Laplacian, since it is model-independent and instead reflects the properties of the charge distribution, correctly predicts the observed structures of the

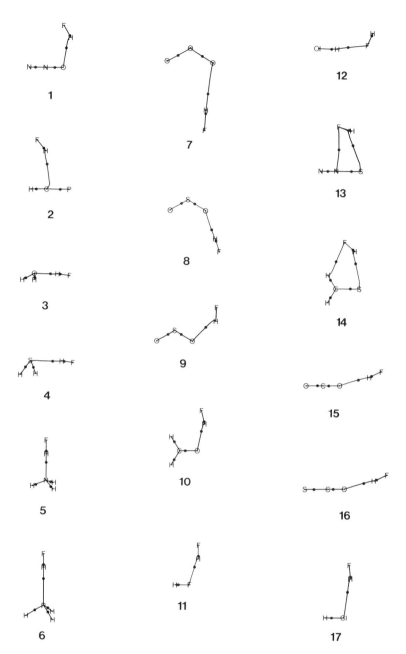

Fig. 7.11. Molecular graphs of non-linear hydrogen-bonded complexes. The dots denote the positions of the bond critical points. This point in HF is close to the proton. The bases in the complexes in numerical order are: (1) N_2O; (2) HCP; (3) H_2O; (4) H_2S; (5) NH_3; (6) PH_3; (7) O_3; (8) SO_2(cis)'; (9) SO_2 (trans); (10) H_2CO; (11) HF; (12) ClH; (13) N_2S; (14) H_2CS; (15) CO_2; (16) SCO; (17) HCl. The latter three are secondary structures, the primary structures for (15) and (16) being linear; for the complex between HCl and HF, the primary structure is (12). Note the cyclic structures obtained for (13) and (14).

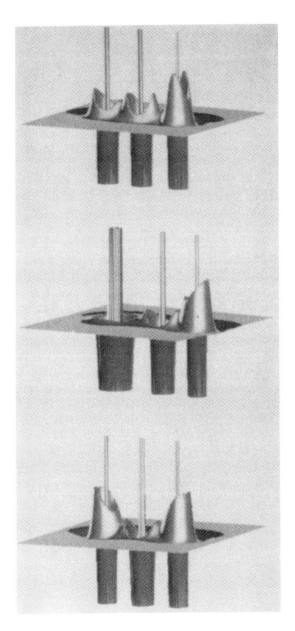

FIG. 7.12.   Relief plots of the negative of the Laplacian distributions for N_2O (top), SCO, and OCO (bottom). An oxygen atom is on the right-hand side in each case.

complexes. The oxygen in NNO possesses a torus of charge concentration (which appears as two maxima in the plane shown in Fig. 7.12) for which $-\nabla^2\rho = 5.63$ au, while the single non-bonded charge concentration on the terminal N has a magnitude of 2.90 au. (This arrangement of non-bonded charges, as does that for the bonded charge concentrations, agrees with the Lewis model, $:N:::N::\ddot{O}$.) The Laplacian predicts the proton to bond with the oxygen and to form a bent structure with a bond angle of 103.1° compared to the calculated value of 104.4°. Contrary to the Lewis model, the oxygen in SCO exhibits only a single non-bonded maximum which is co-linear with the internuclear axis. This maximum is also the largest of the charge concentrations in this molecule and one thus predicts a linear complex SCO–HF, as is both observed and calculated. In OCO, the non-bonded charge concentration on each O forms a torus about the molecular axis, in correspondence with the Lewis model. The value of $-\nabla^2\rho$ for the torus, which forms an angle with the internuclear axis of 149°, is 4.74 au. As is clear from the display of the Laplacian for OCO, however, each oxygen atom is capped by a non-bonded charge concentration which is of almost constant value and the value of $-\nabla^2\rho$ in the VSCC of oxygen on the internuclear axis is 4.72 au. The properties of the Laplacian are thus consistent with the experimental result of a floppy complex with a bond angle within the range $180 \pm 30°$. The calculated equilibrium structure is found to be linear, but its energy differs from that of a secondary bent structure with a bond angle of 165° by only 0.4 kJ/mol.

Further examples of using the Laplacian distribution to predict the geometries of acid–base reactions are provided by the crystal structures of $(I_4^{2+})(AsF_6^-)_2$ and $(S_4^{2+})(AsF_6^-)_2 \cdot 0.6SO_2$ (Gillespie et al. 1983; Passmore et al. 1980). The I_4^{2+} cation is modelled by the isoelectronic Cl_4^{2+} cation. The directions of approach to the holes in the VSCC of the chlorine atoms are indicated by arrows in Fig. 7.13. These predicted orientations of attack by a nucleophile are precisely the positions of the nearest neighbour fluorine atoms of the counter-ions in the I_4^{2+} crystal. The holes in the VSCC of the sulphur atoms in S_4^{2+} are also indicated in Fig. 7.13. They form an acute angle with the S–S bonds and are in the plane of the nuclei. In this case a single nearest-neighbour fluorine atom of a counter-ion lying on the bisector of the S–S bond binds to each such pair of holes. This prediction for S_4^{2+} is in contrast to that of frontier orbital theory. According to this theory a nucleophile should approach the cation from above or below the plane of the nuclei, to maximize overlap with the π^* LUMO (lowest unoccupied molecular orbital) of the cation.

The topology of the VSCC of the carbon atom in the triplet state of difluoromethylene is homeomorphic to that for the oxygen atom in the water molecule. This is illustrated in Fig. 7.14 which displays $-\nabla^2\rho$ for the triplet molecule in planes analogous to those shown for water in Fig. 7.5. In this case

S_4^{2+}

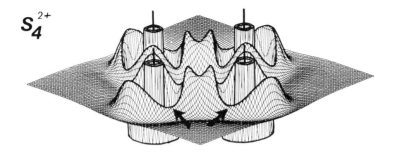

S_2N_2

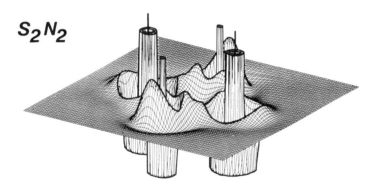

Cl_4^{2+}

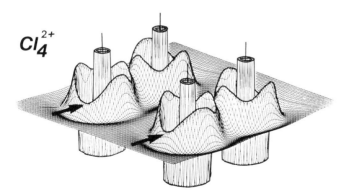

FIG. 7.13.   Relief maps of $-\nabla^2\rho$ for S_4^{2+}, S_2N_2, and Cl_4^{2+}. The arrows indicate the direction of nucleophilic attachment to the cations by the fluorides of the anions in the crystal.

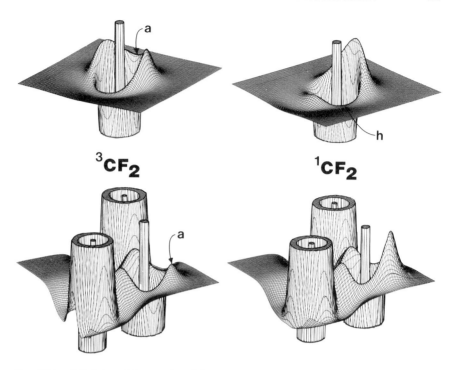

$^3\text{CF}_2$ $^1\text{CF}_2$

FIG. 7.14.   Relief plots of the negative of the Laplacian distributions for triplet and singlet states of CF_2. The lower diagrams are for the plane containing the nuclei, the upper ones for the perpendicular symmetry plane containing the C nucleus, the plane containing the non-bonded charge maxima. There are two non-bonded maxima in the triplet, one in the singlet. The point labelled 'a' is a $(3, -1)$ critical point in the VSCC of triplet carbon. There is no radial maximum or lip at the point labelled 'h' and its mirror image and the VSCC of singlet carbon exhibits 'holes' at these two points. The maxima present in the VSCCs of the F atoms are not shown as they are larger by a factor of 10 than those on the carbons.

the two local maxima in the non-bonded plane correspond to separate concentrations of charge arising from the two unpaired electrons. The plane of the nuclei exhibits two bonded maxima linking the carbon with the fluorines and a second view of the $(3, -1)$ critical point between the non-bonded maxima, a point labelled 'a' in the diagram. In contrast to this, the two corresponding planes for the singlet species of this molecule, while exhibiting two bonded maxima, exhibit a single, symmetrically placed, non-bonded charge concentration. The non-bonded charge concentration in the singlet state corresponds to a single electron pair and, since it is a local maximum in $-\nabla^2\rho$, it appears as a maximum in both planes. This example serves to illustrate that the topology of the Laplacian extends beyond the Lewis model of electron pairs and enables one to determine the location and magnitudes of local concentrations of electronic charge even when the maxima correspond to the presence of a single unpaired electron (MacDougall and Bader 1986). The point in the VSCC of carbon labelled 'h' in Fig. 7.14 and its mirror point in the singlet state denote the presence of holes in the sphere of maximum charge concentration, a point where a nucleophile has direct access to the core of the carbon atom.

The above examples have demonstrated that the relative approach of the reactants in a Lewis acid–base reaction can be predicted by aligning a region of charge concentration with a region of charge depletion, the regions being determined by their Laplacian distributions. In many cases, but not all, the local concentration of charge predicted by the Laplacian to be the site of nucleophilic activity coincides with corresponding properties of HOMO (highest occupied molecular orbital). Similarly, the local depletion of charge predicted by the Laplacian to be the site of the electrophilic activity coincides with the properties of LUMO. In this way the Laplacian of the charge density recovers the predictions of the frontier orbital model of chemical reactivity (Fukui 1971). In this model the relative orientation of reactants is determined by requiring that the overlap of HOMO and LUMO be a maximum. In larger systems where the manifold of one-electron states in the HOMO–LUMO energy region is dense, as in S_4^{2+} for example, the ordering of the orbitals is not well-determined and this model is difficult to apply. The Laplacian of the charge density appears to be more stable in its ability to predict the sites of charge depletion or concentration. This function is determined by a property of the total charge density of a system and is model-independent, assuming a proper representation of the charge density.

Perhaps the most important information contained in the state function that must be sacrificed in its reduction to the charge density is the information regarding its phase. In the orbital model, the relative phases of the orbitals determine their symmetry properties and their effective overlap. The Laplacian of the charge distribution, by recovering predictions based upon orbital phase information, provides a bridge between orbital models and the understanding of reactivity based upon the charge density. This bridge can be further demonstrated and better understood through the use of a model based upon second-order perturbation theory which predicts the essential features of the relaxation in the charge density for a given displacement of the nuclei, the latter being the source of the perturbation. This is done by approximating the vibrationally induced relaxation in ρ by the transition density obtained from a mixing-in of the lowest-energy excited state of appropriate symmetry, the symmetry being determined by the perturbing nuclear displacement. If one further assumes the largest relaxation in ρ to be caused by mixing in the excited state of lowest energy, one is able to predict the nuclear displacement that results in the smallest energy increase (Bader 1960, 1962). This method has been used to predict the signs of interaction constants in vibrational potential functions (a direct test of the assumption that the vibrational mode that causes a mixing-in of the lowest-energy excited state has the smallest force constant) and the dissociative pathways of unimolecular reactions and transition-state complexes. It has also been shown that the relative approach of the reactants, as determined by the properties of their Laplacian distributions, leads to the formation of a complex in which the most facile relaxation of the charge density is the one which leads to further motion along the reaction coordinate defined by this approach of the reactants (Bader and MacDougall 1985). The model demonstrates the manner in which the nature and, in particular, the symmetry of the lowest excited state determines the course of a chemical reaction (Bader 1962). Within the orbital model, the relevant transition density can be approximated by the product of an orbital from the ground state with one from the excited state. In many cases this corresponds to the product of HOMO with LUMO and, in this way, the second-order perturbation approach to reactivity justifies the assumption of the frontier molecular orbital model. In some cases, the symmetry of LUMO is such as to yield a transition density of irrelevant symmetry, that is, one which does not correspond to the symmetry of any nuclear motion of the complex, and it is for this reason that the frontier orbital model sometimes fails.

7.4   The characterization of atomic interactions

7.4.1   Definition of atomic interactions

The gradient vector field of the charge density identifies the set of atomic interaction lines within a molecule. These interactions, which define the molecular structure, can be characterized in terms of the properties of the Laplacian of the charge density. The local expression of the virial theorem, eqn (7.4), relates the sign of the Laplacian of ρ to the relative magnitudes of the local contributions of the potential and kinetic energy densities to their virial theorem averages. By mapping those regions where $\nabla^2\rho < 0$, the regions where electronic charge is concentrated, one is mapping those regions where the potential energy density makes its dominant contributions to the lowering of the total energy of the system, eqn (7.5). It is found that such a mapping of the potential energy density defines a spectrum of possible atomic interactions, with those characterized as shared at one extreme and those characterized as closed-shell at the other (Bader and Essén 1984).

As discussed in Chapter 2, the interaction of two atoms leads to the formation of a critical point in the charge density at which ρ has one positive curvature, labelled λ_3 and two negative curvatures labelled λ_1 and λ_2. The phase portrait of such a $(3, -1)$ critical point is displayed in Fig. 2.6. The position coordinate of the critical point is denoted by $\mathbf{r}_c$. The pair of eigenvectors associated with the negative eigenvalues of the Hessian of ρ at $\mathbf{r}_c$, the two negative curvatures of $\rho(\mathbf{r}_c)$, generate a set of gradient paths of ρ, all of which terminate at the critical point and define the interatomic surface. The unique eigenvector associated with the positive eigenvalue generates a pair of gradient paths, each one of which originates at the critical point and terminates at one of the nuclei, thereby defining a line along which the charge density is a maximum with respect to any neighbouring line, the *atomic interaction line*. In a bound system, such a line is called a *bond path*.

Embodied in the classical notion of structure is the idea of relating the properties of a molecule to a set of pairwise atomic interactions which dominate and characterize the mechanics underlying the structure. The interactions may be attractive or repulsive—the definition of structure in terms of the dominant interatomic interactions is essential to the understanding of either situation. It was shown in Chapter 6 that the quantum force acting on the electronic charge in the basin of an atom Ω, the quantity $\mathbf{F}(\Omega)$ defined in eqns (6.14) and (6.15), is given by the integral of the force $\overset{\leftrightarrow}{\sigma}\,\mathrm{d}S$ exerted on each element $\mathrm{d}S$ of its atomic surface by its neighbouring atoms. (The quantity $\overset{\leftrightarrow}{\sigma}$ is the quantum stress tensor defined in eqn (6.12).) An atomic surface $S(\Omega)$ is, in general, composed of a number of interatomic surfaces $S(\Omega|\Omega')$ and the force $\mathbf{F}(\Omega)$ can be expressed as

$$\mathbf{F}(\Omega) = -\oint \mathrm{d}S\,(\Omega)\overset{\leftrightarrow}{\sigma}\cdot\mathbf{n} = -\sum_{\Omega'\neq\Omega}\oint \mathrm{d}S\,(\Omega|\Omega')\overset{\leftrightarrow}{\sigma}\cdot\mathbf{n}. \qquad (7.8)$$

The sum in eqn (7.8) runs over the atoms, Ω', neighbouring atom Ω—over each atom linked to Ω by an atomic interaction line. Thus eqn (7.8) provides the physical basis for the model in which a molecule is viewed as a set of interacting atoms. It isolates, through the definition of structure, the set of pairwise atomic interactions which determine the force acting on each atom in a molecule for any configuration of the nuclei.

Since the two perpendicular curvatures of ρ whose eigenvectors define a surface $S(\Omega|\Omega')$ are negative, the charge density is a maximum at $\mathbf{r}_c$ in the interatomic surface and charge is locally concentrated there with respect to points in the surface. The curvature of ρ along the interaction line is positive, charge density is locally depleted at $\mathbf{r}_c$ relative to neighbouring points along the line, and ρ is a minimum at $\mathbf{r}_c$ along this line. *Thus, the formation of a chemical bond and its associated interatomic surface is the result of a competition between the perpendicular contractions of ρ towards the bond path which lead to a concentration or compression of charge along this line, and the parallel expansion of ρ away from the surface which leads to its separate concentration in each of the atomic basins.* The sign of the Laplacian of ρ at the bond critical point, the quantity $\nabla^2\rho(\mathbf{r}_c)$, determines which of the two competing effects is dominant and, because of the appearance of $\nabla^2\rho(\mathbf{r}_c)$ in the local expression for the virial theorem, eqn (7.4), its sign also serves to summarize the essential mechanical characteristics of the interaction which creates the critical point.

There is, therefore, an intimate link between the topological properties of $\rho(\mathbf{r})$ and its Laplacian, the trace of the Hessian of ρ, and, through the properties of the Laplacian, one may begin to bridge the gap between the form of the charge distribution and the mechanics which govern it.

The classification of interactions in both bound and unbound states is presented first, followed in Section 7.5 by a discussion relating the parameters of this scheme to the requirements for obtaining a bound state.

7.4.2   Classification of atomic interactions

Properties of the charge density at the $(3, -1)$ critical point for a number of diatomic molecules are given in Table 7.4 and, for some polyatomic molecules, in Table 7.5. Contour and relief maps of the Laplacian distributions for some of these molecules are shown in Figs 7.15–7.17.

When $\nabla^2\rho(\mathbf{r}_c) < 0$ and is large in magnitude, $\rho(\mathbf{r}_c)$ is also large, and electronic charge is concentrated in the internuclear region as a result of the dominance of the perpendicular contractions of ρ towards the interaction line or, equivalently in these bound systems, towards the bond path. The result is a sharing of electronic charge by both nuclei, as is found for interactions usually characterized as covalent or polar and they shall be referred to as *shared interactions*. In shared interactions, as illustrated in Figs 7.15–7.17, the region of space over which the Laplacian is negative and which contains the interatomic critical point is contiguous over the valence regions of both atoms and the VSCCs of the two atoms form one continuous region of charge concentration. The interaction is dominated by the lowering of the potential energy associated with the formation of the $(3, -1)$ critical point. In a shared interaction, the atoms are bound as a consequence of the lowering of the potential energy associated with the concentration of electronic charge between the nuclei (eqns (7.4) and (7.5)).

This concentration of electronic charge in the interatomic surface is reflected in the relatively large values of $\rho(\mathbf{r}_c)$, the value of ρ at the $(3, -1)$ critical point, for the molecules entered under shared interactions in Tables 7.4 and 7.5. In all but one of these examples, the ratio of the perpendicular contractions of ρ to its parallel expansion, as measured by the ratio $|\lambda_1|/\lambda_3$, is greater than unity. In cases of tight binding, as evidenced by the large negative values of $\nabla^2\rho(\mathbf{r}_c)$ in N_2, for example, λ_3 as well as λ_1 and λ_2 are large in magnitude, but the ratio $|\lambda_1|/\lambda_3$ is still greater than unity. Occupation of the antibonding 2π orbital of AB or of the corresponding π_g orbital of A_2, causes an increase in λ_3. Compare, for example, λ_3 for NO, NO^-, and O_2 with that for N_2.

The second limiting type of atomic interaction is that occurring between closed-shell systems, such as found in noble gas repulsive states, in ionic

Table 7.4
*Characterization of atomic interactions**

| Molecule and state | R | $\rho(\mathbf{r}_c)$ | $\nabla^2\rho(\mathbf{r}_c)$ | Eigenvalues of Hessian of $\rho(\mathbf{r}_c)$ | | | Kinetic energy contribution | | | | Net charge on A in AB |
				$\lambda_1=\lambda_2$ perpendicular	λ_3 parallel	$|\lambda_1|/\lambda_3$	$G(\mathbf{r}_c)_\perp$	$G(\mathbf{r}_c)_\parallel$	$G_\perp/G_\parallel$	$G(\mathbf{r}_c)/\rho(\mathbf{r}_c)$	
Shared interaction											
$H_2(^1\Sigma_g^+)$	1.400	0.2728	−1.3784	−0.9917	0.6049	1.64	...	...	...	0.062[†]	
$B_2(^3\Sigma_g^-)$	3.005	0.1250	−0.1983	−0.0988	0.0014	71.3	0.0223	0.0048	4.65	0.396	
$N_2(^1\Sigma_g^+)$	2.068	0.7219	−3.0500	−1.9337	0.8175	2.37	0.3042	0.0170	17.89	0.866	
$NO(^2\Pi)$	2.1747	0.5933	−2.0353	−1.6460	1.2568	1.31	0.2366	0.0463	5.11	0.88	+ 0.495
$NO^-(^3\Sigma^-)$	2.1747	0.5755	−2.0851	−1.5883	1.0914	1.91	0.2284	0.0534	4.22	0.89	− 0.088
$O_2(^3\Sigma_g^-)$	2.282	0.5513	−1.0127	−1.4730	1.9333	0.76	0.2053	0.0721	2.85	0.88	
Closed-shell interactions											
$He_2(^1\Sigma_g^+)$	3.000	0.0367	0.2501	−0.0774	0.4049	0.19	0.0000	0.0540	0.000	1.47	
$Ne_2(^1\Sigma_g^+)$	3.000	0.1314	1.3544	−0.3436	2.0417	0.17	0.0141	0.3024	0.047	2.52	
$Ar_2(^1\Sigma_g^+)$	4.000	0.0957	0.4455	−0.1388	0.7231	0.20	0.0064	0.1144	0.056	1.33	
$LiCl(^1\Sigma^+)$	3.825	0.0462	0.2657	−0.0725	0.4106	0.18	0.0033	0.0577	0.057	1.39	+ 0.924
$NaCl(^1\Sigma^+)$	4.461	0.0358	0.2004	−0.0401	0.2806	0.14	0.0035	0.0396	0.099	1.30	+ 0.911
$NaF(^1\Sigma^+)$	3.629	0.0548	0.4655	−0.0897	0.6449	0.14	0.0081	0.0890	0.090	1.94	+ 0.943
$KF(^1\Sigma^+)$	4.1035	0.0554	0.3132	−0.0717	0.4566	0.16	0.0067	0.0647	0.104	1.41	+ 0.930
$MgO(^1\Sigma^+)$	3.3052	0.0903	0.6506	−0.1331	0.9169	0.15	0.0170	0.1351	0.126	1.87	+ 1.412

* All quantities in atomic units. Results are calculated from functions using a Slater basis of near Hartree–Fock quality. Sources: Gilbert and Wahl (1967); McLean and Yoshimine (1967); Cade and Wahl (1974); Cude and Huo (1975).
[†] A single determinant function for H_2 yields $G(\mathbf{r}_c)=0$. The value of $G(\mathbf{r}_c)/\rho(\mathbf{r}_c)$ is calculated from a correlated wave function for H_2 (Das and Wahl 1966).

Table 7.5
Characterization of atomic interactions*

Molecule and interaction	R	$\rho(r_c)$	$\nabla^2\rho(r_c)$	λ_1	λ_2	λ_3	$\|\lambda_1\|/\lambda_3$	$G(r_c)_\perp$[†]	$G(r_c)_\parallel$	$G_\perp/G_\parallel$	$G(r_c)/\rho(r_c)$
Shared interactions											
CC bond in ethylene	2.489	0.3627	− 1.1892	− 0.8147	− 0.5635	0.1890	4.31	0.0674	0.0040	16.85	0.383
CC bond in benzene	2.619	0.3268	− 1.0134	− 0.7070	− 0.5746	0.2682	2.64	0.0460	0.0037	12.43	0.293
CC bond in ethane	2.886	0.2523	− 0.6615	− 0.4772	− 0.4772	0.2929	1.63	0.0223	0.0048	4.65	0.196
CC bond in cyclopropane	2.830	0.2490	− 0.5331	− 0.4892	− 0.3284	0.2846	1.72	0.0382	0.0073	5.23	0.336
CH bond in CH$_4$	2.048	0.2770	− 0.9784	− 0.7178	− 0.7178	0.4571	1.57	0.0180	0.0030	6.00	0.141
OH bond in H$_2$O	1.782	0.3909	− 2.4416	− 2.0658	− 2.0081	1.6323	1.27	0.0339	0.0141	2.40	0.210
OH bond in proton donor of (H$_2$O)$_2$	1.790	0.3825	− 2.5346	− 2.1187	− 2.0645	1.6486	1.29	0.0310	0.0139	2.40	0.198
FH	1.691	0.4043	− 3.8426	− 2.8882	− 2.8882	1.9338	1.49	0.0401	0.0185	2.17	0.244
FH bond in proton donor of (HF)$_2$	1.712	0.3846	− 3.8643	− 2.8202	− 2.8202	1.7759	1.59	0.0357	0.0196	1.82	0.236
Closed-shell interactions											
Hydrogen bond in (H$_2$O)$_2$	3.853	0.0198	0.0623	− 0.0247	− 0.0240	0.1110	0.223	0.0061	0.0147	0.415	0.806
Hydrogen bond in (HF)$_2$	3.360	0.0262	0.1198	− 0.0406	− 0.0360	0.1994	0.204	0.0013	0.0243	0.053	1.027
Hydrogen bond in HCN–HF	3.554	0.0284	0.0920	− 0.0412	− 0.0412	0.1745	0.236	0.0010	0.0210	0.046	0.807
Hydrogen bond in NN–HF	3.923	0.0169	0.0647	− 0.0216	− 0.0216	0.1079	0.200	0.0005	0.0132	0.037	0.837
Van der Waals bond in											
Ne–HF	3.988	0.0099	0.0484	− 0.0136	− 0.0136	0.0755	0.180	0.0005	0.0099	0.053	1.096
Ar–HF	4.841	0.0077	0.0311	− 0.0080	− 0.0080	0.0470	0.170	0.0003	0.0058	0.052	0.828

All quantities in atomic units. With the exception of the last four systems the data are calculated from single determinant functions using Gaussian basis sets at corresponding optimized geometries: a 6–31G basis was used in the hydrocarbon calculations, a 6–31G** basis for the calculations of H$_2$O and HF. The results for HCNHF, N$_2$HF, NeHF, and ArHF are from MP2 6–311G(2d, 2p) calculations.
[†] This is the average value of the two perpendicular components.

bonds, in hydrogen bonds, and in van der Waals molecules. They are illustrated by the second set of entries in Tables 7.4 and 7.5. One anticipates that such interactions will be dominated by the requirements of the Pauli exclusion principle. Thus, for *closed-shell interactions*, $\rho(\mathbf{r}_c)$ is relatively low in value and the value of $\nabla^2\rho(\mathbf{r}_c)$ is positive. The sign of the Laplacian is determined by the positive curvature of ρ along the interaction line, as the exclusion principle leads to a relative depletion of charge in the interatomic surface. These interactions are dominated by the contraction of charge away from the interatomic surface towards each of the nuclei. The Laplacian of ρ is positive over the entire region of interaction and the kinetic energy contribution to the virial from this region is greater than the contribution from the potential energy. The spatial displays of the Laplacian of ρ given in Figs 7.15–7.17 are atomic-like for the closed-shell interactions. The regions where the Laplacian is negative are, aside from polarization effects to be discussed later, identical in form to those of a free atom or ion. Thus, the spatial regions where the potential energy dominates the total energy are confined separately to each atom, reflecting the contraction of the charge towards each nucleus, away from the region of the interatomic surface. In a closed-shell interaction which gives a bound state, the atoms are bound as a consequence of the charge which is concentrated within the basin of each atom. The ratio $|\lambda_1|/\lambda_3 < 1$ in all the examples of closed-shell interactions.

The almost complete interatomic transfer of one electronic charge indicated in Table 7.4 for the ionic systems is verified by the nodal structure of the corresponding Laplacian maps. The cations, Li^+, Na^+, and K^+, all lack the outer nodes associated with the valence density distribution of the isolated atom. Thus, Li in LiCl has but one negative region rather than two, Na in NaCl has two rather than three, and K in KF has three rather than four. The reader is referred to Fig. E7.2 for displays of the charge distributions and interatomic surfaces for some of these systems. Another characteristic of a closed-shell interaction exemplified by the alkali halides and discussed in Section E7.1 is the separate localization of the electrons within the basin of each atom, as determined by the spatial localization of the Fermi hole.

Data for the hydrogen bonds in the dimers of HF and H_2O are given in Table 7.5. A hydrogen bond results from the interaction of two closed-shell systems and the properties of ρ at the associated bond critical point reflect all of the characteristics associated with such interactions: a low value for $\rho(\mathbf{r}_c)$ and $\nabla^2\rho(\mathbf{r}_c) > 0$. Each HF or H_2O fragment is easily recognizable in the display of the Laplacian for the corresponding dimer. None of the characteristics associated with the sharing and accumulation of charge are evident in the neighbourhood of a hydrogen bond critical point. The negative regions of the Laplacian are separately localized in each HF or H_2O fragment in a pattern similar to that found in the monomers. The O–H and F–H interactions in the monomers and in the monomer fragments present in the

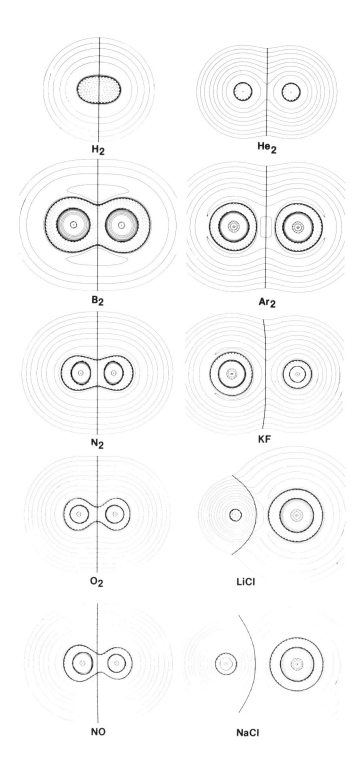

hydrogen-bonded complexes are typical of polar, shared interactions with $\nabla^2\rho(\mathbf{r}_c) < 0$. Thus, a hydrogen bond is defined to be one in which a hydrogen atom is bound to the acid fragment by a shared interaction and to the base by a closed-shell interaction.

The Laplacian map for the dimer of HF is topologically equivalent (i.e. continuously deformable) into that for Ne_2. Why certain closed-shell interactions are stable with respect to separated species while others are not is related to the extent of interatomic charge transfer and to the polarization of the atomic distribution, a topic discussed in Section 7.5.

The Laplacian distributions for S_4^{2+} and S_2N_2 in Fig. 7.13 provide further examples of shared interactions, the unequal sharing of the charge concentration in the latter molecule being an example of a polar system. The final molecule in this figure, Cl_4^{2+}, illustrates both types of interactions. This molecule consists of two identical tightly bound Cl_2^+ ions, with a bond length of 1.88 Å, weakly linked together at an equilibrium separation of 2.77 Å. The two short bonds are shared interactions with the relatively large value of 0.206 au for $\rho(\mathbf{r}_c)$ at the bond critical point and with $\nabla^2\rho(\mathbf{r}_c) = -0.227$ au. The two long bonds have $\rho(\mathbf{r}_c) = 0.032$ au and $\nabla^2\rho(\mathbf{r}_c) = +0.082$ au. In the short bonds, electronic charge is concentrated between the nuclei along the bond path. In the long bonds, electronic charge is concentrated separately in the basins of the atoms linked by the long bond paths. Of the two peaks that appear within the VSCC of each Cl atom and along the axis of a long bond, the peak on the bonded side is of greater magnitude ($\nabla^2\rho(\mathbf{r}_c) = -0.541$ au) than is that on the non-bonded side ($\nabla^2\rho(\mathbf{r}_c) = -0.510$ au). Thus, the charge density of each Cl atom, while localized within the basin of the atom with respect to the long interactions, is polarized in the direction of its bonded neighbour.

Also included in Tables 7.4 and 7.5 are the values of the kinetic energy density $G(\mathbf{r}_c)$ at the interatomic critical point and its parallel and perpendicular components. It is clear from the definition of the density $G(\mathbf{r})$ (eqn (5.48))

FIG. 7.15. Contour maps of the Laplacian of the charge density for molecules with shared and closed-shell interactions. Positive values of $\nabla^2\rho$ are denoted by solid contours, negative values by dashed contours. The Laplacian is also negative within the region bounded by the innermost solid contour enclosing the nucleus for all atoms beyond He in this and the remaining figures. The values of the contours in these and succeeding figures (see Table A2 in Appendix) are not as important as the extent and relative positioning of the regions where $\nabla^2\rho$ is either positive or negative. Some contours overlap one another when $\nabla^2\rho$ undergoes a change in sign. The outermost contour in each plot is $+0.002$ au. The intersection of each interatomic surface with the plane of the figure is also shown. The values of $\nabla^2\rho$ at the $(3, -1)$ critical points, the point where the interatomic surface intersects the atomic interaction line (which in these molecules is coincident with the internuclear axis) are recorded in Table 7.4 under $\nabla^2\rho(\mathbf{r}_c)$. In a shared interaction the region of charge increase is contiguous over the basins of both atoms and is a result of the contractions in ρ perpendicular to the interaction line. In a closed-shell interaction the regions of charge concentration are separately localized within each atom and the interaction is dominated by the contractions in ρ towards each of the nuclei.

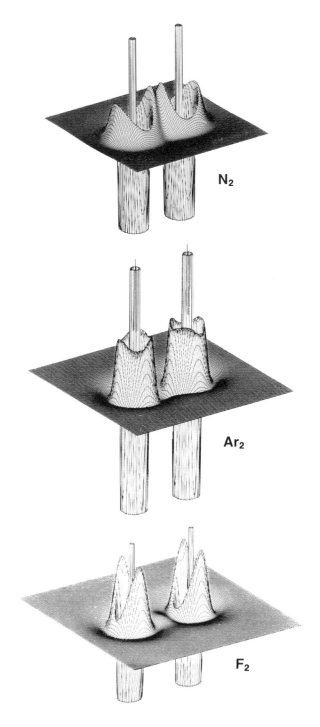

that it is everywhere positive and finite. Its value is discontinuous at a nuclear position because $\nabla\psi$ is discontinuous there. The conditions on the first-order density matrix for nuclear–electron coalescence have been given by Bingle (1963) and by Pack and Brown (1966). These expressions may be used to show that the limiting value of $G(\mathbf{r})$ at a nuclear position is $\frac{1}{2}Z^2\rho(0)$ for a spherically averaged density (Bader and Beddall 1972). The density $G(\mathbf{r})$ is expressible in terms of three contributions along orthogonal axes. The relative values of the component parallel to the internuclear axis $(G_{\parallel})$ and one of the components perpendicular to this axis, $(G_{\perp})$ faithfully reflect the magnitudes of the corresponding curvatures of ρ at $\mathbf{r}_c$. For the shared interactions, $G_{\perp}(\mathbf{r}_c) > G_{\parallel}(\mathbf{r}_c)$, while just the reverse situation is found for the closed-shell interactions. In addition, as anticipated on the basis of the local virial theorem (eqn (7.4)), the kinetic energy per electronic charge, the ratio $G(\mathbf{r}_c)/\rho(\mathbf{r}_c)$, is less than unity for the shared interactions and greater than unity for the closed-shell interactions. Thus, when the positive curvature of ρ is large and dominated by the contraction of the charge towards the nuclei, the kinetic energy per electron is absolutely large and the value of its parallel component exceeds that of a perpendicular component. In the shared interactions, the accumulation of charge in the internuclear region leads to a softening of the gradients of ρ and of the corresponding curvature of ρ along the interaction line, and the parallel component of $G(\mathbf{r})$ is correspondingly less than its perpendicular components. The dominance of these latter components again mirrors the corresponding dominance of the perpendicular contractions of ρ towards the bond axis in shared interactions. Because of the concentration of charge and the concomitant negative values of the Laplacian of ρ over the same region, the potential energy is dominant and the kinetic energy per electron is absolutely small.

Extraordinarily small values of $G(\mathbf{r}_c)/\rho(\mathbf{r}_c)$, of the order of 0.03 au, are exhibited by the non-nuclear maxima, the pseudoatoms, in the metallic clusters of Li and Na atoms illustrated in Fig. 2.11. The same small values of kinetic energy per electron are reflected in the ratio of the average values of $T(\Omega)$ to $N(\Omega)$ for the pseudoatoms and, in accordance with the Heisenberg uncertainty principle, they indicate that the charge density of the pseudo-atoms is loosely bound and unconfined. The Laplacian distributions for these

---

FIG. 7.16.   Relief maps of the negative of the Laplacian distribution of ρ to contrast the distinguishing features of the shared and closed-shell limits of atomic interactions as represented by N_2 and Ar_2, respectively. The map for F_2 is intermediate between the two limits. While $\nabla^2\rho(\mathbf{r}_c)$ is positive for F_2 as found for Ar_2, its value of $\rho(\mathbf{r}_c)$ is three times larger than that for Ar_2. Electronic charge is *accumulated* in the binding region of F_2, as is typical of a shared interaction, but is *concentrated* in the atomic basins, as is typical of a closed-shell interaction. While $\nabla^2\rho(\mathbf{r}_c) > 0$ for both Ar_2 and F_2, the Laplacian distribution is a minimum at $\mathbf{r}_c$ for Ar_2, but a maximum at the same point in F_2. The charge densities are calculated using a 6-311G (2d, 2p) basis in an MP2 calculation.

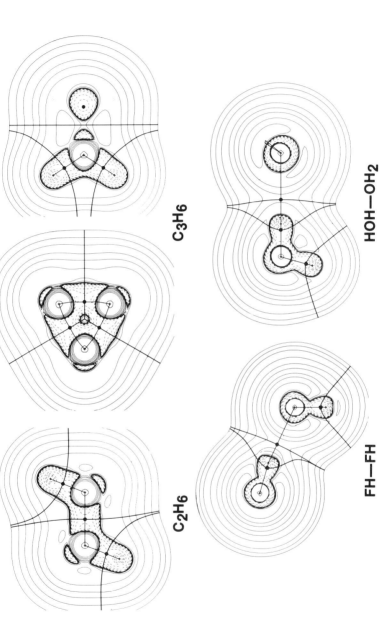

C₃H₆

C₂H₆

HOH—OH₂

FH—FH

FIG. 7.17. Contour maps of the Laplacian of ρ for ethane, cyclopropane, and the hydrogen-bonded dimers of HF and H₂O. The $(3, -1)$ critical points are indicated by black dots and interatomic surfaces are also indicated. The plane for C₂H₆ contains the nuclear framework H–C–C–H. The first diagram for C₃H₆ is for the plane containing three C nuclei. It is followed by a symmetry plane perpendicular to this containing a single C nucleus. The dot shown on the right-hand side of this diagram denotes the $(3, -1)$ critical point defining the interaction line linking the two out-of-plane C nuclei. The line on the right-hand oxygen nucleus in (H₂O)₂ denotes the position of an out-of-plane H nucleus. The Laplacian is positive in the neighbourhood of each hydrogen bond $(3, -1)$ critical point—a closed-shell interaction. The associated interatomic surface defines the boundary of each monomer fragment. Note that the hydrogen-bonded protons are adjacent to a non-bonded charge concentration in the VSCC of the F or O atom of the proton acceptor fragment.

clusters show the metal atoms to be missing their outer charge concentrations, indicating that they are present primarily, as ionic cores (Fig. 7.18). The charge density of the cores is highly localized as indicated by values in excess of 90 per cent for the contained Fermi correlation. The charge on the pseudoatoms, however, is very delocalized with values for the contained Fermi correlation being only 30 per cent of that required for complete localization. Aside from the inner-shell charge concentrations of the metallic cores, the negative values of the Laplacian distribution are confined within the boundaries of the pseudoatoms. The value of $\nabla^2\rho(\mathbf{r_c})$ is positive at the bond critical point linking a pseudoatom to a metallic core, but negative at a critical point linking two pseudoatoms. The properties of ρ at the critical point linking two pseudoatoms are characteristic of a weak, shared interaction: $\rho(\mathbf{r_c}) = 0.0056$ au, $\nabla^2\rho(\mathbf{r_c}) = -2.6 \times 10^{-4}$ au, and $G(\mathbf{r_c})/\rho(\mathbf{r_c}) = 0.10$ au. The ionic cores are linked to one another through the pseudoatoms and the study of these clusters gives a model of Group I metals consisting of positively charged metal atoms with very localized charge distributions immersed in and bound by an intermeshed network of negatively charged pseudoatoms. Metallic binding is a result of the lowering of the potential energy associated with the charge concentrations of the valence density, and these concentrations are confined within the boundaries of the pseudoatoms. These same charge concentrations have a very low kinetic energy per electron and the potential energy lowering associated with the formation of the pseudoatoms is obtained without a large accompanying increase in the kinetic energy. As a consequence, the electron density of the pseudoatoms should be mobile under the influence of an electric field. The quantum mechanical current for a core density is simply the diamagnetic current circulating around the nucleus and the metal atoms in these clusters will not contribute appreciably to the electrical conduction. It is the highly delocalized density of the network of pseudoatoms that is responsible for the binding in a metallic system and for its conducting properties.

Observations regarding the behaviour of the parallel and perpendicular components of the kinetic energy and their relation to the gradients and curvatures of ρ in molecular systems were first made by Bader and Preston (1969) for the molecules H_2 and He_2. They studied the spatial properties of $G(\mathbf{r})$ and $K(\mathbf{r})$ (eqn (5.48)) and their relation to $L(\mathbf{r})$ (eqn (7.1)), the density proportional to the negative of the Laplacian of ρ. The differing behaviour of $G(\mathbf{r})$ in the binding region of a bound and an unbound system is made very clear by comparing the plots of this function these authors give for H_2 and He_2. It was shown that the value of $L(\mathbf{r})$ in a region where the Laplacian of ρ is negative is a measure of the extent to which the charge density can be concentrated in regions of low potential energy beyond that anticipated on classical grounds. The kinetic energy density $G(\mathbf{r})$ and its components may be

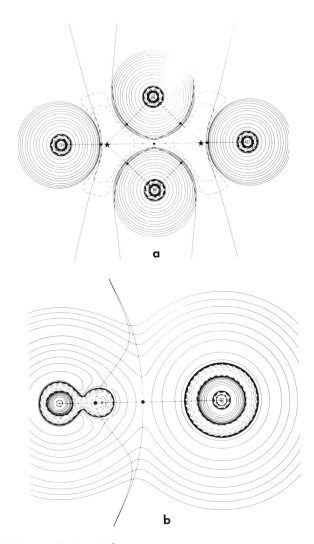

Fig. 7.18. (a) Contour display of $\nabla^2 \rho(\mathbf{r})$ for the Na_4 cluster in the plane of the nuclei. The locations of the two non-nuclear attractors are indicated by stars and of the bond critical points by dots. The contour values are given in the Appendix (Table A2). The inner contour encompassing a sodium nucleus also encompasses the innermost region of charge concentration. The third, or valence shell of charge concentration of the Na atoms has essentially all been transferred to the pseudoatoms and the region where $\nabla^2 \rho < 0$ is almost completely localized within their boundaries. (b) Contour display of $\nabla^2 \rho(\mathbf{r})$ for ArHF. The innermost contour encompassing the Ar or the F nucleus encloses the innermost region of charge concentration. The Ar atom, unlike the Na atoms in (a), possesses all three quantum shells by exhibiting three alternating shells of charge concentration and charge depletion. Contrast the relative planarity of the Ar–H interatomic surface with the curved nature of the H–F surface. The same contrasting behaviour in the shapes of the interatomic surfaces for the closed-shell interaction, $\nabla^2 \rho(\mathbf{r}_c) > 0$, and the shared interaction, $\nabla^2 \rho(\mathbf{r}_c) < 0$, linking the proton are found for all hydrogen-bonded

related directly to the gradients of the orbital densities ρ_i and their occupation numbers n_i,

$$G(\mathbf{r}) = (\hbar^2/8m)\sum_i n_i \nabla\rho_i \cdot \nabla\rho_i/\rho_i. \tag{7.9}$$

For a one-electron system, $G(\mathbf{r})$ is related to the gradients of $\rho(\mathbf{r})$ itself. For H_2, where just one orbital ($1\sigma_g$) and, for He_2, where just two orbitals ($1\sigma_g$ and $1\sigma_u$) make the major contributions to ρ, the above observations regarding the paralleling behaviour of $G_\perp(\mathbf{r})$ and $G_{||}(\mathbf{r})$ with that of the corresponding gradients and curvatures of ρ for the shared and closed-shell interactions are made clear by eqn (7.9). However, the data in Tables 7.4 and 7.5 show that these are general observations and that, even in many-electron systems, the local behaviour of the kinetic energy density may be related to the gradients and curvatures of the total charge density. This correlation is partially accounted for by theory through eqn (7.4). When $\nabla^2\rho(\mathbf{r}) > 0$ and the Laplacian is dominated by the positive curvature of ρ (contraction of ρ towards each nucleus), the larger contribution to the Laplacian comes from the kinetic energy—and, by observation, primarily from its parallel component. Correspondingly, when the Laplacian of ρ is negative and ρ is concentrated as a result of contractions perpendicular to the bond path, not only does the potential energy make the dominant contribution to the virial but one observes that the perpendicular components of $G(\mathbf{r})$ dominate its parellel component. Thus, one concludes that the kinetic energy dominates the contributions to the virial and to the energy in regions of space where its parallel component is dominant. This occurs in regions where $\nabla^2\rho(\mathbf{r}) > 0$. Conversely, in regions where the perpendicular components to $G(\mathbf{r})$ are largest, the potential energy makes the major contribution to the virial and to the energy of the system.

As noted by Bader and Preston (1969), the differing behaviour of $G_{||}(\mathbf{r})$ and $G_\perp(\mathbf{r})$ in H_2 is so extreme as to be reflected in their integrated average values. Thus, the ratio of the average values of the parallel ($T_{||}$) to the sum of the perpendicular ($T_\perp$) components of the kinetic energy for H_2 at its equilibrium separation was found to be 0.3569 as compared to the value of one-half for the separated atoms. As noted by these authors, this result is consistent with the original observation made by Coulson (1941a,b; Coulson and Duncanson 1941a,b) in his study of the electronic momentum distribution for the H_2 molecule. He found that the mean component of the electronic

---

systems (see, for example, the displays for the water and hydrogen fluoride dimers given in Fig. 7.17). A small bonded maximum is formed within the valence shell of charge concentration of the Ar atom in forming the complex with HF. The value of this maximum in the Laplacian distribution, equal to -1.19 au, and its position, 1.088 au from the Ar nucleus, represent only small perturbations of the spherical surface of maximum charge concentration present in the VSCC of the free Ar atom (see Table 7.1).

momentum in the direction of the bond axis is decreased while the mean component perpendicular to the bond axis is increased. For the closed-shell interaction in He_2, just the opposite behaviour is obtained. Bader and Preston noted that, in He_2, $G_{||}(\mathbf{r}) > G_{\perp}(\mathbf{r})$ and they found $T_{||}/T_{\perp} > 0.5$. This dominance of the parallel over perpendicular components of T, both locally and overall, is anticipated on the basis of the Laplacian map for He_2 which is everywhere positive except for small spherical regions encompassing each nucleus (Fig. 7.15). These properties of the kinetic energy are also evident in the electron momentum distribution of He_2. Ramirez (1983) found that this distribution in He_2 is ellipsoidal with the major axis parallel to the line of interactions, just the opposite of the situation found in H_2 where the major axis of the momentum distribution is perpendicular to the bond axis, as found by Coulson (1941a,b).

For systems containing atoms other than H and He, one does not find such large deviations in the ratio of $T_{||}/T_{\perp}$ from the value of one-half, since the average kinetic energy components in these systems are dominated by the core energies. The contributions of the components to the local value of $G(\mathbf{r})$ in the valence region, however, do parallel the behaviour found in H_2 or He_2, as exemplified by the data in Tables 7.4 and 7.5.

7.4.3   Hydrogen bonds and van der Waals molecules

The van der Waals molecules, NeHF and ArHF, although very weakly bound, can also be classified as hydrogen-bonded complexes, since the hydrogen in these molecules is bonded to the acid fragment by a shared interaction and to the base by a closed-shell interaction. The charge density at the critical point linking the hydrogen to the base atom in the van der Waals molecules exhibits the same characteristics as it does in the hydrogen-bonded molecules. This is illustrated by the data in Table 7.5. These are weak interactions: the charge accumulated along the bond path and the stresses induced in ρ caused by a van der Waals interaction are small, as evidenced by the values of $\rho(\mathbf{r}_c)$ and by the magnitudes of the three contributing curvatures to $\nabla^2\rho(\mathbf{r}_c)$. The two kinds of interactions also have common features in terms of the properties of the atoms brought about by the formation of the complex (Table 7.6). (The atom of a polyatomic base engaged in hydrogen bonding is denoted by the symbol B and the remainder of the molecule by X.) In addition to the van der Waals molecules, NeHF and ArHF, which have dissociation energies $(D_e) \approx 1 \, \text{kJ/mol}$, data are given for hydrogen-bonded complexes of HF with hydrogen cyanide and the nitrogen molecule. These two molecules provide examples of normal $(D_e > 15 \, \text{kJ/mol})$ and weak hydrogen bonds, respectively. Similar studies have been published for a large number of hydrogen-bonded systems (Carroll and Bader 1988).

Table 7.6

(a) Properties of the hydrogen bond in complexes of HF with HCN, N_2, Ne, and Ar*

XB–HF	D_e	r_H	r_B	$|\Delta r_H|$	$|\Delta r_B|$	$\sum|\Delta r|$	$\rho^0(r_H)$	$\rho^0(r_B)$	$\sum\rho^0$
HCN–HF	25.65	1.163	2.391	0.92	1.18	2.10	0.0135	0.0137	0.0272
NN–HF	7.87	1.359	2.564	0.72	0.91	1.63	0.0079	0.0082	0.0161
Ne–HF	0.71	1.518	2.469	0.56	0.38	0.94	0.0050	0.0040	0.0090
Ar–HF	0.76	1.669	3.172	0.41	0.55	0.96	0.0033	0.0037	0.0070

(b) Changes in atomic properties

XBHF	Ω	$\Delta N(\Omega)$	$\Delta|\mathbf{M}(\Omega)|$	$\Delta Q_{zz}(\Omega)$
HCNHF	F	0.044	− 0.0140	− 0.0257
	H	− 0.012	− 0.0184	0.0985
	N	0.038	0.0335	0.2277
	C	− 0.046	0.0258	− 0.2089
	H	− 0.022	− 0.0031	− 0.0182
NNHF	F	0.019	− 0.0037	− 0.0091
	H	− 0.004	− 0.0093	0.0640
	N	0.029	− 0.0041	0.1106
	N	− 0.044	− 0.0009	− 0.0239
NeHF	F	0.006	− 0.0002	− 0.0052
	H	− 0.005	− 0.0106	0.0434
	Ne	− 0.001	0.0391	− 0.0487
ArHF	F	0.005	− 0.0004	− 0.0008
	H	0.004	0.0004	0.0316
	Ar	− 0.008	0.0439	0.1218

* D_e in kJ/mol. Other quantities are in atomic units. Results from MP2 6-311G (2d, 2p)/MP2 6-311G (2d, 2p) calculations. The D_e values are corrected for BSSE using the full supermolecule basis set (Boys and Bernardi 1970).

The data are obtained from large basis set calculations (see Table 7.6) which include electron correlation (second-order Moeller–Plesset perturbation theory) and which are corrected for basis set superposition error (BSSE), (Boys and Bernardi 1970). While the results for the stronger hydrogen bonds are qualitatively the same with or without the inclusion of correlation, the weaker interactions require the better description of the wave function.

The D_e value for NeHF in Table 7.6 compares well with the value of 0.78 kJ/mol calculated by ONeil et al. (1989) using a large basis set on a size consistent calculation in the coupled electron pair approximation (CEPA) and corrected for BSSE. Hutson and Howard (1982) have determined an intermolecular potential for ArHF by a least squares fitting of molecular beam spectral data. In the case of HF complexes there is little information in these spectra on the absolute well depth. Their estimate of 2.6 kJ/mol for D_e of ArHF probably does correctly indicate however, that the calculated value is underestimated for this system.

A hydrogen bond results from the mutual penetration of the van der Waals envelopes of the H atom of the acid and of the atom B of the base, the strength of the interaction increasing with the degree of mutual penetration (Carroll and Bader 1988). The bonded radii of the B and H atoms—the distance of a bond critical point to the corresponding nucleus—for the B–H bond in the complexes are given in Table 7.6. Each value is less than the corresponding van der Waals non-bonded radius—the distance of the H or B nucleus to its non-bonded 0.001-au contour in the isolated reactant—and the values $|\Delta r_B|$ and $|\Delta r_H|$ are the measures of the extent of penetration of the van der Waals envelope to yield the bonded radii found in the complex. Also listed in Table 7.6, under the headings $\rho^0(r_H)$ and $\rho^0(r_B)$, are the values of ρ in the isolated acid and base molecules at the point of penetration, i.e. at the position determined by the bonded radius r_H or r_B. The sum of these two values, $\sum \rho^0$, is only slightly smaller in value than $\rho(r_c)$, the value of ρ at the hydrogen bond critical point (Table 7.5). Unlike a shared interaction, there is no large increase in charge density at the bond critical point and in the interatomic surface relative to the unperturbed densities. The final density is instead determined primarily by the extent of penetration—the greater the penetration the larger the value of $\rho(r_c)$ and, in general, the stronger the resulting hydrogen bond. To obtain hydrogen bonds with $D_e > 25$ kJ/mol, the van der Waals envelope of the H atom must be penetrated by ≈ 1 au and that of the B atom by a still larger amount. At these penetrations, both of the unperturbed densities have values in excess of 0.01 au.

Argon, the softer of the two base atoms, is penetrated the most by the hydrogen in the van der Waals complexes, while Ne, the harder of the two bases, penetrates the hydrogen atom's density to a greater extent than does Ar. The sum of the B and H penetrations are nearly equal in these two weakly bound complexes. While the penetration of the argon atom is greater than that of the neon atom, the values of $\rho^0(r_B)$ at these penetrations are nearly equal, as the argon valence density is more diffuse than that of neon. The greater penetration of the hydrogen by the harder base neon gives rise to the largest value of $\rho(r_c)$. The greater penetration of a given acid atom by the harder of two base atoms appears to be general (Carroll and Bader 1988). In the complexes HF–HF and HCl–HF for example, the F atom penetrates the hydrogen of HF to almost twice the extent than does the Cl atom, 0.82 compared to 0.42 au, respectively, while the two base atoms are penetrated by similar amounts. In a like vein, the pair of complexes, HCl–HF and HF–HCl, allows for a comparison of soft and hard bases interacting with hard and soft acids, respectively. The charge distribution of HF and HCl can be compared in Fig. 6.3. The non-bonded radius of H is 2.01 au in the hard acid HF and 2.28 au in the soft acid HCl. The penetration of the soft acid H by the hard base F is more than twice the penetration of the hard H in HF by the soft base Cl—0.96 au compared to 0.42 au—while the base atoms are penetrated to a nearly equal extent by the hard and soft H atoms. Unlike the weakly bound van der Waals complexes, penetration is overriding for a hydrogen bond with the result that HF–HCl is the energetically favoured complex.

A study of the changes in the atomic moments (Table 7.6(b)) indicates how the atomic charge distributions change as a result of the penetration of the H and B non-bonded densities, as occurs in the formation of a stable complex. In all cases of regular hydrogen bond formation, that is, with $D_e > 5$ kJ/mol, there is a relatively small transfer of charge from the base to the acid, equalling 0.03 and 0.02 e for the complexes with HCN and N$_2$, respectively. In the same cases, the hydrogen of the acid loses charge and the A atom of the acid HA gains charge. In a polyatomic base, the base atom B gains charge while a slightly greater amount of charge is lost by the remainder of the base molecule X. These changes are exemplified by the complexes with HCN and N$_2$. Experimentally, it is observed that the dipole moment of a hydrogen-bonded complex is greater than the vector sum of the dipoles of the isolated reactants, an effect termed dipole moment enhancement. Theory demonstrates that this enhancement is a result of a loss of charge by the tail of the base (the X group) and a gain by the head of the acid (the F atom in the present cases). This effective transfer of charge, while small in absolute amount, is across the length of the complex and it creates a moment which adds to the head-to-tail alignment of the acid and base dipoles to yield a moment which is greater than the sum of the reactant moments.

The changes in the atomic populations encountered in the formation of the van der Waals complexes are extremely small. There is a predicted transfer of 0.008 e from the base to the acid in ArHF, while the corresponding transfer of 0.001 e in NeHF is equal to the error in an integrated atomic population. There are significant changes in the first and second moments of the atoms in these complexes, however, and these are discussed next.

The atomic densities of the isolated reactants, as determined by the first moment $\mathbf{M}(\Omega)$ (eqn (6.45)), are polarized in a direction counter to the direction of charge transfer. Thus in HF, F is polarized towards H which is itself polarized away from F and, in HCN, the atoms are polarized in the direction N to C to H. In N$_2$, the atoms are polarized into their non-bonded regions. The atoms in these linear molecules have positive values for $Q_{zz}(\Omega)$, the quadrupole polarization with respect to the internuclear axis (eqn (6.48)), implying a transfer of electronic charge from along the axis to a torus-like distribution about this axis. The largest changes in these moments are found for the H and B atoms and the results for the HCN complex are typical of most-hydrogen-bonded complexes. The charge distributions of the H atom of the acid and of the B atom of the base change so as to remove density from along their axis of approach and thereby facilitate the penetration of their non-bonded charge distributions. The data in Table 7.6(b) show that the polarization of H away from the F is reduced and that of the N atom towards the C of HCN is increased. Thus, the charge distributions of both the H and B atoms polarize away from one another in the formation of the complex. Similarly, the quadrupole polarization of the same two atoms is increased, indicating that an increased amount of electronic charge is removed from along the internuclear axis, the axis of approach, and placed in the torus-like distribution about the axis. This corresponds to a σ to π promotion of the density. The same polarizations occur in the weaker complex with N_2. The N atom acting as a base polarizes slightly away from the H atom on forming the complex and its original atomic polarization is decreased.

The changes in the atomic moments of the Ar and Ne atoms dominate the formation of the van der Waals complexes. The dipole polarizations of the noble gas atoms exceed in magnitude the corresponding changes for the base atom in the regular complexes and in each case they are polarized *towards* the H atom of the acid. The polarization of the Ar atom is greater than that of the Ne atom. In forming the complex with Ne the charge density on hydrogen is polarized away from the base atom while, in the Ar complex, it is polarized to a much lesser extent towards the base atom. These differing polarizations are consistent with the observation made above that Ne, the harder of the two bases, penetrates the H atom to a greater extent than does the Ar atom. The hydrogen atom undergoes the same quadrupolar polarization in the formation of the van der Waals molecules as in the regular hydrogen-bonded complexes but it is of smaller magnitude, being again smallest for the Ar complex. The Ne and Ar atoms exhibit differing behaviour in this respect, Ne increasing the amount of electronic charge along the axis of approach, Ar decreasing it.

The hydrogen atom is destabilized in the formation of a hydrogen bond, the extent of destabilization paralleling the strength of the interaction (Carroll and Bader 1988). This is true for the present examples as well. The A–H bond of the acid is slightly perturbed on hydrogen bond formation and, in general, the A–H bond is lengthened. This lengthening is reflected in a corresponding decrease in the value of $\rho(\mathbf{r}_c)$. These changes are small and the bond of the acid retains the characteristics of a shared interaction. This is illustrated by the data in Table 7.5 for the hydrogen bonds in the water and hydrogen fluoride dimers. In the present examples the H–F bond is lengthened by 0.0160 au in HCNHF and by 0.0044 au in NNHF, with associated decreases in $\rho(\mathbf{r}_c)$ of 0.0154 and 0.0045 au. The H–F bond is perturbed to a much smaller extent in the formation of van der Waals complex, the changes in the H–F bond length and its value of $\rho(\mathbf{r}_c)$ being predicted to be less than one-thousandth of an au for the Ne and Ar complexes. The H–F internuclear separation is actually calculated to decrease by 0.00042 au in the complex with Ne, the most stable of the two complexes.

In van der Waals molecules and regular hydrogen-bonded complexes, a hydrogen atom links an acid to a base fragment with a shared and a closed-shell interaction, respectively. Both types of complexes result from the mutual penetration of the non-bonded van der Waals radii of the H atom of the acid and of the terminal atom B of the base to yield a value for the density at the H–B critical point which is only slightly greater than the sum of the unperturbed densities. For the regular hydrogen-bonded complexes, the B atom is penetrated to a greater extent than the H atom and the strength of the interaction increases with the degree of penetration. In these molecules the H and the B atoms polarize so as to facilitate the mutual penetration of their non-bonded densities. The noble gas atoms are polarized to a greater extent in the formation of the van der Waals complexes than is a base atom of a hydrogen bond. While the dipole and quadrupole polarizations of argon are greater than those of neon, the latter atom penetrates and perturbs the hydrogen atom of the acid to a greater extent. Thus the effect on the binding energy of the greater penetration of hydrogen by neon is offset by the greater polarizability of argon.

Unlike the polarization of the base atom in a regular hydrogen bond interaction, the dipolar polarization of a noble gas atom is towards the hydrogen. In the relatively weak complex of HF with N_2, the change in the polarization of the base N away from H is very small. This interaction is transitional between the two patterns of atomic polarizations that result from the mutual penetration of closed-shell systems with little or no accompanying charge transfer, the features common to van der Waals and hydrogen-bonded interactions.

7.4.4   Intermediate interactions

The examples so far considered illustrate that atomic interactions exhibit two limiting sets of behaviour of the charge density at the $(3, -1)$ critical point. One set is the opposite of the other in terms of the values of $\rho(\mathbf{r}_c)$ and of the regions of charge concentration and depletion and their associated mechanical consequences, as determined by the sign of the Laplacian of ρ. In these examples, the interatomic critical point is situated relatively far from a nodal surface in the Laplacian of ρ. In some molecules, however, the critical point is located close to a nodal surface in $\nabla^2\rho$. The atomic basins neighbouring the

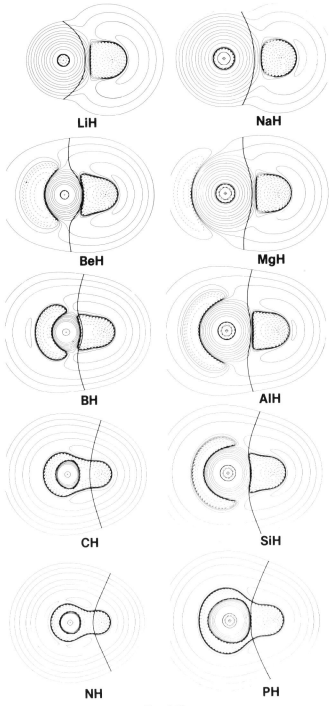

LiH　　　　**NaH**

BeH　　　　**MgH**

BH　　　　**AlH**

CH　　　　**SiH**

NH　　　　**PH**

FIG. 7.19

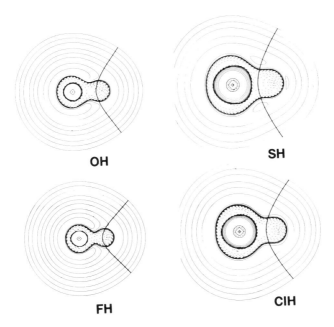

FIG. 7.19.   Contour maps of the Laplacian of ρ for second- and third-row diatomic hydrides. The intersection of the interatomic surface with the plane of the diagram is also shown. These maps illustrate the transition from closed-shell to shared atomic interactions.

interatomic surface exhibit opposite behaviour with respect to the sign of the Laplacian of ρ in these cases and the interactions are transitional in character. The transition from closed-shell to shared interactions is illustrated in Fig. 7.19 for the second- and third-row diatomic hydrides, AH. The properties of $\rho(\mathbf{r})$ at the interatomic critical points for these molecules are given in Table 7.7. The first member of each series, LiH and NaH, exhibits the characteristics of a closed-shell interaction—the value of $\rho(\mathbf{r}_c)$ is small and $\nabla^2\rho(\mathbf{r}_c) > 0$. The charge density of a hydrogen atom is easily polarized, particularly when it is negatively charged. This polarization is evident in the disposition of the negative region of $\nabla^2\rho$ relative to the position of the proton. The Laplacian distributions and the properties of ρ at $\mathbf{r}_c$ for the molecules CH to FH and SH to ClH are typical of shared interactions. In CH there is a close to zero transfer of charge and an equal sharing of the charge increase. In NH, OH, and HF there is an increasing polarization of the shared charge towards A. The transition to shared interaction occurs earlier in the second-row than in the third-row elements as anticipated on chemical grounds (see Fig. 6.2) and is understandable in terms of the greater core size of a third-row element compared to that of a corresponding second-row element.

Table 7.7

Characterization of atomic interactions in diatomic hydrides*

AH	R	$\rho(\mathbf{r}_c)$	$\nabla^2\rho(\mathbf{r}_c)$	$\lambda_1 = \lambda_2$	λ_3	$|\lambda_1|/\lambda_3$	$G_\perp/G_\parallel$†	$G(\mathbf{r}_c)/\rho(\mathbf{r}_c)$	Net charge on A
LiH $^1\Sigma$*	3.015	0.0407	0.1571	−0.0619	0.2808	0.2204	0.0	1.017	+0.911
BeH $^2\Sigma^+$	2.538	0.0965	0.1638	−0.2032	0.5703	0.3563	0.0	0.950	+0.868
BH $^1\Sigma^+$	2.336	0.1843	−0.5847	−0.4622	0.3397	1.3606	0.0	0.396	+0.754
CH $^2\Pi$	2.124	0.2787	−1.0389	−0.7572	0.4754	1.5928	2.231	0.102	+0.032
NH $^3\Sigma^-$	1.961	0.3395	−1.6034	−1.2399	0.8765	1.4146	2.196	0.157	−0.323
OH $^2\Pi$	1.834	0.3756	−2.6482	−1.8968	1.1454	1.6560	2.527	0.208	−0.585
FH $^1\Sigma^+$	1.733	0.3884	−4.3872	−2.8015	1.2159	2.3041	2.990	0.251	−0.760
NaH $^1\Sigma^+$	3.566	0.0337	0.1320	−0.0394	0.2108	0.1869	0.0426	0.979	+0.810
MgH $^2\Sigma^+$	3.271	0.0529	0.1844	−0.0695	0.3233	0.2150	0.0398	1.023	+0.796
AlH $^1\Sigma^+$	3.114	0.0743	0.1883	−0.1054	0.3990	0.2642	0.0368	0.981	+0.825
SiH $^2\Pi$	2.874	0.1134	0.1330	−0.1666	0.4663	0.3573	0.0763	0.9330	+0.795
PH $^3\Sigma^-$	2.708	0.1590	−0.1142	−0.2222	0.3302	0.6729	0.1961	0.7679	+0.579
SH $^2\Pi$	2.551	0.2128	−0.5663	−0.3732	0.1801	2.0722	1.385	0.2617	+0.094
ClH $^1\Sigma^+$	2.409	0.2540	−0.7824	−0.6035	0.4247	1.4210	2.536	0.2327	−0.241

* All quantities in atomic units. The results are calculated from functions using a Slater basis of near Hartree–Fock quality (Cade and Huo 1973).

† In a single determinant state function comprised of σ orbitals only $G_\perp(\mathbf{r}_c) = 0$.

The values of $\rho(\mathbf{r}_c)$ show a monotonic increase through both series, being a minimum at the closed-shell limit. In BeH and BH and in MgH to SiH the valence charge density remaining on A is strongly polarized into the non-bonded region of the A atom (as a consequence of the net negative field exerted on it by the negatively charged hydrogen atom) where it forms a separate region of charge increase. For these molecules and for PH the ratio, $|\lambda_1|/\lambda_3$ exhibits values intermediate between those characteristic of the two limiting types of interactions. The atomic interactions in these molecules have properties which bridge those of the ionic systems at the closed-shell limit and those of the covalent and polar systems of the shared interactions. In molecules, such as BeH, BH, and SiH where the nodal surface in $\nabla^2\rho$ is nearly coincident with the interatomic surface, the atomic interactions correspond to a relatively hard core of density on the A atom which is dominated by contractions towards the A nucleus, interacting with a softer, more polarizable region of charge concentration on the bonded side of the proton which is dominated by contractions of ρ towards the bond path.

The carbon–oxygen and carbon–sulphur bonds provide further examples of intermediate'interactions. The charge distributions and interatomic surfaces for a number of molecules containing C–O and/or C–S bonds are shown in Fig. 6.4. Table 7.8 lists the atomic populations and first moments for the atoms in these bonds, together with the data for their $(3, -1)$ or bond critical points. The position of the C–O interatomic surface relative to the node in the VSCC of the carbon atom is illustrated for the formaldehyde molecule in Fig. 7.20. The positions of the bonded and non-bonded charge concentrations in the VSCCs of the carbon and oxygen atoms are also indicated, and one sees that, while there are two bonded maxima for the C–O bond, both lie within the basin of the oxygen atom because of the transfer of charge from C to O. This is a general observation for all the C–O bonds with the result that the bond critical point lies just within the inner shell of charge depletion. The same observation holds for the C–S bonds because of the charge transfer from S to C.

Transfer of the order of one electronic charge or more does not imply the presence of an ionic interaction. The charge distributions and the properties of the atoms in these molecules do not begin to approach those of closed-shell systems, a fact reflected in the properties of ρ at the bond critical point. The first observation for this set of molecules is that, while $\nabla^2\rho(\mathbf{r}_c) > 0$ for these bonds, the value of $\rho(\mathbf{r}_c)$ is also large in value, much larger than the values associated with the closed-shell or ionic limit. This, in turn, is a result of the large magnitude of the curvatures of ρ at $\mathbf{r}_c$, values which are much larger than those for the closed-shell limit, but comparable to those for the shared limit (Tables 7.4 and 7.5). These values reflect the large stresses that are present in the interatomic charge distribution, stresses which lead to a large accumulation of charge in the interatomic surface and to the formation of a

Table 7.8

Properties of atoms and bonds in compounds of carbon, oxygen and sulphur*

System	Bond	R	$r_c^\dagger$	$q(C)$	$q(O)$ or $q(S)^\ddagger$	$M(C)^\ddagger$	$M(O)$ or $M(S)$	$\rho(\mathbf{r}_c)$	$\nabla^2\rho(\mathbf{r}_c)$	λ_1	λ_2	λ_3	$G(\mathbf{r})/\rho(\mathbf{r})$
H$_2$CO	C–O	2.237	0.7427	+1.2453	–1.2403	–0.912	+0.597	0.4308	0.4593	–1.2499	–1.1815	2.8908	1.988
CO	C–O	2.105	0.7081	+1.3269	–1.3269	–1.718	+0.984	0.5069	0.9796	–1.8323	–1.8323	4.6441	2.287
OCO	C–O	2.160	0.7398	+2.5951	–1.2976	0.000	+0.789	0.4826	0.1508	–1.3969	–1.3969	2.9447	1.918
SCO	C–O	2.137	0.7236	+0.7639	–1.2743	–2.006	+0.825	0.4904	0.5954	–1.5067	–1.5067	3.6087	2.127
	C–S	2.971	1.8426	+0.7639	+0.5104	+2.006	–1.252	0.2215	0.8586	–0.1968	–0.1968	1.2523	2.016
SCS	C–S	2.918	1.8048	–1.3571	+0.6786	0.000	–1.436	0.2448	0.8250	–0.2498	–0.2498	1.3245	1.969
CS	C–S	2.872	1.7786	–0.8870	+0.8870	+0.438	–1.494	0.2843	0.7592	–0.3199	–0.3199	1.3991	1.926
H$_2$CS	C–S	3.022	1.8834	–0.6178	+0.5155	+1.008	–1.479	0.2432	0.3619	–0.2808	–0.2501	0.8929	1.593

* Results obtained from RHF/6–311^{++}G**//6–311^{++}G** calculations. All results in atomic units.

$^\dagger r_c$ is distance from carbon nucleus to bond critical point, its bonded radius.

‡ A positive sign means $M(\Omega)$ is directed towards bonded partner, a negative means that $M(\Omega)$ is directed away from bonded partner, excluding the hydrogens.

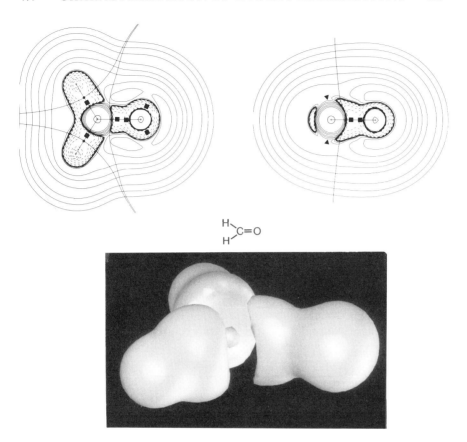

Fig. 7.20.   (top) Contour plots of $\nabla^2 \rho$ in the formaldehyde molecule for the plane containing the nuclei (left-hand side) and the plane perpendicular to this and along the C–O axis (right-hand side). The dashed (solid) lines denote regions of charge concentration (depletion) with contour values given in the Appendix (Table A2). The Laplacian is also negative within the region bounded by innermost solid contour encompassing the C and O nuclei. The former diagram shows the positions of the three bonded charge concentration (solid squares) in the VSCC of carbon and the single bonded and two non-bonded concentrations in the VSCC of oxygen. The latter diagram shows the two $(3, +1)$ critical points in $-\nabla^2 \rho$ (solid triangles) which determine the sites of nucleophilic attack at carbon. The positions of the zero-flux surfaces are also indicated. (bottom) The zero envelope 'reactive surface' for formaldehyde.

concentration of charge on the electronegative side of the interatomic surface and to a corresponding depletion of charge on the electropositive side. The large positive curvature, λ_3, found for the C–O bond dominates the interaction and the properties of the bond. Such a bond is perturbed only slightly by substitution, of a hydrogen in formaldehyde for example. When λ_3 is low in value, as for an ethylenic bond, atomic substitution can more readily lead to a shift in the interatomic surface, corresponding to a transfer of charge across the surface and to a change in the characteristics of the bond (Slee, 1986a).

The atomic interactions in C–O and C–S are clearly intermediate in character: $\rho(\mathbf{r}_c)$ is large as there is a large accumulation of charge in the interatomic surface resulting from the considerable contractions of ρ towards the bond path, but the interaction is dominated by the still larger positive stress in ρ along the interaction line giving a tight distribution with a large kinetic energy per electron. The ability to summarize the details of the mechanics of an interaction in terms of the properties of ρ at $\mathbf{r}_c$ and the properties of the atoms whose nuclei are linked by the associated interaction line precludes the need to assign oversimplifying labels to describe the interaction. It is, however, important to realize that the parameters which summarize an interaction do exhibit a continuous spectrum of values linking limiting situations that can be classified as shared and closed-shell interactions.

It is also possible to observe behaviour transitional between shared and closed-shell interactions in the absence of charge transfer. In the homonuclear series $B_2 \rightarrow F_2$, both $\rho(\mathbf{r}_c)$ and $|\nabla^2\rho(\mathbf{r}_c)|$ increase to a maximum at N_2 and decrease to minimum values at F_2. Their behaviour parallels the binding energies of these molecules, as it does the occupation of the $1\pi_u$ bonding orbital (fully occupied at N_2) and of the $1\pi_g$ antibonding orbital (fully occupied at F_2). Occupation of the $1\pi_g$ orbital leads to an increase in the localization of the charge on each atom and this has the expected consequences on the properties of ρ at $\mathbf{r}_c$, namely, a decrease in $\rho(\mathbf{r}_c)$ and an increase in the positive curvature λ_3 and, hence, in $\nabla^2\rho(\mathbf{r}_c)$. In F_2 the contraction of ρ towards each nucleus dominates the interaction and the VSCC of each F atom is localized in a near-spherical atomic-like shell, as is typical of the closed-shell limit (Fig. 7.16). As demonstrated in Fig. 7.16, however, the behaviour of the Laplacian at $\mathbf{r}_c$ for F_2 is just opposite to that found for Ar_2. The characteristics of ρ at $\mathbf{r}_c$ for F_2 are intermediate between those for O_2 and Ne_2 (Table 7.4); $\rho(\mathbf{r}_c) = 0.296$ au, $\nabla^2\rho(\mathbf{r}_c) = 0.233$ au, $\lambda_1 = \lambda_2 = -0.731$ au, $\lambda_3 = 1.694$ au, and the ratio $G(\mathbf{r}_c)/\rho(\mathbf{r}_c) = 0.86$ au. The ratio of the perpendicular to parallel kinetic energy contributions at $\mathbf{r}_c$ is 0.751, a value which, while less than unity as typical of closed-shell interactions, is considerably greater than the value for Ne_2.

As discussed in detail in Section E7.1, both the charge density and the pair density of the electrons of a F atom are very localized within its basin, as is characteristic of a closed-shell interaction. There is only a 7 per cent exchange of the electrons of a F atom in F_2 with the electrons on the neighbouring atom. The contribution to the exchange integrals arising from the interatomic electron exchange (a measure of the covalent character in the molecular orbital model of electronic structure) is correspondingly small. In contrast, the corresponding extents of electron localization in C_2 and N_2 are 73 and 78 per cent, respectively. For these two molecules the interatomic surfaces defining the atoms maximize the extent of interatomic exchange, as is characteristic of a shared interaction. Thus, the strong potential field exerted on the valence electrons in a fluorine atom, while providing the major source of binding in fluorides via charge transfer to the fluorine atom, is also the cause of the weak binding found in the F_2 molecule.

There is no suggestion that the atomic interaction in the F_2 molecule is at the closed-shell limit which requires that $\rho(\mathbf{r}_c)$ be small in value in addition to $\nabla^2\rho(\mathbf{r}_c)$ being greater than zero. It is, however, demonstrated that the binding in this molecule is qualitatively different from that found in N_2 (see Fig. 7.16) and that the differences are made quantitative by the properties of each system at the critical point in its charge density, formed as a consequence of the interaction of the two atoms.

7.5 Atomic interactions in bound and unbound states

We now consider the problem of further classifying a given interaction as belonging to a bound or unbound state of a system, i.e. bound or unbound with respect to a separation of the system into the two fragments linked by the corresponding interaction line. This problem is first considered from the electrostatic point of view using the Hellmann–Feynman theorem to relate the form of the charge distribution to the forces exerted on the nuclei. This is followed by a discussion of the energetics of an interaction in terms of the virial theorem. This discussion of the energy is linked to the properties of the charge density via the Laplacian of ρ using the local expression of the virial theorem, eqn (7.4).

7.5.1 The electrostatic theorem and chemical binding

The Hellmann–Feynman theorem determines the derivative of the energy E with respect to a parameter s appearing in the Hamiltonian (Feynman 1939). This theorem is derived in eqn (5.36) as a special case of the hypervirial theorem. It was pointed out there via eqn (5.37) that, when s is taken to be a nuclear position coordinate, one obtains the electrostatic theorem. This theorem states that the force exerted on nucleus α, the quantity $\mathbf{F}_\alpha$, is $Z_\alpha e$ times the electrostatic field exerted at the nucleus by the other nuclei and by the distribution of electronic charge. An explicit expression for $\mathbf{F}_\alpha$ is given in eqn (6.60). Accepting the quantum mechanical expression for the distribution of electronic charge as given by $\rho(\mathbf{r})$, the theorem is a statement of classical electrostatics and therein lies both its appeal and usefulness. The theorem also applies beyond the Born–Oppenheimer approximation, García-Sucre and Castells (1989).

Following Berlin (1951) the expression for the axial component of the force on either nucleus in a diatomic molecule (with reference to the coordinate system shown in Fig. 7.21) can be expressed as

$$\mathbf{F}_\alpha = \frac{e^2}{2} \int \left(\frac{Z_\alpha \cos \theta_\alpha}{r_\alpha^2} + \frac{Z_\beta \cos \theta_\beta}{r_\beta^2} \right) \rho(\mathbf{r}) \, d\tau - Z_\alpha Z_\beta / R^2 = -\mathbf{F}_\beta. \qquad (7.10)$$

When the element of electronic charge $- e\rho(\mathbf{r}) \, d\tau$ is placed so that the first of the two following conditions is fulfilled,

$$\left(Z_\alpha \frac{\cos \theta_\alpha}{r_\alpha^2} + Z_\beta \frac{\cos \theta_\beta}{r_\beta^2} \right) \quad \left\{ \begin{array}{l} > 0 \ \text{binding} \\ < 0 \ \text{antibinding,} \end{array} \right. \qquad (7.11)$$

then its effect is to draw the two nuclei together, or to exert a binding force. When the charge element is placed so as to satisfy the second condition, its effect is to exert a larger force on one nucleus than on the other and draw the nuclei apart. Such a charge element exerts an antibinding force on the nuclei which acts in concert with the nuclear force of repulsion. In this manner the space of a diatomic system can be partitioned into a *binding region*, the open region in Fig. 7.21(a), and *antibinding regions*, the hatched areas in the same figure (Berlin 1951). They are separated by two surfaces of revolution defined by the condition that the expression in eqn (7.11) equal zero. Charge elements at a point on such a surface exert equal forces on both nuclei and tend to neither increase nor decrease the separation between the nuclei. Early studies on molecular binding used density difference distributions, $\Delta \rho = \rho(\text{molecule}) - \rho(\text{atoms})$, in conjunction with the concepts of binding and antibinding regions to study the relation between the redistribution of electronic charge which results from the formation of a molecule and chemical binding (Bader et al. 1967a,b; Bader 1981). It was proposed that the density difference distribution, or $\Delta \rho$ map, may be taken as the pictorial representation of the 'bond density'.

The forces acting on the nuclei in a diatomic molecule and directed along the internuclear axis, expressed in terms of the internuclear separation R, the quantity $F(R)$, is given by $- dE(R)/dR$, Fig. 7.21b. $E(R)$ is the total energy of the molecule as determined by the fixed-nucleus Hamiltonian given in eqn (6.8). The change in the total energy $E(R)$ as the internuclear separation R is *decreased* from R_1 to R_2 is given by

$$\Delta E_{12} = - \int_{R_1}^{R_2} F(R) \, dR. \qquad (7.12)$$

Two necessary conditions for a state to be bound are that the forces on the nuclei vanish at some $R_2 \equiv R_e$ and that $\Delta E_{12} < 0$. Thus, in addition to the forces vanishing, there must be net attractive forces acting on the nuclei, that is, forces drawing the nuclei together and decreasing the separation R, over

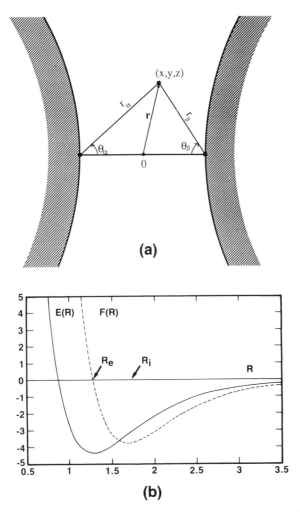

FIG. 7.21. (a) The coordinate system used to define Berlin's binding and antibinding regions and a display of these regions for the case of a homonuclear diatomic molecule. (b) Plots of the total energy of a diatomic molecule $E(R)$ versus the internuclear separation R—the potential energy curve for motion of the nuclei. Superimposed on this curve is the curve for $F(R)$ defined as $F(R) = -dE(R)/dR$. From the E versus R curve, the forces on the nuclei are attractive when $F(R) < 0$. The area lying between the force curve and the R axis between the limits $R = \infty$ and $R = R_e$ equals the binding energy of the molecule. The larger the attractive forces on the nuclei, forces which attain their maximum value at the inflexion point R_i, the larger is the binding energy.

some range of internuclear separations $R \geqslant R_e$ for $\Delta E_{12} < 0$. The energy increases for an increase or decrease in R about R_e for a bound state. These statements are illustrated in Fig. 7.21b for the formation of a bound state from the separated atoms.

In order for the Hellmann–Feynman forces on the nuclei to be attractive and so yield a bound state, the electronic charge must be distributed so that its contribution to the force from the binding region exceeds both the nuclear force of repulsion and the contributions to the electronic force from charge density in the antibinding regions. This can be accomplished by accumulating electronic charge in the binding region and/or by concentrating it along the internuclear axis within the binding region ($\cos \theta_\alpha = \cos \theta_\beta = 1$ in eqn (7.11)). It is for this reason that the formation of a $(3, -1)$ critical point in the charge density is a necessary condition for the formation of a bound state, since such a critical point implies the existence of an atomic interaction line—a line linking the nuclei along which the charge density is a maximum with respect to any neighbouring line. As emphasized in Sections 2.4.2 and 3.2.4, the presence of such a line indicates that electronic charge is accumulated between the nuclei, with $\rho(\mathbf{r})$ reaching its maximum values along the bond path. *If the forces resulting from the accumulation of electronic charge in the binding region are sufficient to exceed the antibinding forces over a range of separations to yield an equilibrium configuration, then the state is bound and the atomic interaction line is called a bond path.*

The Laplacian distributions for the shared interactions illustrated in Figs 7.15 to 7.17 clearly indicate the predominance of the concentration of electronic charge over the binding region for such interactions. In all of these molecules $\nabla^2\rho < 0$ over the binding region for each interacting pair of nuclei. Thus, in a shared interaction, electronic charge is concentrated in the binding region as a result of the contraction of ρ towards the bond path and this concentration extends over the basins of both atoms. It is the shared concentration of charge which exerts the net attractive forces on the nuclei for $R > R_e$ and balances the forces of repulsion at $R = R_e$. While $\nabla^2\rho < 0$ in the antibinding regions of nuclei as well, the amount of charge accumulated there is less than that accumulated in the binding region, and the net force exerted on the nuclei is one of attraction. The atomic graph of the N atom in N_2, for example, consists of two maxima in $-\nabla^2\rho$, one bonded, the other non-bonded with magnitudes 3.15 au and 2.52 au, respectively (values from an MP2 calculation, see Fig. 7.16). The maxima are linked by a torus of $(2, 0)$ critical points in $-\nabla^2\rho$ which encircle the bond axis and have a magnitude of 0.74 au. The bonded maxima of the two nitrogen VSCCs are linked by a $(3, -1)$ critical point in $-\nabla^2\rho$ of magnitude 2.61 au and the region of charge concentration is contiguous over both atomic basins.

In a shared interaction it is the electronic charge concentrated in the neighbourhood of the interatomic surface and extending over both atomic basins which exerts the attractive forces on the nuclei. In the closed-shell interactions the region in the neighbourhood of the interatomic critical point is one of charge depletion and charge is instead concentrated separately in each atomic basin in shell-like regions similar to those found in an isolated atom or ion (Figs 7.15–7.17). Clearly, such an interaction can result in a bound or unbound state depending upon the direction of the polarization of each atomic charge distribution. The directions of these atomic polarizations can be evident in the Laplacian distributions. Consider, as an example of an unbound state, the Laplacian distribution for the Ar_2 molecule. The atomic graph for an argon atom in this molecule, like that for nitrogen in N_2, consists of two axial critical points in $-\nabla^2\rho$. Unlike nitrogen, however, where both critical points denote local charge concentrations, the critical point in the binding region of Ar is a $(3, +1)$ critical point, denoting a local minimum in the sphere of charge concentration, while that in its antibinding region is a $(3, -3)$ maximum, a local charge concentration. The values of $\nabla^2\rho$ at these two points are -1.26 and -1.34 au, respectively (values from an MP2 calculation). Two tori of critical points of rank two encircle the internuclear axis, one of signature -2, the other of signature 0, and link the two axial critical points to complete the atomic graph. The two atomic graphs are themselves linked by a broad central minimum for which $\nabla^2\rho = +0.45$ au. Thus, for this unbound state, there is a region of maximum *depletion* in electronic charge in the central binding region and the VSCC of each atom is polarized into its antibinding region (see Fig. 7.16). The major difference in the Laplacian distributions of bound N_2 and unbound Ar_2 is the presence of a considerable concentration of charge in the central binding region of the former linking the two atomic VSCCs, and its replacement in the latter molecule by a region of charge depletion.

Interactions at the closed-shell limit can result in a bound state if accompanied by charge transfer, examples being ionic and hydrogen-bonded systems. The charge distributions shown in Fig. E7.2 for diatomic molecules formed from Group I and Group VII elements and their Laplacian distributions shown in Fig. 7.15 demonstrate that such systems approximate the model of two nearly spherical charge distributions. The atomic charges given in Tables 7.4 and E7.1 bear out the ionic model, as do the measures of the degree of localization of the pair density within each of the separate atomic boundaries given in Table E7.1. As a consequence of Gauss's law, the binding in a model ionic system A^+B^- is achieved by a polarization of both ions, considered to be initially spherical, in a direction counter to the direction of charge transfer. It is shown that the charge distributions of the systems considered here recover the dictates of this model.

Gauss's law of electrostatics states that the field exerted at a point external to a spherical charge distribution is equal to that which would be obtained if all the charge were concentrated at its centre as an equivalent point charge. Thus, the cationic nucleus is attracted by the net negative field exerted on it by the anion, and the charge distribution of the cation must polarize away from the anion to balance this net attractive force on its nucleus and achieve electrostatic equilibrium. The nucleus of the anion is repelled by the net positive field of the cation, and the charge distribution of the anion must polarize towards the cation to balance the repulsive force exerted on its nucleus. Thus, in an ionic system, both nuclei are bound by the forces exerted on them by the electronic charge localized on the anion and polarized towards the cation. That is, the charge accumulation in the binding region is localized within the basin of the anionic nucleus (Bader and Henneker 1965). The polarization of the VSCC of the Cl atom in LiCl and NaCl towards the cation is evident in the displays of $\nabla^2 \rho$ for these molecules in Fig. 7.15.

The contributions to the forces on the nuclei in such molecules show that they achieve electrostatic equilibrium by a polarization of the atomic charge distributions in a direction counter to the transfer of approximately one electronic charge. The electronic contribution to the force on nucleus α in the molecule AB is given by the expression

$$\mathbf{F}_\alpha^e = \mathbf{F}_\alpha^e(A) + \mathbf{F}_\alpha^e(B) \tag{7.13}$$

where $\mathbf{F}_\alpha^e(A)$ is the force exerted on nucleus α by the electronic charge in its own basin, the basin of atom A and $\mathbf{F}_\alpha^e(B)$ is the force exerted on nucleus α by the electronic charge in the basin of atom B. For the cation, $\mathbf{F}_\alpha^e(A)$ is antibinding and $\mathbf{F}_\alpha^e(B)$ is binding, while for the anion the corresponding contributions are both binding as predicted by the model. Not only are the directions of the forces as predicted by the model, so are their magnitudes. In the idealized limit of A^+B^-, the electronic contributions to the forces on the two nuclei may be expressed in terms of equivalent charges $Q_\alpha(\Omega)$ as

$$\begin{aligned}
\mathbf{F}_\alpha^e &= (Z_\alpha/R^2)(Q_\alpha(A) + Q_\alpha(B)), \\
\mathbf{F}_\beta^e &= (Z_\beta/R^2)(Q_\beta(B) + Q_\beta(A)).
\end{aligned} \tag{7.14}$$

The charge $Q_\alpha(A)$ or $Q_\beta(B)$ is the charge which, when placed at the second nucleus, reproduces the force on the cationic or anionic nucleus, respectively, to account for the polarization of their atomic densities, antibinding for the cation and binding for the anion. Ideally, $Q_\alpha(A) = +1$ and $Q_\beta(B) = -1$. The charges $Q_\alpha(B)$ and $Q_\beta(A)$ should equal $-(Z_\beta + 1)$ and $-(Z_\alpha - 1)$, respectively. For spherical ions, these latter effective charges should equal the negative of the corresponding electronic populations $N(B)$ and $N(A)$. The charge equivalents of the electronic contributions to the forces for some ionic systems are given in Table 7.9. The charge equivalents of the forces $\mathbf{F}_\alpha^e(B)$ and $\mathbf{F}_\beta^e(A)$ are indeed close in value to the actual populations of the B and A atoms which are given in brackets. The contributions to the forces from polarizations of the atomic densities also approach their limiting charge equivalents of ± 1. The sums of the charge equivalents for nucleus α and of those for nucleus β must equal $-Z_\beta$ and $-Z_\alpha$, respectively, for electrostatic equilibrium. Forces on nuclei are notoriously difficult to calculate quantitatively and only in the case of LiF do the numbers in Table 7.9 sum to yield

Table 7.9

Charge equivalents* of electronic contribution to nuclear forces

AB	$Q_\alpha(A)$	$Q_\alpha(B)$		$Q_\beta(B)$	$Q_\beta(A)$
LiF	0.77	− 9.76 (9.94)		− 0.99	− 2.04 (2.06)
LiCl	0.60	− 17.82 (17.92)		− 0.91	− 2.07 (2.07)
KF	0.73	− 9.86 (9.93)		− 0.71	− 17.84 (18.07)
A_2			A_2		
C_2	− 0.86	− 5.13 (6.00)	F_2	− 0.42	− 8.52 (9.00)
N_2	− 1.01	− 5.98 (7.00)	He_2	+ 0.71	− 1.94 (2.00)
O_2	− 0.78	− 7.21 (8.00)	Ar_2	+ 0.33	− 17.61 (18.00)

*$Q_\alpha(A) = \mathbf{F}_\alpha^e(A) \cdot R^2/Z_\alpha$, $Q_\alpha(B) = \mathbf{F}_\alpha^e(B) \cdot R^2/Z_\alpha$ with corresponding definitions for nucleus β. A negative sign means that force is attractive, i.e. directed towards the other nucleus.

small net forces. The experimental bond length was employed in the calculation of the two remaining systems (McLean and Yoshimine 1967) and the errors, while larger, are smaller than any of the individual contributions to the forces.

Also included in Table 7.9 are charge equivalents of the forces in homonuclear diatomic molecules for bound and unbound systems. In the bound systems, which result from shared interactions, both contributions to the electronic force on a nucleus are binding. The descreening of the second nucleus and the force resulting from the atomic polarization are both a maximum for the strongest bond with the greatest charge accumulation between the nuclei, N_2. Because of the contraction of the charge density away from the interatomic surface and its separate localization in the basin of each atom in the closed-shell interactions, the descreening of the second nucleus is much less for the unbound systems. The charge equivalents of $\mathbf{F}_\alpha^e(B)$ approach $- N(B)$ in value for He_2 and Ar_2, as found for the ionic closed-shell systems. The atomic contributions for the unbound systems are, however, strongly antibinding as a result of the polarization of each atomic density away from the internuclear region. The F_2 molecule, which is intermediate between the two limits of interaction and possesses the smallest dissociation energy of the bound systems, is seen to possess the largest screening of the second nucleus and the smallest binding contribution from the atomic polarizations for the bound systems.

The operator for the electronic force exerted on a nucleus is proportional to $1/r_\alpha^2$, the inverse of the square of the distance from the nucleus (eqn (7.10)), and is, therefore, dominated by the polarization of the charge density close to the nucleus. In the expression for an atomic first moment $\mathbf{M}(\Omega)$, however, the charge density is weighted by the operator $\mathbf{r}_\alpha$ (eqn (6.45)), and this property is dominated by the polarization of the valence density, particularly its diffuse component. It is possible, therefore, for the atomic force $\mathbf{F}_\alpha^e(A)$ and the atomic first moment $\mathbf{M}(A)$ to be oppositely directed, as indeed occurs for the bound ground states of the homonuclear diatomics C_2 to F_2. As evidenced by the data in Table 7.9, the atomic force in this series is binding and the inner density is polarized towards the second nucleus. The atomic first moment in each of these molecules is, however, directed away from the second nucleus, the magnitudes of this moment for C_2, N_2, O_2, and F_2 being, respectively, 0.172, 0.590, 0.438, and 0.197 au. Both moments attain their maximum magnitude for the nitrogen molecule, the most tightly bound of these systems. In an ionic molecule both the atomic force and atomic first moment of the anion are directed towards the cation.

A study of the contributions of the individual molecular orbital densities to the total force exerted on a nucleus in homonuclear diatomic molecules was made by Bader *et al.* (1967*a*). This study shows that the atomic polarizations binding the nuclei in these molecules do not arise primarily from the $1\sigma_g$ and $1\sigma_u$ molecular orbitals whose densities correspond to slightly polarized 1s-like atomic distributions. Indeed, in C_2 these orbital densities exert small antibinding forces on the nuclei. It is the density of the $2\sigma_g$ orbital in these molecules which is primarily responsible for the atomic forces which bind the nuclei, while the densities of the $2\sigma_u$ and $3\sigma_g$ orbitals are responsible for the oppositely directed atomic first moments.

The requirements of binding, as viewed through the electrostatic theorem, emphasize the existence of an atomic interaction line as a necessary condition for a state to be bound, whether it be at the shared or closed-shell limit of interaction. The differing properties associated with the distributions of electronic charge at the shared and closed-shell limits of interaction are reflected in the differing mechanisms by which the forces on the nuclei are balanced to achieve electrostatic equilibrium in the two cases.

7.5.2   The virial theorem and chemical bonding

The Laplacian of ρ determines the regions of space where electronic charge is concentrated. This function, coupled with the electrostatic theorem, enables one to qualitatively characterize a charge distribution as binding or antibinding with respect to a given interaction in a molecule. These same regions of space, those where $\nabla^2\rho < 0$, are the regions in which the potential energy makes its dominant contributions to the virial of a system. Thus, the Laplacian of ρ provides a link between the forces acting on the nuclei and the energetics of an interaction, a link we now pursue.

The molecular statement of the virial theorem (Slater 1933) is given in eqn (6.65). For a diatomic molecule, the virial of the external forces acting on the nuclei assumes the simple form $-R(dE/dR)$ and in this case the virial theorem becomes

$$2T = -V - R(dE/dR). \tag{7.15}$$

T is the average kinetic energy of the electrons, $V = \langle \hat{V} \rangle$, where $\hat{V}$, the complete potential energy operator, is defined in eqn (6.9) and $E = T + V$ is the eigenvalue of the fixed nucleus Hamiltonian defined in eqn (6.8). The force is zero for the separated atoms and for the equilibrium separation R_e. Defining ΔT and ΔV to be the differences in these energies for the molecular values minus the separated atom values, the virial theorem yields the following well-known constraints on the energy changes accompanying the formation of a bound state, i.e. one for which $\Delta E < 0$, $\Delta T > 0$, $\Delta V < 0$, and $2\Delta T = -\Delta V$. Thus, the virial theorem ascribes molecular stability to a decrease in the potential energy, a change which necessarily demands that the kinetic energy increase. Since, next to the nuclear positions, the internuclear region exerts the lowest potential, such an interpretation is consistent with the requirement of the electrostatic theorem of binding, that electronic charge be accumulated in the binding region (Slater 1963). It is also consistent with the *local statement* of the virial theorem, eqn (7.4), which states that the potential energy makes its dominant stabilizing contributions to the virial and to the lowering of the total energy in those regions where electronic charge is concentrated, that is where the Laplacian distribution is negative. The preceding discussion has demonstrated that this corresponds to a concentration of charge in the binding region. The relation between the local and integrated forms of the virial theorem provided by the Laplacian of ρ is subject to further investigation below. However, it is clear at this point that the electrostatic view that binding is obtained as a consequence of the

Table 7.10

General behaviour of dT/dR and dV/dR in the formation of a bound molecular state

Range of value of R		$F(R)$	$dF(R)/dR$	$dV(R)/dR$	$dT(R)/dR$
I	$\infty > R > R_i$	< 0	> 0	$\leqq 0^*$	$\leqq 0^\dagger$
	$R = R_i$	< 0	0	> 0	< 0
II	$R_i > R > R_e$	< 0	< 0	> 0	< 0
	$R = R_e$	0	< 0	> 0	< 0
III	$R < R_e$	> 0	< 0	> 0	< 0
IV	$R \ll R_e$	> 0	< 0	< 0	> 0

$^* < 0$ for $R \gg R_i$, $^\dagger > 0$ for $R \gg R_i$.

eventual balancing of attractive forces on the nuclei and the view of the virial theorem that bonding is a result of the lowering of the potential energy are, in fact, ascribing a common origin to a bound state: that electronic charge be concentrated in the internuclear or binding region (Slater 1963, pp. 39–40). The study of the properties of the Laplacian distribution for bound systems demonstrates that this is where electronic charge is concentrated in bound systems to achieve electrostatic equilibrium and simultaneously bring about the stabilizing decrease in the potential energy of the system.* The local and integral virial theorems place restrictions on the accompanying changes in the kinetic energy as well, and these will also be illustrated and discussed.

The virial and electrostatic theorems are connected by the relation

$$dE = -F(R)dR \qquad (7.16)$$

which shows, as demonstrated in Fig. 7.21, that the energy will decrease over a decreasing range of R values only if the forces acting on the nuclei are attractive—that is, electronic charge is accumulated in the binding region. The two theorems may be combined to yield constraints on the signs of dT/dR and dV/dR for ranges of R values as defined by extrema in the $F(R)$ and $E(R)$ curves (Bader 1981). The subtraction of T from both sides of eqn (7.15) followed by differentiation with respect to R yields

$$dT(R)/dR = 2F(R) + R\,dF(R)/dR \qquad (7.17)$$

with $F(R)$ defined as in eqn (7.16). The addition of $2V$ to each side of eqn (7.15) followed by differentiation with respect to R yields a corresponding expression for dV/dR,

$$dV(R)/dR = -3F(R) - R\,dF(R)/dR. \qquad (7.18)$$

---

* It is perhaps important to mention that one can find articles in the literature which dispute these conclusions. Since the conclusions are based upon physical laws, one is assured that the criticisms employ non-physical arguments, as is indeed the case. For example, the argument that a build-up of electronic charge in the internuclear or 'bond' region causes an increase rather than a decrease in the potential energy of a system is obtained from the analysis of a system which does not obey the virial theorem. That is, the result is obtained for a very approximate (unscaled) wave function, one that predicts the total energy to decrease as a result of $\Delta T < 0$ and $\Delta V > 0$, a result in direct violation of the virial theorem which it is used to criticize! Another argument, which ascribes molecular stability to a decrease in kinetic energy, measures the changes in energies accompanying the formation of a bound state to those of an arbitrarily defined promoted state, which, since it is not a realizable quantum state, does not obey the virial or electrostatic theorems. In addition, the promoted state is defined so as to yield the desired result. The conclusion of the electrostatic theorem regarding the necessity of the accumulation of electronic charge in the binding region has been criticized on the basis of maps obtained by subtracting from a total molecular density distribution, an arbitrarily defined reference density, which again requires that one invoke a physically unrealizable state of the system. In brief, if one stays within the bounds of physics and avoids opinions as to what arbitrarily defined reference states should be used to obtain a desired result, the theorems and the mutually consistent interpretations they provide are inviolate. If one does desire an opinion regarding these two theorems, one can refer to Slater (1972) who regarded them as 'two of the most powerful theorems applicable to molecules and solids'.

The points at which $F(R)$ equals zero and where it attains its minimum value enable one to define ranges for values of R in which the signs of $F(R)$ and its derivatives are uniquely determined (Table 7.10). The inflexion point in the curve for $E(R)$ corresponds to the point at which $F(R)$ is a minimum, a point labelled R_i. At this point the magnitude of the binding forces exerted on the nuclei by the charge density is a maximum.

In range I (Table 7.10), where $\infty > R > R_i$, $F(R)$ and $dF(R)/dR$ are of opposite sign and the signs of the derivatives of T and V are not uniquely determined. However, for large values of R, these signs should be determined by the sign of just $dF(R)/dR$, since this term is multiplied by R in the expressions for the derivatives. One thus predicts the kinetic energy to initially decrease, $dT/dR > 0$, and the potential energy to increase for the *approach* of two atoms. This is the observed result for the approach of two neutral atoms in S states, and is, in general, true for the approach of neutral molecules for these changes in T and V are characteristic of the so-called London or van der Waals' forces. These forces appear as the leading term $-6C_6/R^7$, in the perturbation expansion for long-range interactions.

Feynman, in his original paper on the electrostatic theorem (1939), suggested that the long-range force on each nucleus should result from its attraction to the centroid of its own charge distribution. This view is borne out by the diagram of the change in the atomic charge densities shown in Fig. 7.22 for the approach of two ground-state hydrogen atoms at a separation of 8 au.

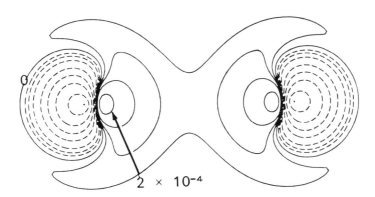

2×10^{-4}

FIG. 7.22. Density difference map calculated from a correlated wave function for H_2 (Das and Wahl 1966) for a separation of 8 au. The density of the unperturbed hydrogen atoms, each in its ^2S ground state, is subtracted from the molecular density using the same nuclear position coordinates. The overlap of the atomic densities at this internuclear separation is negligible and the difference map illustrates the polarization of each of the atomic densities as caused by the correlative interaction of the two electrons. The solid and dashed contours, increase $(+)$ or decrease $(-)$, respectively, from the zero contour in the order $\pm 2 \times 10^{-n}$, $\pm 4 \times 10^{-n}$, $\pm 8 \times 10^{-n}$ with n beginning at 5 and decreasing in steps of unity.

This diagram is obtained from a highly correlated wave function by subtracting the unperturbed atomic densities from the correlated molecular density. The transfer of electronic charge from the antibinding regions to the binding region, which results from the polarization of each atomic density towards the other, is a consequence of the correlative interactions between the electrons on the two centres. One can do no better at explaining this picture than to give Feynman's prediction of it by quoting his 1939 paper.

The Schrödinger perturbation theory for two interacting atoms at a separation R, large compared to the radii of the atoms, leads to the result that the charge distribution of each is distorted from central symmetry, a dipole moment of order $1/R^7$ being induced in each atom. The negative charge distribution of each atom has its center of gravity moved slightly toward the other. It is not the interaction of these dipoles which leads to van der Waals' force, but rather the attraction of each nucleus for the distorted charge distribution of its own electrons that give the attractive $1/R^7$ force.

Thus the long-range forces are boot-strap forces—each nucleus is pulled by its own distorted charge distribution. The form of the charge rearrangement found to result from this correlative interaction between the electrons and shown in Fig. 7.22 provides a ready explanation for the initial decrease in T and increase in V. Charge density is removed from the region of each nucleus where $G(\mathbf{r})$, the kinetic energy density, has its maximum value and is placed in a diffuse distribution in the binding region. This accounts for $\Delta T < 0$. The charge density is also tightly bound in the region of the nucleus and its removal from this region to a region further removed from the nucleus raises the potential energy of the system. Thus, an initial attractive interaction can result from a decrease in T and an increase in V, but never for the energy change for a bound state at electrostatic equilibrium, not even a van der Waals' minimum, where one must have $\Delta T > 0$ and $\Delta V < 0$.

At $R = R_i$, where the binding force attains its largest magnitude, the signs of $dT/dR (< 0)$ and $dV/dR (> 0)$ are uniquely determined. Thus, the decrease in energy for $R \gg R_i$ can result from a decrease in T and in spite of an increase in V. However, for $R < R_i$, up to $R = R_e$, T and V must begin to increase and decrease, respectively. In H_2, for example, as R decreases from infinity, initially $dV/dR < 0$ and $\Delta V > 0$. However, at $R = R_i = 2$ au, $\Delta V = -0.0625$ au and, hence, dV/dR changes sign before $R = R_i$. Thus for H_2, the present analysis shows that the charge rearrangements for $R \leqslant 2$ au lead to a decrease in the potential energy of the system and to an increase in its kinetic energy. The contributions from $F(R)dR$ to the energy in the range $R_i \geqslant R \geqslant R_e$ must dominate over those in the range $\infty > R > R_i$, since the total changes in T and V determining $\Delta E(R_e)$ must be such that $\Delta T > 0$ and $\Delta V < 0$.

The signs of dT/dR and dV/dR at $R = R_e$ will persist for some $R < R_e$, since, initially, $F(R)$ is small and dF/dR is large. However, $F(R)$ rapidly increases in this range (tending to infinity) while dF/dR remains finite. Thus, eventually, for some $R < R_e$, the signs of dT/dR and dV/dR will change, and V and T will increase and decrease, respectively, with R.

The ideas underlying the equating of E, the *total* energy of a molecule, to the sum of electronic and nuclear energies E_e and E_n were developed in Section 6.3. The electronic energy E_e is given by

$$E_e = T + \mathscr{V} \tag{7.19}$$

where $\mathscr{V}$, the electronic potential energy, is the virial of the Ehrenfest forces exerted on the electrons

$$-\int \mathbf{r} \cdot \nabla \cdot \overleftrightarrow{\sigma}(\mathbf{r}) d\mathbf{r} = \mathscr{V} \tag{7.20}$$

where $\overleftrightarrow{\sigma}$ is the quantum stress tensor, eqn (6.13). The electronic potential energy is related to the total potential energy V appearing in $E = T + V$ by the expression

$$V = V_{en} + V_{ee} + V_{nn} = \mathscr{V} - \sum_\alpha \mathbf{X}_\alpha \cdot \nabla_\alpha E. \tag{7.21}$$

Since the nuclei are stationary, E_n is purely potential and is equal to the virial of the electrostatic forces acting on the nuclei

$$E_n = -\sum_\alpha \mathbf{X}_\alpha \cdot \nabla_\alpha E. \tag{7.22}$$

The virial theorem may be stated in terms of the electronic contribution to E as

$$T = -E_e = -\tfrac{1}{2}\mathscr{V} \tag{7.23}$$

or, in terms of the total energy, as

$$T = -E - \sum_\alpha \mathbf{X}_\alpha \cdot \nabla_\alpha E = -\tfrac{1}{2}(V + \sum_\alpha \mathbf{X}_\alpha \cdot \nabla_\alpha E). \tag{7.24}$$

For an isolated atom, or for a molecule in an equilibrium configuration one has $T = -E_e = -E$. In general, however, for a molecule

$$T = -E_e > -E \qquad \text{or} \qquad T = -E_e < -E \tag{7.25}$$

depending on whether there are repulsive $(T > |E|)$ or attractive $(T < |E|)$ forces acting on the nuclei.

The Laplacian of the charge density as it appears in the local expression of the virial theorem, eqn (7.4), compares twice the kinetic energy density $G(\mathbf{r})$ not with contributions to the potential V, but with $\mathscr{V}$, the electronic potential energy. This is an important distinction from the point of view of whether or not a system is bound. Consider an interaction where $\nabla^2 \rho$ is predominantly negative over the binding region and *net* forces of attraction act on the nuclei. In this case the local contribution to the virial of the Hellmann–Feynman forces exerted on the electrons, which, as explained following eqn (6.60), is given by

$$-\sum_\alpha \mathbf{X}_\alpha \cdot \mathbf{F}_{\alpha e} = V_{nn} + \sum_\alpha \mathbf{X}_\alpha \cdot \nabla_\alpha E, \qquad (7.26)$$

is greater than V_{nn}. Thus, because of the concentration of charge in the binding region, the total potential energy V is more negative than $\mathcal{V}$ (eqn (7.21)), $|V| > 2T$, and $|E| > T$. Thus, E decreases with a decrease in R, until V_{nn} increases to a value such that equilibrium is attained and $V = \mathcal{V}$. It is, however, the concentration of negative charge in the binding region resulting from contractions in ρ perpendicular to the bond path which causes the local contributions to V to exceed the contributions from the kinetic energy for $R > R_e$, thereby leading to the creation of attractive forces and, hence, to a bound, shared interaction. In a system where $\nabla^2\rho$ is predominantly positive over a binding region and net forces of repulsion act on the nuclei, the kinetic energy contributions to the virial dominate over the internuclear region as a result of the contraction of the electronic charge density towards the nuclei and its consequent depletion in this region. In this situation $\nabla^2\rho$ is most negative in the antibinding regions, V is less negative than $\mathcal{V}$, $2T > |V|$, repulsive forces are exerted on the nuclei, and the system is unbound. In an ionic bond the transfer of charge to the anion for $R > R_e$ and the polarization of its charge towards the cation yield net attractive contributions to $\mathcal{V}$; V exceeds $\mathcal{V}$ in magnitude (eqn (7.21)) because the forces are attractive and $2T < |V|$.

A comparison of the Laplacian distribution for a shared interaction with its shell-like structure in the separated atoms shows that the major change in this function on the formation of such an interaction is the creation of a region over which $\nabla^2\rho < 0$ extending over the basins of both atoms. This concentration of charge results from a contraction of ρ perpendicular to the interaction line and it leads to a local lowering of the potential energy. The magnitude of this lowering in the potential energy is greater than the contributions to the kinetic energy from the same region. In particular, the component of the kinetic energy parallel to the bond path is decreased in value as a result of the decreases in the corresponding gradients and curvatures of ρ.

The differing properties of the charge distributions for bound and unbound states are made very clear by comparing the Laplacian distributions of ρ for H_2 and He_2 (Fig. 7.15). In the former, the atomic interaction is dominated by the contractions in ρ perpendicular to the bond path while in the latter it is dominated by the contractions in ρ towards each of the nuclei.

It has been suggested that H_2^+ and H_2 are bound, not because of the accumulation of electronic charge in the 'bond' region, but because of the contraction of the charge towards the nuclei. In actual fact, the major difference between the Laplacian distribution for bound H_2 and unbound He_2 is that the former but not the latter possesses accumulated, shared charge in the binding region, while the distribution of the latter but not the former is dominated by the contractions of the charge density towards the nuclei. It is important to read all the information that is contained in these maps of the Laplacian distribution. Because of the appearance of $\nabla^2\rho(\mathbf{r})$ in the local expression of the virial theorem, these maps also reveal the role of the local contributions of the kinetic energy to the total energy of the system. As noted in Section 7.4.2, the kinetic energy components parallel and perpendicular to the interaction line mirror the behaviour of the corresponding gradients and curvatures of ρ. For H_2, $\nabla^2\rho < 0$ in the binding region because of the contraction of charge towards the interaction line, and the parallel gradient and curvature of ρ are softened. Consequently, the kinetic energy per electron is small in value, a result in accord with the local virial theorem. Thus for H_2 at R_e, the ratio of the two perpendicular components to the parallel component of the average kinetic energy, the ratio $T_\perp/T_{\parallel}$, is greater than 2.0 and equals 2.80 (Bader and Preston 1969). For He_2, $\nabla^2\rho > 0$ in the region of the interatomic surface and the interaction is dominated by the contraction of ρ towards each nucleus. The parallel gradient and curvature of ρ are large, as is the parallel component of the kinetic energy which exceeds the perpendicular components. The kinetic energy per electron is also absolutely large. These observations are again in accord with the local statement of the virial theorem which states that the kinetic energy is dominant in regions where $\nabla^2\rho > 0$. For He_2 at $R = 2.0$ au, the ratio $T_\perp/T_{\parallel}$ is less than two and equals 1.81 (Bader and Preston 1969).

The binding of the nuclei in ring or cage structures is made particularly clear by the Laplacian distribution of the charge density. In cyclopropane the magnitudes of the two negative curvatures of ρ at a carbon–carbon bond critical point are unequal (Table 7.5). The principal axis of the curvature of smallest magnitude is directed at the ring or $(3, +1)$ critical point at the centre of the ring surface. Thus, a carbon–carbon bond of cyclopropane has a noticeable ellipticity ($\varepsilon = 0.490$) and electronic charge is preferentially concentrated in the ring surface (Fig. 7.17). The Laplacian of ρ shows a concentration of valence charge density over the surface of the ring with the exception of a small region in the neighbourhood of the $(3, +1)$ critical point. (The principal axes of the two positive curvatures of a ring critical point lie in and define the ring surface.) It is this concentration of electronic charge in the interior of the ring structure which binds the ring nuclei. The large extent of this concentration of charge is reflected in a noticeable curvature of the bond paths away from the geometrical perimeter of the ring (Fig. 7.17).

Electron-deficient molecules, such as the boranes, form ring and cage structures to gain maximum stability from a minimum amount of electronic glue by a very pronounced delocalization of charge over ring surfaces. The molecular graph for $B_6H_6^{-2}$ shows each boron atom to be linked to one hydrogen atom and four other boron atoms to form an octahedral cage structure bounded by eight ring surfaces. The Laplacian of the charge density for one such three-membered boron ring surface is shown in Fig. 7.23. The boron–boron bonds exhibit extreme ellipticities equal to 3.58 and, as illustra-

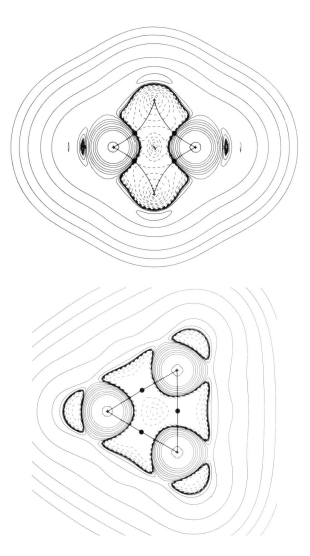

FIG. 7.23.   Contour maps of the Laplacian of ρ for the plane containing the two nuclei and the two bridging protons in B_2H_6 and for a face containing three boron nuclei in the octahedral molecule $B_6H_6^{-2}$. The bond paths and bond critical points are also shown. The inner contour around each boron nucleus encircles the inner region of charge concentration, a region which is in turn encompassed by the inner shell of charge depletion.

ted in Fig. 7.23, electronic charge is concentrated almost evenly over the whole of the ring surface. The contraction of ρ along the axis perpendicular to the ring surface at the ring critical point is also large and it dominates the two positive curvatures of ρ in the ring surface so that $\nabla^2\rho$ is negative even at the ring critical point. The value of ρ at a ring critical point is 0.120 au, only slightly less than its value of 0.127 au at a boron–boron bond critical point. The same observations hold true for four-membered rings involving one or two bridging hydrogen atoms in molecules such as B_2H_6 or B_5H_9. Figure 7.23 illustrates the Laplacian of ρ in the plane of the bridging hydrogen nuclei of B_2H_6. In this four-membered ring the bond paths are inwardly curved and electronic charge is again delocalized over the valence region of the ring surface. The ellipticities of the B–H bridging bonds equal 0.561 and their values of $\rho(\mathbf{r}_c) = 0.119$ au exceed the value of ρ at the ring critical point by only 0.013 au. Thus, in three- and four-membered rings of electron-deficient structures, electronic charge is preferentially delocalized over and concentrated in the ring surface where it serves to bind all nuclei of the ring. The description of such molecules as globally delocalized is particularly apt (King and Rouvray 1977).

7.5.3 A summary

The electrostatic and virial theorems, when coupled through the use of the Laplacian of the electronic charge distribution, provide an understanding and classification of the mechanics underlying atomic interactions leading to both bound and repulsive states. The regions of charge concentration as defined by the Laplacian, $\nabla^2\rho < 0$, can be used in conjunction with the electrostatic theorem to determine whether the forces on the nuclei will be attractive or repulsive. The electronic charge in these same regions makes the dominant stabilizing contributions to the potential and total energies of the system and, in this way, the Laplacian provides a link between force and energy. If the charge is concentrated in the binding region, the forces are attractive, the magnitude of the total potential energy V exceeds that of $\mathscr{V}$, and the virial of the nuclear forces contributes to the binding of the system. If charge is concentrated in the antibinding regions, the forces are repulsive, the magnitude of $\mathscr{V}$ exceeds that of V and the virial of the nuclear forces destabilizes the system. As explained in Section 6.4.4, this is what occurs when ethane in its minimum-energy staggered conformation undergoes a rigid rotation into its 'frozen' eclipsed conformation. Charge concentrations contribute to a lowering of $E_e = T + \mathscr{V}$, the electronic energy of the system, wherever they occur. Even non-bonded charge concentrations contribute to the lowering of the electronic energy. However, the effect of local charge concentrations on the total energy E is determined by whether they exert binding or antibinding forces on the nuclei, leading to a stabilization of the total energy in the former case and to its destabilization in the latter.

The Laplacian maps, again through the local virial theorem, also provide information regarding the local contributions to the kinetic energy. The low value of the gradient in directions parallel to the axis leads to small contributions to the kinetic energy density $G(\mathbf{r})$ and hence to T, while the contraction of $\rho(\mathbf{r})$ perpendicular to the axis gives large contributions. The charge density in the binding region of a stable molecule is, therefore, distributed in such a way as to keep the accompanying increase in kinetic energy to a minimum. The otherwise large increase in $G(\mathbf{r})$ resulting from the contraction of $\rho(\mathbf{r})$ perpendicular to the bond, the effect responsible for the reduction in the potential energy, is partially offset by the smaller contributions to $G(\mathbf{r})$ and hence to T as a result of the softening of the gradient of $\rho(\mathbf{r})$ along the bond path (Bader and Preston 1969).

In addition to these local properties of ρ and its Laplacian distribution which serve to provide a broad basis for the classification of atomic interactions, the theory of atoms in molecules makes possible the determination of the average values of all the mechanical properties of an interacting atom. The use of an atomic population and of the fluctuation in this value (which determines the degree to which the electrons are localized within the boundaries of the atom) have also been used in the classification and understanding of atomic interactions. One can also determine the changes in the energies undergone by the individual atoms in each interaction, as illustrated for the origin of the barrier to rotation in C_2H_6. Depending upon the nature of the problem, the theory enables one to obtain as detailed a view of an atomic interaction as one desires, stressing any particular set of properties as one will, all in terms of the mechanical properties of the atoms involved.

E7.1   The pair density and the localization of electrons

This section details the mathematics and the physics which enable one to determine the extent to which some average number of electrons are localized to a portion of the total space available to them. The localization of electronic charge is determined by the quantum mechanical distribution function $\psi^* \psi \, dx_1 \, dx_2 \ldots dx_N$ (eqn (1.1)). Consider the partitioning of a total system containing N electrons into two mutually exclusive regions Ω and Ω'. The probability of the event that n electrons will be found in Ω when the remaining $(N - n)$ are confined to Ω' is

$$P_n(\Omega) = \frac{N!}{n! \, (N - n)!} \int_\Omega dx_1 \ldots \int_\Omega dx_n \int_{\Omega'} dx_{n+1}$$

$$\ldots \int_{\Omega'} dx_N \, \Gamma^{(N)} (x_1, x_2, \ldots x_N) \tag{E7.1}$$

where $\Gamma^{(N)}$, the many-particle probability density, is a diagonal element of the spinless N-particle density matrix,

$$\Gamma^{(N)}(\mathbf{x}_1, \mathbf{x}_2, \ldots \mathbf{x}_N) = \sum_{\text{spins}} \psi^*(\mathbf{x}_1, \mathbf{x}_2, \ldots \mathbf{x}_N) \psi(\mathbf{x}_1', \mathbf{x}_2', \ldots \mathbf{x}_N')\big|_{\mathbf{x}_i = \mathbf{x}_i'}$$

where $\mathbf{x}_i$ denotes a product of space and spin variables, $\mathbf{x}_i = \mathbf{r}_i \sigma_i$. The binomial coefficient in eqn (E7.1) accounts for the indistinguishability of the electrons, so that it matters not which n electrons are considered to be in Ω when the remaining $(N - n)$ are in Ω'. Since the wave function is assumed to be normalized, one has

$$\sum_n P_n(\Omega) = 1. \tag{E7.2}$$

The average number of particles in a given region Ω is

$$\bar{N}(\Omega) = \sum_n n P_n(\Omega) = \int_\Omega \rho(\mathbf{r})\,d\mathbf{r}, \tag{E7.3}$$

a result which, as indicated in eqn (E7.3), must be equivalent to the integration of the charge density $\rho(\mathbf{r})$ over the same region of space. If a single event has a dominant probability, it is referred to as the leading event. If $P_n(\Omega)$ is a leading event and if the probability distribution of the other events is symmetrical about $P_n(\Omega)$, then $\bar{N}(\Omega)$ will equal n. One must in general, however, distinguish between the number of electrons localized in a given region to yield a leading event and the average occupation number of the region.

Daudel and co-workers (Daudel 1953, 1968; Daudel et al. 1955; Aslangul et al. 1972) were the first to provide a definition of the localization of electrons based solely upon information contained in the state function. They proposed that information theory (Shannon 1948) be used to define the 'best' division of a system into regions or 'loges'. According to this theory our knowledge about a given system is a maximum when the 'missing information function' $I(P_n, \Omega)$ is minimized where

$$I(P_n, \Omega) = -\sum_n P_n(\Omega) \ln P_n(\Omega). \tag{E7.4}$$

If the probability of one particular event (n particles in Ω) approaches a certainty,

$$P_n(\Omega) \to 1 \quad \text{and} \quad \sum_{m \neq n} P_m(\Omega) \to 0. \tag{E7.5}$$

In this instance our knowledge of the system would be maximal as we would know with certainty that n electrons are in Ω and $(N - n)$ in Ω'. Correspondingly, in this case $I(P, \Omega)$ approaches zero. At the other extreme, when all events are equally probable, our knowledge of the specific state of the system is minimal and the missing information function is maximized.

Using the definition of the 'best loges' as the ones which minimize $I(P_n, \Omega)$, Daudel et al. (1974) applied this theory to a number of diatomic and triatomic molecules containing up to six electrons, namely, LiH, BeH^+, BH, and BeH_2.

It was indeed possible in these molecules to partition total space into regions so that the missing information function was simultaneously minimized for each region. The quantum probabilities were in each case dominated by $P_2(\Omega)$ and the most probable partitioning was found to be the one in which pairs of electrons were localized in well-defined regions of space.

Figure E7.1 illustrates the search for the best spherical loge in a two-loge partitioning of the six-electron $BH(X\,{}^1\Sigma^+)$ molecule. Plotted in this figure are the values of $P_n(\Omega)$ and $I(P_n, \Omega)$, where the volume Ω refers to a spherical loge centred on the boron nucleus of variable radius r. Aside from the trivial peaks in $P_0(\Omega)$ for $r = 0$ and in $P_6(\Omega)$ for $r = \infty$, the single most probable event is obtained for a loge of radius 0.7 au which simultaneously maximizes $P_2(\Omega)$ and minimizes $I(P_n, \Omega)$. The satellite events $P_1(\Omega)$ and $P_3(\Omega)$ have small and almost equal probabilities and thus $\bar{N}(\Omega)$, the average number of electrons in this most probable core loge, is two, as determined by the leading event

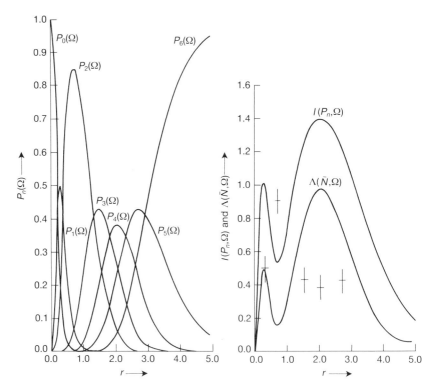

FIG. E7.1.   The variation in the probabilities $P_n(\Omega)$, the missing information function $I(P, \Omega)$ and the fluctuation in $\bar{N}(\Omega)$, $\Lambda(\bar{N}, \Omega)$ with the radius r (in au) of a spherical loge centred on the boron nucleus in $BH\,(X^1\Sigma^+)$. The positions of the maxima in the $P_n(\Omega)$ values for each value of n are also indicated on the plot for $\Lambda(\bar{N}, \Omega)$.

$P_2(\Omega)$. In this case there is an 85 per cent probability of finding two electrons in a sphere of radius 0.7 au. Note from Fig. E7.1 that maximizing a single event, $P_1(\Omega)$ for example, does not necessarily lead to a minimum in $I(P_n, \Omega)$. What is important to note is that $I(P_n, \Omega)$ can be extremized and that it exhibits only the single (non-trivial) minimum; thus there is but a single 'best' core loge. The reader is referred to the original paper (Daudel *et al.* 1974*a,b*) for details regarding the multi-loge partitioning.

The study by Daudel *et al.* (1974*a,b*) also showed that the fluctuation in the average number of particles in a region paralleled closely the behaviour of the missing information function (see Fig. E7.1, for example). The fluctuation in the average number of electrons for a region Ω is defined as

$$\Lambda(\bar{N}, \Omega) = \overline{N^2}(\Omega) - [\bar{N}(\Omega)]^2 = \sum_n n^2 P_n(\Omega) - [\sum_n n P_n(\Omega)]^2. \qquad \text{(E7.6)}$$

Clearly, the fluctuation in the average population of a region Ω (denoted here by $\bar{N}(\Omega)$) also approaches zero when a single event dominates the distribution. A difficulty with using quantum probabilities to determine the extent of spatial localization is that their evaluation requires the use of the full Nth-order density matrix. Because of this, the calculation of $P_n(\Omega)$ rapidly becomes prohibitive with increasing N. The fluctuation $\Lambda(\bar{N}, \Omega)$ however, can be expressed entirely in terms of the diagonal elements of just the second-order density matrix (Bader and Stephens 1974).

A diagonal element of the second-order density matrix or *pair density*, $\rho(\mathbf{r}_1, \mathbf{r}_2)$, as it is referred to here, is defined as

$$\rho(\mathbf{r}_1, \mathbf{r}_2) = (N(N-1)/2) \int d\sigma_1 \int d\sigma_2 \int dx_3 \dots \int dx_N \psi^* \psi \qquad \text{(E7.7)}$$

where, as before, $\mathbf{x}_i$ denotes a product of spin and space coordinates and $d\mathbf{x}_i = d\sigma_i d\mathbf{r}_i$. The expression for $\overline{N^2}(\Omega)$ can be expressed in terms of $\rho(\mathbf{r}_1, \mathbf{r}_2)$ as

$$\overline{N^2}(\Omega) = 2 \int_\Omega d\mathbf{r}_1 \int_\Omega d\mathbf{r}_2 \rho(\mathbf{r}_1, \mathbf{r}_2) + \int_\Omega d\mathbf{r}_1 \rho(\mathbf{r}_1) \qquad \text{(E7.8)}$$

where the charge density $\rho(\mathbf{r}_1)$, or *number density* as it is referred to here, is normalized to N electrons. Thus, the fluctuation around $\bar{N}(\Omega)$ for a region Ω can be written

$$\Lambda(\bar{N}, \Omega) = 2 \int_\Omega d\mathbf{r}_1 \int_\Omega d\mathbf{r}_2 \rho(\mathbf{r}_1, \mathbf{r}_2) + \bar{N}(\Omega) - [\bar{N}(\Omega)]^2. \qquad \text{(E7.9)}$$

There is a great conceptual advantage in abandoning the missing information $I(P_n, \Omega)$ in favour of the fluctuation $\Lambda(\bar{N}, \Omega)$ to determine the localization of electrons. Lennard-Jones (1952) pointed out that the extent to which a set of indistinguishable particles is spatially localized is determined by the system's pair density, the same distribution function which determines the fluctuation in $\bar{N}$. For a system of electrons, the relevant properties of the pair density are a consequence of the so-called Fermi correlation which results from the antisymmetrization of the state function as required by the Pauli principle (Bader and Stephens 1975).

The relationship between the Fermi correlation and the spatial localization of electrons is developed by defining, as McWeeny (1960) does, a correlation function $f(\mathbf{r}_1, \mathbf{r}_2)$ in an expression which determines the extent to which the pair density deviates from a simple product of number densities,

$$\rho(\mathbf{r}_1, \mathbf{r}_2) = \tfrac{1}{2}\rho(\mathbf{r}_1)\rho(\mathbf{r}_2)[1 + f(\mathbf{r}_1, \mathbf{r}_2)]. \tag{E7.10}$$

The pair density can never equal the product of number densities as given by the first term on the right-hand side of eqn (E7.10) since the product $\tfrac{1}{2}\rho(\mathbf{r}_1)\rho(\mathbf{r}_2)$ integrates to $\tfrac{1}{2}N^2$ or $\tfrac{1}{2}N$ pairs in excess of the total number of distinct pairs $\tfrac{1}{2}N(N-1)$ that can be formed from N particles and the number obtained by integration of the pair density (eqn (E7.7)). Hence, the correlation factor $f(\mathbf{r}_1, \mathbf{r}_2)$ must correct for the improper counting of pairs as given by a product of number densities. Because of this requirement one finds that (McWeeny 1960)

$$\int d\mathbf{r}_2 \rho(\mathbf{r}_2) f(\mathbf{r}_1, \mathbf{r}_2) = -1 \tag{E7.11}$$

when eqn (E7.10) is integrated over the coordinate $\mathbf{r}_2$ (the left-hand side of eqn (E7.10) yielding $((N-1)/2)\rho(\mathbf{r}_1)$ in this step) for any fixed value of $\mathbf{r}_1$. This result corresponds to the removal of one electron and the total effect of the correlation over all space is to remove the $\tfrac{1}{2}N$ self-pairs improperly counted by the product of number densities, that is,

$$\tfrac{1}{2}\int d\mathbf{r}_1 \int d\mathbf{r}_2 \, \rho(\mathbf{r}_1)\rho(\mathbf{r}_2) f(\mathbf{r}_1, \mathbf{r}_2) = -N/2. \tag{E7.12}$$

Thus, $\rho(\mathbf{r}_2) f(\mathbf{r}_1, \mathbf{r}_2)$ is a distribution function that, for a fixed value of $\mathbf{r}_1$, describes the decrease in the density of other electrons by the equivalent of one unit of charge. Its effect is, therefore, to create a hole, a correlation hole, in the pair distribution function.

In this work we are particularly concerned with the spatial extent of the effects of this self-pairing correlation on the motions of the electrons and with whether or not the net effect of this correlation for any one particle, the correlation hole, may be localized to one particular region of space.

In an electronic system it is essential to distinguish between the correlation of electrons with identical spin, Fermi correlation, from that of electrons with opposite spins, Coulomb correlation. The correlation term $f(\mathbf{r}_1, \mathbf{r}_2)$ as defined in eqn (E7.10) will measure both types of correlation (McWeeny 1960). However, the limiting value of the correlation hole as expressed in eqn (E7.11) arises only from the correlation between electrons of the same spin, the Fermi hole. That part of $f(\mathbf{r}_1, \mathbf{r}_2)$ that refers to the correlation between electrons of opposite spin contributes zero when integrated over all space. If the coordinates of an α-spin electron are held fixed during the integration over all space in eqn (E7.11), then only the removal of α-spin density contributes to the limiting value of -1 as a consequence of the orthogonality of the spin functions. Conceptually, it arises because the α and β electrons form two distinct sets, and self-pairing can occur only within a set of identical particles.

The pair density or second-order density matrix, obtained from a single determinantal function composed of orthogonal spin functions ϕ_i is given in eqn (E1.4). Comparison of that expression for the pair density with that given in eqn (E7.10) yields for the Fermi hole for a reference electron of α spin at $\mathbf{r}_1$

$$h^{\alpha}(\mathbf{r}_1, \mathbf{r}_2) = - \sum_i^{\alpha} \sum_j^{\alpha} \phi_i(\mathbf{r}_1)\phi_j^*(\mathbf{r}_1)\phi_i^*(\mathbf{r}_2)\phi_j(\mathbf{r}_2) / \sum_i^{\alpha} \phi_i^*(\mathbf{r}_1)\phi_i(\mathbf{r}_1).$$

$$(E7.13)$$

Only orbitals of α spin contribute as integration (or summation) over the spin coordinates will cause any contributions from β-spin orbitals to vanish. The summation in the denominator of eqn (E7.13) is the α-spin density,

$$\rho^{\alpha}(\mathbf{r}_1) = \sum_i^{\alpha} \phi_i^*(\mathbf{r}_1)\phi_i(\mathbf{r}_1). \qquad (E7.14)$$

Corresponding expressions obtain for the electrons of β spin. Integration of $h^{\alpha}(\mathbf{r}_1, \mathbf{r}_2)$ over $\mathbf{r}_2$ for fixed $\mathbf{r}_1$ yields -1 corresponding to the removal of one electronic charge of α spin. One also notes that

$$h^{\alpha}(\mathbf{r}_1, \mathbf{r}_2) = -\rho^{\alpha}(\mathbf{r}_1) \qquad \text{when} \quad \mathbf{r}_2 = \mathbf{r}_1, \qquad (E7.15)$$

corresponding to the complete removal of α-spin density from the position of the reference electron. This result is a direct consequence of the antisymmetry imposed on Ψ and hence on $\rho(\mathbf{r}_1, \mathbf{r}_2)$ by the Pauli exclusion principle (McWeeny 1960). Because of this antisymmetry the exchange of space coordinates of two electrons of identical spin must change the sign of the corresponding pair density. Thus,

$$\rho^{\alpha}(\mathbf{r}_1, \mathbf{r}_2) = -\rho^{\alpha}(\mathbf{r}_2, \mathbf{r}_1), \qquad (E7.16)$$

a result which demands that the pair density vanish when the space coordinates of two electrons of identical spin are coincident, or that $f^{\alpha\alpha}(\mathbf{r}_1, \mathbf{r}_2) = -1$ for $\mathbf{r}_1 = \mathbf{r}_2$. This limiting extreme in correlation is not obtained for electrons of opposite spin and there is a finite probability of electrons of opposite spin 'colliding' in real space. The state function must, however, exhibit a cusp for the coalescence of two electronic coordinates just as it does for the coalescence of an electron with a nucleus (Kato 1957; Roothaan and Weiss 1960).

The effect of the Fermi correlation on the pair density as given in eqn (E7.16) is so pronounced that even a single determinantal function yields a good description of the local properties of the Fermi hole. The addition of Coulomb correlation does not change its significant spatial features. One can view the net result of the Pauli principle as correcting for the self-pairing of electrons of identical spin in a non-trivial way, as determined by the spatial properties of the Fermi hole (see eqn (E7.13), for example). Slater (1951) was led to propose that the exchange charge potential be proportional to $\rho^{1/3}$ on the basis of a modelling of the Hartree–Fock expression for the Fermi hole. We now show that the extent to which sets of electrons may be localized in real space is determined solely by the spatial properties of the Fermi hole (Bader and Stephens 1974, 1975).

The average number of pairs of electrons in a region Ω, a quantity denoted by $D_2(\Omega, \Omega)$, is determined by integrating both coordinates in $\rho(\mathbf{r}_1, \mathbf{r}_2)$ over Ω, a result which is equivalent to expressing the result in terms of the event probabilities $P_n(\Omega)$,

$$D_2(\Omega, \Omega) = \tfrac{1}{2}\sum_n P_n(\Omega)n(n-1) = \tfrac{1}{2}(\overline{N^2}(\Omega) - \bar{N}(\Omega)). \qquad \text{(E7.17)}$$

Consider now the properties of the region (Ω) in the limiting situation when $P_n(\Omega)$, the probability of one particular event, equals unity and the probabilities of all other events are thus zero. In this situation $\bar{N}(\Omega)$, the average number of electrons in Ω, equals n. The number of pairs which can be formed is $n(n-1)/2 = \bar{N}(\Omega)(\bar{N}(\Omega) - 1)/2$, the number of distinct pairs which can be formed from an isolated and distinct set of n objects. When the pair population of a region Ω is expressible as $\bar{N}(\Omega)(\bar{N}(\Omega) - 1)/2$ it is referred to as being a 'pure pair' population. A set of electrons with such properties would be perfectly localized in the region Ω, i.e. they would behave as a separate and distinguishable subset of electrons and, since the self-pairing is totally accounted for within the region, the wave function for the total system could be expressed as a simple product of separately antisymmetrized wave functions for the n electrons in Ω and the $(N - n)$ in Ω'. This is the limiting case of localization in an electronic system and we show that it occurs when the total Fermi hole for each of the n electrons in Ω is confined totally to the region Ω.

Using eqn (E7.10), the expression for the average number of pairs in a region Ω may be expressed so as to incorporate the correlation term to yield

$$D_2(\Omega, \Omega) = \int_\Omega d\mathbf{r}_1 \int_\Omega d\mathbf{r}_2 \rho(\mathbf{r}_1, \mathbf{r}_2) = (\bar{N}(\Omega)^2 + F(\Omega, \Omega))/2 \qquad \text{(E7.18)}$$

where

$$F(\Omega, \Omega) = \int_\Omega d\mathbf{r}_1 \int_\Omega d\mathbf{r}_2 \rho(\mathbf{r}_1)\rho(\mathbf{r}_2)f(\mathbf{r}_1, \mathbf{r}_2). \qquad \text{(E7.19)}$$

$F(\Omega, \Omega)$ is a measure of the total correlation which is contained within the region (Ω). For a Hartree–Fock wave function, which includes only Fermi correlation, the magnitude of $F(\Omega, \Omega)$ is a measure of the total (integrated) Fermi hole of the $N(\Omega)$ particles that lies within the region Ω. From eqns (E7.11) and (E7.19), the limiting value of the correlation is seen to be $-\bar{N}(\Omega)$.

Thus, only if the correlation hole for each of the $\bar{N}(\Omega)$ particles in Ω is totally contained within Ω for all values of $\mathbf{r}_1$ in Ω will $D_2(\Omega, \Omega)$ correspond to the exact number of pairs which can be formed from a distinct set of $\bar{N}(\Omega)$ particles. That is, in this limiting situation the $\bar{N}(\Omega)$ electrons (where $\bar{N}(\Omega)$ is necessarily an integer) generate a pure pair population with

$$D_2(\Omega, \Omega) = \bar{N}(\Omega)(\bar{N}(\Omega) - 1)/2. \tag{E7.20}$$

Otherwise, $\bar{N}(\Omega) < [\overline{N^2}(\Omega)]^{1/2}$ and, from eqn (E7.17), $D_2(\Omega, \Omega) > \bar{N}(\Omega)(\bar{N}(\Omega) - 1)/2$. Only when the relationship expressed in eqn (E7.20) applies is the number of pairs in a region determined by (or in any way related to) the average number of particles within this same region. In this limit both $\bar{N}(\Omega)$ and $D_2(\Omega, \Omega)$ yield pure populations as their values are determined by a single event.

It is now easy to demonstrate that the fluctuation in the average electron population of a region Ω vanishes when the Fermi correlation for the $\bar{N}(\Omega)$ electrons in Ω is totally contained in Ω. Equation (E7.9) for $\Lambda(\bar{N}, \Omega)$ can be re-expressed in terms of the average number of pairs in Ω as

$$\Lambda(\bar{N}, \Omega) = 2D_2(\Omega, \Omega) + \bar{N}(\Omega) - [\bar{N}(\Omega)]^2. \tag{E7.21}$$

Substitution of eqn (E7.18) for $D_2(\Omega, \Omega)$ into eqn (E7.21) yields

$$\Lambda(\bar{N}, \Omega) = \bar{N}(\Omega) + F(\Omega, \Omega). \tag{E7.22}$$

The fluctuation attains its minimum value of zero under the same condition that $D_2(\Omega, \Omega)$ is reduced to a pure pair population, that is, when one particular event has a probability of unity. From eqn (E7.22) the limiting value of $F(\Omega, \Omega)$ is equal to $-\bar{N}(\Omega)$, a value which, as stated previously, implies that the total Fermi correlation hole for each electron in Ω is totally contained within the region Ω.

The magnitude of the limiting value of $F(\Omega, \Omega)$, which equals $\bar{N}^\alpha(\Omega) + \bar{N}^\beta(\Omega)$, corresponds to the Fermi holes for both sets of electrons being totally contained within Ω for any motion within the boundaries of Ω. Thus, all other electrons of either spin are totally excluded from Ω and there is no exchange of electrons in Ω with electrons outside of the region Ω. The ratio $|F(\Omega, \Omega)/\bar{N}(\Omega)|$ therefore, provides a quantitative measure of the extent to which the average number of electrons $\bar{N}(\Omega)$ in Ω are localized to that particular region of space. The quantity,

$$l(\Omega) = |F(\Omega, \Omega)/\bar{N}(\Omega)| \times 100, \tag{E7.23}$$

is a measure of the per cent localization of the electrons to a region Ω. *A localization of 100 per cent corresponds to the Fermi hole of each electron being totally contained in the region Ω.*

The expression for $F(\Omega, \Omega)$ is particularly simple for a Hartree–Fock or single determinantal wave function. Multiplication of the expression for the

Fermi hole given in eqn (E7.13) by $\rho^\alpha(\mathbf{r}_1)$ and integration over the coordinates of both electrons yields, together with a corresponding treatment for the electrons of β spin,

$$F(\Omega, \Omega) = -\sum_{ij}^\alpha S_{ij}^2(\Omega) - \sum_{ij}^\beta S_{ij}^2(\Omega) = F^{\alpha\alpha}(\Omega, \Omega) + F^{\beta\beta}(\Omega, \Omega) \qquad (E7.24)$$

where $S_{ij}(\Omega)$ is the overlap of the spin orbitals ϕ_i and ϕ_j over the region Ω.

Complete localization is, of course, possible only for an isolated system. What is remarkable, however, is the extent to which the electrons of atoms in an ionic molecule approach this limit of perfect localization, with $l(\Omega)$ values in excess of 95 per cent not being uncommon. In systems, such as the fluorides and chlorides of lithium and sodium displayed in Fig. E7.2, the atomic surface of zero flux is found to minimize the fluctuation in the atomic populations and, thus, the magnitude of the correlation hole per particle is an extremum for such atoms. The properties of the number and pair densities for these

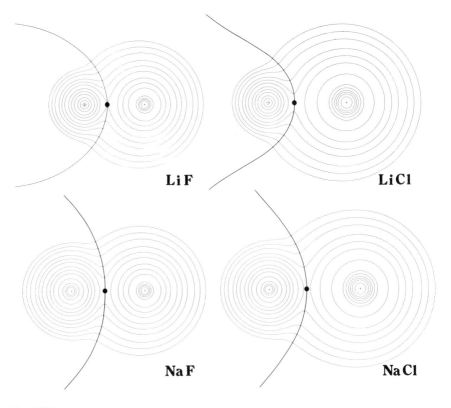

LiF   LiCl   NaF   NaCl

FIG. E7.2. Contour maps of the charge distribution showing the position of the interatomic surfaces for ionic diatomic molecules. The outer contour is 0.002 au with succeeding values as given in the Appendix (Table A).

Table E7.1

*Properties of number and pair densities in ionic systems**

AB	R	ρ_b	$\nabla^2\rho_b$	$q(A) = -q(B)$	$l(A)$ (%)	$l(B)$ (%)	$D_2(A, A)$	$D_2(B, B)$	$D_2(A, B)$
LiF	2.955	0.0802	0.7018	+ 0.937	95.5	99.1	1.14	44.4	20.4
LiCl	3.825	0.0462	0.2657	+ 0.930	96.0	99.5	1.15	151.7	37.1
NaF	3.629	0.0543	0.4655	+ 0.944	99.5	98.7	45.6	44.5	99.9
NaCl	4.461	0.0358	0.2004	+ 0.911	98.9	99.4	45.9	151.5	180.6

* Calculated from functions close to the Hartree–Fock limit (McLean and Yoshimine 1967). All molecules in $^1\Sigma^+$ ground state. All dimensioned quantities in atomic units.

molecules AB are summarized in Table E7.1. The atomic populations, together with their derived charges, and the atomic pair populations are close to the limiting values anticipated for the ionic model A^+B^- for these systems—net charges of ± 1 with the electrons localized in two separate closed-shell structures. The localization of the electrons in the separate atomic basins, as measured by the localization of the Fermi hole for both α and β electrons, is in excess of 95 per cent in every case. Correspondingly, the atomic pair populations are close in value to those associated with pure pair populations resulting from complete localization of the electrons in each atomic basin. These values are 1.0 for Li^+, 45.0 for Na^+ and F^-, and 153 for Cl^-. The actual pair populations are, in each case, slightly in excess of the values determined by the pure pair limit $N(\Omega)(N(\Omega) - 1)/2$ using the observed values of $N(\Omega)$. In the limit of complete localization, the number of pairs formed *between* the electrons on the anion and cation is maximized and should equal simply $N(A)N(B)$. These values, followed by those for the ionic model A^+B^- in brackets, are to be compared with those listed under $D_2(A, B)$ in Table E7.1. They are 20.5 (20) for LiF, 37.1 (36) for LiCl, 100 (100) for NaF, and 180.7 (180) for NaCl. To the extent that the values of $D_2(A, B)$ match the limiting values, there is no exchange of the electrons on either ion with those of its neighbour.

The spatial properties of the Fermi hole that give rise to this physical localization of electronic charge within the basins of the separate atoms is illustrated in Fig. E7.3. *The display of the Fermi hole, since it represents the removal of spin density equivalent to one electron, is a picture of the spreading out or delocalization of the charge density of the reference electron when placed at* $\mathbf{r}_1$.

While it is possible to find regions of space bounded by surfaces such that the contained Fermi correlation is maximized to yield localized groupings of electrons, these regions correspond to atomic cores or localized atomic populations and not, in general, to localized pairs of bonded and non-bonded

LiF

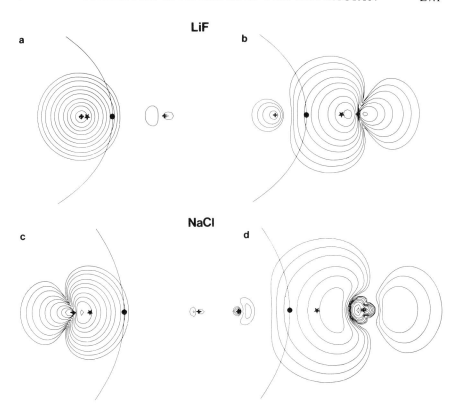

FIG. E7.3.   Contour maps of the density of the Fermi hole for various positions of the reference electron (denoted by a star) in LiF and NaCl. The positions of the interatomic surfaces are also shown. The contours used in this and the following figure are in au and decrease in value from the outermost contour inwards in steps of -2×10^n, -4×10^n, and -8×10^n with n beginning at -3 and increasing in steps of unity. Reference electron is: (a) 0.20 au from Li nucleus; (b) 0.58 au from F nucleus; (c) 0.60 au from Na; and (d) 1.66 au from Cl nucleus. In (a) the reference electron is necessarily within the s-like core of Li. In the remaining plots the reference electron lies beyond the inner s-like core. The magnitude of the contours in (a) are close to one-half the value of corresponding spatial contours of the total density of the Li atom in Fig. E7.2. That is, the Fermi density approaches $\rho^\alpha = \rho^\beta = \rho/2$, the density of the α or β electrons over the basin of the Li atom, leading to the almost total exclusion of all other α and β electrons from this basin for this position of the reference electron.

electrons as anticipated on the basis of the Lewis model (Bader and Stephens 1975). In the hydrides past BH_3 for example, there are no regions of space outside of the core for which the fluctuation can be minimized, i.e. in which the electrons are physically localized. Methane is borderline in its behaviour, but the motions of the valence electrons in the remaining hydrides, NH_3, H_2O, and HF, are so strongly intercorrelated that no best partitioning is found and the localized pair concept loses its meaning. A bond loge in HF, for example, which has a number population of two, contains only 53 per cent of the Fermi correlation required for complete localization, while $l(\Omega)$ equals only 49 per cent for a non-bonded loge in the same molecule. Attempts to define bonded loges in N_2 or F_2 with average populations of two fail completely as the contained Fermi correlation amounts to only 28 and 17 per cent, respectively, of that required for complete localization. The atomic surface of zero flux in N_2 yields regions which contain 78 per cent of the Fermi correlation required for complete localization of the seven electrons to the basin of the atom. In this example, which one anticipates to be typical of a delocalized system, the charge density of either atom is delocalized into the region of the other to the extent of 22 per cent as a result of molecule formation. This percentage delocalization increases if one excludes the highly localized core. In N_2 the valence density of a N atom is 28 per cent delocalized over the second valence region and 1 per cent over the second core. In F_2 the electrons on each F atom are 93 per cent localized. Even the valence density is delocalized to only 9 per cent over the adjacent atom. The electronic charge of a fluorine atom is both tightly bound and highly localized within the basin of the atom. The difficulty in disrupting the very pronounced intraatomic Fermi correlation in a F atom and the resulting small degree of delocalization of its electrons provides an explanation for the weak binding found in F_2 compared to that in N_2.

In summary, electrons can be localized within an atomic boundary in a closed-shell interaction. They can also be separately localized within the inner or valence shells, but not to bonded regions which cross interatomic surfaces, nor, in general, to non-bonded regions which occupy only a portion of a valence shell, to yield the Lewis model of the localized electron pair.

In terms of the orbital model, it is clear from eqn (E7.24), that it is not sufficient that orbitals be localizable to achieve physical localization of pairs of electrons. Instead, each orbital, for both α and β spins, must be localized to a separate region of space, i.e. $S_{ii}(\Omega) = 1$ and $S_{i \neq j}(\Omega) = 0$. Such separate localization is approached for the same systems, LiH, BeH_2, and BH_3, for example, where the intrapair correlation energy dominates the total correlation energy, but it is not a general result. Localized orbitals are not that localized and they are, in general, not localized to separate regions of space, the requirement for physical localization of electron pairs.

E7.2   Local charge concentrations and partial localization of the Fermi hole

It is demonstrated through a study of the Fermi hole density that the local decrease in the potential resulting from the approach of each ligand to the central atom does result in a partial condensation of the pair density to yield patterns of localization as anticipated by the VSEPR model. It is demonstrated that there is a correspondence between this partial localization of the pair density and the appearance of local charge concentrations in the VSCC of the central atom.

The properties of the Fermi hole for a physically localized pair of electrons are illustrated by the core electrons in methane. In this molecule a sphere of radius 0.53 au centred on the carbon nucleus defines a region of space which maximizes the contained Fermi correlation to yield a pair of electrons which are 88 per cent localized. The localized nature of the Fermi hole is illustrated in Fig. E7.4. In spite of the unsymmetrical position of the reference electron, the hole is nearly symmetrical with respect to the nucleus, and it corresponds closely to a plot of the α density distribution in the region of the core. With the exception of the contours of very low value extending out to the proton, this plot remains essentially unchanged for any position of the reference electron within this localized core region. Thus the core region is maximally isolated from the remainder of the system. The fluctuation in its average population of two is minimized, as is the exchange of its electrons with those in the remainder of the system.

While there is localization into shells in a free atom, as reflected in the Laplacian distribution as well, there is no possibility of defining a localized pair within the L or M shells, as the Fermi hole for each electron is spread over its respective shell. It has already been noted that the uniformity of the outer shell of charge concentration is disrupted with the approach of ligands to the central atom with the resultant formation of local concentrations of charge. The formation of bonded charge maxima in the VSCC of the central atom is understandable in view of a decrease in the electron–nuclear potential in each of the resulting internuclear regions of the molecule. However, the decrease in potential is, in general, not sufficient to condense the Fermi hole of a single electron to a degree sufficient to localize a pair of electrons from the remainder of the system as is found for the core region of an atom. This situation is also illustrated for the methane molecule. While it is not possible to define a valence region which maximizes the contained Fermi correlation in this molecule (Bader and Stephens 1975), the division of space excluding the core into four equivalent bonded domains does represent the best partitioning of its valence space in that it minimizes the sum of the fluctuations in the populations of the individual regions. The pair of electrons in each such bonded region is 69 per cent localized. It appears from many examples that it is not possible to maximize the contained Fermi correlation for a region of space containing on the average a pair of electrons when the total Fermi correlation falls below 70 per cent. The diffuse nature of the Fermi hole in the valence region of this molecule which prevents the physical localization of a pair of electrons is illustrated in Fig. E7.4 for a reference electron located at the bonded maximum in the VSCC of carbon. This hole is

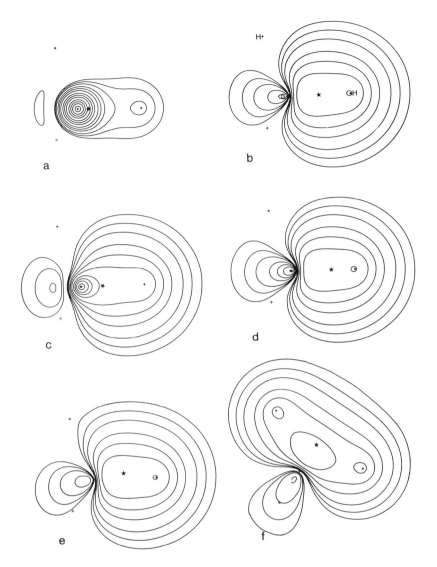

Fig. E7.4. Contour maps of the Fermi hole density in methane for various positions of the reference electron, as indicated by a star. Each map is for a plane containing the carbon and two hydrogen nuclei as labelled in map (b). (a) Reference electron, 0.35 au from C nucleus, is within the localized core of radius 0.53 au. (b) Reference electron is at position of bonded maximum in VSCC of carbon, 1.02 au from nucleus. (c) Reference electron is 0.69 au from nucleus. The Fermi density is a maximum at the C nucleus and is less localized in the valence region of the C–H bond than in map (b). (d) Reference electron is 1.28 au from C nucleus. The area within the inner (0.08 au) C–H 'bond' contour is decreased relative to map (b) and more density is placed within the core and non-bonded regions. (e) Reference electron is moved off bonded maximum towards second proton. It is 0.99 au from nucleus and still within neighbourhood of bonded maximum. Map (e) is essentially unchanged from map (b). (f) Reference electron, at same distance from nucleus as in map (e), is moved further, to a point on boundary of two bonded regions. This map represents the maximum possible delocalization of the Fermi hole for this distance of the reference electron from the C nucleus.

much less localized than that for the core electrons, extending into the non-bonding region as well. Its shape remains insensitive to the position of the reference electron only for motion of the reference electron within the neighbourhood of the bonded maximum. As the reference electron is removed from this region, the hole becomes more delocalized and its overlap with the core and neighbouring bonded regions increases. For this reason one cannot maximize the contained Fermi correlation nor, equivalently, minimize the fluctuation in its population; any variation in the boundary of a valence domain causes the average electron population to increase faster than the contained Fermi correlation.

While the valence density is not physically localized into isolated pairs of electrons as is the core density, there is partial localization of the pair density as reflected in the value of 0.69 for the fraction of the total possible Fermi correlation for an average population of two electrons. Rather than a single electron pair, each bonded domain in methane contains an average of 1.3 distinct pairs of electrons as opposed to the 1.75 pairs which would be obtained if the electrons were fully delocalized over the entire valence region. Of equal importance is the observation that the Fermi hole for a reference electron at, or in the neighbourhood of, a bonded maxima in the VSCC of the carbon atom attains its minimum values along the axes of the other C–H bonds. Recalling that the Fermi hole is a display of the spread or delocalization of the density of the reference electron, one sees that this density is concentrated along a tetrahedral or threefold axis to maximally avoid the three other corresponding bonded regions.

The Fermi hole for the reference electron at a bonded maxima in the VSCC of the carbon atom has the appearance of the density of a directed sp^3 hybrid orbital of valence bond theory or of the density of a localized bonding orbital of molecular orbital theory. Luken (1982, 1984) has also discussed and illustrated the properties of the Fermi hole and noted the similarity in appearance of the density of a Fermi hole to that for a corresponding localized molecular orbital. We emphasize here again that localized orbitals like the Fermi holes shown above for valence electrons are, in general, not sufficiently localized to separate regions of space to correspond to physically localized or distinct electron pairs. The fact that the Fermi hole resembles localized orbitals in systems where physical localization of pairs is not found further illustrates this point.

It should be borne in mind that the resemblance of a Fermi hole density to that of a localized valence orbital is obtained only when the reference electron is placed in the neighbourhood of a local maximum in the VSCC. The Fermi hole and hence the density of the reference electron are much more delocalized for general positions throughout the valence region (see Fig. E7.4(f)). Localized molecular orbitals thus overemphasize electron localizability and do not provide true representations of the extent to which electrons are spatially localized.

There is partial localization of the valence density in methane. The condensation into four partially localized pairs of electrons arranged along four tetrahedral axes is a result of the combined effects of the ligand field and the Pauli exclusion principle described above. Most important is that this partial localization of the pair density is reflected in the properties of the VSCC of the carbon atom which undergoes a corresponding condensation into four local concentrations of electronic charge. These properties of the pair density are not just the result of the tetrahedral symmetry of the ligand field in methane because, as we will now see, the Fermi hole exhibits the same behaviour in the ammonia molecule.

The Fermi holes obtained for a reference electron at the positions of a bonded and the non-bonded maxima in the VSCC of the nitrogen atom in ammonia are illustrated in Fig. E7.5. This figure shows that there is a partial condensation of the pair density in ammonia to yield four partially localized pairs of electrons arranged along four axes which are approximately tetrahedrally directed. The pair density is less localized than in methane with the bonded and non-bonded domains containing, respectively, 61 and 55 per cent of the total possible Fermi correlation for average electron populations of two. There are, on average, 1.38 distinct pairs of electrons in a bonded domain and 1.45 in a non-bonded domain of ammonia, and the pair density is further removed from the limit of localized pairs than it is in methane. As in methane, the pattern of delocalization of the density of a bonded or of the non-bonded reference electron is such as to maximally avoid the three other axes of pair localization. Thus, in agreement with the postulated behaviour of the electron pairs in the VSEPR model, the most probable arrangement of the partially localized pairs of electrons in ammonia is approximately tetrahedral, even though the ligand field is only threefold. Also, in accord with the VSEPR model, the Fermi hole for the non-bonded reference electron is laterally more diffuse than is that for a bonded reference electron, a property also reflected in the properties of the corresponding maxima in the VSCC of the nitrogen atom.

Of equal importance to the VSEPR model is the observation that, for geometries which, on the basis of the VSEPR model, do not maximize the distance between the electron pairs, the Fermi holes are less localized and exhibit greater overlap with the holes for neighbouring domains. As an example, the Fermi hole for a reference electron on the threefold axis in planar ammonia, situated at one of the two symmetrically related non-bonded charge concentrations, is strongly delocalized over both sides of the plane of the nuclei (Fig. E7.5). The density of the non-bonded reference electron clearly overlaps the holes for reference electrons placed at the bonded maxima more in this geometry than in the most stable pyramidal geometry. *The Fermi holes for reference electrons placed at the corresponding bonded and non-bonded maxima in the VSCC of the central atom are most localized and least mutually overlapping for those geometries which in the*

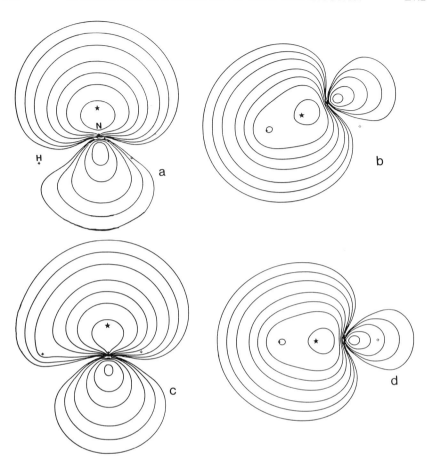

FɪG. E7.5.   Contour maps of the Fermi hole density for pyramidal (a, b) and planar (c, d) ammonia. In maps (a) and (b) the reference electron is positioned at the non-bonded and bonded maxima, respectively, in the VSCC of the nitrogen atom. Note that the Fermi density is more contracted towards the core in NH_3 than it is in CH_4, as are the maxima in its VSCC. Maps (c) and (d) are corresponding plots for planar ammonia. The density of the non-bonded Fermi hole, map (c), is more delocalized than that for the pyramidal geometry, map (a). In the planar geometry, contours of the non-bonded Fermi hole density encompass the N–H internuclear axis. Clearly, maps (c) and (d) overlap one another to a greater extent than do maps (a) and (b)—the electron pairs are more localized in pyramidal than in planar ammonia.

VSEPR model maximize the interpair separations or, equivalently, maximize the separations between the local maxima in the VSCC of the central atom.

It is important to emphasize that the correlation which exists between the properties of the Fermi hole and the VSEPR model is made most evident through the use of the properties of the Laplacian of the charge density. The correspondence is greatest when the position of the reference electron coincides with a local maximum in the VSCC of the central atom. The Fermi holes are individually most localized for such a coincidence in position, and they are least overlapping for the geometry which maximizes the distances between the local maxima in the VSCC of the central atom.

References

Aslangul, R., Constanciel, R., Daudel, R., and Kottis, P. (1972). *Adv. quantum Chem.* **6**, 93.

Bader, R. F. W. (1960). *Mol. Phys.* **3**, 137.

Bader, R. F. W. (1962). *Can. J. Chem.* **40**, 1164.

Bader, R. F. W. (1981). In *The force concept of chemistry* (ed. B. M. Deb), pp. 39–136. Van Nostrand Reinhold, New York.

Bader, R. F. W. and Beddall, P. M. (1972). *J. Chem. Phys.* **56**, 3320.

Bader, R. F. W. and Chang, C. (1989). *J. Phys. Chem* **93**, 2946.

Bader, R. F. W. and Essén, H. (1984). *J. Chem. Phys.* **80**, 1943.

Bader, R. F. W. and Henneker, W. H. (1965). *J. Am. chem. Soc.* **87**, 3063.

Bader, R. F. W. and MacDougall, P. J. (1985). *J. Am. chem. Soc.* **107**, 6788.

Bader, R. F. W. and Preston, H. J. T. (1969). *Int. J. quantum Chem.* **3**, 327.

Bader, R. F. W. and Stephens, M. E. (1974). *Chem. Phys. Lett.* **26**, 445.

Bader, R. F. W. and Stephens, M. E. (1975). *J. Am. chem. Soc.* **97**, 7391.

Bader, R. F. W., Henneker, W. H., and Cade, P. E. (1967a). *J. Chem. Phys.* **46**, 3341.

Bader, R. F. W., Keaveny, I., and Cade, P. E. (1967b). *J. Chem. Phys.* **47**, 3381.

Bader, R. F. W., MacDougall, P. J., and Lau, C. D. H. (1984). *J. Am. chem. Soc.* **106**, 1594.

Bader, R. F. W., Gillespie, R. J., and MacDougall, P. J. (1988). *J. Am. chem. Soc.* **110**, 7329.

Berlin, T. (1951). *J. Chem. Phys.* **19**, 208.

Bingel, W. A. (1963). *Z. Naturforsch.* **18a**, 1249.

Birchall, T. and Gillespie, R. J. (1963). *Can. J. Chem.* **41**, 2642.

Boys, S. F. and Bernardi, F. (1970). *Mol. Phys.* **19**, 553.

Burgi, H. B. and Dunitz, J. D. (1983). *Acc. Chem. Res.* **16**, 153.

Cade, P. E. and Huo, W. M. (1973). *At. Data nucl. Data Tables* **12**, 415.

Cade, P. E. and Huo, W. M. (1975). *At. Data nucl. Data Tables* **15**, 1.

Cade, P. E. and Wahl, A. C. (1974). *At. Data nucl. Data Tables* **13**, 339.

Carroll, M. T. and Bader, R. F. W. (1988). *Mol. Phys.* **65**, 695.

Carroll, M. T., Chang, C., and Bader, R. F. W. (1988). *Mol. Phys.* **63**, 387.

Carroll, M. T., Cheeseman, J. R., Osman, R., and Weinstein, H. (1989). *J. Phys. Chem.* **93**, 5120.

Coulson, C. A. (1941a). *Proc. Cambridge phil. Soc.* **37**, 55.

Coulson, C. A. (1941b). *Proc. Cambridge phil. Soc.* **37**, 74.

Coulson, C. A. and Duncanson, W. E. (1941a). *Proc. Cambridge phil. Soc.* **37**, 67.

Coulson, C. A. and Duncanson, W. E. (1941b). *Proc. Cambridge phil. Soc.* **37**, 406.

Das, G. and Wahl, A. C. (1966). *J. Chem. Phys.* **44**, 87.

Daudel, R. (1953). *C. R. Acad. Sci., Paris* **237**, 601.

Daudel, R. (1968). *The fundamentals of theoretical chemistry*. Pergamon Press, Oxford.

Daudel, R., Brion, H., and Odiot, S. (1955). *J. Chem. Phys.* **23**, 2080.

Daudel, R., Bader, R. F. W., Stephens, M. E., and Borrett, D. S. (1974*a*). *Can. J. Chem.* **52**, 1310.

Daudel, R., Bader, R. F. W., Stephens, M. E., and Borrett, D. S. (1974*b*). *Can. J. Chem.* **52**, 3077.

Daudel, R., Stephens, M. E., Kapuy, E., and Kozmutza, C. (1976). *Chem. Phys. Lett.* **40**, 194.

Davidson, E. R. (1972). *Rev. mod. Phys.* **44**, 451.

Ebbing, D. D. and Henderson, R. C. (1965). *J. Chem. Phys.* **42**, 2225.

Feynman, R. P. (1939). *Phys. Rev.* **56**, 340.

Fukui, K. (1971). *Acc. Chem. Res.* **4**, 57.

García-Sucre, M. and Castells, V. (1989). *Int. J. Quantum Chem. Symp.* **23**, 79.

Gilbert, T. L. and Wahl, A. C. (1967). *J. Chem. Phys.* **47**, 3425.

Gillespie, R. J. (1972). *Molecular geometry*. Van Nostrand Reinhold, London.

Gillespie, R. J., Kapoor, R., Faggiani, R., Lock, C. J. L., Murchie, M., and Passmore, J. (1983). *J. chem. Soc. Commun.* 8.

Hutson, J. M. and Howard, B. J. (1982). *Mol. Phys.* **45**, 791.

Kato, W. A. (1957). *Commun. pure appl. Math.* **10**, 151.

King, R. B. and Rouvray, D. H. (1977). *J. Am. chem. Soc.* **99**, 7834.

Kutzelnigg, W. (1973). *Fortschr. Chem. Forsch.* **41**, 31.

Lennard-Jones, J. E. (1952). *J. Chem. Phys.* **20**, 1024.

Lewis, G. N. (1916). *J. Am. chem. Soc.* **38**, 762.

Lewis, G. N. (1933). *J. Phys. Chem.* **1**, 17.

Luken, W. L. (1984). *Croat. Chem. Acta* **57**, 1283.

Luken, W. L. (1982). *Theor. Chim. Acta* **61**, 265.

MacDougall, P. J. and Bader, R. F. W. (1986). *Can. J. Chem.* **64**, 1496.

McLean, A. D. and Yoshimine, M. (1967). *IBM J. Res. Dev.* **12**, 206.

McWeeny, R. (1960). *Rev. mod. Phys.* **32**, 335.

Morse, P. and Feshbach, H. (1953). *Methods of theoretical physics*, Part 1. McGraw-Hill, New York.

ONeil, S. V., Nesbitt, D. J., Rosmus, P., Werner, H.-J., and Clary, D. C. (1989). *J. Chem. Phys.* **91**, 711.

Pack, R. T. and Brown, W. B. (1966). *J. Chem. Phys.* **45**, 556.

Passmore, J., Sutherland, G., and White, P. S. (1980). *J. chem. Soc. Chem. Commun.* 330.

Politzer, P., Landry, S. J., and Warnheim, T. (1982). *J. Phys. Chem.* **86**, 4767.

Ramirez, B. I. (1983). *Chem. Phys. Lett.* **94**, 180.

Roothaan, C. C. J. and Weiss, A. W. (1960). *Rev. mod. Phys.* **32**, 194.

Sagar, R. P., Ku, A. C., Smith, V. H., and Sinas, A. M. (1988). *J. chem. Phys.* **88**, 4367.

Scrocco, E. and Tomasi, J. (1978). *Adv. quantum Chem.* **11**, 116.

Shannon, E. (1948). *Bell. Syst. tech. J.* **27**, 379.

Shi, Z. and Boyd, R. J. (1988). *J. Chem. Phys.* **88**, 4375.

Sinanoğlu, O. (1962). *J. Chem. Phys.* **36**, 706.

Sinanoğlu, O. and Skutnik, B. (1968). *Chem. Phys. Lett.* **1**, 699.

Slater, J. C. (1933). *J. Chem. Phys.* **1**, 687.

Slater, J. C. (1951). *Phys. Rev.* **81**, 385.

Slater, J. C. (1963). *Quantum theory of molecules and solids*, Vol. I. McGraw-Hill, New York.

Slater, J. C. (1972). *J. Chem. Phys.* **57**, 2389.

Slee, T. (1986*a*). *J. Am. chem. Soc.* **108**, 606.

Slee, T. (1986*b*). *J. Am. chem. Soc.* **108**, 7541.

Slee, T., Larouche, A., and Bader, R. F. W. (1988). *J. Phys. Chem.* **92**, 6219.

THE ACTION PRINCIPLE FOR A QUANTUM SUBSYSTEM

Quantum mechanics involves two distinct sets of hypotheses—the general mathematical scheme of linear operators and state vectors with its associated probability interpretation and the commutation relations and equations of motion for specific dynamical systems. It is the latter aspect that we wish to develop, by substituting a single quantum dynamical principle for the conventional array of assumptions based on classical Hamiltonian dynamics and the correspondence principle.

Julian Schwinger (1951)

8.1 A common basis for classical and quantum mechanics

8.1.1 Introduction

It has been demonstrated in Chapter 5 that Schrödinger's original derivation of the wave equation can be generalized to provide a quantum description of an atom in a molecule. His derivation uses the calculus of variations as applied to an energy functional of ψ and its extension to an atom in a molecule entails two essential steps: the introduction of a particular type of operator through its identification with the variation of the state function and a variation of the surface bounding the atomic subsystem. The replacement of the variations in the state function by the action of an operator of the form $-(i\varepsilon/\hbar)\hat{G}\psi$ is, as recognized by Schwinger, a fundamental step with far-reaching consequences. Such operators are known as generators of infinitesimal unitary transformations. Dirac (1958) pointed out that they are the quantum counterparts of the generators of infinitesimal contact transformations in classical mechanics and the correspondence between the properties of these two sets of generators is responsible for the fundamental similarity in the structure of classical and quantum mechanics. Because of this correspondence, Schwinger was able to state a single principle—the principle of stationary action—from which either classical or quantum mechanics can be derived in its entirety. It is through this principle that one is able to extend the predictions of quantum mechanics to the properties of a subsystem of some total system, that is, to an atom in a molecule, an open quantum system. Because of its central role in the theory of atoms in molecules, and no less because of the corresponding role it plays in providing a unifying structure for physics, a résumé of what the principle of stationary action is and of how it is obtained is given here.

The presentation is preceded by a summary of the principal mathematical aspects of quantum mechanics as required for the development of the theory, including, in particular, the properties of infinitesimal unitary transformations. This review is not abstract but is paralleled by a demonstration of the elegant manner in which these purely mathematical constructs account for the properties that are unique to a quantum system. Section 8.1 concludes with a review of infinitesimal contact transformations in classical mechanics as required to develop the analogy between the classical and quantal equations of motion, a development made possible by infinitesimal canonical transformations—the generic name for infinitesimal transformations in both classical and quantum mechanics. A reader is referred to books by Messiah (1958) or Roman (1965) for more detailed accounts of the mathematics underlying the framework of quantum mechanics. In particular, Roman gives a detailed discussion of canonical transformations and their role in Schwinger's action principle. A highly recommended treatise on classical mechanics is the one by Goldstein (1980).

8.1.2   State vectors, state functions, and transformation functions

The mathematical formalism of quantum mechanics is expressed in terms of linear operators, which represent the observables of a system, acting on a state vector which is a linear superposition of elements of an infinite-dimensional linear vector space called Hilbert space. We require a knowledge of just the basic properties and consequences of the underlying linear algebra, using mostly those postulates and results that have direct physical consequences. Each state of a quantum dynamical system is exhaustively characterized by a state vector denoted by the symbol $|\Psi\rangle$. This vector and its complex conjugate vector $\langle\Psi|$ which describes the same state of the system, are expressed in Hilbert space. The product $c|\Psi\rangle$, where c is a number which may be complex, describes the same state.

In general, for any pair of such vectors denoted by $|a\rangle$ and $|b\rangle$ we can define a number $\langle b|a\rangle$ which is called the scalar product of $\langle b|$ with $|a\rangle$. This product satisfies eqn (8.1) when acting on a linear combination of state vectors,

$$\langle b|\{\textstyle\sum_i c_i|a_i\rangle\} = \textstyle\sum_i c_i\langle b|a_i\rangle \tag{8.1}$$

as well as the relation

$$\langle b|a\rangle = (\langle a|b\rangle)^*. \tag{8.2}$$

That is, the scalar product equals the complex conjugate of its transpose. From this it follows that $\langle a|a\rangle = \langle a|a\rangle^*$ and is a real number.

A measurable physical quantity or observable is represented by a linear Hermitian operator $\hat{\omega}$. The action of a linear operator on a state vector $|a\rangle$ is to transform it into another vector $\hat{\omega}|a\rangle$. It is linear since its action on a linear combination of state vectors is given by

$$\hat{\omega}\{\textstyle\sum_i c_i|a_i\rangle\} = \textstyle\sum_i c_i\hat{\omega}|a_i\rangle. \tag{8.3}$$

The complex conjugate of the vector $\hat{\omega}|a\rangle$ is $\langle a|\hat{\omega}^\dagger$ where $\omega^\dagger$ is the Hermitian conjugate or adjoint of $\hat{\omega}$. Replacement of $|b\rangle$ by $\hat{\omega}|b\rangle$ in eqn (8.2) yields

$$\langle b|\hat{\omega}^\dagger|a\rangle = \{\langle a|\hat{\omega}|b\rangle\}^*, \tag{8.4}$$

an equation which relates the action of $\hat{\omega}$ to that of its Hermitian adjoint $\hat{\omega}^\dagger$. An operator is said to be Hermitian if

$$\hat{\omega} = \hat{\omega}^\dagger, \tag{8.5}$$

a property which yields the important result

$$\langle b|\hat{\omega}|a\rangle = \{\langle a|\hat{\omega}|b\rangle\}^*. \tag{8.6}$$

Use of eqn (8.2) again, with $\langle a|$ replaced by $\langle a|\hat{\lambda}$ and with $|b\rangle$ replaced by $\hat{\mu}|b\rangle$, yields through the use of eqn (8.4) with $\hat{\omega} = \hat{\lambda}\hat{\mu}$, the additional result

$$\hat{\mu}^\dagger\hat{\lambda}^\dagger = (\hat{\lambda}\hat{\mu})^\dagger. \tag{8.7}$$

The only observable values for a property Ω are one of the eigenvalues ω_i of the associated linear Hermitian operator $\hat{\omega}$ as defined in the eigenvalue equation

$$\hat{\omega}|\omega_i\rangle = \omega_i|\omega_i\rangle \tag{8.8}$$

where, since $\hat{\omega}$ is Hermitian, the eigenvalues are real. An eigenvector $|\omega_i\rangle$ is labelled by its associated eigenvalue ω_i. By assumption, all of the eigenvectors which satisfy eqn (8.8) form a complete set. That is, they span the Hilbert space and, as illustrated below, any arbitrary state vector can be expanded in terms of this set of functions. Thus, the eigenvectors of a Hermitian operator either form a complete orthogonal set or may be combined to yield such a set. When normalized to unity they thus satisfy the expression

$$\langle \omega_i|\omega_j\rangle = \delta_{ij}. \tag{8.9}$$

The orthogonality extends to the linearly independent members of a degenerate set. An eigenvector is labelled with its eigenvalue for the property Ω because, when the system is in a state described by this vector, one is certain that the measurement of the property Ω will yield the particular eigenvalue ω_i. Since the ω_i are the only observable values of the property Ω, actual measurement of this property will carry the state vector $|\Psi\rangle$ of a system over into one of the eigenvectors $|\omega_i\rangle$. In this way the measurement of a physical property places the system in a specific state. Once the measurement has been carried out to yield the value ω_i, one can be certain that the state vector

$|\Psi\rangle = |\omega_i\rangle$. Without such previous preparation, one cannot predict which of the eigenvalues will be observed in a given measurement.

This latter situation is handled within the theory through the principle of superposition wherein the state vector is expanded in the complete set of eigenvectors associated with the property Ω, an expansion which ultimately allows for the observation of any one of the possible values of the property. Thus, the state vector is expanded as

$$|\Psi\rangle = \sum_j c_j |\omega_j\rangle. \tag{8.10}$$

A coefficient c_k in this expansion is evaluated by forming the scalar product of $|\Psi\rangle$ with $\langle\omega_k|$ and using the orthonormality property expressed in eqn (8.9) to yield

$$c_k = \langle\omega_k|\Psi\rangle. \tag{8.11}$$

The physical interpretation of this expansion and the manner in which it can be used to predict the outcome of a measurement of the property Ω are obtained by considering the expression for the average value of the property Ω obtained for a system in the state $|\Psi\rangle$. This average value, by postulate, is given by

$$\langle\hat{\omega}\rangle \equiv \langle\Psi|\hat{\omega}|\Psi\rangle = \sum_j \sum_k c_j^* c_k \langle\omega_j|\hat{\omega}|\omega_k\rangle \tag{8.12}$$

when the state vector is normalized to unity. Using eqn (8.8) and the orthogonality condition of eqn (8.9) this average value reduces to

$$\langle\hat{\omega}\rangle = \sum_j |c_j|^2 \omega_j. \tag{8.13}$$

The value $\langle\hat{\omega}\rangle$ is the one that would be obtained for the property Ω by averaging the results of many measurements and it is equal to the value of each ω_k weighted by the probability of observing that particular value, where the weighting factor is given by $|c_k|^2$. For $|\Psi\rangle$ normalized to one, the sum of the weighting factors is unity, as can be seen by setting ω equal to the unit operator in eqn (8.12). Thus $|c_k|^2$ is the probability of obtaining the value ω_k in a single measurement. Using eqn (8.11) this probability is given by

$$|c_k|^2 = |\langle\omega_k|\Psi\rangle|^2 \tag{8.14}$$

and the quantity $\langle\omega_k|\Psi\rangle$ is thus the *amplitude of this probability*. Equation (8.10) is the representation of the state vector in terms of the ω-basis and the probability amplitude $\langle\omega|\Psi\rangle$ is called the *representative*.

Two linear operators possess a common set of eigenvectors only if the two operators commute. Suppose two observables $\hat{\omega}$ and $\hat{\tau}$ associated with the properties Ω and T, respectively, possess a common set of eigenvectors,

$$\hat{\omega}|\omega_i, \tau_j\rangle = \omega_i|\omega_i, \tau_j\rangle$$

and

$$\hat{\tau}|\omega_i, \tau_j\rangle = \tau_j|\omega_i, \tau_j\rangle.$$

Measurement of the property Ω carries the state vector over into an eigenvector of $\hat{\omega}$ with the particular value for this property equal to ω_i. Since an eigenvector of $\hat{\omega}$ is also an eigenvector of $\hat{\tau}$, the same measurement has also prepared the system to yield a particular value for the property T and it is clear that subsequent measurement of this latter property will yield the value τ_j. Thus, the values of the properties Ω and T may be simultaneously known and measured for a given state of a system. It is also clear that, if two observables do not commute, their values cannot be measured simultaneously, for measurement of the second property will force the system into a new state which is no longer an eigenvector of the first observable and, hence, no longer possessive of a sharp value for the first property.

In this manner the mathematical formalism incorporates the important experimental result that not all physical observables can be known simultaneously for a quantum system, the incompatibility of a knowledge of the position and its conjugate momentum for a particle being an example of this behaviour. The non-compatibility of these two observables is enshrined in the Heisenberg commutation rules

$$[\hat{q}_i, \hat{p}_j] = i\hbar\delta_{ij},$$

$$[\hat{q}_i, \hat{q}_j] = 0, \tag{8.15}$$

$$[\hat{p}_i, \hat{p}_j] = 0.$$

These rules are a part of the formalism of quantum mechanics that is recovered from Schwinger's action principle.

For any given system there exists a maximal number of mutually commuting or compatible observables. The maximum information about a given quantum system is obtained by the simultaneous measurement of this maximal set of compatible observables, a process called the *maximal measurement*. The resulting set of eigenvalues is used to classify the state of the system. For example, within the Russell–Saunders coupling scheme for a many-electron atom, the observables for the energy, for the square of both the orbital and spin angular momenta, along with those for a single component of both angular momenta, the set $\{\hat{H}, \hat{L}^2, \hat{S}^2, \hat{L}_z, \hat{S}_z\}$, form a mutually commuting set of operators. Thus, each state of a many-electron atom is characterized by giving the results of the maximal measurement as $|E, L, M_L, S, M_S\rangle$ where the values of the angular momenta are expressed in terms of their associated quantum numbers.

We are now ready to address the consequences of a change in the representation of the state function through a change in the set of basis vectors used in the expansion, eqn (8.10). We will incorporate the result that a given set of functions will, in general, be simultaneous eigenfunctions of a set of commuting operators. Assume a set of mutually commuting operators $\hat{\omega}_1$, $\hat{\omega}_2, \ldots, \hat{\omega}_r$ and denote a possible collected set of eigenvalues by the single symbol ω_r and label the associated eigenvector correspondingly, $|\omega_r\rangle$. The state vector expressed in the ω-representation is

$$|\Psi\rangle = \sum_r \langle\omega_r|\Psi\rangle|\omega_r\rangle. \tag{8.16}$$

The representative $\langle\omega_r|\Psi\rangle$ may alternatively be replaced by the *state function* in the ω-representation and denoted by $\psi(\omega_r)$ to yield

$$|\Psi\rangle = \sum_r \psi(\omega_r)|\omega_r\rangle. \tag{8.17}$$

The state vector $|\Psi\rangle$ is a vector spanning Hilbert space and is a symbol representing all possible values of the set of eigenvalues ω_r. The state function $\psi(\omega_r)$ is the projection of $|\Psi\rangle$ on to a particular base vector of this space and signifies a state with a particular set of values for the eigenvalues ω_r. The same state vector can be expressed in terms of another representation, the τ-representation, where τ_s denotes a set of eigenvalues obtained from the action of the mutually commuting set of observables $\hat{\tau}_1, \hat{\tau}_2, \ldots, \hat{\tau}_s$ on one of its simultaneous eigenvectors $|\tau_s\rangle$,

$$|\Psi\rangle = \sum_s \psi(\tau_s)|\tau_s\rangle, \tag{8.18}$$

where $\psi(\tau_s)$ is the state function in the τ-representation. The connection between the two representations is easily obtained by forming the scalar product of $|\Psi\rangle$ with $\langle\omega_r|$ in eqn (8.18),

$$\langle\omega_r|\Psi\rangle = \psi(\omega_r) = \sum_s \langle\omega_r|\tau_s\rangle\psi(\tau_s). \tag{8.19}$$

The quantity $\langle\omega_r|\tau_s\rangle$, the scalar product of the base vectors used in the two representations, is the probability amplitude of finding the ω_r base state in the τ_s base state and is called the *transformation function*. The state function in the ω-representation, which represents the probability amplitude of observing a particular set of ω eigenvalues for the system described by $|\Psi\rangle$, is obtained from the τ-representation by summing over all of its base states. The possibility of changing representations is fundamental to quantum mechanics and the work of both Feynman and Schwinger is concerned with the evaluation of the representative which enables one to pass from a representation at one time to another at a later time.

If one rewrites eqn (8.16) in the form

$$|\Psi\rangle = \sum_r |\omega_r\rangle\langle\omega_r|\Psi\rangle, \tag{8.20}$$

it is evident that the effect of the summation term is to multiply $|\Psi\rangle$ by the unit operator, $\hat{I}$. Equation (8.20) is a result of the $|\omega_r\rangle$ forming a complete set and hence the identity

$$\sum_r |\omega_r\rangle\langle\omega_r| = \hat{I} \tag{8.21}$$

is called the completeness condition.

The concept of a representation and the associated definitions have been introduced for the case of a discrete spectrum of eigenvalues. This concept is readily extended to the situation where the commuting set of observables possesses a continuous eigenvalue spectrum. Denote by α_r a set of eigenvalues for the set of mutually commuting operators, $\hat{\alpha}_1, \hat{\alpha}_2, \ldots, \hat{\alpha}_r$ with corresponding eigenfunctions $|\alpha_r\rangle$, which possess a continuous set of eigenvalues. The analogue of eqn (8.20) for the expansion of $|\Psi\rangle$ in the α-representation is

$$|\Psi\rangle = \int |\alpha_r\rangle \, d\alpha_r \langle \alpha_r | \Psi \rangle \tag{8.22}$$

where the 'r' summation is replaced by a corresponding integration over the continuous eigenvalue spectrum. The quantity $\langle \alpha_r | \Psi \rangle$ is again the representative or state function in the α-representation. Let $\hat{\beta}_s$ denote a second set of mutually commuting observables with basis $|\beta_s\rangle$. The two representations of $|\Psi\rangle$ using the two different bases are related by (compare eqn (8.19))

$$\langle \alpha_r | \Psi \rangle = \int \langle \alpha_r | \beta_s \rangle \, d\beta_s \langle \beta_s | \Psi \rangle \tag{8.23}$$

where $\langle \alpha_r | \beta_s \rangle$ is again called the transformation function. If one introduces a third set of mutually commuting observables γ_t and their associated basis $|\gamma_t\rangle$, then we have as a special case of eqn (8.23)

$$\langle \alpha_r | \gamma_t \rangle = \int \langle \alpha_r | \beta_s \rangle \, d\beta_s \langle \beta_s | \gamma_t \rangle. \tag{8.24}$$

The relation expressed in eqn (8.24) is known as the *multiplicative law of transformation functions*.

An important example of a maximal set of commuting observables with a continuous spectrum of eigenvalues is provided by the operators $\hat{q}_r$ representing the position coordinates of a set of particles. The state function in terms of the coordinate representation is given by

$$|\Psi\rangle = \int |q_r\rangle \, dq_r \langle q_r | \Psi \rangle = \int |q_r\rangle \, dq_r \psi(q_r) \tag{8.25}$$

where the ket $|q_r\rangle$ denotes the simultaneous eigenvector corresponding to the set of eigenvalues q_r $(r = 1, \ldots, n)$ for a system with n particles. In this representation the representative or state function $\langle q_r | \Psi \rangle$ is *Schrödinger's wave function* $\psi(q_r)$. The state vector denotes all possible functional values of $\psi(q_r)$, while the state function itself represents the value of this function for one particular set of coordinates q_r, the commuting set of eigenvalues. Consider two sets of the commuting set of coordinate operators, one at time t, the other at time t'. Using eqn (8.25), one obtains for the relation between these two bases in analogy with eqn (8.23) the expression

$$\langle q_{r'}, t' | \Psi \rangle = \psi(q_{r'}, t') = \int \langle q_{r'}, t' | q_r, t \rangle \, dq_r \psi(q_r, t). \tag{8.26}$$

Equation (8.26) is an important result. It shows that a knowledge of the transformation function or transition amplitude $\langle q_{r'}, t' | q_r, t \rangle$ suffices to determine the dynamical behaviour of the system with time, since it relates the state function at time t' to that at time t. Knowledge of how a system passes from its state at one time to that at a later time implies knowledge of the system's equation of motion. The path integral equation of Feynman and the differential form of this equation as given by Schwinger are equations for the determination of the transformation function in eqn (8.26).

Von Neumann has demonstrated that, for a system with a finite number of degrees of freedom, the transformation from one representation to another discussed above is what is termed a unitary transformation. It is in terms of unitary transformations that one is able to demonstrate a correspondence between classical and quantum mechanics and it is upon this correspondence that Schwinger based his formulation of quantum mechanics.

8.1.3   Unitary transformations

The state of a classical system is specified in terms of the values of a set of coordinates q and conjugate momenta p at some time t, the coordinates and momenta satisfying Hamilton's equations of motion. It is possible to perform a coordinate transformation to a new set of ps and qs which again satisfy Hamilton's equation of motion with respect to a Hamiltonian expressed in the new coordinates. Such a coordinate transformation is called a canonical transformation and, while changing the functional form of the Hamiltonian and of the expressions for other properties, it leaves all of the numerical values of the properties unchanged. Thus, a canonical transformation offers an alternative but equivalent description of a classical system. One may ask whether the same freedom of choosing equivalent descriptions of a system exists in quantum mechanics. The answer is in the affirmative and it is a unitary transformation which is the quantum analogue of the classical canonical transformation.

A unitary transformation is one which, when applied to both the state function and the observables of a system, leaves the description of the system unchanged. Denote by $\hat{U}$ an operator with the property that its Hermitian conjugate or adjoint is equal to its inverse, that is

$$\hat{U}\hat{U}^\dagger = \hat{U}^\dagger\hat{U} = \hat{I} \qquad \text{or} \qquad \hat{U}^\dagger = \hat{U}^{-1}. \tag{8.27}$$

Equation (8.27) defines a unitary operator and $\hat{U}$ is used to perform simultaneous transformations on both the state vector $|\Psi\rangle$ and on the observables $\hat{\omega}$ such that

$$|\Psi\rangle \to |\Psi'\rangle = \hat{U}|\Psi\rangle, \tag{8.28}$$

$$\hat{\omega} \to \hat{\omega}' = \hat{U}\hat{\omega}\hat{U}^{-1}. \tag{8.29}$$

It is easy to show that this transformation leaves the description of the system unchanged. The transformed operator $\hat{\omega}'$ has the same eigenvalue spectrum as does $\hat{\omega}$. If initially one has the eigenvalue eqn (8.8), the action of $\hat{\omega}'$ on $|\omega_i'\rangle$ yields the same eigenvalue ω_i,

$$\hat{\omega}'|\omega_i'\rangle = \hat{U}\hat{\omega}\hat{U}^{-1}\hat{U}|\omega_i\rangle = \hat{U}\hat{\omega}|\omega_i\rangle = \omega_i\hat{U}|\omega_i\rangle = \omega_i|\omega_i'\rangle. \qquad (8.30)$$

The expectation value of an observable is also left unchanged

$$\langle\Psi'|\hat{\omega}'|\Psi'\rangle = \langle\hat{U}\Psi, \hat{\omega}'\hat{U}\Psi\rangle = \langle\Psi, \hat{U}^{-1}\hat{U}\hat{\omega}\hat{U}^{-1}\hat{U}\Psi\rangle$$
$$= \langle\Psi|\hat{\omega}|\Psi\rangle \qquad (8.31)$$

as is the normalization of $|\Psi\rangle$

$$\langle\Psi'|\Psi'\rangle = \langle\hat{U}\Psi, \hat{U}\Psi\rangle = \langle\Psi, \hat{U}^{-1}\hat{U}\Psi\rangle = \langle\Psi|\Psi\rangle. \qquad (8.32)$$

It is also easily demonstrated that operator equations for sums and products of operators are cast into equivalent expressions by a unitary transformation.

We shall be particularly interested in the effects of a unitary transformation which is infinitely close to unity as applied separately to the observables or state vectors of a system. Consider the operator

$$\hat{U} = \hat{I} - (i\varepsilon/\hbar)\hat{G} \qquad (8.33)$$

where ε denotes a real infinitesimal quantity and $\hat{G}$ is a linear Hermitian operator. If we retain terms only to first order in ε, the inverse of $\hat{U}$ is

$$\hat{U}^{-1} = \hat{I} + (i\varepsilon/\hbar)\hat{G}. \qquad (8.34)$$

That is, $\hat{U}\hat{U}^{-1} \approx \hat{I}$ ignoring the terms in ε^2. Since $\hat{G}$ is Hermitian, the adjoint of $\hat{U}$ is also given by eqn (8.34). Hence $\hat{U}^\dagger = \hat{U}^{-1}$ and $\hat{U}$ is unitary. The infinitesimal operator $\varepsilon\hat{G}$ when used in this manner is referred to as the generator of the infinitesimal transformation.

The effect of such a transformation on an observable $\hat{\alpha}$ is

$$\hat{\alpha} \rightarrow \hat{\alpha}' = \{\hat{I} - (i\varepsilon/\hbar)\hat{G}\}\hat{\alpha}\{\hat{I} + (i\varepsilon/\hbar)\hat{G}\} = \hat{\alpha} + (i\varepsilon/\hbar)[\hat{\alpha}, \hat{G}]. \qquad (8.35)$$

Following Schwinger, the infinitesimal transformation induced on the observable $\hat{\alpha}$ by the generator $\varepsilon\hat{G}$ is defined to be

$$\delta\hat{\alpha} = \hat{\alpha} - \hat{\alpha}' = (i\varepsilon/\hbar)[\hat{G}, \hat{\alpha}] = (i/\hbar)[\hat{F}, \hat{\alpha}] \qquad (8.36)$$

where the infinitesimal Hermitian operator $\hat{F} = \varepsilon\hat{G}$. The same transformation, when applied to the state vector, yields

$$|\alpha\rangle \rightarrow |\alpha'\rangle = \hat{U}|\alpha\rangle = |\alpha\rangle - (i/\hbar)\hat{F}|\alpha\rangle, \qquad (8.37)$$

and the generator $\hat{F}$ induces an infinitesimal change in the state vector given by

$$\delta|\alpha\rangle = |\alpha'\rangle - |\alpha\rangle = -(i/\hbar)\hat{F}|\alpha\rangle. \qquad (8.38)$$

The concept of an infinitesimal unitary transformation can be used to provide a differential characterization for the representative of a state, or for a transformation function. The change in the representative in eqn (8.23) when the commuting set of operators $\hat{\alpha}_r$ is altered by the unitary transformation generated by the infinitesimal Hermitian operator $\hat{F}$, is given by

$$\delta\langle\alpha_r|\Psi\rangle = \langle\delta\alpha_r|\Psi\rangle = (i/\hbar)\langle\alpha_r|\hat{F}|\Psi\rangle \qquad (8.39)$$

or

$$\delta\langle\alpha_r|\Psi\rangle = (i/\hbar)\int\langle\alpha_r|\hat{F}|\beta_s\rangle\,d\beta_s\langle\beta_s|\Psi\rangle. \qquad (8.40)$$

Similarly, the transformation function $\langle\alpha_s|\gamma_t\rangle$ of eqn (8.24) can be characterized by the effect of altering the two commuting sets $\hat{\alpha}_r$ and $\hat{\gamma}_t$ into $\hat{\alpha} - \delta\hat{\alpha}$ and $\hat{\gamma} - \delta\hat{\gamma}$ as induced by the two infinitesimal generating operators $\hat{F}_\alpha$ and $\hat{F}_\gamma$. One obtains

$$\delta\langle\alpha_r|\gamma_t\rangle = \langle\delta\alpha_r|\gamma_t\rangle + \langle\alpha_r|\delta\gamma_t\rangle = (i/\hbar)\langle\alpha_r|\hat{F}_\alpha - \hat{F}_\gamma|\gamma_t\rangle \qquad (8.41)$$

or

$$\delta\langle\alpha_r|\gamma_t\rangle = (i/\hbar)\int\langle\alpha_r|\hat{F}_\alpha|\beta_s\rangle\,d\beta_s\langle\beta_s|\gamma_t\rangle - (i/\hbar)\int\langle\alpha_r|\beta'_s\rangle\,d\beta'_s\langle\beta'_s|\hat{F}_\gamma|\gamma_t\rangle. \quad (8.42)$$

It was emphasized above that the transformation function determines the temporal evolution of a system. Equation (8.41) or (8.42) relates the differential of a transformation function to the action of infinitesimal generators acting on the two associated base states. In his action principle, Schwinger postulates an alternative expression for the infinitesimal change in the transformation function, proposing that this change be given by a variation of the quantum action integral between the same two states.

The above has demonstrated the important result that an infinitesimal unitary transformation can be used to generate changes in the dynamical variables of a quantum system. Corresponding changes can be made in a classical system through the use of generators of infinitesimal canonical transformations. Illustrating this correspondence will make clear that all possible changes, spatial or temporal, in a classical or quantal system are describable in terms of generators of corresponding infinitesimal 'canonical transformations'. The deep understanding of physics afforded by Schwinger's principle derives from the common basis it provides for both classical and quantum mechanics in terms of such transformations. For this reason it is desirable to develop, in parallel, the classical case as is done in the next section. A reader already familiar with classical Hamiltonian mechanics or one eager to proceed with the quantum development may omit this section, at least on first reading for the latter readers.

8.1.4   Canonical transformations and classical mechanics

The concept of a canonical transformation is fundamental to the Hamiltonian formulation of classical mechanics, the formulation which provides the usual bridge from classical to quantum mechanics. The essentials of this formulation and of the role played by a canonical transformation are introduced here.

The classical equations of motion for any generalized set of coordinates can be obtained from Hamilton's principle which states that the motion of a system from time t_1 to time t_2 is such that the line integral

$$W_{12} = \int_{t_1}^{t_2} L\,dt \qquad (8.43)$$

where the Lagrangian $L = T - V$, is an extremum for the path of motion. (There is a growing movement (Feynman 1964) to call the quantity W_{12} the action integral and to change the name of this principle to the principle of least action, names which historically referred to the variation of the time integral of a sum of terms $p_i q_i$ for all degrees of freedom. We shall follow these suggested changes in nomenclature.) In Chapter 5 it was shown that Schrödinger's equation for a stationary state is obtained as the Euler–Lagrange equation in the extremization of the integral $\mathscr{G}[\psi]$ by finding that state function ψ which extremized this integral. According to the principle of least action, the classical equations of motion can be obtained in a similar manner by finding that trajectory connecting the times t_1 and t_2 which extremizes the action integral. The Lagrangian is a function of the coordinates and their time derivatives

$$L(q, \dot{q}, t) = T - V = \sum_i \tfrac{1}{2} m\dot{q}_i^2 - V(q) \qquad (8.44)$$

where q and $\dot{q}$ without the subscript denote corresponding sets of coordinates for all particles. Suppose the proper sets of generalized coordinates at t_1 and t_2 to be given by $q(t_1)$ and $q(t_2)$, respectively, and then imagine connecting the initial and final sets of coordinates with a space–time trajectory in configuration space as pictured in Fig. 8.1. There is only one true trajectory and its form is predicted by the equations of motion. Alternatively, we may derive the equations of motion by using the principle of least action to find that trajectory connecting t_1 and t_2 such that the action integral attains its minimum value. This is clearly once again a problem for the calculus of variations.

One denotes a point on the true path at any time t by the symbol $q(t)$ and a point on the varied path by $q'(t)$. Then, as for the variation of a state function in eqn. (5.69), one has

$$q'(t) = q(t) + \delta q(t) \qquad (8.45)$$

where $\delta q(t)$ indicates an arbitrary change in the path at time t. The variations in the path are to vanish at the two time end-points and the end-points themselves are not changed. A possible such varied path is shown in Fig. 8.1. As detailed in Chapter 5 for the variation of a general integral I (eqn (5.68)),

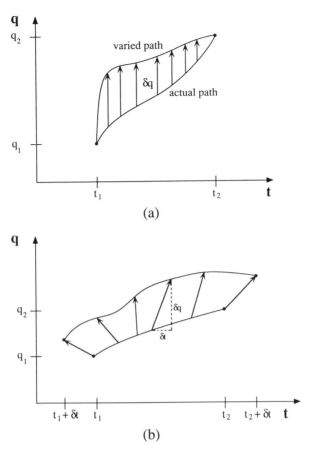

(a)

(b)

FIG. 8.1.   Schematic representations of actual and varied paths linking initial and final states. In (a) the time end-points are fixed and the variations in the path vanish at these points. In (b) the time end-points are varied, as is the path at these points.

the first-order change in W_{12} caused by this variation of q is given by

$$\delta W_{12} = \int_{t_1}^{t_2} \sum_i \{(\partial L/\partial q_i)\delta q_i + (\partial L/\partial \dot{q}_i)\delta \dot{q}_i\} \mathrm{d}t. \qquad (8.46)$$

Using the methods developed in Chapter 5, one rids the expression of variations in $\dot{q}_i$ using an integration by parts

$$\mathrm{d}\{(\partial L/\partial \dot{q}_i)\delta q_i\}/\mathrm{d}t = (\partial L/\partial \dot{q}_i)\delta \dot{q}_i + \{\mathrm{d}(\partial L/\partial \dot{q}_i)/\mathrm{d}t\}\delta q_i \qquad (8.47)$$

to obtain

$$\delta W_{12} = \int_{t_1}^{t_2} \sum_i \{(\partial L/\partial q_i) - \mathrm{d}(\partial L/\partial \dot{q}_i)/\mathrm{d}t\}\delta q_i \mathrm{d}t + \sum_i (\partial L/\partial \dot{q}_i)\delta q_i|_{t_1}^{t_2}. \qquad (8.48)$$

Since each δq_i is assumed to vanish at the time end-points, the condition that the variation in the action integral vanish, i.e. that $q_i(t)$ is such as to minimize the action, is given by

$$\delta W_{12} = \int_{t_1}^{t_2} \sum_i \{(\partial L/\partial q_i) - \mathrm{d}(\partial L/\partial \dot{q}_i)/\mathrm{d}t\}\delta q_i \mathrm{d}t = 0. \qquad (8.49)$$

For this result to be true for any and all variations in q_i which vanish at the time end-points, it is necessary and sufficient that one obtain the Euler equation

$$\partial L/\partial q_i - \mathrm{d}(\partial L/\partial \dot{q}_i)/\mathrm{d}t = 0. \qquad (8.50)$$

For a system with N degrees of freedom, q_i, $i = 1$ to N, this equation is obtained for each of the N coordinates q_i. These are Lagrange's equations of motion, the equations of motion for a system obeying classical mechanics. Thus, the Lagrangian, which minimizes the value of the action integral along the true trajectory between the times t_1 and t_2, is also the function which yields the equations of motion when inserted into the Euler equation (8.50).

A conceptually richer statement of classical mechanics is obtained using Hamilton's formulation. This entails a change in variables from the set of position coordinates and their dependent velocities, which, together with t, determine the Lagrangian, to a set consisting of the coordinates q and their conjugate momenta p, the momentum p_j conjugate to the coordinate q_j being defined as

$$p_j = \partial L/\partial \dot{q}_j. \qquad (8.51)$$

The ps and qs are treated as independent coordinates in the definition of the Hamiltonian, $H(p, q, t)$. The coordinate transformation from $(q, \dot{q}, t)$ to (q, p, t) is accomplished using a Legendre transform of the form

$$H(q, p, t) = \sum_i \dot{q}_i p_i - L(q, \dot{q}, t). \qquad (8.52)$$

For a conservative system for which the kinetic energy T is a homogeneous function of the $\dot{q}_j$s, the summation in eqn (8.52) yields $2T$ and the Hamiltonian reduces to the total energy of the system

$$H = 2T - (T - V) = T + V = E. \qquad (8.53)$$

The classical equations of motion in Hamilton's form can be obtained from a modification of Hamilton's principle as outlined above. The Lagrangian in the action integral is expressed in terms of the Hamiltonian using eqn (8.52) to yield the integral

$$I = \int_{t_1}^{t_2} \{\sum_i p_i \dot{q}_i - H(p, q, t)\}\mathrm{d}t. \qquad (8.54)$$

The extremization of I,

$$\delta \int_{t_1}^{t_2} \{\textstyle\sum_i p_i \dot{q}_i - H(p, q, t)\} dt = 0, \tag{8.55}$$

is carried out between fixed time end-points by making independent variations in both the p_i and q_i. This is easily done and, after using an integration by parts to rid the expression of the terms involving $\delta \dot{q}_j$, one obtains

$$\delta I = \int_{t_1}^{t_2} \textstyle\sum_i \{\delta p_i(\dot{q}_i - \partial H/\partial p_i) - \delta q_i(\dot{p}_i + \partial H/\partial q_i)\} dt = 0 \tag{8.56}$$

which, for arbitrary values of the independent variations δp_i and δq_i, yields the set of Euler equations

$$\dot{q}_i = \partial H/\partial p_i, \qquad \dot{p}_i = -\partial H/\partial q_i. \tag{8.57}$$

These are Hamilton's canonical equations. One may determine the temporal behaviour of a classical system with N degrees of freedom by solving Lagrange's N second-order differential equations with the constants of integration being fixed by the $2N$ initial values of the coordinates and velocities which determine the initial state of the system, or by solving Hamilton's $2N$ first-order equations for the same initial state.

It is within the Hamiltonian formulation of classical mechanics that one introduces the concept of a *canonical transformation*. This is a transformation from some initial set of ps and qs, which satisfy the canonical equations of motion for $H(p, q, t)$ as given in eqn (8.57), to a new set Q and P, which depend upon both the old coordinates and momenta with defining equations,

$$Q_i = Q_i(p, q, t),$$
$$P_i = P_i(p, q, t), \tag{8.58}$$

and which also satisfy the canonical equations of motion for a new Hamiltonian $K(P, Q, t)$. The name *contact transformation* is also applied to such a change of coordinates. The original coordinates satisfy the variation principle stated in eqn (8.55) and, if the P and Q are to be canonical coordinates, they must simultaneously satisfy a corresponding variation principle stated in terms of P, Q, and K. This does not imply that the integrands for the two sets of coordinates must be equal, but that they differ at most by a total time derivative of an arbitrary function F. The function is called the generating function of the transformation and as demonstrated in Goldstein (1980), it totally determines the transformation equations.

The generating functions are functions of one of the old coordinates and one of the new, as well as of the time. There are, therefore, four such generating functions and we are interested in the one of the form $F(q, P, t)$. In terms of this generator, the equations for obtaining the old momenta and new coordinates are

$$p_i = \partial F/\partial q_i \quad \text{and} \quad Q_i = \partial F/\partial P_i. \tag{8.59}$$

With the particular choice of $\sum_i q_i P_i$ for F, one obtains the identity transformation for which $p_i = P_i$ and $Q_i = q_i$. It is to emphasized that any of the canonical transformations, which possibly change the functional form of a property and are classical analogues of a unitary transformation in quantum mechanics, leaves the values of the properties for a system unchanged. This situation is to be contrasted with the effects of an infinitesimal canonical transformation, introduced next, which is the analogue of an infinitesimal unitary transformation of quantum mechanics.

Consider a coordinate transformation which differs from the identity transformation discussed above by an infinitesimal amount for each coordinate, i.e.

$$\begin{aligned} Q_i &= q_i + \delta q_i, \\ P_i &= p_i + \delta p_i. \end{aligned} \tag{8.60}$$

The δq_i and δp_i do not represent virtual changes here as did the variations in the coordinates δq_i in the variation of the action integral to describe a purely mathematical change in the system's trajectory between two times. Here they represent real infinitesimal changes in the coordinates and momenta. The generating function in this case will differ only by an infinitesimal amount from the one describing the identity transformation and hence can be written as

$$F = \sum_i q_i P_i + \varepsilon G(q, P) \tag{8.61}$$

where again ε denotes an infinitesimal parameter of the transformation. Through the use of the defining equations for this generating function, eqn (8.59), one obtains

$$\delta p_i = P_i - p_i = -\varepsilon \partial G/\partial q_i \quad \text{and} \quad \delta q_i = Q_i - q_i = \varepsilon \partial G/\partial p_i. \tag{8.62}$$

Since P_i differs from p_i only by an infinitesimal and the final derivative in eqn (8.62) already involves ε, P_i in the derivative of G has been replaced by p_i to retain only first-order terms. It is usual practice to refer to G as well as F as the generator, and eqns (8.62) define infinitesimal canonical or contact transformations with the generator G.

To illustrate that these generators cause real displacements in a system we take the important example where $G(p, q)$ is set equal to the Hamiltonian $H(p, q)$ and ε to a small time interval dt. The changes induced in the coordinates and momenta by this transformation are found to be

$$\begin{aligned} \delta p_i &= -(\partial H/\partial q_i)dt = \dot{p}_i \, dt = dp_i, \\ \delta q_i &= (\partial H/\partial p_i)dt = \dot{q}_i \, dt = dq_i. \end{aligned} \tag{8.63}$$

Thus, the transformation changes the coordinates and momenta at the time t into the values they have at the time $t + dt$ as predicted by Hamilton's equations of motion, eqn (8.57). We shall demonstrate shortly that $-\hat{H}\delta t$ is the temporal generator of an infinitesimal unitary transformation in quantum mechanics. After noting that the variation with time of an observable in the Heisenberg representation of quantum mechanics may be looked upon as the continuous unfolding of a unitary transformation, Dirac (1930) went on to state, 'In classical mechanics the dynamical variables at time $t + \delta t$ are connected with their values at time t by an infinitesimal contact transformation and the whole motion may be looked upon as the continuous unfolding of a contact transformation. We have here the mathematical foundation of the analogy between the classical and quantum equations of motion . . .'

There remains but one important concept to complete our summary of the role of canonical transformations in classical mechanics, that of the Poisson bracket. Let $F(p, q)$ and $G(p, q)$ denote two mechanical properties of the system. Their Poisson bracket is defined as

$$\{F, G\} = \sum_i [(\partial F/\partial q_i)(\partial G/\partial p_i) - (\partial F/\partial p_i)(\partial G/\partial q_i)]. \qquad (8.64)$$

The so-called fundamental Poisson bracket relations are given by

$$\{p_i, p_j\} = 0, \qquad \{q_i, q_j\} = 0, \qquad \{q_i, p_j\} = \delta_{ij}. \qquad (8.65)$$

In terms of these results it can be shown that the value of a Poisson bracket is invariant under a canonical transformation of the coordinates. Like the corresponding commutator relationships in quantum mechanics, to which it is related by the expression

$$\{F, G\} = (-i/\hbar)[\hat{F}, \hat{G}],$$

the Poisson bracket notation may be used to summarize the fundamental relations of classical physics. The equations of motion may be expressed as

$$\{q_i, H\} = \partial H/\partial p_i = \dot{q}_i \qquad \text{and} \qquad \{p_i, H\} = -\partial H/\partial q_i = \dot{p}_i. \qquad (8.66)$$

These are special cases of the general expression for the time derivative of a property. The total time derivative of a property $F(p, q)$ which does not possess an explicit time dependence is

$$dF/dt = \sum_i [(\partial F/\partial q_i)\dot{q}_i + (\partial F/\partial p_i)\dot{p}_i],$$

which, using the equations of motion (8.57), can be re-expressed as

$$dF/dt = \sum_i [(\partial F/\partial q_i)(\partial H/\partial p_i) - (\partial F/\partial p_i)(\partial H/\partial q_i)]$$

or

$$dF/dt = \{F, H\}. \qquad (8.67)$$

Thus, one has the important result that all functions whose Poisson brackets with the Hamiltonian vanish will be constants of the motion and, conversely, the Poisson brackets of all constants of the motion with H must be zero.

The properties of the angular momentum are usefully summarized by the relations (which are more familiar in their corresponding quantum mechanical form than they are in classical mechanics): the components of the angular momentum vector $\mathbf{L}$ are related by

$$\{L_i, L_j\} = L_k \qquad i, j, k \text{ in cyclic order,} \tag{8.68}$$

and L^2 and one of the components of $\mathbf{L}$, $\mathbf{L} \cdot \mathbf{n}$ are in turn related by

$$\{L^2, \mathbf{L} \cdot \mathbf{n}\} = 0. \tag{8.69}$$

Equation (8.65) shows that the Poisson bracket of two canonical momenta vanishes. Thus, combining the results of the above two equations, one has the important result that, if one of the components of $\mathbf{L}$ and its magnitude are chosen as canonical variables, the two perpendicular components cannot simultaneously be canonical variables.

Of particular importance to the forthcoming development is the statement of the change induced in a property $A(p, q)$ by the generator of an infinitesimal canonical transformation G as expressed in the Poisson bracket notation. The change in the property A obtained as a result of an infinitesimal canonical transformation of the coordinates is given by

$$\delta A = A(q_i + \delta q_i, p_i + \delta p_i) - A(q_i, p_i).$$

A Taylor's series expansion to first-order in the infinitesimals yields

$$\delta A = \sum_i [(\partial A/\partial q_i)\delta q_i + (\partial A/\partial p_i)\delta p_i],$$

which, using the equations of transformation (8.62), can be re-expressed as

$$\delta A = \varepsilon \sum_i \{(\partial A/\partial q_i)(\partial G/\partial p_i) - (\partial A/\partial p_i)(\partial G/\partial q_i)\}$$

or

$$\delta A = \varepsilon\{A, G\}. \tag{8.70}$$

Thus, in analogy with the quantum result for the change induced in an observable by an infinitesimal unitary transformation (eqn (8.36)), the change in a property A caused by an infinitesimal contact transformation is given by the Poisson bracket of A with the generator G. The Poisson bracket of a constant of the motion with H vanishes, as does its corresponding commutator in quantum mechanics. Thus, the equation

$$\delta H = \varepsilon\{H, G\} = -(\mathrm{i}/\hbar)\varepsilon[\hat{H}, \hat{G}] = 0$$

states the important result that the constants of the motion are the generating functions of those infinitesimal canonical transformations which leave the Hamiltonian invariant in both quantum and classical mechanics.

The analoguous behaviour of the Poisson bracket and the commutator has been used to establish a correspondence between classical and quantum mechanics. It is, however, shown in the next section following the derivation of Schwinger's quantum action principle that the correspondence goes deeper and that the analogous behaviour of the Poisson bracket and commutator is a consequence of the properties of infinitesimal canonical transformations which are common to both mechanics.

8.2   The quantum action principle

8.2.1   The principle of stationary action

The development given here follows closely that given by Schwinger (1951). A base vector system $|\alpha_r, t\rangle$ is specified by the eigenvalues α_r of some complete set of commuting operators $\hat{\alpha}_r$ at the time t. One can consider having two similarly constructed operator sets at times t_1 and t_2, $\hat{\alpha}_{r1}$ and $\hat{\alpha}_{r2}$, respectively, which possess the same eigenvalue spectrum α_r and are, therefore, connected by a unitary transformation as outlined in eqns (8.28) and (8.29). Thus,

$$\hat{\alpha}_{r2} = \hat{U}_{12}\hat{\alpha}_{r1}\hat{U}_{12}^{-1},$$

$$|\alpha_r, t_2\rangle = \hat{U}_{12}|\alpha_r, t_1\rangle, \qquad \text{where} \qquad \alpha_{r1} = \alpha_{r2} = \alpha_r.$$

The transformation function connecting a base state at time t_2 with that of another representation derived from the commuting set of observables β_s at time t_1 may be viewed as the matrix of $\hat{U}_{12}^{-1}$ in the original eigenvector system $\hat{\alpha}_{r1}$. That is,

$$\langle \alpha_r, t_2|\beta_s, t_1\rangle = \langle \alpha_r, t_1|\hat{U}_{12}^{\dagger}|\beta_s, t_1\rangle = \langle \alpha_r, t_1|\hat{U}_{12}^{-1}|\beta_s, t_1\rangle. \qquad (8.71)$$

As previously discussed, a description of the temporal evolution of a system is accomplished by stating the relationship between eigenvectors associated with different times or, in other words, by exhibiting the transformation function in eqn (8.71). One may expect that the quantum dynamical laws will find their proper expression in terms of the transformation function and we now present Schwinger's development (1951) of a differential formulation of this type.

Schwinger's reasoning is, with minor modification, as follows. The operator $\hat{U}_{12}^{-1}$ describes the development of the system from time t_1 to time t_2 and involves not only the detailed dynamical characteristics of the system in this space–time region, but also the choice of commuting operators $\hat{\alpha}_{r1}$ and $\hat{\alpha}_{r2}$ at the times t_1 and t_2. Any infinitesimal change in the quantities on which the transformation function depends induces a corresponding alteration in $\hat{U}_{12}^{-1}$,

$$\delta\langle \alpha_r, t_2|\beta_s, t_1\rangle = \langle \alpha_r, t_1|\delta\hat{U}_{12}^{-1}|\beta_s, t_1\rangle. \qquad (8.72)$$

The quantity $i\hat{U}_{12}\delta\hat{U}_{12}^{-1}$ is Hermitian as a result of the unitary property $\hat{U}^{-1} = \hat{U}^\dagger$ or $\hat{U}\hat{U}^\dagger = 1$. Because of this property one has

$$\delta\hat{U}\hat{U}^\dagger = -\hat{U}\delta\hat{U}^\dagger$$

and, by using the identity in eqn (8.7) followed by the use of the above result,

$$(i\hat{U}\delta\hat{U}^\dagger)^\dagger = -i\delta\hat{U}\hat{U}^\dagger = +i\hat{U}\delta\hat{U}^\dagger$$

as required for a Hermitian operator (eqn (8.5)). Accordingly, one can define the Hermitian infinitesimal operator $\delta\hat{W}_{12}$

$$\delta\hat{W}_{12} = -i\hbar\hat{U}_{12}\delta\hat{U}_{12}^{-1},$$

which has the dimensions of action as indicated by the inclusion of $\hbar$ on the right-hand side of the definition. Multiplication of the above by $(i/\hbar)\hat{U}_{12}^{-1}$

$$\delta\hat{U}_{12}^{-1} = (i/\hbar)\hat{U}_{12}^{-1}\delta\hat{W}_{12} \tag{8.73}$$

and the use of this result for $\delta\hat{U}_{12}^{-1}$ in eqn (8.72) give

$$\delta\langle\alpha_r, t_2|\beta_s, t_1\rangle = (i/\hbar)\langle\alpha_r, t_2|\delta\hat{W}_{12}|\beta_s, t_1\rangle. \tag{8.74}$$

The Hermitian operator $\delta\hat{W}$ is the generating operator of infinitesimal transformations. The composition law of transformation functions, eqn (8.24), imposes restrictions on this generator. Because of this law, the generating operators must satisfy an additive law of composition, that is,

$$\langle\alpha_r, t_1|\delta\hat{W}_{13}|\gamma_t, t_3\rangle = \int \langle\alpha_r, t_1|\delta\hat{W}_{12}|\beta_s, t_2\rangle \, d\beta_s\langle\beta_s, t_2|\gamma_t, t_3\rangle$$
$$+ \int \langle\alpha_r, t_1|\beta_s, t_2\rangle \, d\beta_s\langle\beta_s, t_2|\delta\hat{W}_{23}|\gamma_t, t_3\rangle \tag{8.75}$$

or

$$\delta\hat{W}_{13} = \delta\hat{W}_{12} + \delta\hat{W}_{23}. \tag{8.76}$$

The basic assumption made by Schwinger is that the infinitesimal generating operator $\delta\hat{W}_{12}$ is obtained by a variation of the quantities contained in a Hermitian operator $\hat{W}_{12}$ which, because of the additive requirement given in eqn (8.76), must have the general form

$$\hat{W}_{12} = \int_{t_1}^{t_2} \hat{\mathscr{L}}[t]\,dt \tag{8.77}$$

where $\hat{W}$ and $\hat{\mathscr{L}}$ are referred to as action integral and Lagrange function operators, respectively. The Lagrange operator is an invariant Hermitian function of the field Ψ and its first derivatives with respect to space and time.

We may state the resulting dynamical principle in the form of a differential statement for the transformation function connecting two coordinate eigenstates at times t_1 and t_2,

$$\delta \langle q_{r2}, t_2 | q_{r1}, t_1 \rangle = (i/\hbar) \langle q_{r2}, t_2 | \delta \hat{\mathscr{W}}_{12} | q_{r1}, t_1 \rangle$$

$$= (i/\hbar) \left\langle q_{r2}, t_2 \left| \delta \int_{t_1}^{t_2} \mathscr{L}[t] dt \right| q_{r1}, t_1 \right\rangle. \quad (8.78)$$

If variations are effected in a quantum mechanical system, the corresponding change in the transformation function between the eigenstates $|q_{r1}, t_1\rangle$ and $|q_{r2}, t_2\rangle$ is $i/\hbar$ times the matrix element of the variation of the action integral connecting the two states. Equation (8.78) will be referred to as the *quantum action principle*.

Assuming that the parameters of the system are not altered, the variation of the action integral in eqn (8.78) arises only from infinitesimal changes of the sets of commuting observables at the two times t_1 and t_2. We have previously seen (eqn (8.41)) that such transformations may be characterized in terms of the generators of infinitesimal unitary transformations $\hat{F}(t_1)$ and $\hat{F}(t_2)$, which act on the eigenvectors $|q_{r1}, t_1\rangle$ and $|q_{r2}, t_2\rangle$. Comparing eqns (8.41) and (8.78) one obtains for such variations

$$\delta \hat{\mathscr{W}}_{12} = \hat{F}(t_2) - \hat{F}(t_1), \quad (8.79)$$

which is the *operator principle of stationary action*. It states that the action integral operator is unaltered by infinitesimal variations in state functions between the times t_1 and t_2, being affected only by the action of generators at the two time end-points. The principle of stationary action is the operational statement of the theory. It states that the variation of the action integral does not vanish as in Hamilton's principle, but instead equals the difference in the effects of infinitesimal operators acting at the two time end-points. This result demands that the variation of the action integral be generalized to include variations of the state functions and of the time at the time end-points. The principle of stationary action then implies the equations of motion of the system as obtained in Hamilton's principle, and the end-point variations define the generators of the infinitesimal canonical transformations which induce changes in the dynamical properties of the system. In this way a single dynamical principle, the principle of stationary action, recovers not only the equation of motion, but defines the observables, their equations of motion, and the commutation relations.

8.2.2 Applying the principle of stationary action

The first step in applying the principle of stationary action is to generalize the variation of the action integral to include a variation of the time end-points and to retain the variations at the end-points in order to define the generator $\hat{F}(t)$. We shall express the Lagrangian operator in terms of the complete set of commuting generalized coordinate operators denoted by $\hat{q}_{rt}$ introduced in eqn (8.25), and their time derivatives $\dot{\hat{q}}_{rt}$,

$$\hat{\mathscr{W}}_{12} = \int_{t_1}^{t_2} \hat{\mathscr{L}}(\hat{q}_{rt}, \dot{\hat{q}}_{rt}, t)\mathrm{d}t. \tag{8.80}$$

While this is a formal approach, it maximizes the correspondence with the classical case. In fact, if one assumes the coordinate operators and their time derivatives to be instead their classical analogues, one obtains the corresponding results for a classical system, thereby illustrating the generality of the method and the similarity in the structure underlying the two mechanics that is made evident using this approach. The same variation is performed within the Schrödinger representation in the following section in the extension of the principle to a subsystem.

The difference between the generalized variation of the action integral and the restricted one is illustrated by comparing the form of the varied paths for the classical space–time trajectories depicted in Fig. 8.1(a), (b). The varied path may end at different times and the variation in the coordinates does not vanish at the time end-points. The generalized variation corresponds to retaining the first-order changes between the Lagrangian evaluated along the varied path, integrated between the varied time limits, and its unvaried value integrated between the unvaried time end-points. That is,

$$\delta\hat{\mathscr{W}}_{12} = \int_{t_1+\delta t_1}^{t_2+\delta t_2} \hat{\mathscr{L}}(\hat{q}_{rt} + \delta\hat{q}_{rt}, \dot{\hat{q}}_{rt} + \delta\dot{\hat{q}}_{rt}, t + \delta t)\mathrm{d}(t + \delta t)$$

$$- \int_{t_1}^{t_2} \hat{\mathscr{L}}(\hat{q}_{rt}, \dot{\hat{q}}_{rt}, t)\mathrm{d}t. \tag{8.81}$$

To first-order in the infinitesimals this reduces to the change in $\hat{\mathscr{L}}$ along the varied path between the unvaried time end-points and the unvaried integrand times the variations in the time at the two time end-points,

$$\delta\hat{\mathscr{W}}_{12} = \int_{t_1}^{t_2} \delta\hat{\mathscr{L}}(\hat{q}_{rt}, \dot{\hat{q}}_{rt}, t)\mathrm{d}t + \hat{\mathscr{L}}(\hat{q}_{rt}, \dot{\hat{q}}_{rt}, t)\delta t|_{t_1}^{t_2}. \tag{8.82}$$

The variation of the Lagrangian operator is formally identical to the variation of the classical action integral as developed in eqns (8.45)–(8.48). We may take over the final result in eqn (8.48) completely in its corresponding operator form, retaining in this case the terms involving $\delta\hat{q}_{rt}$ at the time end-points as these variations are no longer required to vanish. The addition of this result to the end-point variations in eqn (8.82) yields, for the general variation of the action integral operator,

$$\delta \hat{\mathscr{W}}_{12} = \int_{t_1}^{t_2} \{\partial \mathscr{L}/\partial \hat{q}_{rt} - \mathrm{d}(\partial \mathscr{L}/\partial \hat{\dot{q}}_{rt})/\mathrm{d}t\} \delta \hat{q}_{rt} \mathrm{d}t$$
$$+ \{(\partial \mathscr{L}/\partial \hat{\dot{q}}_{rt})\delta \hat{q}_{rt} + \hat{\mathscr{L}}(\hat{q}_{rt}, \hat{\dot{q}}_{rt}, t)\delta t\}|_{t_1}^{t_2}. \tag{8.83}$$

The end-point variations can be recast into a more useful form. One first notes that the complete change in a coordinate operator at a time end-point, denoted by the symbol Δq_{rt}, can arise as a result of a variation in the operator or as a result of a variation in the time

$$\Delta \hat{q}_{rt} = \delta \hat{q}_{rt} + \hat{\dot{q}}_{rt}\delta t. \tag{8.84}$$

An end-point variation in eqn (8.83), expressed in terms of Δq_{rt}, is given by

$$(\partial \mathscr{L}/\partial \hat{\dot{q}}_{rt})\Delta \hat{q}_{rt} - \{(\partial \mathscr{L}/\partial \hat{\dot{q}}_{rt})\hat{\dot{q}}_{rt} - \mathscr{L}(\hat{q}_{rt}, \hat{\dot{q}}_{rt}, t)\}\delta t. \tag{8.85}$$

In accordance with the classical definition of the Hamiltonian, eqn (8.52), and recalling that the derivative of the Lagrangian with respect to a velocity (eqn (8.51)) is the momentum conjugate to the corresponding coordinate, the Hamiltonian operator is defined as

$$\hat{H}(\hat{q}_{rt}, \partial \mathscr{L}/\partial \hat{\dot{q}}_{rt}) = (\partial \mathscr{L}/\partial \hat{\dot{q}}_{rt})\hat{\dot{q}}_{rt} - \hat{\mathscr{L}}(\hat{q}_{rt}, \hat{\dot{q}}_{rt}, t). \tag{8.86}$$

By using this result in eqn (8.85), an end-point variation arising from the generalized variation of the action integral operator at a time t, a quantity denoted by the operator $\hat{F}(t)$, is given by

$$\hat{F}(t) = \{(\partial \mathscr{L}/\partial \hat{\dot{q}}_{rt})\Delta \hat{q}_{rt} - \hat{H}(\hat{q}_{rt}, \partial \mathscr{L}/\partial \hat{\dot{q}}_{rt})\delta t\}. \tag{8.87}$$

Incorporating this result into eqn (8.83) yields, for the generalized variation,

$$\delta \hat{\mathscr{W}}_{12} = \int_{t_1}^{t_2} \{(\partial \mathscr{L}/\partial \hat{q}_{rt}) - \mathrm{d}(\partial \mathscr{L}/\partial \hat{\dot{q}}_{rt})/\mathrm{d}t\} \delta \hat{q}_{rt} \mathrm{d}t + \hat{F}(t_2) - \hat{F}(t_1). \tag{8.88}$$

Comparison of this result with the operator statement of stationary action given in eqn (8.79) shows that the quantity multiplied by the arbitrary variations in $\hat{q}_{rt}$ must vanish as it does in the restricted variation of the action integral, thereby yielding the equation of motion

$$\partial \mathscr{L}/\partial \hat{q}_{rt} = \mathrm{d}(\partial \mathscr{L}/\partial \hat{\dot{q}}_{rt})/\mathrm{d}t$$

and the statement of stationary action

$$\delta \hat{\mathscr{W}}_{12} = \hat{F}(t_2) - \hat{F}(t_1). \tag{8.89}$$

The comparison also leads to the identification of the generators of the infinitesimal unitary transformations of the principle of stationary action with the terms derived from the end-point variations in $\hat{q}_{rt}$ and in the time as given in eqn (8.87) for the operator $\hat{F}(t)$.

A generator $\hat{F}(t)$ may be decomposed into two parts: one which determines the purely temporal changes in a system, the generator $-\hat{H}\delta t$, the other determining all possible spatial changes in a system, the generator $(\partial\mathscr{L}/\partial\hat{q}_{rt})\delta\hat{q}_{rt}$. The use of these generators in the derivation of the Heisenberg equations of motion for the observables and of the commutation relations are illustrated next. These relations, together with the equation of motion for the system, provide a complete description of the properties of a quantum dynamical system, and all are derived from a single dynamical principle.

For a purely temporal change the complete variation in an observable $\hat{\omega}$, $\Delta\hat{\omega}$, equals zero and, from eqn (8.84), the whole of its change results from the change in the time end-point. By eqn (8.84) this change is given by $\delta\hat{\omega} = -\dot{\hat{\omega}}\delta t$. According to eqn (8.36), the change in the observable $\hat{\omega}$ caused by the operator $-\hat{H}\delta t$ acting as the generator of an infinitesimal unitary transformation is

$$\delta\hat{\omega} = -(i/\hbar)[\hat{H}, \hat{\omega}]\delta t = -\dot{\hat{\omega}}\delta t \qquad (8.90)$$

or

$$\dot{\hat{\omega}} = (i/\hbar)[\hat{H}, \hat{\omega}], \qquad (8.91)$$

which is Heisenberg's equation of motion for the observable $\hat{\omega}$. If one takes $\hat{H}$ in this equation to be the classical Hamiltonian and multiplies the commutator by $(i\hbar)$, one obtains the expression for the time rate of change of the observable $\hat{\omega}$ in the Poisson bracket notation (eqn (8.67)). As in a classical system, the Hamiltonian is the variable conjugate to the time and the action of the generator $-\hat{H}\delta t$ is to propagate the system through time.

When there is no change in the time end-point, $\Delta\hat{q}_{rt} = \delta\hat{q}_{rt}$ and, from eqn (8.87), the generator of a pure displacement is given by $(\partial\mathscr{L}/\partial\hat{q}_{rt})\delta\hat{q}_{rt}$. We may derive the commutation relations previously introduced in eqn (8.15) by considering the effect of this generator on the coordinate operator $\hat{q}_{st}$ and the conjugate momentum $\hat{p}_{st}$. According to eqn (8.36) the change in $\hat{q}_{st}$ induced by this generator is given by

$$\delta\hat{q}_s = (i/\hbar)[(\partial\mathscr{L}/\partial\hat{q}_r)\delta\hat{q}_r, \hat{q}_s]$$
$$= (i/\hbar)(\partial\mathscr{L}/\partial\hat{q}_r)[\delta\hat{q}_r, \hat{q}_s] + (i/\hbar)[(\partial\mathscr{L}/\partial\hat{q}_r), \hat{q}_s]\delta\hat{q}_r \qquad (8.92)$$

where the subscript t has been dropped, since all operators refer to the same time, and the product in the commutator has been expanded. The variation $\delta\hat{q}_r$ is arbitrary and can be taken to be $\varepsilon\hat{q}_r$ and the first of the commutators in the expanded product yields, as a component of the simplest solution to eqn (8.92), the result

$$[\hat{q}_r, \hat{q}_s] = 0, \qquad (8.93a)$$

which is the first of the commutator relations. This leaves

$$\delta\hat{q}_s = (i/\hbar)\,[\partial\hat{\mathcal{L}}/\partial\hat{q}_r, \hat{q}_s]\delta\hat{q}_r. \tag{8.94}$$

Equation (8.94) requires the satisfaction of the second commutator relation

$$[\partial\hat{\mathcal{L}}/\partial\hat{q}_r, \hat{q}_s] = [\hat{p}_r, \hat{q}_s] = -i\hbar\delta_{rs} \tag{8.93b}$$

where $\partial\hat{\mathcal{L}}/\partial\hat{q}_r$ has been identified with the conjugate momentum $\hat{p}_r$. Following Roman (1965), we apply the same displacement of the coordinate to the momentum operator which must yield zero and one has

$$\delta\hat{p}_s = (i/\hbar)\,[\hat{p}_r\delta\hat{q}_r, \hat{p}_s] = 0$$
$$= (i/\hbar)\hat{p}_r[\delta\hat{q}_r, \hat{p}_s] + (i/\hbar)[\hat{p}_r, \hat{p}_s]\delta\hat{q}_r = 0. \tag{8.95}$$

Again taking δq_r equal to $\varepsilon\hat{q}_r$, the first of the commutators in the expanded product yields $-\varepsilon\hat{p}_s$. The resulting operator equation must be valid for any state including one for which $\hat{p}_s|\Psi\rangle = 0$, which, for arbitrary $\delta\hat{q}_r$, yields the last of the Heisenberg commutator relations

$$[\hat{p}_r, \hat{p}_s] = 0. \tag{8.93c}$$

Thus, Schwinger's principle of stationary action yields the equation of motion for the system, Heisenberg's equation of motion for each observable, and the commutation relations. The Heisenberg equations and the commutator relations are obtained through the use of generators of infinitesimal canonical transformations, a fundamental set of operations which are introduced by identifying them with the end-point variations in a generalized variation of the action principle. The generators define the conjugate pairs of variables, momentum $\partial\hat{\mathcal{L}}/\partial\hat{q}_r$ with coordinate $\hat{q}_r$, and the Hamiltonian $\hat{H}$ with time. All physical observables are constructed from the canonically conjugate pairs of variables and the displacement and temporal operators thus generate all possible changes in the dynamical variables of a system and predict the consequences of these changes. All of this is obtained for a classical as well as for a quantum system with the fundamental Poisson bracket relations (eqn (8.65)) replacing the commutation relations. It is clear that Schwinger's variational principle, by providing a common mathematical foundation through its use of infinitesimal canonical transformations, provides a single dynamical principle from which classical and quantum mechanics can be derived.

A form of the principle of stationary action that will prove of great use in later applications is obtained by expressing the principle of stationary action (eqn (8.79)) for an infinitesimal time interval, that is, as a variation of the Lagrange function operator. Equation (8.79) is re-expressed as

$$\delta\hat{\mathcal{W}}_{12} = \int_{t_1}^{t_2} (d\hat{F}/dt)\,dt. \tag{8.96}$$

By dividing both sides of eqn (8.96) by $t_2 - t_1$ and subjecting the result to the limit of $\Delta t \to 0$, one obtains an expression for the principle of stationary action in terms of a variation of the Lagrange function

$$\delta \hat{\mathscr{L}}[t] = \mathrm{d}\hat{F}/\mathrm{d}t. \tag{8.97}$$

By using Heisenberg's eqn (8.91) for the time derivative of an observable, the result can be equivalently expressed as

$$\delta \hat{\mathscr{L}}[t] = (\mathrm{i}/\hbar)[\hat{H}, \hat{F}] + \partial \hat{F}/\partial t \tag{8.98}$$

where the term $\partial \hat{F}/\partial t$ is included for those cases in which $\hat{F}$ possesses an explicit time dependence. *The effect of an infinitesimal canonical transformation on the Lagrange function is equal to the time rate of change of the generator of the transformation.*

The stage is now set for the demonstration that the principle of stationary action can be extended to a particular class of subsystems of some total system, that is, to a particular class of open systems. It will be shown that the variation of the action integral for such a subsystem recovers the primary result of Schwinger's principle—that the change in its action integral is totally determined by the generators of infinitesimal unitary transformations acting at the two time end-points, with no contributions from the intervening times.

8.3   Atomic action principle

8.3.1   Properties of the quantum mechanical Lagrangian

The Schrödinger representation will be used in the development of the subsystem action principle (Bader *et al.* 1978; Srebrenik *et al.* 1978; Bader and Nguyen-Dang 1981). The action integral for a total system, $\mathscr{W}_{12}[\Psi]$, is defined as the time integral of the Lagrangian $\mathscr{L}[\Psi, t]$, which is in turn obtained by integration of the Lagrangian density $L[\Psi, \nabla\Psi, \dot{\Psi}, t]$ over the coordinates of all of the particles in the system

$$\mathscr{W}_{12}[\Psi] = \int_{t_1}^{t_2} \mathscr{L}[\Psi, t]\mathrm{d}t = \int_{t_1}^{t_2} \mathrm{d}t \int \mathrm{d}\tau L[\Psi, \nabla\Psi, \dot{\Psi}, t]. \tag{8.99}$$

The quantum mechanical Lagrangian density for a system of many particles interacting via a many-particle potential energy operator $\hat{V}$ is

$$L[\Psi, \nabla\Psi, \dot{\Psi}, t] = (\mathrm{i}\hbar/2)(\Psi^*\dot{\Psi} - \dot{\Psi}^*\Psi)$$
$$- (\hbar^2/2m)\sum_i \nabla_i\Psi^* \cdot \nabla_i\Psi - \hat{V}\Psi^*\Psi. \tag{8.100}$$

That this is the correct Lagrangian density is demonstrated by showing that the variation of the resulting action integral with respect to Ψ and Ψ^* yields

Schrödinger's equations,

$$i\hbar\dot\Psi = \hat{H}\Psi \quad \text{and} \quad -i\hbar\dot\Psi^* = \hat{H}\Psi^*, \quad \hat{H} = -(\hbar^2/2m)\sum_i\nabla_i^2 + \hat{V}.$$
$$(8.101)$$

There is no loss in generality by considering a Lagrangian density for a single particle in this demonstration. The variation of $\mathcal{W}_{12}[\Psi]$, using the methods developed in eqns (5.66)–(5.69), after integration by parts to rid the expression of terms in $\delta\nabla\Psi$ and $\delta\dot\Psi$, is

$$\delta\mathcal{W}_{12}[\Psi] = \int_{t_1}^{t_2} dt \int d\tau \left\{ \partial L/\partial\Psi - \nabla\cdot(\partial L/\partial\nabla\Psi) - \frac{d}{dt}\frac{\partial L}{\partial\dot\Psi} \right\} \delta\Psi$$

$$+ \int_{t_1}^{t_2} dt \oint dS(\partial L/\partial\nabla\Psi)\cdot\mathbf{n}\delta\Psi$$

$$+ \int d\tau(\partial L/\partial\dot\Psi)\delta\Psi|_{t_1}^{t_2} + \text{cc.} \qquad (8.102)$$

The Lagrangian, as are the functionals considered in the derivation of Schrödinger's stationary-state equation, is a functional of $\nabla\Psi$, and the associated surface term in eqn (8.102) (compare eqn (5.59)) vanishes because of the natural boundary condition (eqn (5.62)) that $\nabla\Psi\cdot\mathbf{n} = 0$ on the boundary at infinity or by demanding that $\delta\Psi$ vanish on the same boundaries. The variations at the time end-points also vanish in this restricted variation which is the analogue of Hamilton's principle or the principle of least action and the resulting Euler–Lagrange equation, obtained by demanding that the variation in $\mathcal{W}_{12}[\Psi]$ vanish for arbitrary $\delta\Psi$, yields Schrödinger's eqn (8.101). This is easily verified by evaluating the partial derivatives of L in eqn (8.102) in those terms multiplied by $\delta\Psi$.

We may retain a formal analogy with the variation of the classical Lagrangian (eqns (8.48) and (8.49)) and with the variation of the quantum Lagrangian integral operator (eqn (8.83)) by defining the functional derivative of $\hat{\mathcal{L}}$ with respect to Ψ (and correspondingly Ψ^*) to be

$$\delta\hat{\mathcal{L}}/\delta\Psi = \{\partial L/\partial\Psi - \nabla\cdot(\partial L/\partial\nabla\Psi)\}. \qquad (8.103)$$

The variation of $\mathcal{L}$ with respect to $\dot\Psi$ is already of the same form for both the classical and quantum cases.

The Lagrangian density, eqn (8.100), exhibits a most important property when the state function Ψ satisfies Schrödinger's eqns (8.101). By using eqns (8.101) to rid eqn (8.100) of the terms in $\dot\Psi$ and $\dot\Psi^*$, the Lagrangian density reduces to the difference between the two kinetic energy expressions as introduced in eqn (5.46) in terms of the identity involving the quantity $\nabla^2(\Psi^*\Psi)$. Denoting by L^0 the particular form assumed by the Lagrangian density when Schrödinger's equation is assumed to hold, as opposed to its general functional dependence on Ψ, $\nabla\Psi$, and $\dot{\Psi}$, one has

$$L^0 = -(\hbar^2/4m)\sum_i\{\Psi^*\nabla_i^2\Psi + \Psi\nabla_i^2\Psi^* + 2\nabla_i\Psi^*\cdot\nabla_i\Psi\}, \quad (8.104)$$

which, using eqn (5.46), can be re-expressed as

$$L^0 = -(\hbar^2/4m)\sum_i\nabla_i^2(\Psi^*\Psi). \quad (8.105)$$

The same identity involving the Laplacian of $\Psi^*\Psi$ was used to transform the integrand in the variation of the functional $\mathscr{G}[\Psi, \Omega]$. Equation (5.83) details the result of integrating one of the terms appearing in the sum found in eqn (8.105); the use of Gauss's theorem transforms each term into a surface integral of the quantity $\Psi^*\nabla_i\Psi\cdot\mathbf{n}$ and its complex conjugate, terms which vanish at the boundary of the system at infinity. Thus, at the point of variation where Schrödinger's equation applies, the Lagrangian integral $\mathscr{L}^0$ vanishes,

$$\mathscr{L}^0[\Psi, t] = -(\hbar^2/4m)\int d\tau_1 \ldots \int d\tau_N \sum_i\nabla_i^2(\Psi^*\Psi) = 0. \quad (8.106)$$

Because the many-particle Lagrangian density L^0 reduces to a sum of single-particle operators, one may define an effective single-particle Lagrangian density $\mathscr{L}^0(\mathbf{r}, t)$ by the usual recipe of summing over all spins and integrating over the coordinates of all the electrons but one, a step which expresses the result in terms of the charge density $\rho(\mathbf{r}, t)$ as

$$\mathscr{L}^0(\mathbf{r}, t) = \int d\tau' L^0 = -(\hbar^2/4mN)\nabla^2\rho(\mathbf{r}, t). \quad (8.107)$$

The vanishing of the Lagrangian integral $\mathscr{L}^0[\Psi, t]$ is seen to be a consequence of the vanishing of the flux in the gradient vector field of the charge density at the infinite boundary of the system

$$\mathscr{L}^0[\Psi, t] = \int d\tau \mathscr{L}^0(\mathbf{r}, t)$$
$$= -(\hbar^2/4mN)\oint dS(\mathbf{r}, t)\nabla\rho(\mathbf{r}, t)\cdot\mathbf{n}(\mathbf{r}, t) = 0. \quad (8.108)$$

Thus, the action integral for a closed quantum system described by Schrödinger's equation vanishes for any time interval Δt because of the zero-flux surface condition that

$$\nabla\rho(\mathbf{r})\cdot\mathbf{n}(\mathbf{r}) = 0 \qquad \text{at every point on the boundary.} \quad (8.109)$$

It is this property that is common to the corresponding Lagrangian and action integrals for a quantum subsystem, the chemical atom.

8.3.2   Atomic action and Lagrangian integrals

The subsystem or atomic Lagrangian integral is defined by the standard mode of integration used in the definition of the subsystem functional $\mathscr{G}[\Psi, \Omega]$ for a stationary state and for the definition of subsystem properties.

That is,

$$\mathscr{L}[\Psi, \Omega, t] = \int_{\Omega} d\mathbf{r} \int d\tau' L(\Psi, \nabla\Psi, \dot{\Psi}, t) \tag{8.110}$$

Correspondingly, the atomic action integral is

$$\mathscr{W}_{12}[\Psi, \Omega] = \int_{t_1}^{t_2} dt \, \mathscr{L}[\Psi, \Omega, t]. \tag{8.111}$$

When Ψ and Ψ^* satisfy Schrödinger's equations, the atomic Lagrangian integral reduces to

$$\mathscr{L}^0[\Psi, \Omega, t] = -(\hbar^2/4mN) \int_{\Omega} d\mathbf{r} \nabla^2 \rho(\mathbf{r}, t)$$

$$= -(\hbar^2/4mN) \oint dS(\mathbf{r}, t) \nabla\rho(\mathbf{r}, t) \cdot \mathbf{n}(\mathbf{r}, t). \tag{8.112}$$

Since the zero-flux boundary condition (eqn (8.109)) is also satisfied by an atom, that is, by a quantum subsystem, the atomic action and Lagrangian integrals vanish as they do for a total closed system. Indeed, one may view the vanishing of the action integral over some total system as being the result of the action integral vanishing separately over the space of each atom in the system.

As a system in a given quantum state changes and evolves with time under the action of the generator $-\hat{H}\delta t$, the atomic surfaces evolve in a continuous manner and the property of exhibiting a zero flux in $\nabla\rho(\mathbf{r})$ is continuously maintained. Thus, the atomic action integral will always vanish

$$\mathscr{W}_{12}^0[\Psi, \Omega] = -(\hbar^2/4mN) \int_{t_1}^{t_2} dt \oint dS \nabla\rho(\mathbf{r}, t) \cdot \mathbf{n}(\mathbf{r}, t) = 0 \tag{8.113}$$

for any time interval, finite or infinitesimal. One may imagine an open system with an arbitrarily defined surface whose Lagrangian integral might vanish at some time t because of the vanishing in the *net* flux of $\nabla\rho(\mathbf{r}, t)$ across its surface (eqn (8.112)). However, in general, as the system changed with time, a net flux in $\nabla\rho(\mathbf{r}, t)$ through the surface would arise and the action integral for some time interval would not vanish as it must for the total system. Thus, the satisfaction of eqn (8.113) for all time intervals, as found for the total system, demands that the open system be defined by a surface that itself evolves with time in a manner determined by a property of the open system, this property being one of a zero flux in $\nabla\rho$ at every point on its surface (eqn (8.109)).

The condition, started in eqn (8.113), that the atomic action integral vanish for all time intervals may be taken as the quantum definition of an atom. It is a direct consequence of the topological definition of an atom as the union of an attractor and its basin and is the basis for the particular variational properties possessed by an atom considered as an open quantum system.

8.3.3   Variation of the atomic action integral

We are now in a position to perform a generalized variation of the action integral for an open system to demonstrate that Schwinger's principle of stationary action can be extended in such a manner as to provide a quantum definition of an atom in a molecule. We shall be considering the change in the atomic action integral $W_{12}[\Psi, \Omega]$ of eqn (8.111) ensuing from variations $\delta\Psi$ in the state function Ψ such that the following conditions are fulfilled.

1. The trial function Φ, where $\Phi = \Psi + \delta\Psi$, and its first and second derivatives vanish whenever an èlectronic position vector is of infinite length.

2. In terms of Φ, a region $\Omega(\Phi, t)$ can be defined that is bounded by a zero-flux surface in $\nabla\rho_\Phi$ where

$$\nabla\rho_\Phi(\mathbf{r}, t) \cdot \mathbf{n}(\mathbf{r}, t) = 0 \qquad \text{for all points on the surface of } \Omega \qquad (8.114)$$

where the trial density is defined as

$$\rho_\Phi(\mathbf{r}, t) = N\int d\tau' \Phi^*(\mathbf{r}, \tau', t)\Phi(\mathbf{r}, \tau', t). \qquad (8.115)$$

Moreover, it is required that, as Φ tends to Ψ at any time t, $\Omega(\Phi, t)$ is continuously deformable into the region $\Omega(t) = \Omega(\Psi, t)$ associated with the atom. The region $\Omega(\Phi, t)$ thus represents the atom in the varied total system, which is described by the trial function $\Phi(\mathbf{r}, \tau', t)$ just as $\Omega(t)$ represents the atom when the total system is in the state described by $\Psi(\mathbf{r}, \tau', t)$.

Requiring the fulfilment of condition (2) amounts to imposing the variational constraint that the divergence of $\nabla\rho_\Phi$ integrate to zero at all stages of the variation, i.e. that

$$I_\Phi = \int_{\Omega(\Phi, t)} \nabla^2\rho_\Phi d\mathbf{r} = 0 \qquad (8.116)$$

for all admissible Φ and for all t, which implies

$$\delta I_\Psi = \delta\left\{ \int_{\Omega(t)} \nabla^2\rho(\mathbf{r}, t)d\mathbf{r} \right\} = 0. \qquad (8.117)$$

To impose the variational constraint given in eqn (8.117) and thereby define a particular class of open system, one must vary the surface of the subsystem. This requirement necessarily leads to the relaxation of the usual variational constraint that the $\delta\Psi$ vanish at all boundaries of the system and at the time end-points as is done in the restricted variation of the action integral in Hamilton's principle to obtain the equations of motion. Such a variation was performed above in Section 8.3.1 in the restricted variation of the quantum mechanical action integral to demonstrate that the quantum mechanical Lagrangian density defined in eqn (8.100) does indeed lead to Schrödinger's equation of motion. The reader is aware that it is the generalization of the variation of the action integral obtained by the removal of the constraints that the variations in Ψ vanish at the time end-points and by the additional variation of the time end-points themselves that leads to Schwinger's principle of stationary action. Thus, the variation of the atomic action integral with the necessary retention of the $\delta\Psi$ on the boundaries and, hence, at the time end-points perforce leads to the same generalization of the variation principle. What is now demonstrated is that, if the subsystem action integral is restricted to regions satisfying the above variational constraint, this generalized variation also leads to the same physical principle of stationary action.

The techniques of the calculus of variations, now familiar to the reader, are applied to the expression for $\mathscr{W}_{12}[\Psi, \Omega]$ (eqn (8.111)) with the Lagrangian density as defined in eqn (8.100). Included is a variation of the time end-points as detailed in eqns (8.81) and (8.82), and a variation of the surface of the subsystem in the manner previously detailed in eqn (5.74). Variations in Ψ and Ψ^* at the time end-points are also allowed in this generalized variation of the action integral. Following these steps, one obtains the following explicit results for variations in Ψ and the complex conjugates of these (cc) for variations in Ψ^*

$$\delta\mathscr{W}_{12}[\Psi, \Omega] = \int_{t_1}^{t_2} dt \int_{\Omega} d\mathbf{r} \int d\tau' \{(-i\hbar\dot{\Psi}^* - \hat{H}\Psi^*)\delta\Psi + cc\}$$

$$+ \int_{t_1}^{t_2} dt \oint dS(\Omega, \mathbf{r}) \int d\tau'(-\hbar^2/2m)\{\nabla\Psi^*\delta\Psi \cdot \mathbf{n}(\mathbf{r}) + cc\}$$

$$+ \int_{t_1}^{t_2} dt \oint dS(\Omega, \mathbf{r})\mathscr{L}(\mathbf{r})\delta_\psi S(\Omega, \mathbf{r}) + cc$$

$$+ \int_{t_1}^{t_2} dt \oint dS(\Omega, \mathbf{r}) \int d\tau' \{(-i\hbar/2)(\partial S(\Omega, \mathbf{r})/\partial t)\Psi^*\delta\Psi + cc\}$$

$$+ \left\{(i\hbar/2)\int_{\Omega} d\mathbf{r} \int d\tau' \Psi^*\delta\Psi + cc + \mathscr{L}[\Psi, \Omega, t]\delta t\right\}\Big|_{t_1}^{|t_2}$$

$$(8.118)$$

As in previous chapters, the position vector of the electron whose coordinates are integrated over Ω is denoted by $\mathbf{r}$. The first surface term in eqn (8.118) arises from the integration by parts used to rid the expression of variations in

$V_i\Psi$, terms of the form

$$(\partial L/\partial V_i\Psi)\delta V_i\Psi = -\frac{\hbar^2}{2m}V_i\Psi^*\delta V_i\Psi. \tag{8.119}$$

This step was previously detailed in the variation of $\mathscr{G}[\psi,\Omega]$ in eqn (5.76) and the identity preceding it. As described there, all the resulting surface terms of $V_i\Psi^*\cdot\mathbf{n}$ vanish save the one arising from the gradient operator for the electron whose coordinates are integrated over the (finite) surface coordinates of the region Ω, as only for this electron is the position vector in the surface not of infinite length. The integration by parts used to rid the expression of the term involving $\delta\dot\Psi$

$$(\partial L/\partial\dot\Psi)\delta\dot\Psi = (i\hbar/2)\Psi^*\delta\dot\Psi \tag{8.120}$$

requires special comment. This step is accomplished using the identity

$$\int_{t_1}^{t_2}dt\,\frac{d}{dt}\left\{\int_\Omega d\mathbf{r}\int d\tau'(i\hbar/2)\Psi^*\delta\Psi\right\}$$

$$= (i\hbar/2)\int_{t_1}^{t_2}dt\left\{\int_\Omega d\mathbf{r}\int d\tau'(\dot\Psi^*\delta\Psi + \Psi^*\delta\dot\Psi)\right.$$

$$\left. + \oint dS(\Omega,\mathbf{r})\int d\tau'(\partial S(\Omega,\mathbf{r})/\partial t)\Psi^*\delta\Psi\right\}. \tag{8.121}$$

The time derivative of the surface appearing in this expression denotes the time rate of change in the surface bounding the unvaried region $\Omega(\Psi,t)$. That is, the time derivative of an integral with a time-dependent boundary must include a contribution arising from the change in the boundary with time. It is the use of the identity in eqn (8.121) that leads to the inclusion of the terms involving $\Psi^*\delta\Psi$ at the two time end-points. Since the time end-points are themselves varied, one also obtains the term $\mathscr{L}[\Psi,\Omega,t]\delta t$ at the time end-points, as noted previously in the generalized variation of the action integral operator.

8.3.4   Principle of stationary action in the Schrödinger representation

Before pursuing the variation of the atomic action integral, it is helpful to first recover the statement of the principle of stationary action in the Schrödinger representation for the total system. If one sets the boundary of the region Ω at infinity in eqn (8.118) to obtain the variation of the total system action integral $\mathscr{W}_{12}[\Psi]$, and restricts the variation so that $\delta\Psi$ vanishes at the time end-points and the end-points themselves are not varied, then only the terms multiplied by the variations $\delta\Psi$ in the first integral on the right-hand side remain. The Euler equation obtained by the requirement that this restricted variation of the action integral vanish is Schrödinger's eqn (8.101). If the end-point variations are restored for the variation of $\mathscr{W}_{12}[\Psi]$, one obtains, in addition to the equation of motion, Schwinger's principle of stationary action stated in the Schrödinger representation

$$\delta\mathscr{W}_{12}[\Psi] = \{(i\hbar/2)\int d\tau\Psi^*\delta\Psi + cc + \mathscr{L}[\Psi, t]\delta t\}|_{t_1}^{t_2}. \tag{8.122}$$

Equation (8.122) states that the action integral responds only to changes induced in the system at the time end-points. In analogy with the classical result given in eqn (8.52) which enables a change from the coordinates q, $\dot{q}$, t to q, p, t as employed in eqn (8.87) in the formulation of the end-point operators in the variation of the action integral operator, one may express the Schrödinger-based average Hamiltonian $\mathscr{H}[\Psi]$ as the Legendre transform of $\mathscr{L}[\Psi, t]$,

$$\mathscr{H}[\Psi] = \tfrac{1}{2}\{\int d\tau(\Psi^*\hat{H}\Psi + (H\Psi^*)\Psi)\} = (i\hbar/2)\int d\tau(\Psi^*\dot{\Psi} - \dot{\Psi}^*\Psi)$$
$$- \mathscr{L}[\Psi, t] - (\hbar^2/4m)\int d\mathbf{r}\nabla^2\rho(\mathbf{r}). \tag{8.123}$$

This transformation enables one to rewrite eqn (8.122) in the form

$$\delta\mathscr{W}_{12}[\Psi] = \{(i\hbar/2)\int d\tau\Psi^*\Delta\Psi + cc - \mathscr{H}[\Psi]\delta t\}|_{t_1}^{t_2}$$
$$= F(t_2) - F(t_1) \tag{8.124}$$

where the symbol $\Delta\Psi$ denotes the general variation

$$\Delta\Psi = \delta\Psi + \dot{\Psi}\delta t.$$

The principle of stationary action identifies the end-point variations with the generators of temporal and spatial unitary transformations. In the Schrödinger representation one obtains the average value of the generator $\hat{F}(t)$, as denoted by the symbol $F(t)$. The expression given here for $F(t)$ can be compared with that obtained by the variation of the action integral operator in eqn (8.87). For a temporal change, $\Delta\Psi = 0$, δt is arbitrary and the change in $\mathscr{W}_{12}[\Psi]$ reduces to the difference in the average values of $-\hat{H}\delta t$ at the two time end-points, thereby identifying the generator of temporal development with the operator $-\hat{H}\delta t$. Similarly, for a purely spatial change, $\delta t = 0$ and

$$\delta\mathscr{W}_{12}[\Psi] = \tfrac{1}{2}\{\int d\tau[\Psi^*\hat{F}\Psi + (\hat{F}\Psi)^*\Psi]\}|_{t_1}^{t_2} \tag{8.125}$$

where in this situation $\Delta\Psi = \delta\Psi = (-i/\hbar)\hat{F}\Psi$ and $\hat{F}$ is an infinitesimal Hermitian operator. It may be a function of the coordinates (inducing a gauge transformation) and/or of the momenta (inducing a translation of coordinates). If both sides of eqn (8.125) are divided by $t_2 - t_1$ and the result is subjected to the limit $\Delta t \to 0$, one obtains, to first-order, an expression for $\delta\mathscr{L}[\Psi, t]$,

$$\delta\mathscr{L}[\Psi, t] = \tfrac{1}{2}\{d\langle\hat{F}\rangle/dt + cc\}, \tag{8.126}$$

which, through Heisenberg's relation for the time rate of change of the average value of an observable, may be re-expressed as

$$\delta\mathscr{L}[\Psi, t] = (\tfrac{1}{2})\{(i/\hbar)\langle[\hat{H}, \hat{F}]\rangle + \langle\partial\hat{F}/\partial t\rangle + \text{cc}\}. \tag{8.127}$$

Equations (8.126) and (8.127) are identical in form to the corresponding results obtained for the variation of the Lagrange function operator in eqns (8.97) and (8.98). They are variational statements of the Heisenberg equation of motion for the observable $\hat{F}$ in the Schrödinger representation. When Ψ describes a stationary state

$$i\hbar\,\partial\Psi/\partial t = \hat{H}\Psi = E\Psi(\tau, t) = \exp(iEt/\hbar)E\psi(\tau), \tag{8.128}$$

the Lagrangian integral reduces to the form

$$\mathscr{L}[\Psi, t] = -\langle\Psi, \Psi\rangle(\mathscr{E}[\psi] - E) \tag{8.129}$$

where $\mathscr{E}[\psi]$ is the energy functional as defined in eqn (5.63), but used here for the many-electron case. It differs from the functional $\mathscr{G}[\psi]$ in eqn (5.58) only in that the normalization constraint is handled by the factor $\langle\psi, \psi\rangle^{-1}$ rather than by an undetermined multiplier. The variations in this case will be of the form $\delta\Psi(\tau, t) = \exp(iEt/\hbar)\delta\psi(\tau)$ and it follows that

$$\delta\mathscr{E}[\psi] = -\langle\psi, \psi\rangle^{-1}\delta\mathscr{L}[\Psi, t]. \tag{8.130}$$

Thus, for a stationary state with a normalized state function ψ, eqn (8.127) assumes the form

$$\delta\mathscr{E}[\psi] = -\tfrac{1}{2}\{(i/\hbar)\langle[\hat{H}, \hat{F}]\rangle + \text{cc}\} = 0, \tag{8.131}$$

which may be equivalently stated in terms of the functional $\mathscr{G}[\psi]$ as

$$\delta\mathscr{G}[\psi] = -\tfrac{1}{2}\{(i/\hbar)\langle[\hat{H}, \hat{F}]\rangle + \text{cc}\} = 0 \tag{8.132}$$

as previously given in eqn (5.101). The derivation of these latter two equations from the principle of stationary action substantiates the claim made at the conclusion of Chapter 5 that eqn (5.101), and hence eqn (8.131), are statements of this principle for a stationary state. As in the time-dependent case, the principle yields the equation of motion for the state function in addition to the variational statement of the equation of motion for an observable $\hat{F}$, which, for a stationary state, is the hypervirial theorem.

8.3.5   Atomic statement of the principle of stationary action

We return to eqn (8.118) for the case where Ω refers to a region of space bounded by a surface of zero flux in the gradient vector field of the charge density. In addition to retaining the terms at the time end-points which arise from the generalization of the variation and which are present for a total system as well, we must consider the terms arising from the non-vanishing of

$\delta\Psi$ on the surface of the subsystem, the change of this surface with time, and, finally, the term arising from the variation of the surface of the subsystem. The symbol $\delta S(\Omega, \mathbf{r})$ in this latter term denotes the infinitesimal normal shift of a surface element centred on $\mathbf{r}$, as conditioned by the continuous variation in the admissible trial function $\Phi = \Psi + \delta\Psi$, in a small neighbourhood of the state function Ψ. This continued development of eqn (8.118) is taken when the variation is complete. That is, we are at the point of variation, Schrödinger's equation applies, and the expression for $\delta W_{12}[\Psi, \Omega]$ is consequently reduced to the surface and time end-point terms. It was shown in eqn (8.108) that the Lagrangian integral vanishes when Schrödinger's equation is satisfied. We have already demonstrated in a formal sense that the presence of this term does indeed define $-\hat{H}\delta t$ as the generator of a temporal transformation. Operationally, this role is taken over by eqn (8.127) in the Schrödinger representation and there is no loss in generality to simply set the Lagrangian integral at the two time end-points equal to zero for the remainder of the discussion.

Another consequence of the applicability of Schrödinger's equation to the expression for $\delta\mathscr{W}_{12}[\Psi, \Omega]$ at the point of variation is that the Lagrangian density appearing in the term arising from the variation of the surface is, as shown in eqn (8.107), proportional to $\nabla^2\rho(\mathbf{r})$, the Laplacian of the charge density. Thus, at the point of variation, the expression for the variation of the atomic action integral is

$$
\begin{aligned}
\delta\mathscr{W}_{12}[\Psi, \Omega] = \int_{t_1}^{t_2} dt &\left\{ \oint dS(\Omega, \mathbf{r}) \int d\tau' [(-\hbar^2/2m)(\nabla\Psi^*\delta\Psi \cdot \mathbf{n}(\mathbf{r}) + cc) \right. \\
&- (i\hbar/2)(\partial S(\Omega, \mathbf{r})/\partial t)\Psi^*\delta\Psi + cc] \\
&\left. + \oint dS(\Omega, \mathbf{r})(-\hbar^2/4mN)\nabla^2\rho(\mathbf{r})\delta_\psi S(\Omega, \mathbf{r}) + cc \right\} \\
&+ \left\{ (i\hbar/2) \int_\Omega d\mathbf{r} \int d\tau' \Psi^*\delta\Psi + cc \right\}\Big|_{t_1}^{t_2}.
\end{aligned}
\tag{8.133}
$$

It is through this property of the Lagrangian density at the point of variation, as it appears in the surface variation, that the consequences of the constraint given in eqns (8.116) and (8.117) are introduced into the variational expression. The same procedure was followed in the derivation of the principle for a stationary state in which case the integrand of the functional $\mathscr{G}[\psi, \Omega]$ was also found to be proportional to the Laplacian of the charge density at the point of variation. Here, as there, the imposition of the variational constraint is accomplished using the identity given in eqn (8.134). This result is obtained when one carries out the variation of the Laplacian integral for the region Ω, including a variation of its surface, in the manner prescribed in eqn (5.74)

$$\delta_\psi \left\{ (\hbar^2/4mN) \int_\Omega dr \nabla^2 \rho(\mathbf{r}) \right\} = (\hbar^2/4mN) \int_\Omega dr \delta_\psi \{ \nabla^2 \rho(\mathbf{r}) \}$$

$$+ (\hbar^2/4mN) \oint dS(\Omega, \mathbf{r}) \nabla^2 \rho(\mathbf{r}) \delta_\psi S(\Omega, \mathbf{r}). \tag{8.134}$$

The variation of the integrand indicated in eqn (8.134) was considered previously and the result is given in eqn (5.91). As emphasized there, the Laplacian is a divergence expression—its variation yields only a surface term; thus the imposition of the constraint does not affect the Euler–Lagrange equations that have been obtained at this point. According to the constraint, eqn (8.117), the sum of variations of the integrand and surface must equal zero and, thus, the result of imposing the constraint is to replace the surface variation with the term arising from the variation of $\nabla^2 \rho(\mathbf{r})$, i.e.

$$- (\hbar^2/4mN) \oint dS(\Omega, \mathbf{r}) \nabla^2 \rho(\mathbf{r}) \delta_\psi S(\Omega, \mathbf{r}) = (\hbar^2/4m) \oint dS(\Omega, \mathbf{r}) \int d\tau' \{ (\nabla \Psi^*) \delta \Psi$$

$$+ \Psi^* \delta \nabla \Psi \} \cdot \mathbf{n}(\mathbf{r}) + cc. \tag{8.135}$$

Substitution of this result into eqn (8.133) yields

$$\delta \mathscr{W}_{12}[\Psi, \Omega] = - (\hbar^2/4m) \left\{ \int_{t_1}^{t_2} dt \oint dS(\Omega, \mathbf{r}) \int d\tau' [(\nabla \Psi^*) \delta \Psi \right.$$

$$\left. - \Psi^* \nabla (\delta \Psi)] \cdot \mathbf{n}(\mathbf{r}) + cc \right\}$$

$$+ \left\{ - (i\hbar/2) \int_{t_1}^{t_2} dt \oint dS(\Omega, \mathbf{r}) \int d\tau' (\partial S/\partial t) \Psi^* \delta \Psi + cc \right\}$$

$$+ \left\{ (i\hbar/2) \int_\Omega dr \int d\tau' \Psi^* \delta \Psi + cc \right\} \Big|_{t_1}^{t_2}. \tag{8.136}$$

One recognizes the first term in eqn (8.136) as the variation in the quantum mechanical current density (eqns (5.94) and (5.95)). It is obtained by combining the surface term arising from the variation with respect to $\nabla \Psi$ with the surface term arising from the imposition of the variational constraint, eqn (8.135). *Thus the variation of the surface of the subsystem together with the restriction that the subsystem be an atom bounded by a zero-flux surface causes the quantum mechanical current density to appear in the variation of the action integral, a term whose presence is a necessary requirement for the proper description of the properties of an open system.* It is now demonstrated that eqn (8.136) is the atomic equivalent of the principle of stationary action.

The variation of Ψ is identified with the action of a generator of an infinitesimal unitary transformations on the corresponding state function, a fundamental identification in the development of the principle of stationary action. Following the same procedure that was used to obtain the principle of stationary action for the total system in the Schrödinger representation (eqn (8.125)), we set $\delta\Psi = (-i/\hbar)\hat{F}\Psi$ where $\hat{F}$ is an infinitesimal Hermitian operator. The operator $\hat{F}$ refers to the electron whose coordinates are integrated over the region Ω. This substitution yields

$$\Delta\mathcal{W}_{12}[\Psi, \Omega] = F(\Omega, t_2) - F(\Omega, t_1)$$

$$- \int_{t_1}^{t_2} dt \oint dS(\Omega, \mathbf{r})\{(\partial S/\partial t)\rho_F(\mathbf{r}, t) - \tfrac{1}{2}(\mathbf{J}_F \cdot \mathbf{n} + cc)\}.$$

$$(8.137)$$

All of the quantities appearing in eqn (8.137) have been previously defined: the term $F(\Omega, t)$ arising from the end-point variations is the atomic average of the generator $\hat{F}$ (defined in eqn (8.138); compare eqn (6.34) for a stationary state),

$$F(\Omega, t) = \int_\Omega d\mathbf{r}\rho_F(\mathbf{r}, t) = (N/2)\int_\Omega d\mathbf{r} \int d\tau'\{\Psi^*\hat{F}\Psi + (\hat{F}\Psi)^*\Psi\}. \qquad (8.138)$$

$\rho_F(\mathbf{r}, t)$ is the corresponding property density (compare eqn (6.35)),

$$\rho_F(\mathbf{r}, t) = (N/2)\int d\tau'\{\Psi^*\hat{F}\Psi + (\hat{F}\Psi)^*\Psi\}, \qquad (8.139)$$

and $\mathbf{J}_F(\mathbf{r}, t)$ the associated vector current density,

$$\mathbf{J}_F(\mathbf{r}, t) = (N\hbar/2mi)\int d\tau'\{\Psi^*\mathbf{V}(\hat{F}\Psi) - (\mathbf{V}\Psi^*)\hat{F}\Psi\}, \qquad (8.140)$$

which is N times the current density previously defined in eqn (5.98). The variation $\Delta\mathcal{W}_{12}[\Psi, \Omega] = N\delta\mathcal{W}_{12}[\Psi, \Omega]$ because the atomic averages are defined for a density which integrates to N.

It was emphasized in Chapter 6 that the definition of an atomic stationary state property is determined by the form of the atomic stationary state functional $\mathcal{G}[\psi, \Omega]$. In precisely the same manner, the definition of an atomic property in the general time-dependent case is determined by the form of the atomic Lagrangian integral $\mathcal{L}[\Psi, \Omega, t]$. In both the stationary-state and time-dependent cases, the modes of integration used in the definition of the atomic functionals that are required to give atomic statements of the principle of stationary action are those which yield atomic averages of an associated effective single-particle density. The property density $\rho_F(\mathbf{r})$ is obtained in the same manner as is the charge density itself, by summing over all spins and averaging over the coordinates of all the electrons but one as denoted by the symbol $\int d\tau'$. The quantity $\rho_F(\mathbf{r})d\mathbf{r}$, the contribution to the average value of $\hat{F}$ from the volume element $d\mathbf{r}$, is thus N times the value of the property F for a single electron as determined by its interactions with the averaged motion of all remaining particles in the system.

The variation of the atomic action integral differs from the result obtained for the total system (eqn (8.125)) in that, in addition to the difference in the average values of the generators at the two time end-points, a time integral of surface contributions arising from the flux in the current density and from the change in the surface with time appears in the atomic expression. We now demonstrate that both equations are equivalent expressions of the principle of stationary action—that the change in action is determined by the contributions of the generators acting at the two time end-points. To see this result in the equation for the variation of the atomic action integral, consider the expression for the time rate of change of an atomic expectation value which is easily obtained by using Schrödinger's time-dependent eqns (8.101) in the expression for the time derivative of an atomic average (see eqn (5.44)). This yields

$$dF(\Omega, t)/dt = (1/2)\{(i/\hbar)\langle[\hat{H}, \hat{F}]\rangle_\Omega + \text{cc}\}$$

$$+ \oint dS(\Omega, \mathbf{r})\{(\partial S(\Omega, \mathbf{r})/\partial t)\rho_F(\mathbf{r}, t)$$

$$- \tfrac{1}{2}(\mathbf{J}_F \cdot \mathbf{n}(\mathbf{r}) + \text{cc})\}. \tag{8.141}$$

The first term on the right-hand side of eqn (8.141) represents, by Heisenberg's equation of motion, the total change in the expectation value of an observable for a system with infinite boundaries. It describes the contribution to the time rate of change of an observable arising from the interior of the system. An open system, a system with finite boundaries, has further contributions corresponding to a change in its surface with time and to a net flux in the property current density across the surface. These two terms appear with different signs in eqn (8.141) because, when $\partial S/\partial t > 0$ corresponding to an expansion of the volume Ω, the average value of $\hat{F}$ is increased, while, when $\mathbf{J}_F \cdot \mathbf{n} > 0$ corresponding to a net outflow of density ρ_F from Ω, the result is a decrease in the average value of $\hat{F}$.

Now it is the essence of the principle of stationary action that the total change in action is equal to the difference in the values of the generator $\hat{F}(t)$ evaluated at the two time end-points (eqn (8.79)). In the Schrödinger representation this corresponds to $\delta\mathcal{W}_{12}[\Psi]$ equalling the difference in the average values of the generator at the time end-points (eqn (8.125)). According to eqn (8.141) the difference in the average values of the generator at two times for an open system is

$$F(\Omega, t_2) - F(\Omega, t_1) = \int_{t_1}^{t_2} dt(dF(\Omega, t)/dt)$$

$$= \int_{t_1}^{t_2} dt\left[\tfrac{1}{2}\{(i/\hbar)\langle[\hat{H}, \hat{F}]\rangle_\Omega + \text{cc}\}\right.$$

$$\left. + \oint dS\{(\partial S/\partial t)\rho_F - \tfrac{1}{2}(\mathbf{J}_F \cdot \mathbf{n} + \text{cc})\}\right]. \tag{8.142}$$

From this equation it is clear that the difference in the average values of a property for an open system between two times implicitly includes, in addition to the difference in values determined within the boundaries of the system as described by the commutator average, an integrated contribution over the time-like surface, arising from the flux in the property across the surface at each time t and from the change in the surface with time. This situation is illustrated in Fig. 8.2 for a two-dimensional open system. Thus, to obtain a statement of the principle of stationary action for an open system equivalent to that for the total system, one must subtract from the difference in the open-system average values the integral over the time-like surface of the boundary contributions to this difference. This is precisely the expression given for the change in the atomic action integral in eqn (8.137). Comparison of this expression for $\Delta \mathscr{W}_{12}[\Psi]$ with eqn (8.142) shows that the terms subtracted from the difference in the end-point average values of the generator are just those which account for the surface contributions to this difference integrated over the time-like surface connecting the two time end-points. Thus what remains is the difference in the values of the generator at

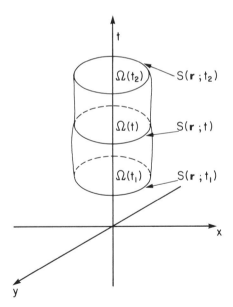

FIG. 8.2. Pictorial representation of the space–time development of a two-dimensional sub-system $\Omega(t)$ of a total system. The space-like surfaces are denoted by $\Omega(t)$, each being bounded by a surface $S(\mathbf{r}, t)$. The collection of the latter constitute a time-like surface, which, together with $\Omega(t_1)$, and $\Omega(t_2)$, define the space–time integration volume in $\mathscr{W}_{12}[\Psi, \Omega]$. The principle of stationary action states that the total change in action is equal to the difference in the values of the generators $\hat{F}(t)$ evaluated in the two *space-like* surfaces at t_1 and t_2. There is no contribution to $\delta \mathscr{W}_{12}$ from the interior of the space–time volume swept out as a consequence of the system evolving in time or from the time-like surface.

the two time end-points averaged over the interior of the subsystem, the essence of the principle of stationary action. The change in the atomic action integral can also be expressed as

$$\Delta \mathscr{W}_{12}[\Psi, \Omega] = \int_{t_1}^{t_2} dt\{(i/\hbar)\langle[\hat{H}, \hat{F}]\rangle_\Omega + cc\}/2, \qquad (8.143)$$

an expression which is equivalent in form and content to the corresponding expression for the total system

$$\Delta \mathscr{W}_{12}[\Psi] = \int_{t_1}^{t_2} dt\{(i/\hbar)\langle[\hat{H}, \hat{F}]\rangle + cc\}/2. \qquad (8.144)$$

Equation (8.144) is an alternative form of the expression given in eqn (8.125) for the total system. The principle of stationary action for a subsystem can be expressed for an infinitesimal time interval in terms of a variation of the Lagrangian integral, similar to that given in eqn (8.127) for the total system. For the atomic Lagrangian, assuming $\hat{F}$ to have no explicit time dependence, this statement is

$$\delta \mathscr{L}[\Psi, \Omega, t] = (1/2)\{(i/\hbar)\langle[\hat{H}, \hat{F}]\rangle_\Omega + cc\}. \qquad (8.145)$$

This result may be restated in the form appropriate for a stationary state by using the relationship

$$\mathscr{L}[\Psi, \Omega, t] = -\langle\Psi, \Psi\rangle_\Omega(\mathscr{E}[\psi, \Omega] - E), \qquad (8.146)$$

corresponding to the total system expression given in eqn (8.129). This yields the statements

$$\delta\mathscr{E}[\psi, \Omega] = -\tfrac{1}{2}\{(i/\hbar)\langle[\hat{H}, \hat{F}]\rangle_\Omega + cc\}/\langle\psi, \psi\rangle_\Omega \qquad (8.147)$$

and

$$\delta\mathscr{G}[\psi, \Omega] = -\tfrac{1}{2}\{(i/\hbar)\langle[\hat{H}, \hat{F}]\rangle_\Omega + cc\} \qquad (8.148)$$

as the analogues of the statements for the total system given in eqns (8.131) and (8.132). Equation (8.148) is, of course, identical to that obtained directly from the variation of Schrödinger's functional for a stationary state as described in Chapter 5 and stated in eqn (5.100).

The derivation of the principle of stationary action for an atom in a molecule (eqn (8.143)) yields Schrödinger's equation of motion for the total system, identifies the observables of quantum mechanics with the variations of the state function, defines their average values, and gives their equations of motion. We have demonstrated in Chapter 6 how one can use the atomic statement of the principle of stationary action given in eqn (8.148) to derive the theorems of subsystem quantum mechanics and thereby obtain the mechanics of an atom in a molecule. The statement of the atomic action principle given in eqn (8.145) plays the same role in the general time-dependent case, as will be illustrated in the following section. Equations (8.143) and (8.145) represent an extension of quantum mechanics. They enable one to obtain a quantum mechanical description of the properties of any region of space bounded by a surface of zero flux in the gradient vector field of the charge density. In this sense, the mechanics of a total system is obtained as a special case of these more general equations.

8.3.6   Examples of unitary transformations

The variation of the functional $\mathscr{G}[\psi, \Omega]$ for a stationary state can be equated to an infinitesimal times the flux in the current density of the generator through the surface of the atom (eqn (5.99)). The corresponding statement for the variation of the atomic Lagrangian integral, obtained by using eqns (8.141) and (8.145), is

$$\delta\mathscr{L}[\Psi, \Omega, t] = (1/2N)\{dF(\Omega)/dt - \oint dS[(\partial S/\partial t)\rho_F$$
$$- \tfrac{1}{2}(\mathbf{J}_F \cdot \mathbf{n} + cc)]\}. \tag{8.149}$$

One may use this expression to evaluate $\delta\mathscr{L}[\Psi, \Omega, t]$ for any particular generator $\hat{F}$, corresponding to the use of eqn (5.99) for the evaluation of $\delta\mathscr{G}[\psi, \Omega]$ as was done in Chapter 6. However, we shall take the opportunity of evaluating $\delta\mathscr{L}$ for various generators to illustrate the properties of unitary transformations.

The action of the generator of a unitary transformation, as represented by the Hermitian operator $\hat{F}$, can be determined through the use of the operator $\hat{U}$ defined as

$$\hat{U} = e^{-(i/\hbar)\hat{F}} \qquad \text{and} \qquad \hat{U}^\dagger = e^{+(i/\hbar)\hat{F}} \tag{8.150}$$

The same expression applies if $\hat{F}$ is an infinitesimal Hermitian operator equal to $\varepsilon\hat{G}$, for, as $\varepsilon \to 0$, the series expansion for an exponential yields, to first-order in ε,

$$\hat{U} = \hat{I} - (i/\hbar)\hat{F} = \hat{I} - (i/\hbar)\varepsilon\hat{G}. \tag{8.151}$$

As demonstrated in Chapter 6 for a stationary state, the atomic force law is obtained by setting the generator equal to the momentum conjugate to the coordinate $\mathbf{r}$, the electronic coordinate that is integrated over the basin of the atom Ω. It follows immediately from eqn (8.35), which shows the effect of an infinitesimal unitary transformation on an operator, that the generator $\hat{F} = \varepsilon \cdot \mathbf{p}$ induces an infinitesimal uniform translation of the electronic coordinate $\mathbf{r}$ by $-\varepsilon$. Each component of the vector ε is an arbitrary, infinitesimal real number, fixed for all $\mathbf{r}$ and, from eqn (8.35),

$$\mathbf{r}' = \hat{U}\mathbf{r}\hat{U}^\dagger = \mathbf{r} - (i\varepsilon/\hbar)\cdot[\mathbf{p}, \mathbf{r}] = \mathbf{r} - \varepsilon. \tag{8.152}$$

It is easily shown that the same result is obtained for finite ε through the use of the generator $\mathbf{p}$ in a unitary transformation. This is most easily done by parametrizing $\hat{U}$ as

$$\hat{U}(\zeta) = e^{-(i/\hbar)\zeta\varepsilon\cdot\mathbf{p}}. \tag{8.153}$$

For $\zeta = 1$, $\hat{U}(1) = \hat{U}$ and, for $\zeta = 0$, $\hat{U}(0) = 1$. Differentiation of the transformed coordinate denoted by $\mathbf{r}(\zeta)$ by the parameter ζ yields

$$\partial_\zeta \mathbf{r}(\zeta) = \partial_\zeta(\hat{U}(\zeta)\mathbf{r}\hat{U}^\dagger(\zeta)) = -(i/\hbar)\hat{U}(\zeta)[\varepsilon\cdot\mathbf{p}, \mathbf{r}]\hat{U}^\dagger(\zeta) = -\varepsilon. \tag{8.154}$$

The integration of this result over ζ between the limits 0 and 1 yields $\mathbf{r}(1) = \mathbf{r}(0) - \varepsilon$. It is clear that the generator $\mathbf{p}$ has no effect on a momentum coordinate. In parallel with the results for $\hat{F} = \varepsilon\cdot\mathbf{p}$, a uniform translation in momentum space is instead obtained through the use of the generator $\hat{F} = \varepsilon\cdot\mathbf{r}$ in a unitary transformation.

A perhaps more interesting example of a unitary transformation is provided by the 'virial operator' $\mathbf{r}\cdot\mathbf{p}$. The Hermitian form of this operator denoted by $\hat{\mathscr{V}}$, where $\hat{\mathscr{V}}$ is given by the equivalent expressions

$$\hat{\mathscr{V}} = \tfrac{1}{2}(\mathbf{r}\cdot\mathbf{p} + \mathbf{p}\cdot\mathbf{r}) = \mathbf{r}\cdot\mathbf{p} - (3/2)i\hbar \tag{8.155}$$

when multiplied by the real number ε and used as the generator of a unitary transformation, causes the electronic position coordinate to be multiplied by the factor $\zeta = e^\varepsilon$. Hence, the virial operator is called the scaling operator, since it generates a scaling of the electronic coordinates. This accounts for the well-known result that the variational property of the energy and the satisfaction of the virial theorem are related by the operation of scaling the electronic coordinates of the wave function. The use of the single product $\mathbf{r}\cdot\mathbf{p}$ rather than the Hermitian average shown in eqn (8.155), as was done in the derivation of the atomic virial theorem in Chapter 6 by determining the atomic average of its commutator with $\hat{H}$, is permissible, since the difference between these two forms is simply the constant term $3i/2$.

To obtain a scaling in a positive sense we shall take $\hat{F} = -\varepsilon\hat{\mathscr{V}}$ to define the unitary operator

$$\hat{U}(\varepsilon) = e^{(i/\hbar)\varepsilon\hat{\mathscr{V}}} \tag{8.156}$$

Proceeding as in eqn (8.154) by taking the derivative of the transformed coordinate $\mathbf{r}$ with respect to the parameter, one has

$$\partial_\varepsilon \mathbf{r}(\varepsilon) = \partial_\varepsilon\{\hat{U}(\varepsilon)\mathbf{r}\hat{U}^\dagger(\varepsilon)\} = (i/\hbar)\hat{U}(\varepsilon)[\hat{\mathscr{V}}, \mathbf{r}]\hat{U}^\dagger(\varepsilon) = \mathbf{r}(\varepsilon). \tag{8.157}$$

The integrated form of this result between the limits 0 and ε is

$$\mathbf{r}(\varepsilon) = \mathbf{r}e^\varepsilon = \mathbf{r}\zeta, \tag{8.158}$$

that is, the coordinate $\mathbf{r}$ is scaled by the factor $\zeta = e^\varepsilon$. If ε is taken to be an infinitesimal and $\hat{F}$, therefore, becomes the generator of an infinitesimal unitary transformation, the coordinate $\mathbf{r}$ is transformed into $\mathbf{r}(1 + \varepsilon)$, keeping only the first-order term in the expansion of the exponential. The same result is of course obtained using eqn (8.35).

We next demonstrate that the action of $\hat{U}(\varepsilon)$ on the state function yields a properly normalized function with the coordinate $\mathbf{r}$ scaled by ζ (Epstein 1974). That is,

$$\hat{U}(\varepsilon)\Psi(\mathbf{r}, \tau') = \zeta^{3/2} \Psi(\zeta\mathbf{r}, \tau') \tag{8.159}$$

where, as before, τ' denotes all the electronic coordinates save $\mathbf{r}$. This result is proved by noting that with $X = \hat{U}(\varepsilon)\Psi$ one has the identity,

$$\frac{\hbar}{i}\frac{dX}{d\varepsilon} = \hat{\mathscr{V}}X, \qquad X = \Psi \qquad \text{when} \qquad \varepsilon = 0, \tag{8.160}$$

and then showing that this relation is satisfied when X is given by the right-hand side of eqn (8.159). The differentiation gives

$$\frac{\hbar}{i}\frac{dX}{d\varepsilon} = \frac{\hbar}{i}\left\{(3/2)\zeta^{1/2}\Psi(\zeta\mathbf{r}, \tau') + \zeta^{3/2}\frac{\partial\Psi(\mathbf{r}(\varepsilon))}{\partial\mathbf{r}(\varepsilon)}\frac{\partial\mathbf{r}(\varepsilon)}{\partial\varepsilon}\right\}, \tag{8.161}$$

which, together with the result that

$$\frac{\partial\Psi(\mathbf{r}(\varepsilon))}{\partial\mathbf{r}(\varepsilon)}\frac{\partial\mathbf{r}(\varepsilon)}{\partial\varepsilon}\bigg|_{\varepsilon=0} \equiv \mathbf{r}\cdot\nabla\Psi(\mathbf{r}, \tau'), \tag{8.162}$$

yields the desired result when $\varepsilon = 0$.

8.4   Variational basis for atomic properties

8.4.1   Variational derivation of the atomic force law

The statement of the atomic action principle given in eqn (8.145) is a variational principle which enables one to derive the properties of an atom in a molecule—it is an atomic variation principle. We shall use it first to derive the atomic statement of the Ehrenfest force law, the equation of motion of an atom in a molecule. This is accomplished through a variation of $\mathscr{L}[\Psi, \Omega]$ with generator $\hat{F} = \varepsilon\cdot\mathbf{p}$. As detailed above, such a generator induces an infinitesimal uniform translation of the electronic coordinate $\mathbf{r}$ by $-\varepsilon$. Denoting by Ψ_ε the image of the state function Ψ in the infinitesimal unitary transformation generated by $\hat{F}$, one has

$$\Psi_\varepsilon(\mathbf{r}, \tau', t) = (\hat{I} - (i/\hbar)\varepsilon\cdot\hat{\mathbf{p}})\Psi(\mathbf{r}, \tau', t) = \Psi(\mathbf{r} - \varepsilon, \tau', t). \tag{8.163}$$

The variation in $\mathscr{L}[\Psi, \Omega]$ is given by the first-order difference between $\mathscr{L}(\varepsilon)$ and $\mathscr{L}(\varepsilon = 0)$ or

$$\delta\mathscr{L}[\Psi, \Omega] = \varepsilon\cdot[\nabla_\varepsilon\mathscr{L}(\varepsilon)]_{\varepsilon=0}. \tag{8.164}$$

Thus the atomic variation principle states that

$$\boldsymbol{\varepsilon} \cdot \mathbf{V}_\varepsilon \mathscr{L}[\Psi_\varepsilon, \Omega]|_{\varepsilon=0} = \tfrac{1}{2}\boldsymbol{\varepsilon} \cdot \{(i/\hbar)\langle[\hat{H}, \hat{p}]\rangle_\Omega + \text{cc}\}. \tag{8.165}$$

Substituting Ψ_ε for Ψ in the expression for $\mathscr{L}[\Psi, \Omega]$ (eqns (8.100) and (8.110)), differentiating with respect to ε, and noting that

$$\mathbf{V}_\varepsilon \Psi_\varepsilon|_{\varepsilon=0} = -\nabla\Psi,$$

one obtains for eqn (8.165) (including a variation of the surface $\delta S = \boldsymbol{\varepsilon} \cdot \mathbf{n}$)

$$
\begin{aligned}
\mathbf{V}_\varepsilon \mathscr{L}[\Psi_\varepsilon, \Omega]|_{\varepsilon=0} = & -\int_\Omega d\mathbf{r} \int d\tau' \Big\{ \tfrac{1}{2} i\hbar((\nabla\Psi^*)\dot{\Psi} + \Psi^*\nabla\dot{\Psi} \\
& - (\nabla\dot{\Psi}^*)\Psi - \dot{\Psi}^*\nabla\Psi) \\
& - \frac{\hbar^2}{2m}\sum_i ((\nabla\nabla_i\Psi^*)\cdot\nabla_i\Psi) \\
& + (\nabla\nabla_i\Psi)\cdot\nabla_i\Psi^*) - \hat{V}(\nabla\Psi^*\Psi + \Psi^*\nabla\Psi) \Big\} \\
& - \frac{\hbar^2}{4mN}\oint dS\nabla^2\rho\mathbf{n}.
\end{aligned}
\tag{8.166}
$$

The dyadic notation has been used to express the terms arising from the kinetic energy part of the Lagrangian. The explicit form for one such term is

$$(\nabla\nabla_i\Psi^*)\cdot\nabla_i\Psi = \sum_k \mathbf{e}_k\left(\sum_j \frac{\partial^2\Psi^*}{\partial\mathbf{r}_k\partial\mathbf{r}_j^{(i)}}\cdot\frac{\partial\Psi}{\partial\mathbf{r}_j^{(i)}}\right) \tag{8.167}$$

where $\mathbf{r}_k(\mathbf{r}^{(i)})$ is the component of $\mathbf{r}(\mathbf{r}_i)$ along the direction of the unit vector $\mathbf{e}_k$, $k = 1, 2, 3$. It follows from eqn (8.167) that

$$(\nabla\nabla_i\Psi^*)\cdot\nabla_i\Psi = \nabla_i\cdot(\nabla_i\Psi\nabla\Psi^*) - \nabla\Psi^*\nabla_i^2\Psi. \tag{8.168}$$

Using eqn (8.168) and its complex conjugate together with Schrödinger's equation, eqn (8.166) becomes

$$
\begin{aligned}
\mathbf{V}_\varepsilon \mathscr{L}[\Psi_\varepsilon, \Omega]|_{\varepsilon=0} = & -\int_\Omega d\mathbf{r} \int d\tau' \{\tfrac{1}{2} i\hbar(\dot{\Psi}^*\nabla\Psi + \Psi^*\nabla\dot{\Psi} \\
& - (\nabla\dot{\Psi}^*)\Psi - (\nabla\Psi^*)\dot{\Psi})\} \\
& + (\hbar^2/2m)\oint dS \int d\tau' \{\nabla\Psi^*\nabla\Psi + \nabla\Psi\nabla\Psi^*\}\cdot\mathbf{n} \\
& - (\hbar^2/4mN)\oint dS\,\nabla^2\rho\mathbf{n},
\end{aligned}
$$

a result which may be equivalently expressed in terms of $\mathbf{J}(\mathbf{r})$ as given in eqn (8.140) for $\hat{F} = \hat{I}$ as

$$\boldsymbol{\varepsilon} \cdot \nabla_\varepsilon \mathscr{L}[\Psi_\varepsilon, \Omega]|_{\varepsilon=0} = \boldsymbol{\varepsilon} \cdot \frac{m}{N} \int_\Omega d\mathbf{r} \frac{\partial \mathbf{J}}{\partial t} - \boldsymbol{\varepsilon} \cdot \frac{\hbar^2}{4mN} \oint d\mathbf{S} \, \nabla^2 \rho \mathbf{n}$$

$$+ \boldsymbol{\varepsilon} \cdot \frac{\hbar^2}{2m} \oint d\mathbf{S} \int d\tau' \{\nabla\Psi^* \cdot \nabla\Psi + \nabla\Psi\nabla\Psi^*\} \cdot \mathbf{n}. \quad (8.169)$$

Since the subsystem is bounded by a zero-flux surface at all stages of the variations generated by $\hat{F}$, the zero-flux variational constraint (eqn (8.117)) applies. In the present case it assumes the form

$$\boldsymbol{\varepsilon} \cdot \frac{\hbar^2}{4m} \nabla_\varepsilon \left(\int_\Omega \nabla^2 \rho_{\Psi_\varepsilon} d\mathbf{r} \right) \bigg|_{\varepsilon=0} = 0. \quad (8.170)$$

Using the same procedure as above, we readily find the left-hand side of eqn (8.170) to be given by

$$\boldsymbol{\varepsilon} \cdot \frac{\hbar^2}{4m} \nabla_\varepsilon \left(\int_\Omega \nabla^2 \rho_{\Psi_\varepsilon} d\mathbf{r} \right) \bigg|_{\varepsilon=0} = - \boldsymbol{\varepsilon} \cdot \frac{\hbar^2 N}{4m} \oint d\mathbf{S} \int d\tau' [(\nabla\nabla\Psi^*)\Psi + \nabla\Psi^* \cdot \nabla\Psi$$

$$+ \nabla\Psi\nabla\Psi^* + \Psi^*(\nabla\nabla\Psi)] \cdot \mathbf{n}$$

$$+ \boldsymbol{\varepsilon} \cdot \frac{\hbar^2}{4m} \oint d\mathbf{S} \nabla^2 \rho \mathbf{n} = 0. \quad (8.171)$$

The addition of this expression for the constraint to the right-hand side of eqn (8.169) yields

$$\boldsymbol{\varepsilon} \cdot \nabla_\varepsilon \mathscr{L}[\Psi_\varepsilon, \Omega]|_{\varepsilon=0} = \boldsymbol{\varepsilon} \cdot \frac{1}{N} \left\{ m \int_\Omega d\mathbf{r} \frac{\partial \mathbf{J}}{\partial t} - \oint d\mathbf{S} \overleftrightarrow{\sigma} \cdot \mathbf{n} \right\} \quad (8.172)$$

where

$$\overleftrightarrow{\sigma}(\mathbf{r}) = \frac{N\hbar^2}{4m} \int d\tau' \{\nabla(\nabla\Psi^*)\Psi + \Psi^*\nabla(\nabla\Psi) - \nabla\Psi^* \nabla\Psi - \nabla\Psi\nabla\Psi^*\} \quad (8.173)$$

is the quantum stress tensor previously defined in eqn (6.12). As emphasized there, the stress tensor plays a dominant role in the description of the mechanical properties of an atom in a molecule. It has the dimensions of an energy density or, equivalently, of force per unit area. Its dot product with $\mathbf{n}(\mathbf{r})$, the outwardly directed normal to $S(\Omega, \mathbf{r})$, gives the force exerted on the associated element of the atomic surface. As shown in eqn (6.13), $\overleftrightarrow{\sigma}(\mathbf{r})$ is totally determined by the one-electron density matrix $\Gamma^{(1)}(\mathbf{r}, \mathbf{r}')$.

Evaluation of the commutator appearing in eqn (8.165) demonstrates that this variation of $\mathscr{L}[\Psi, \Omega]$ yields the atomic average of the total force acting on the electron described by the position vector $\mathbf{r}$,

$$\varepsilon \cdot \tfrac{1}{2}\{(i/\hbar)\langle[\hat{H}, \hat{\mathbf{p}}]\rangle_\Omega + \text{cc}\} = \varepsilon \cdot \langle - \boldsymbol{\nabla}\hat{V}\rangle_\Omega = \varepsilon \cdot (1/N)\mathbf{F}(\Omega, t). \tag{8.174}$$

By combining the results given in eqns (8.172) and (8.174) one obtains from the atomic variational principle the integrated atomic force law

$$\mathbf{F}(\Omega, t) = m \int_\Omega \mathrm{d}\mathbf{r}(\partial\mathbf{J}(\mathbf{r})/\partial t) - \oint \mathrm{d}S(\mathbf{r})\overleftrightarrow{\sigma}(\mathbf{r})\cdot\mathbf{n}(\mathbf{r}). \tag{8.175}$$

Equation (8.175) is a generalization of Ehrenfest's theorem (Ehrenfest 1927). This theorem relates the forces acting on a subsystem or atom in a molecule to the forces exerted on its surface and to the time derivative of the momentum density $m\mathbf{J}(\mathbf{r})$. It constitutes the quantum analogue of Newton's equation of motion in classical mechanics expressed in terms of a vector current density and a stress tensor, both defined in real space.

8.4.2   Differential force law

Within a Lagrangian formulation of the mechanics of a field as used here one may define an energy–momentum tensor whose components summarize the principal properties of the field (Morse and Feshbach 1953; Landau and Lifshitz 1975). We show that the divergence equations satisfied by the spatial components of this tensor for the Schrödinger field yield the differential form of the atomic force law, eqn (8.175) (Bader 1980).

Define a component of the energy–momentum tensor as

$$W_{\mu v} = \Psi_\mu(\partial L/\partial\Psi_v) + \text{cc} - \delta_{\mu v}L, \tag{8.176}$$

where μ and v run over the $3N$ spatial degrees of freedom and the time, and $\Psi_v \equiv \partial\Psi/\partial x_v$. The spatial components W_{jv} (j referring only to the $3N$ spatial coordinates) satisfy the set of divergence equations

$$\sum_v \frac{\partial W_{jv}}{\partial x_v} = -\frac{\partial L}{\partial x_j} \tag{8.177}$$

where $\partial L/\partial x_j$ vanishes unless L possesses an explicit dependence on x_j. In the case of the Lagrangian density defined here (eqn (8.100)),

$$\partial L/\partial x_j = -\Psi^*(\partial\hat{V}/\partial x_j)\Psi$$

and eqn (8.177) yields

$$-\Psi^*\boldsymbol{\nabla}_j\hat{V}\Psi = \frac{i\hbar}{2}\frac{\partial}{\partial t}[(\boldsymbol{\nabla}_j\Psi^*)\Psi - \Psi^*\boldsymbol{\nabla}_j\Psi] - \frac{\hbar^2}{4m}\sum_k\boldsymbol{\nabla}_k\cdot[(\boldsymbol{\nabla}_k\boldsymbol{\nabla}_j\Psi^*)\Psi$$

$$+ \Psi^*(\boldsymbol{\nabla}_k\boldsymbol{\nabla}_j\Psi) - \boldsymbol{\nabla}_k\Psi^*\boldsymbol{\nabla}_j\Psi - \boldsymbol{\nabla}_k\Psi\boldsymbol{\nabla}_j\Psi^*]. \tag{8.178}$$

Equation (8.178) has been previously obtained by Pauli (1958) and by Epstein (1975), who termed it the differential force law. The integration of this expression over the coordinates of all electrons but one by the usual recipe yields (Bader 1980) an equation governing a force density

$$\mathbf{F}(\mathbf{r}, t) = m\, \partial \mathbf{J}(\mathbf{r})/\partial t - \mathbf{V} \cdot \overleftrightarrow{\sigma}(\mathbf{r}). \tag{8.179}$$

Equation (8.179) is the differential form of the integrated atomic force law given in eqn (8.175),

$$\mathbf{F}(\Omega, t) = \int_\Omega d\mathbf{r}\, \mathbf{F}(\mathbf{r}, t). \tag{8.180}$$

As previously discussed in Chapter 6 in eqn (6.17) and following for a stationary state, the force density $\mathbf{F}(\mathbf{r}, t)$ is the total instantaneous force exerted on the electron at $\mathbf{r}_1$ by the nuclei and the remaining electrons

$$-\mathbf{V}_1 \hat{V} = \hat{\mathbf{F}}_1 = -e^2 \sum_\alpha Z_\alpha \frac{(\mathbf{r}_1 - \mathbf{X}_\alpha)}{|\mathbf{r}_1 - \mathbf{X}_\alpha|^3} + e^2 \sum_{j>1} \frac{(\mathbf{r}_1 - \mathbf{r}_j)}{|\mathbf{r}_1 - \mathbf{r}_j|^3}$$

$$= \sum_\alpha \hat{\mathbf{F}}_{e\alpha} + \sum_{j>1} \hat{\mathbf{F}}_{e1ej} \tag{8.181}$$

averaged over the motions of the remaining electrons in the system to yield

$$\mathbf{F}(\mathbf{r}, t) = \sum_\alpha \hat{\mathbf{F}}_{e\alpha} \rho(\mathbf{r}) - 2\int d\mathbf{r}_2\, \hat{\mathbf{F}}_{e1e2}\, \Gamma^{(2)}(\mathbf{r}_1, \mathbf{r}_2) \tag{8.182}$$

where $\Gamma^{(2)}(\mathbf{r}_1, \mathbf{r}_2)$ is the diagonal element of the two-density matrix, eqn (E1.3). While $\mathbf{F}(\mathbf{r}, t)$ clearly involves two-body forces, according to eqn (8.179) it is totally determined by $\mathbf{J}(\mathbf{r})$ and $\overleftrightarrow{\sigma}(\mathbf{r})$, which are functionals of the one-density matrix, $\Gamma^{(1)}(\mathbf{r}, \mathbf{r}')$ evaluated at the point $\mathbf{r} = \mathbf{r}'$ in real space. The force density $\mathbf{F}(\mathbf{r}, t)$ may be interpreted as the total force exerted on the element of charge density located at $\mathbf{r}$ and obeying an equation of motion in real space, eqn (8.179). This equation is identical in form and physical content to Cauchy's first equation of motion of classical continuum mechanics (Malvern 1969).

The variational derivation of the integral atomic force law, eqn (8.175), is applicable only to a region of space bounded by a zero-flux surface in $\mathbf{V}\rho(\mathbf{r})$, i.e. to an open system whose Lagrangian integral vanishes at the point of variation. Thus the variational derivation of the atomic force,

$$\varepsilon \cdot \mathbf{V}_\varepsilon \mathscr{L}[\Psi, \Omega]\big|_{\varepsilon=0} = \varepsilon \cdot \mathbf{F}(\Omega, t)/N, \tag{8.183}$$

and its law of motion (eqn (8.175)) may be regarded as a quantum mechanical definition of an atom in a molecule.

For a stationary state the corresponding expressions are

$$\varepsilon \cdot \mathbf{V}_{\varepsilon} \mathscr{E}\left[\Psi, \Omega\right]\big|_{\varepsilon = 0} = (\varepsilon \cdot \mathbf{F}(\Omega)/N)/\langle \psi, \psi \rangle_{\Omega} \tag{8.184}$$

where

$$\mathbf{F}(\Omega) = -\oint dS(\mathbf{r}) \vec{\sigma}(\mathbf{r}) \cdot \mathbf{n}(\mathbf{r}). \tag{8.185}$$

In general physical terms, the average force exerted on some enclosed volume is equal to the negative of the integral of the pressure acting on each element of its surface—a statement of the physical content of eqn (8.185). This same equation demonstrates that the effect of the environment on an open quantum system Ω is determined by just the flux in the forces across the boundary of Ω (Bader 1980). (See eqn (6.18) and accompanying discussion.)

8.4.3   Variational derivation of the atomic virial theorem

The general time-dependent virial theorem for an atom in a molecule is derived from the atomic variational principle. We shall find a close connection between the expressions so obtained for the virial and those derived in the previous section for the force. In particular, the differential force law leads directly to a corresponding local expression for the virial theorem.

The generator of the infinitesimal unitary transformation, whose atomic projection determines the variation in $\mathscr{L}[\Psi, \Omega]$, is in this case $\hat{F} = -\varepsilon \hat{\mathscr{V}}$ where ε is an infinitesimal real number and where $\hat{\mathscr{V}}$ is the virial operator, eqn (8.155). As discussed in Section 8.3.6, the action of this generator is equivalent to a scaling of the electronic coordinate $\mathbf{r}$ by the factor $\zeta = 1 + \varepsilon$. We shall use the latter of the two forms given for $\hat{\mathscr{V}}$ in eqn (8.155). Since the constant term $-(3/2)\mathrm{i}$ does not contribute to the variation in $\mathscr{L}$ or to the commutator average in eqn (8.145), the generator $\hat{F}$ is set equal to $-\varepsilon \mathbf{r} \cdot \mathbf{p}$, the term responsible for generating the scaling of the coordinates. Thus the transformed state function $\Psi_{\varepsilon}(\mathbf{r}, \tau', t)$ is given as

$$\Psi_{\varepsilon}(\mathbf{r}, \tau', t) = (1 + (\mathrm{i}/\hbar)\varepsilon \mathbf{r} \cdot \hat{\mathbf{p}})\Psi(\mathbf{r}, \tau', t) = \zeta^{3/2}\Psi(\zeta \mathbf{r}, \tau', t)$$

and the variational principle now states that

$$\frac{\partial}{\partial \varepsilon}\mathscr{L}[\Psi_{\varepsilon}, \Omega]\big|_{\varepsilon = 0} = -\frac{1}{2}\left\{\left(\frac{\mathrm{i}}{\hbar}\right)\langle[\hat{H}, \hat{\mathbf{r}} \cdot \hat{\mathbf{p}}]\rangle_{\Omega} + \mathrm{cc}\right\}. \tag{8.186}$$

Writing the expression for $\mathscr{L}[\Psi, \Omega]$ explicitly in terms of ε, differentiating with respect to ε, and noting that

$$(\partial/\partial \varepsilon)\big|_{\varepsilon \to 0} = \mathbf{r} \cdot \mathbf{V},$$

one obtains (with $\delta S = -\varepsilon \mathbf{r} \cdot \mathbf{n}$)

$$\frac{\partial}{\partial \varepsilon} \mathscr{L}[\Psi_\varepsilon, \Omega]|_{\varepsilon=0} = \int_\Omega d\mathbf{r} \int d\tau' \mathbf{r} \cdot \left[\frac{i\hbar}{2} (\nabla \Psi^* \dot{\Psi} + \Psi^* \nabla \dot{\Psi} - (\nabla \dot{\Psi}^*)\Psi - \dot{\Psi}^* \nabla \Psi) \right.$$

$$-\frac{\hbar^2}{2m} \sum_i ((\nabla \nabla_i \Psi^*) \cdot \nabla_i \Psi + (\nabla \nabla_i \Psi) \cdot \nabla_i \Psi^*)$$

$$\left. - \hat{V}(\nabla \Psi^* \Psi + \Psi^* \nabla \Psi) \right] + \frac{\hbar^2}{4mN} \oint dS \nabla^2 \rho \mathbf{r} \cdot \mathbf{n}$$

$$+ \int_\Omega d\mathbf{r} \int d\tau' \left(-\frac{\hbar^2}{2m} \right) \nabla \Psi^* \cdot \nabla \Psi. \tag{8.187}$$

Equation (8.187) can be rearranged using eqn (8.168), Schrödinger's equation, and the identities,

$$\nabla_i \cdot (\nabla_i \Psi^* \nabla \Psi) \cdot \mathbf{r} = \nabla_i \cdot [(\nabla_i \Psi^* \nabla \Psi) \cdot \mathbf{r}] - (\nabla_i \Psi^* \nabla \Psi) \cdot \nabla_i \mathbf{r},$$

$$\nabla_i \cdot (\nabla_i \Psi \nabla \Psi^*) \cdot \mathbf{r} = \nabla_i \cdot [(\nabla_i \Psi \nabla \Psi^*) \cdot \mathbf{r}] - (\nabla_i \Psi \nabla \Psi^*) \cdot \nabla_i \mathbf{r},$$

to yield

$$\frac{\partial}{\partial \varepsilon} \mathscr{L}[\Psi_\varepsilon, \Omega]|_{\varepsilon=0} = -\frac{m}{N} \int_\Omega d\mathbf{r} \left(\mathbf{r} \cdot \frac{\partial \mathbf{J}}{\partial t} \right) + \frac{\hbar^2}{4mN} \oint dS \nabla^2 \rho \mathbf{r} \cdot \mathbf{n}$$

$$-\frac{\hbar^2}{2m} \oint dS \int d\tau' \mathbf{r} \cdot [\nabla \Psi^* \nabla \Psi + \nabla \Psi \nabla \Psi^*] \cdot \mathbf{n}. \tag{8.188}$$

As the atom is to be bounded by a zero-flux surface at all stages of the variation, the variational constraint of eqn (8.117) must be imposed. This constraint in the present case demands that the variation

$$\frac{\hbar^2}{4m} \frac{\partial}{\partial \varepsilon} \left(\int_\Omega \nabla^2 \rho_{\Psi_\varepsilon} d\mathbf{r} \right) \bigg|_{\varepsilon=0} = \frac{\hbar^2}{4m} \left\{ \oint dS \int d\tau' \mathbf{r} \cdot [\nabla(\nabla \Psi^*)\Psi \right.$$

$$+ \Psi^* \nabla(\nabla \Psi) + \nabla \Psi^* \nabla \Psi + \nabla \Psi \nabla \Psi^*] \cdot \mathbf{n}$$

$$\left. + \int_\Omega d\mathbf{r} \nabla^2 \rho(\mathbf{r}) - \oint dS \nabla^2 \rho \mathbf{r} \cdot \mathbf{n} \right\} = 0 \tag{8.189}$$

vanishes in the limit $\varepsilon \to 0$. The addition of the right-hand side of this constraint to eqn (8.188) yields, for the constrained variation,

$$\frac{\partial}{\partial \varepsilon} \mathscr{L}[\Psi_\varepsilon, \Omega]|_{\varepsilon=0} = -\frac{m}{N} \int_\Omega d\mathbf{r} \, \mathbf{r} \cdot \frac{\partial \mathbf{J}}{\partial t} + \oint dS(\mathbf{r}) \mathbf{r} \cdot \overset{\leftrightarrow}{\boldsymbol{\sigma}}(\mathbf{r}) \cdot \mathbf{n}(\mathbf{r})$$

$$+ \frac{\hbar^2}{4m} \int_\Omega d\mathbf{r} \nabla^2 \rho(\mathbf{r}) \tag{8.190}$$

where $\overleftrightarrow{\sigma}(\mathbf{r})$ is the stress tensor previously defined in eqn (8.173). Evaluating the commutator of the right-hand side of eqn (8.186),

$$\langle (i/\hbar)[\hat{H}, \hat{\mathbf{r}} \cdot \hat{\mathbf{p}}] \rangle_\Omega = 2\langle \Psi^*, -(\hbar^2/2m)\nabla^2 \Psi \rangle_\Omega + \langle \Psi^*, -\mathbf{r} \cdot \nabla \hat{V} \Psi \rangle_\Omega,$$

thereby yielding

$$N\frac{\partial}{\partial \varepsilon} \mathscr{L}[\Psi_\varepsilon, \Omega]|_{\varepsilon=0} = -2T_k(\Omega) - \mathscr{V}_b(\Omega). \tag{8.191}$$

The quantity $T_k(\Omega)$ is the average kinetic energy of the subsystem defined specifically in terms of the density $K(\mathbf{r})$ (eqns (5.48) and (5.49)) and $\mathscr{V}_b(\Omega)$ is the virial of the forces exerted on an electron in the subsystem resulting from its instantaneous interactions with the nuclei and other electrons in the system (see eqn (6.19))

$$\mathscr{V}_b(\Omega) = \int_\Omega d\mathbf{r} \int d\tau' \{-\mathbf{r} \cdot \nabla \hat{V}\} \Psi^* \Psi. \tag{8.192}$$

Equating the results given in eqns (8.190) and (8.191), one obtains an expression for the virial theorem of an open system which satisfies the zero-flux boundary condition (8.109)

$$-2T_k(\Omega) = \mathscr{V}_b(\Omega) + \mathscr{V}_s(\Omega) + \mathscr{V}_p(\Omega) + (\hbar^2/4m)\int_\Omega d\mathbf{r} \nabla^2 \rho(\mathbf{r}).$$

The final term on the right-hand side of this equation vanishes because of the zero-flux surface condition thereby yielding

$$-2T_k(\Omega) = \mathscr{V}_b(\Omega) + \mathscr{V}_s(\Omega) + \mathscr{V}_p(\Omega). \tag{8.193}$$

The term $\mathscr{V}_s(\Omega)$ is the virial of the forces exerted on the surface of the subsystem, a term expressible in terms of the stress tensor previously defined in eqns (8.173) and (6.12),

$$\mathscr{V}_s(\Omega) = \oint dS(\mathbf{r})\mathbf{r} \cdot \overleftrightarrow{\sigma}(\mathbf{r}) \cdot \mathbf{n}(\mathbf{r}). \tag{8.194}$$

($\overleftrightarrow{\sigma} \cdot \mathbf{n}$ is the outwardly directed force exerted per unit area of surface and $\mathbf{r} \cdot \overleftrightarrow{\sigma} \cdot \mathbf{n}$ is the virial of this force.) The term $\mathscr{V}_p(\Omega)$, specific to a time-dependent system, is the virial of the forces arising from the time rate of change of the momentum density $m\mathbf{J}$ throughout the subsystem,

$$\mathscr{V}_p(\Omega) = -m\int_\Omega d\mathbf{r} \, \mathbf{r} \cdot \partial \mathbf{J}/\partial t. \tag{8.195}$$

It was pointed out in the discussion of the atomic virial theorem for a stationary state (eqn (6.23)) that, while the values of the individual contributions to the virial of the subsystem are dependent upon the choice of origin for the vector $\mathbf{r}$, their sum, which determines the total subsystem virial, is independent of the choice of origin. It is clear from eqn (8.193) that the same origin independence of the total virial

$$\mathcal{V}(\Omega) = \mathcal{V}_b(\Omega) + \mathcal{V}_s(\Omega) + \mathcal{V}_{\dot{p}}(\Omega) \tag{8.196}$$

is obtained for the general time-dependent case. The basin and surface virials are identical with their stationary-state counterparts, eqns (6.19) and (6.21), respectively; the general statement of the atomic virial theorem given in eqn (8.193) reduces to the stationary-state expression, eqn (6.23), when $\mathcal{V}_{\dot{p}}(\Omega) = 0$.

8.4.4   Local virial relationship

The average of the atomic virial $\mathcal{V}_b(\Omega)$, eqn (8.192), as defined by the atomic variational principle, is the virial of the quantum mechanical force density as defined in the differential force law, averaged over the atomic volume. By taking the virial of $\mathbf{F}(\mathbf{r}, t)$ in eqn (8.179), one obtains

$$\mathbf{r} \cdot \mathbf{F}(\mathbf{r}, t) = m\mathbf{r} \cdot \partial \mathbf{J}/\partial t - \mathbf{r} \cdot \nabla \cdot \overleftrightarrow{\sigma}. \tag{8.197}$$

One proceeds as in the case of a stationary state (eqn (6.25) and following). Using the identity

$$\nabla \cdot (\mathbf{r} \cdot \overleftrightarrow{\sigma}) = \operatorname{Tr} \overleftrightarrow{\sigma} + \mathbf{r} \cdot \nabla \cdot \overleftrightarrow{\sigma}$$

where $\operatorname{Tr} \overleftrightarrow{\sigma}$ denotes the trace or spur of the stress tensor, eqn (8.197) gives

$$\mathbf{r} \cdot \mathbf{F}(\mathbf{r}, t) = m\mathbf{r} \cdot \partial \mathbf{J}/\partial t + \operatorname{Tr} \overleftrightarrow{\sigma} - \nabla \cdot (\mathbf{r} \cdot \overleftrightarrow{\sigma}). \tag{8.198}$$

From eqn (8.173) for the definition of $\overleftrightarrow{\sigma}(\mathbf{r})$, $\operatorname{Tr} \overleftrightarrow{\sigma}$ may be variously expressed as (see eqns (6.26) and (6.27))

$$\operatorname{Tr} \overleftrightarrow{\sigma}(\mathbf{r}) = -K(\mathbf{r}) - G(\mathbf{r}) = -2K(\mathbf{r}) - (\hbar^2/4m)\nabla^2\rho(\mathbf{r})$$

$$= -2G(\mathbf{r}) + (\hbar^2/4m)\nabla^2\rho(\mathbf{r}). \tag{8.199}$$

Its integral over an atomic volume yields the unique value

$$\int_\Omega d\mathbf{r} \operatorname{Tr} \overleftrightarrow{\sigma}(\mathbf{r}) = -2T(\Omega). \tag{8.200}$$

Using the final identity given in eqn (8.199), eqn (8.198) may be expressed as

$$-2G(\mathbf{r}) = \mathbf{r} \cdot \mathbf{F}(\mathbf{r}, t) + \nabla \cdot (\mathbf{r} \cdot \overleftrightarrow{\sigma}(\mathbf{r})) - m\mathbf{r} \cdot \partial \mathbf{J}(\mathbf{r})/\partial t$$

$$- (\hbar^2/4m)\nabla^2\rho(\mathbf{r}). \tag{8.201}$$

Integration of eqn (8.201) over an atomic volume for which the integral of $\nabla^2\rho(\mathbf{r})$ vanishes yields, term for term, the atomic virial theorem for a time-dependent system (eqn (8.193)) or for a stationary state (eqn (6.23)). Thus, eqn (8.201) is, in terms of its derivation and its integrated form, a local expression of the virial theorem. The atomic virial theorem provides the basis for the definition of the average energy of an atom, as discussed in Chapter 6.

8.4.5 Summary of the atomic variational principle

The principle of stationary action provides a complete quantum mechanical description of a system in terms of the changes induced in the system by infinitesimal unitary transformations. The generators of such transformations describe all possible system changes, caused both by displacements or by a temporal development (Roman 1965). It has been shown that this principle, which is a variational principle, applies uniquely to a connected region of real space bounded by a surface of zero flux in $\nabla\rho(\mathbf{r})$. Thus, the principle of stationary action defines an atom as being a bounded region of space, the variational properties of which, as determined by the action of infinitesimal unitary transformations, are identical to those obtained for the total system. The generators of such transformations describe real physical changes in the system. *Thus, a single variational principle serves to describe the properties of a total system and of an atom within the system.*

The operational statement of this principle is the atomic variational principle

$$\delta\mathscr{L}[\Psi,\Omega] = (\tfrac{1}{2})\{(i/\hbar)\langle[\hat{H},\hat{F}]\rangle_\Omega + \mathrm{cc}\}$$

or, for a stationary state,

$$\delta\mathscr{E}[\Psi,\Omega] = -(\tfrac{1}{2})\{(i/\hbar)\langle[\hat{H},\hat{F}]\rangle_\Omega + \mathrm{cc}\}/\langle\psi,\psi\rangle_\Omega.$$

The generator $\hat{F}$ causes an infinitesimal transformation in the state function describing the total system. The physical nature of this transformation is identified through the commutator of $\hat{H}$ with $\hat{F}$. The atomic value of the change caused by the transformation is given by the projection of the total system value over the atomic volume. Since a generator can be found to describe any change in a system, all atomic properties and the laws governing these properties can be derived from this principle. These derivations of atomic properties parallel the derivations of the corresponding properties for the total system, since both are described by the same variational principle.

For example, $\hat{F} = \varepsilon\hat{N} \equiv \varepsilon\hat{I}N$, where N is the total number of electrons in the system, generates a simple infinitesimal transformation, which leaves the Lagrangian $\mathscr{L}[\Psi,\Omega]$ invariant. In addition, since N is a constant of the motion for the total system, $[\hat{H},\hat{N}] = 0$. However, the time rate of change of the average electronic population of an atom, $N(\Omega)$, is not zero in general and the equation of continuity governing the time evolution of $N(\Omega)$ is obtained directly from the equivalent statement of the atomic variational principle, eqn (8.149), as

$$\frac{\mathrm{d}}{\mathrm{d}t}\int_\Omega \mathrm{d}\mathbf{r}\,\rho(\mathbf{r}) = N(\Omega) = -\oint\mathrm{d}S\left\{(\mathbf{J}\cdot\mathbf{n} + \mathrm{cc}) - \frac{\partial S}{\partial t}\rho\right\}. \qquad (8.202)$$

As illustrated earlier, setting the generator equal to $\varepsilon \cdot \hat{\mathbf{p}}$ defines the atomic force and the variational principle leads to the integral atomic force law, or the equation of motion for an atom in a molecule. Finally, it was shown that, when $\hat{F} = -\varepsilon \mathbf{r} \cdot \hat{\mathbf{p}}$, the commutator defines the electronic kinetic energy and virial for an atom, and the variational principle yields the relationship between these quantities, the atomic virial theorem. These three relationships—the equation of continuity, the equation of motion, and the virial theorem—form the basis for the understanding of the mechanics of an atom in a molecule.

Within the Lagrangian formalism one obtains a definition of the energy–momentum tensor for the Schrödinger field. The many-particle divergence equation satisfied by the spatial components of this tensor, when integrated in the standard manner, yields a differential expression of the atomic force law. The virial of the force density defined in this manner gives a local expression for the atomic virial theorem. Thus, from a single formalism one obtains a description of the local mechanical properties of an atom as well as of their average values.

This chapter concludes with the demonstration that the atomic statement of the principle of stationary action obtains in the presence of an electromagnetic field. Thus, atoms continue to exist in the presence of applied electric and/or magnetic fields and the response of each atom to such fields and its contribution to the magnetic and electric properties of the total system can be determined.

8.5   Atoms in an electromagnetic field

8.5.1   The Lagrangian and Hamiltonian

We shall first find that Lagrangian for a system of charged particles in an electromagnetic field which, through the principle of least action, gives the correct equation of motion. The electric and magnetic fields of the electromagnetic field, $\mathscr{E}$ and $\mathbf{B}$, respectively, are related to the scalar and vector potentials, ϕ and $\mathbf{A}$, by the equations

$$\mathscr{E} = -\frac{1}{c}\frac{\partial \mathbf{A}}{\partial t} - \nabla \phi \tag{8.203}$$

and

$$\mathbf{B} = \nabla \times \mathbf{A}. \tag{8.204}$$

It is clear that ϕ and $\mathbf{A}$ are not uniquely determined by these equations, as one can modify both potentials through the addition of the scalar $\chi(\mathbf{r}, t)$, i.e.

$$\mathbf{A}'(\mathbf{r}, t) = \mathbf{A}(\mathbf{r}, t) + \nabla \chi(\mathbf{r}, t) \tag{8.205}$$

and

$$\phi'(\mathbf{r}, t) = \phi(\mathbf{r}, t) - (1/c)(\partial \chi(\mathbf{r}, t)/\partial t) \tag{8.206}$$

without affecting the fields $\mathscr{E}$ and $\mathbf{B}$ in any way. Since the state function depends upon the potentials, this so-called change in gauge of the potentials represented in eqns (8.205) and (8.206) changes the state function as well. The wave function in the new gauge is related to the function in original gauge by a gauge transformation, a unitary transformation represented by

$$\Psi'(\mathbf{r}, t) = e^{ie\chi(\mathbf{r}, t)/\hbar c} \Psi(\mathbf{r}, t). \tag{8.207}$$

If the physical results are to remain unchanged under a unitary transformation, it is necessary to transform the operators as well as the state functions (eqns (8.28) and (8.29)). A gauge transformation has no effect on a coordinate operator but the momentum operator $\hat{\mathbf{p}}$ is changed into $\hat{\mathbf{p}} + (e/c)\nabla\chi$. Thus, to maintain gauge invariance in properties determined by the momentum operator, $\hat{\mathbf{p}}$ must be replaced by a new operator. In the presence of an electromagnetic field or just a magnetic field, the momentum operator $\hat{\mathbf{p}}$ is replaced by the expression

$$\hat{\boldsymbol{\pi}} = \hat{\mathbf{p}} - (e/c)\mathbf{A}(\mathbf{r}, t) \tag{8.208}$$

for a particle of charge e. The use of $\hat{\boldsymbol{\pi}}$ will insure that all averages over a momentum operator will be gauge-invariant as the extra term $(e/c)\nabla\chi$ is exactly cancelled by the change in $\mathbf{A}$ on going to the new gauge. With the momentum expressed as in eqn (8.208), one is free to choose any gauge and the one most convenient for our purposes is called the transverse gauge. In this gauge one sets $\nabla \cdot \mathbf{A} = 0$, a choice which has the consequence of causing the quantities $\hat{\mathbf{p}}$ and $\mathbf{A}$ to commute, $\mathbf{p} \cdot \mathbf{A} = \mathbf{A} \cdot \mathbf{p}$, since their commutator is $-i\hbar\nabla \cdot \mathbf{A}$.

Through the use of eqn (8.208) and the inclusion of the scalar potential ϕ, $L(\Psi, \nabla\Psi, \dot{\Psi})$, the many-particle Lagrangian given in eqn (8.100), can be modified for the presence of external electric and magnetic fields. The modified Lagrangian $L(\Psi, \nabla\Psi, \dot{\Psi}, \mathbf{A}, \phi)$ can be expressed as

$$L(\Psi, \nabla\Psi, \dot{\Psi}, \mathbf{A}, \phi) = L(\Psi, \nabla\Psi, \dot{\Psi}) + \frac{i\hbar e}{2mc} \sum_i \{\nabla_i \Psi^* \cdot \mathbf{A}(\mathbf{r}_i)\Psi$$
$$- \mathbf{A}(\mathbf{r}_i)\Psi^* \cdot \nabla_i \Psi\} - \frac{e^2}{2mc^2} \sum_i \mathbf{A}^2(\mathbf{r}_i)\Psi^*\Psi$$
$$- \sum_j e_j \phi(\mathbf{r}_j)\Psi^*\Psi \tag{8.209}$$

where the sum j runs over all the charged particles in the system. The associated action integral is defined as

$$\mathscr{W}_{12}[\Psi, \mathbf{A}, \phi] = \int_{t_1}^{t_2} \int d\tau \, L(\Psi, \nabla\Psi, \dot{\Psi}, \mathbf{A}, \phi)dt. \tag{8.210}$$

Variation of the action integral with respect to Ψ^*, subject to the natural boundary condition that $\mathbf{V}_i\Psi \cdot \mathbf{n} = 0$ on the surface at infinity and the condition that the variations in Ψ^* vanish at the time end-points, yields as the equation of motion the Euler–Lagrange equation of the variation. This statement of Schrödinger's equation, the one appropriate for use when the system is in the presence of electric and magnetic fields, is

$$i\hbar\dot{\Psi} = \frac{1}{2m}\sum_i\left(\frac{\hbar}{i}\mathbf{V}_i - \frac{e}{c}\mathbf{A}(\mathbf{r}_i)\right)^2\Psi + (\hat{V} + \sum_j e_j\phi(\mathbf{r}_j))\Psi. \qquad (8.211)$$

This result may be re-expressed in terms of $\hat{H}^0$, the usual field-free Hamiltonian, and $\hat{H}^i$, an interaction Hamiltonian, as

$$i\hbar\dot{\Psi} = \hat{H}^0\Psi + \hat{H}^i\Psi \qquad (8.212)$$

where $\hat{H}^i$ is given by

$$\hat{H}^i = -\frac{e}{2mc}\sum_i(\hat{\mathbf{p}}_i \cdot \mathbf{A}(\mathbf{r}_i) + \mathbf{A}(\mathbf{r}_i) \cdot \hat{\mathbf{p}}_i)$$

$$+ \frac{e^2}{2mc^2}\sum_i\mathbf{A}^2(\mathbf{r}_i) + \sum_j e_j\phi(\mathbf{r}_j). \qquad (8.213)$$

It will be recalled (eqn (5.22)) that the first of Ehrenfest's relationships, obtained by taking the commutator of $\hat{H}$ with the position vector $\hat{\mathbf{r}}$, states that the momentum operator $\hat{\mathbf{p}}$ is given by $md\hat{\mathbf{r}}/dt$. If one uses the Hamiltonian $\hat{H}^0 + \hat{H}^i$ in the commutator with $\hat{\mathbf{r}}$, then one correctly determines that the momentum is now given by $\hat{\pi}$ in eqn (8.208), that is, $md\hat{\mathbf{r}}/dt = \hat{\mathbf{p}} - (e/c)\mathbf{A}$.

Before proceeding with the definition and variation of the corresponding atomic Lagrangian and action integrals, it is important to verify that, when Schrödinger's eqn (8.211) is satisfied, the many-particle Lagrangian in the presence of the external fields (eqn (8.209)) still reduces to a sum of terms, each of which is proportional to the Laplacian operator for a single electron acting on the product $\Psi^*\Psi$. It is through this property of the Lagrangian that the consequences of the variational constraint which demands that the subsystem be bounded by a surface of zero flux in the gradient vector field of ρ is incorporated into the variation of the atomic action integral. Substituting for $\dot{\Psi}$ and $\dot{\Psi}^*$ in eqn (8.209) through the use of eqn (8.211) and its complex conjugate, one finds as before that

$$L^0(\Psi, \nabla\Psi, \dot{\Psi}, \mathbf{A}, \phi) = \sum_i\left\{\frac{-\hbar^2}{4m}(\Psi^*\nabla_i^2\Psi + \Psi\nabla_i^2\Psi^*) - \frac{\hbar^2}{2m}\nabla_i\Psi^* \cdot \nabla_i\Psi\right\}$$

$$= \frac{-\hbar^2}{4m}\sum_i\nabla_i^2(\Psi^*\Psi) \qquad (8.214)$$

where the final equality is obtained using the identity developed in eqn (5.46).

In analogy with eqn (8.110), the atomic Lagrangian integral in the presence of external fields is defined as

$$\mathscr{L}[\Psi, \Omega, \mathbf{A}, \phi, t] = \int_\Omega d\mathbf{r} \int d\tau' L(\Psi, \nabla\Psi, \dot\Psi, \mathbf{A}, \phi)$$

$$= \int_\Omega d\mathbf{r}\, \mathscr{L}(\mathbf{r}, \mathbf{A}, \phi) \tag{8.215}$$

and the associated action integral as

$$\mathscr{W}_{12}[\Psi, \Omega, \mathbf{A}, \phi] = \int_{t_1}^{t_2} dt\, \mathscr{L}[\Psi, \Omega, \mathbf{A}, \phi, t]. \tag{8.216}$$

At the point of variation, where Ψ and Ψ^* satisfy Schrödinger's equation, the atomic Lagrangian integral, because of eqn (8.214), reduces to a surface integral of the flux in $\nabla\rho$ through the surface of the subsystem

$$\mathscr{L}^0[\Psi, \Omega, \mathbf{A}, \phi, t] = \int_\Omega d\mathbf{r}\, \mathscr{L}^0(\mathbf{r}, \mathbf{A}, \phi) = -(\hbar^2/4mN) \int_\Omega d\mathbf{r} \nabla^2\rho(\mathbf{r}, t)$$

$$= -(\hbar^2/4mN)\oint dS(\Omega, t)\nabla\rho(\mathbf{r}, t) \cdot \mathbf{n}(\mathbf{r}, t). \tag{8.217}$$

Thus, the atomic Lagrangian and action integrals in the presence of an electromagnetic field, like their field-free counterparts, vanish as a consequence of the zero-flux surface condition (eqn (8.109)). These properties are common to the corresponding integrals for the total system and it is a consequence of this equivalence in properties that the action integrals for the total system and each of the atoms which comprise it have similar variational properties.

8.5.2   Definition of an atom in an external field

The definition of an atom in a molecule in the presence of an electromagnetic field is accomplished by defining its properties using the principle of stationary action. This principle for a total system and for an atom within the system is obtained as in the field-free case, through a variation of the corresponding action integrals. The variation is generalized to include a variation of the time end-points and by the retention of the variations in the state function at the time end-points. It is the end-point variations that are identified with the generators of infinitesimal unitary transformations. In this manner, the change in the action integral is equated to the action of infinitesimal unitary transformations acting on the state functions at the two time end-points, a relationship known as the principle of stationary action. At every stage of the variation of the atomic action integral, which includes a variation of its surface, the variation is subject to the constraint given in eqns (8.116) and (8.117). This constraint is equivalent to demanding that the atom be bounded by a surface of zero flux in the gradient vector field of the charge density at every stage of the variation (eqn (8.114)). The result of such a generalized variation of the atomic action integral (given in eqn (8.216)) is (compare the field-free result, eqn (8.118))

$$\delta \mathscr{W}_{12}[\Psi, \Omega, \mathbf{A}, \phi]$$

$$= \int_{t_1}^{t_2} \mathrm{d}t \int_{\Omega} \mathrm{d}\mathbf{r} \int \mathrm{d}\tau' \{(-i\hbar\dot{\Psi}^* - \hat{H}^0\Psi^* - \hat{H}^i\Psi^*)\delta\Psi + \mathrm{cc}\}$$

$$+ \int_{t_1}^{t_2} \mathrm{d}t \oint \mathrm{d}S(\Omega, \mathbf{r}) \int \mathrm{d}\tau'(-\hbar^2/2m)\left\{\left(\boldsymbol{\nabla}\Psi^* + \frac{ie}{\hbar c}\mathbf{A}(\mathbf{r})\Psi^*\right)\delta\Psi \cdot \mathbf{n}(\mathbf{r}) + \mathrm{cc}\right\}$$

$$+ \int_{t_1}^{t_2} \mathrm{d}t \oint \mathrm{d}S(\Omega, \mathbf{r})\left\{\mathscr{L}^0(\mathbf{r}, \mathbf{A}, \phi)\delta_\psi S(\Omega, \mathbf{r}) + \mathrm{cc}\right.$$

$$\left. + \int \mathrm{d}\tau'(-i\hbar/2)(\partial S(\Omega, \mathbf{r})/\partial t)\Psi^*\delta\Psi + \mathrm{cc}\right\}$$

$$+ \left\{(i\hbar/2)\int_{\Omega} \mathrm{d}\mathbf{r} \int \mathrm{d}\tau'\Psi^*\delta\Psi + \mathrm{cc}\right\}\bigg|_{t_1}^{t_2}. \tag{8.218}$$

The first set of terms multiplying the variations in Ψ and the corresponding complex conjugate set yield Schrödinger's equation for Ψ and Ψ^* (eqn (8.211)) for the restricted variation where the natural boundary conditions are imposed, the end-point variations are set equal to zero, and Ω equals all space. The equations of motion are assumed from this point on (they are in any case implied by the principle of stationary action, eqn (8.79)). Since the Lagrangian integral vanishes at the point of variation, its contribution to the change in action at the time end-points has been omitted. The Lagrangian density at the point of variation $\mathscr{L}^0(\mathbf{r}, \mathbf{A}, \phi)$, appearing in the second of the surface terms multiplied by the variation in the surface of the subsystem, is defined in eqn (8.217) and is seen to be proportional to the Laplacian of the charge density. It is through this term that the variational constraint, as given in eqn (8.117), is incorporated into the expression for the variation of the atomic action integral. As previously detailed in eqn (8.134), the satisfaction of the constraint given in eqn (8.117) enables one to replace the surface term, involving the variation of the surface of the subsystem, with an integral of the variation of $\nabla^2\rho$. Since the Laplacian is a divergence expression, its variation as detailed in eqn (8.135) yields only a surface term and, thus, the Euler equation obtained in the variation is left unchanged. Substitution of the result given for the variation of the Laplacian in eqn (8.135) for the surface variation term in eqn (8.218) together with the use of Schrödinger's eqn (8.211) yields the following expression for the variation of the atomic action integral

$$\delta \mathscr{W}_{12}[\Psi, \Omega, \mathbf{A}, \phi] = \int_{t_1}^{t_2} dt \oint dS(\Omega, \mathbf{r})(i\hbar/2)\left\{\delta_\Psi \mathbf{j_B}(\mathbf{r}) \cdot \mathbf{n}(\mathbf{r}) + \text{cc}\right.$$

$$- \int d\tau'(\partial S/\partial t)\Psi^* \delta\Psi + \text{cc}\bigg\}$$

$$+ (i\hbar/2)\left\{\left[\int_\Omega d\mathbf{r} \int d\tau'\Psi^* \delta\Psi + \text{cc}\right]\right\}\Bigg|_{t_1}^{t_2}. \qquad (8.219)$$

The quantity $\mathbf{j_B}$ is the vector current density for a particle in the presence of an electromagnetic field and is defined as

$$\mathbf{j_B}(\mathbf{r}) = \mathbf{j}(\mathbf{r}) - (e/mc)\rho'(\mathbf{r})\mathbf{A}(\mathbf{r})$$

$$= (\hbar/2mi)\int d\tau'[\Psi^*\nabla\Psi - \nabla\Psi^*\Psi] - (e/mc)\int d\tau'\Psi^*\Psi\mathbf{A}(\mathbf{r}). \qquad (8.220)$$

The first term $\mathbf{j}(\mathbf{r})$ is the normal vector current density previously defined (eqn (5.11)) and present in the expressions for the variation of the energy and action functionals for the field-free case (eqns (5.96) and (8.136)). The second term owes its existence to the presence of a magnetic field and is simply the charge density $\rho(\mathbf{r})$ multiplied by the vector potential $\mathbf{A}$. The two terms are called the paramagnetic and diamagnetic currents, respectively. The presence of the two contributions to the total current in the presence of an electromagnetic field is a consequence of the velocity being determined by $\hat{\mathbf{p}}/m - (e/cm)\mathbf{A}$ rather than just by $\hat{\mathbf{p}}/m$. Thus, as in the variation of the field-free functionals, the variation of the surface of the subsystem, together with the restriction that the subsystem be bounded by a zero-flux surface, causes the quantum mechanical current density to appear in the variation, thereby transforming a general mathematical relation into a physical result. The presence of the full magnetic current $\mathbf{j_B}$ is essential to the description of the properties of an open system in a magnetic field.

Proceeding as before in the field-free case, the variations in the state function are replaced by operators which act as generators of infinitesimal unitary transformations. That is, $\delta\Psi = (-i/\hbar)\hat{F}\Psi$ where $\hat{F}$ is an infinitesimal Hermitian operator ($\hat{F} = \varepsilon\hat{G}$). Introducing the notion of generators into the result for the variation of the atomic action integral yields

$$\Delta\mathscr{W}_{12}[\Psi, \Omega, \mathbf{A}, \phi] = F(\Omega, t_2) - F(\Omega, t_1) - \int_{t_1}^{t_2} dt \oint dS(\Omega, \mathbf{r})\{(\partial S/\partial t)\rho_F$$

$$- [\mathbf{J}_F^d(\mathbf{r}) + \tfrac{1}{2}(\mathbf{J}_F(\mathbf{r}) + \text{cc})] \cdot \mathbf{n}(\mathbf{r})\}. \qquad (8.221)$$

The result is expressed in terms of property averages for N electrons, so $\Delta\mathscr{W} = N\delta\mathscr{W}$. The atomic averages of the generator at the time end-points $F(\Omega, t)$ and the corresponding property density ρ_F have been previously defined, eqns (8.138) and (8.139), respectively, as has the paramagnetic contribution to the magnetic current density for property F, eqn (8.140). The diamagnetic contribution to the total magnetic current density for property F is given by

$$\mathbf{J}_F^d(\mathbf{r}) = -(e/mc)\rho_F(\mathbf{r})\mathbf{A}(\mathbf{r}). \tag{8.222}$$

To re-express this result in the form given in eqn (8.143) for the field-free case, we need the Heisenberg equation of motion for $F(\Omega, t)$ for a system in the presence of a magnetic field. This is obtained in the same manner as eqn (8.141) in the field-free case, using eqn (8.211) for the Schrödinger equation of motion to give

$$dF(\Omega, t)/dt = (\tfrac{1}{2})\{(i/\hbar)\langle[\hat{H}, \hat{F}]\rangle_\Omega + cc\}$$
$$+ \oint dS(\Omega, \mathbf{r})\{(\partial S/\partial t)\rho_F$$
$$- [\mathbf{J}_F^d(\mathbf{r}) + \tfrac{1}{2}(\mathbf{J}_F(\mathbf{r}) + cc)]\cdot\mathbf{n}(\mathbf{r})\}. \tag{8.223}$$

Comparison of this expression with that for the change in action in eqn (8.221) shows that the terms subtracted from the end-point averages of the generator are just those which account for the surface contribution to this difference integrated over the time-like surface connecting the two time end-points. Thus, as in the field-free case, the change in the atomic action integral can be expressed entirely in terms of the interior averages of the generator as

$$\Delta\mathscr{W}_{12}[\Psi, \Omega, \mathbf{A}, \phi] = \int_{t_1}^{t_2} dt\{(i/\hbar)\langle[\hat{H}, \hat{F}]\rangle_\Omega + cc\}/2, \tag{8.224}$$

a result equivalent to the statements of stationary action obtained for the total system (eqn (8.144)) and for an atom (eqn (8.143)) in the absence of an electromagnetic field. Statements corresponding to eqn (8.145) for $\delta\mathscr{L}[\Psi, \Omega, \mathbf{A}, \phi]$ and to eqns (8.147) and (8.148) for the stationary-state functionals $\mathscr{E}[\psi, \Omega, \mathbf{A}, \phi]$ and $\mathscr{G}[\psi, \Omega, \mathbf{A}, \phi]$ are also obtained, with the understanding that the Hamiltonian appearing in these expressions includes the interaction Hamiltonian defined in eqn (8.213).

Thus, the definition of an atom as a region of space bounded by a zero-flux surface in the gradient vector field of the charge density with properties predicted by quantum mechanics persists when the system is in the presence of an electromagnetic field and the charge density is correspondingly perturbed. We shall demonstrate this by determining the Ehrenfest force acting on an atom in a molecule in the presence of an electromagnetic field and the corresponding statement of the atomic virial theorem. This is followed by an illustration of how a field-dependent property, namely the polarizability, can be defined and determined for an atom in a molecule.

8.5.3 Atomic force and virial theorems in the presence of external fields

The statement of the atomic action principle which enables one to calculate the properties of an atom in a molecule in the presence of an electromagnetic field is (compare eqn (8.145) for the field-free case)

$$\delta \mathscr{L}[\Psi, \Omega, \mathbf{A}, \phi] = \tfrac{1}{2}\{(i/\hbar)\langle[\hat{H}, \hat{F}]\rangle_\Omega + \mathrm{cc}\}. \tag{8.225}$$

The Hamiltonian appearing in eqn (8.225) for a system of electrons, charge $= -e$, and nuclei with general charge $+ Z_\alpha e$ is

$$\hat{H} = \sum_i \{\hat{\pi}_i^2/2m - \sum_\alpha Z_\alpha e^2(|\mathbf{X}_\alpha - \mathbf{r}_i|)^{-1} - e\phi(\mathbf{r}_i)\}$$

$$+ \sum\sum_{i<j} e^2(|\mathbf{r}_i - \mathbf{r}_j|)^{-1}$$

$$+ \sum\sum_{\alpha<\beta} Z_\alpha Z_\beta e^2(|\mathbf{X}_\alpha - \mathbf{X}_\beta|)^{-1} + \sum_\alpha Z_\alpha e\phi(\mathbf{X}_\alpha)$$

$$= \sum_i \hat{T}_{\pi i} + \hat{V} + \Phi \tag{8.226}$$

with the momentum operator $\hat{\pi}_i$ given by

$$\hat{\pi}_i = \hat{\mathbf{p}}_i + (e/c)\mathbf{A}(\mathbf{r}_i); \tag{8.227}$$

the potential energy contributions have been separated into the normal field-free term $\hat{V}$ and another, Φ, containing the effects of the external field ϕ. One may determine the variation in the atomic Lagrangian by subjecting the state function to the infinitesimal unitary transformation generated by the operator $\hat{F}$. This procedure was illustrated in Section 8.4 in the derivation of the atomic force law and the atomic virial theorem, the same derivations that are to be considered here. However, as pointed out in the case of the stationary-state variations, the variation of the functional can be evaluated in a simpler manner by using the expression obtained prior to the substitution of the Heisenberg equation of motion. Thus, the expression for the variation of the atomic action integral given in eqn (8.221), when divided by the time difference $\Delta t = t_2 - t_1$ and taken to the limit of $\Delta t \to 0$, yields an alternative expression for the variation of the atomic Lagrangian

$$\delta \mathscr{L}[\Psi, \Omega, \mathbf{A}, \phi] = dF(\Omega)/dt - \oint dS(\Omega, \mathbf{r})\{(\partial S/\partial t)\rho_F$$

$$- [\mathbf{J}_F^d(\mathbf{r})\cdot\mathbf{n}(\mathbf{r}) + \tfrac{1}{2}(\mathbf{J}_F(\mathbf{r})\cdot\mathbf{n}(\mathbf{r}) + \mathrm{cc})]\}. \tag{8.228}$$

Note that the result in eqn (8.228) is given for the general case of a particle of charge e, as is done throughout Sections 8.5.1 and 8.5.2. When used in conjunction with the electronic Hamiltonian in eqn (8.226) to obtain expressions for the electronic force and virial, the charge must be replaced by $-e$. Equating the results for $\delta \mathscr{L}$ in eqns (8.225) and (8.228) is equivalent to using the Heisenberg equation of motion (eqn (8.223)) for the generator $\hat{F}$. However, the equating of the two separate expressions emphasizes that the results are obtained from a variational expression, the atomic statement of the principle of stationary action, one that is restricted to subsystems which satisfy the zero-flux boundary condition. Equation (8.225) is, in effect, a variational statement of the Heisenberg equation of motion for the generator $\hat{F}$. As has been stressed in previous applications of the atomic statements of the action principle, the mode of integration of the variational functional determines the definition of the average value of an atomic property. The coordinates appearing in the generator $\hat{F}$ of the action principle and denoted by the position vector $\mathbf{r}$ are those of the electron whose coordinates are integrated over the basin of the atom, the coordinates of the remaining electrons being integrated over all space and the spins of all electrons summed.

Ehrenfest's first relation, the definition of the velocity and its associated momentum, is obtained using the commutator of $\hat{H}$ and the position vector $\mathbf{r}$. This commutator, using the above Hamiltonian and multiplied by m, does indeed yield $\hat{\pi}$ as the momentum of an electron in an electromagnetic field

$$m(i/\hbar)[\hat{H}, \hat{\mathbf{r}}] = \tfrac{1}{2}(i/\hbar)[\hat{\pi}^2, \hat{\mathbf{r}}] = \hat{\pi}. \qquad (8.229)$$

The evaluation of the right-hand side of eqn (8.228) for $\hat{F} = m\hat{\mathbf{r}}$ and equating the result to the right-hand side of eqn (8.225) then yields an expression for the atomic average of the momentum $\hat{\pi}$ in the presence of a varying electromagnetic field

$$\pi(\Omega) = m \int_{\Omega} d\mathbf{r} \, \mathbf{r} \partial \rho(\mathbf{r})/\partial t + \frac{1}{2}m \oint dS(\Omega, \mathbf{r}) \{\mathbf{J}_r(\mathbf{r}) \cdot \mathbf{n}(\mathbf{r}) + cc$$
$$- (e/mc)\rho_r \mathbf{A}(\mathbf{r}) \cdot \mathbf{n}(\mathbf{r})\}. \qquad (8.230)$$

The atomic statement of the Ehrenfest force law is obtained by setting the generator $\hat{F}$ equal to the momentum $\hat{\pi}$. The commutator of the many-particle Hamiltonian eqn (8.226) and $\hat{\pi}$ can be expressed as

$$(i/\hbar)[\hat{H}, \hat{\pi}] = (i/\hbar)[\hat{V} - e\phi(\mathbf{r}), \hat{\mathbf{p}}] + (i/2m\hbar)[\hat{\pi}^2, \hat{\pi}] \qquad (8.231)$$

where $\hat{V}$ represents the potential energy contributions from just the internal electric fields, as defined in eqn (8.226). All of the potential energy terms commute with the vector potential $\mathbf{A}(\mathbf{r})$. The first of the commutators on the right-hand side of eqn (8.231) gives the force exerted on the electron at $\mathbf{r}_1 = \mathbf{r}$ by the nuclei, the other electrons (see eqn (8.181)), and the external field $\phi(\mathbf{r}_1)$,

$$(i/\hbar)[\hat{V} - e\phi(\mathbf{r}), \hat{\mathbf{p}}] = -\nabla\hat{V} + e\nabla\phi(\mathbf{r}), \qquad (8.232)$$

while the second, which does not vanish as one might guess at first sight, gives the force exerted on the electron by the magnetic field

$$(i/2m\hbar)[\hat{\pi}^2, \hat{\pi}] = -(e/2mc)\{\hat{\pi} \times \mathbf{B} - \mathbf{B} \times \hat{\pi}\}. \qquad (8.233)$$

This latter commutator does not vanish because the components of $\hat{\boldsymbol{\pi}}$ do not commute amongst themselves. Consider the commutators of $\hat{\pi}^2$ formed with the component $\hat{\pi}_x$. One has $[\hat{\pi}_x^2, \hat{\pi}_x] = 0$ and

$$[\hat{\pi}_y^2, \hat{\pi}_x] = \hat{\pi}_y[\hat{\pi}_y, \hat{\pi}_x] + [\hat{\pi}_y, \hat{\pi}_x]\hat{\pi}_y$$

$$= -i\hbar\left(\frac{e}{c}\right)\left\{\hat{\pi}_y\left(\frac{\partial A_x}{\partial y} - \frac{\partial A_y}{\partial x}\right) + \left(\frac{\partial A_x}{\partial y} - \frac{\partial A_y}{\partial x}\right)\hat{\pi}_y\right\}. \tag{8.234}$$

Each term in the curly brackets in eqn (8.234) is the negative of the z component of the magnetic field $\mathbf{B} = \mathbf{V} \times \mathbf{A}$; hence,

$$[\hat{\pi}_y^2, \hat{\pi}_x] = i\hbar\left(\frac{e}{c}\right)\{\hat{\pi}_y B_z + B_z \hat{\pi}_y\}. \tag{8.235}$$

Similarly,

$$[\hat{\pi}_z^2, \hat{\pi}_x] = -i\hbar\left(\frac{e}{c}\right)\{\hat{\pi}_z B_y + B_y \hat{\pi}_z\} \tag{8.236}$$

and, thus, as required to obtain eqn (8.233),

$$[\hat{\pi}^2, \hat{\pi}_x] = -i\hbar\left(\frac{e}{c}\right)\{B_y \hat{\pi}_z - B_z \hat{\pi}_y + \hat{\pi}_z B_y - \hat{\pi}_y B_z\}$$

$$= -i\hbar\left(\frac{e}{c}\right)\{(\mathbf{B} \times \hat{\boldsymbol{\pi}})_x - (\hat{\boldsymbol{\pi}} \times \mathbf{B})_x\} \tag{8.237}$$

where $(\mathbf{B} \times \hat{\boldsymbol{\pi}})_x$ denotes the x-component of $\mathbf{B} \times \hat{\boldsymbol{\pi}}$. The averaging of the commutators in eqns (8.232) and (8.233), as required by eqn (8.225), and multiplication by N gives the force exerted on the electrons in the atom Ω, a quantity denoted by $\mathbf{F}(\Omega, \mathbf{B}, \phi, t)$

$$\mathbf{F}(\Omega, \mathbf{B}, \phi, t) = \frac{N}{2}\{(i/\hbar)\langle[\hat{H}, \hat{\boldsymbol{\pi}}]\rangle_\Omega + cc\} = \langle -\mathbf{V}\hat{V}\rangle_\Omega + \langle e\mathbf{V}\phi(\mathbf{r})\rangle_\Omega$$

$$- (e/4mc)\{\langle(\hat{\boldsymbol{\pi}} \times \mathbf{B} - \mathbf{B} \times \hat{\boldsymbol{\pi}})\rangle_\Omega + cc\} \tag{8.238}$$

where $\langle -\mathbf{V}\hat{V}\rangle_\Omega$ (see eqns (6.10) and (6.17)) is the force in the absence of external fields and the remaining terms give the atomic average of the contributions of the external electric and magnetic fields to the Ehrenfest force.

This result is to be equated to the time rate of change of $\hat{\boldsymbol{\pi}}$ and the flux in its paramagnetic and diamagnetic currents through the surface of the atom using eqn (8.228). Thus the forces exerted on the electrons in an atom Ω arising from the electric fields within and external to the molecule and from an external magnetic field are equivalently expressed as

$$\mathbf{F}(\Omega, \mathbf{B}, \phi, t) = m\langle(\partial \mathbf{J}_\pi/\partial t)\rangle_\Omega - \oint dS(\Omega, \mathbf{r}) \overset{\leftrightarrow}{\sigma}(\mathbf{r}) \cdot \mathbf{n}(\mathbf{r})$$

$$+ \frac{e}{c}\oint dS(\Omega, \mathbf{r})\left[\{\mathbf{A}(\mathbf{r})\mathbf{J}(\mathbf{r}) + \mathbf{J}(\mathbf{r})\mathbf{A}(\mathbf{r})\} \right.$$

$$\left. + \frac{e}{mc}\rho(\mathbf{r})\mathbf{A}(\mathbf{r})\mathbf{A}(\mathbf{r}) \right] \cdot \mathbf{n}(\mathbf{r}) \qquad (8.239)$$

where $\overset{\leftrightarrow}{\sigma}(\mathbf{r})$ is the quantum mechanical stress tensor, eqn (8.173). In the absence of external fields the surface integral of the stress tensor and the atomic average of the time rate of change of the current density $\mathbf{J}(\mathbf{r})$ determine the force, eqn (8.175). The stress tensor has the dimensions of an energy density, that is, of a force per unit area or a pressure, and the force acting on the interior of the atom is given by the integral of the pressure acting on each of its surface elements. A current density and $(e/mc)\mathbf{A}\rho$ both have the dimensions of a velocity density and their multiplication by $(e/c)\mathbf{A}$, as occurs in eqn (8.239), converts them into energy densities or pressures. Thus, the second surface term appearing in the atomic force law measures the contribution to the force acting on the interior of the atom that results from the paramagnetic and diamagnetic pressures acting on its surface.

For a stationary state of a system in the presence of uniform external fields, the expression for the force becomes

$$\mathbf{F}(\Omega, \mathbf{B}, \phi) = \langle -\nabla\hat{V}\rangle_\Omega + \langle e\nabla\phi(\mathbf{r})\rangle_\Omega$$

$$- (e/4mc)\{\langle(\hat{\pi} \times \mathbf{B} - \mathbf{B} \times \hat{\pi})\rangle_\Omega + \text{cc}\}. \qquad (8.240)$$

These interior forces are generated in their entirety by the mechanical and magnetic pressures acting on the atom's surface, i.e.

$$\mathbf{F}(\Omega, \mathbf{B}, \phi) = -\oint dS(\Omega, \mathbf{r})\overset{\leftrightarrow}{\sigma}(\mathbf{r}) \cdot \mathbf{n}(\mathbf{r})$$

$$+ \frac{e}{c}\oint dS(\Omega, \mathbf{r})\left[\{\mathbf{A}(\mathbf{r})\mathbf{J}(\mathbf{r}) + \mathbf{J}(\mathbf{r})\mathbf{A}(\mathbf{r})\} \right.$$

$$\left. + \frac{e}{mc}\rho(\mathbf{r})\mathbf{A}(\mathbf{r})\mathbf{A}(\mathbf{r}) \right] \cdot \mathbf{n}(\mathbf{r}). \qquad (8.241)$$

As for a stationary state in the absence of fields (eqns (6.14) and (6.18)), eqn (8.241) states that the force acting on an atom in a molecule is totally determined by the flux in the forces acting on its surface. The same result is true in the general case of time-varying fields (eqn (8.239)) if one includes the basin term arising from the change in the current with time with other basin contributions to the force acting on the interior of the atom.

The atomic virial theorem for a system in the presence of an electromagnetic field is obtained by setting the generator $\hat{F}$ equal to $\hat{\mathbf{r}} \cdot \hat{\boldsymbol{\pi}}$. The commutators required for the evaluation of $\delta\mathscr{L}$, eqn (8.225), are readily evaluated giving

$$(i/\hbar)[\hat{H}, \hat{\mathbf{r}} \cdot \hat{\boldsymbol{\pi}}] = (i/2m\hbar)\{[\hat{\pi}^2, \hat{\mathbf{r}}] \cdot \hat{\boldsymbol{\pi}} + \hat{\mathbf{r}} \cdot [\hat{\pi}^2, \hat{\boldsymbol{\pi}}]\}$$
$$+ (i/\hbar)[\hat{V} - e\phi(\mathbf{r}), \hat{\mathbf{r}} \cdot \hat{\boldsymbol{\pi}}]. \qquad (8.242)$$

The results obtained for these commutators parallel those for the field-free case, eqn (8.191). Thus the first commutator yields twice the operator for the kinetic energy, expressed here in terms of $\hat{\boldsymbol{\pi}}$ (see eqn (8.226))

$$(i/2m\hbar)[\hat{\pi}^2, \hat{\mathbf{r}}] \cdot \hat{\boldsymbol{\pi}} = \hat{\pi}^2/m = 2\hat{T}_\pi, \qquad (8.243)$$

while the last yields the virial of the gradient of the potential energy operator $\hat{V}$ and external field $\phi(\mathbf{r})$ with respect to the electron of position vector $\hat{\mathbf{r}}$, the virial of the force operator given in eqn (8.232)

$$(i/\hbar)[\hat{V} - e\phi(\mathbf{r}), \hat{\mathbf{r}} \cdot \hat{\boldsymbol{\pi}}] = -\mathbf{r} \cdot \nabla\hat{V} + e\mathbf{r} \cdot \nabla\phi(\mathbf{r}). \qquad (8.244)$$

The commutator of $\hat{\pi}^2$ with $\hat{\boldsymbol{\pi}}$ gives the magnetic force acting on an electron (eqn (8.237)) and the remaining commutator in eqn (8.242) gives the virial of this magnetic force

$$(i/2m\hbar)\mathbf{r} \cdot [\hat{\pi}^2, \hat{\boldsymbol{\pi}}] = -\frac{e}{2mc}\mathbf{r} \cdot \{\hat{\boldsymbol{\pi}} \times \mathbf{B} - \mathbf{B} \times \hat{\boldsymbol{\pi}}\}. \qquad (8.245)$$

The averaging of these commutators in the manner determined by the atomic action principle (eqn (8.225)) serves to define the atomic averages of the kinetic energy and of the virials of the internal and external forces acting on the basin of the atom, i.e.

$$\frac{N}{2}\{(i/\hbar)\langle[\hat{H}, \hat{\mathbf{r}} \cdot \hat{\boldsymbol{\pi}}]\rangle_\Omega + \mathrm{cc}\} = \left\langle\left(\hat{\mathbf{p}} + \frac{e}{c}\mathbf{A}(\mathbf{r})\right)^2\right\rangle_\Omega/m + \langle-\mathbf{r} \cdot \nabla\hat{V}\rangle_\Omega$$
$$+ e\langle\mathbf{r} \cdot \nabla\phi(\mathbf{r})\rangle_\Omega$$
$$- \frac{e}{4mc}\{\langle\mathbf{r} \cdot (\hat{\boldsymbol{\pi}} \times \mathbf{B} - \mathbf{B} \times \hat{\boldsymbol{\pi}})\rangle_\Omega + \mathrm{cc}\}$$
$$= 2T_\pi(\Omega) + \mathscr{V}_b(\Omega) + \mathscr{V}_\phi(\Omega) + \mathscr{V}_B(\Omega). \qquad (8.246)$$

These contributions are equal to the terms given by the alternative expression for the variation in the atomic Lagrangian for the generator $\hat{\mathbf{r}} \cdot \hat{\boldsymbol{\pi}}$, eqn (8.228), which are

$$\delta \mathscr{L}[\Psi, \Omega, \mathbf{B}, \phi] = \langle m(\mathbf{r} \cdot \partial \mathbf{J}_{\mathbf{r} \cdot \pi}(\mathbf{r})/\partial t) \rangle_\Omega$$

$$+ \oint dS(\Omega, \mathbf{r}) \{ (-\hbar^2/4m)\nabla\rho(\mathbf{r}) \cdot \mathbf{n}(\mathbf{r}) - \mathbf{r} \cdot \overleftrightarrow{\sigma}(\mathbf{r}) \cdot \mathbf{n}(\mathbf{r}) \}$$

$$+ \frac{e}{c}\oint dS(\Omega, \mathbf{r}) \mathbf{r} \cdot \left[\{ \mathbf{A}(\mathbf{r})\mathbf{J}(\mathbf{r}) + \mathbf{J}(\mathbf{r})\mathbf{A}(\mathbf{r}) \} \right.$$

$$\left. + \frac{e}{mc}\rho(\mathbf{r})\mathbf{A}(\mathbf{r})\mathbf{A}(\mathbf{r}) \right] \cdot \mathbf{n}(\mathbf{r})$$

$$= -\mathscr{V}_{\dot{\pi}}(\Omega) - \mathscr{V}_{\mathrm{S}}(\Omega) - \mathscr{V}_{\mathrm{S}}^{\mathbf{B}}(\Omega). \tag{8.247}$$

The first term ($\mathscr{V}_{\dot{\pi}}(\Omega)$) is the virial of the force over the basin of the atom arising from a varying magnetic field. The first of the terms in the first surface integral vanishes because of the zero-flux surface condition defining the atom. The second is the virial of the mechanical pressures acting on the atomic surface as determined by the stress tensor $\overleftrightarrow{\sigma}$. The second surface integral is the virial of the magnetic pressures acting on the atomic surface. Equating the two expressions for the variation in the atomic Lagrangian given in eqns (8.246) and (8.247) followed by some rearrangement yields the virial theorem for an atom in a molecule in the presence of an electromagnetic field,

$$-2T_\pi(\Omega) = \mathscr{V}(\Omega, \mathbf{B}, \phi) = \mathscr{V}_{\mathrm{b}}(\Omega) + \mathscr{V}_\phi(\Omega) + \mathscr{V}_{\mathbf{B}}(\Omega)$$

$$+ \mathscr{V}_{\dot{\pi}}(\Omega) + \mathscr{V}_{\mathrm{S}}(\Omega) + \mathscr{V}_{\mathrm{S}}^{\mathbf{B}}(\Omega). \tag{8.248}$$

In addition to the basin virial $\mathscr{V}_{\mathrm{b}}(\Omega)$ arising from the internal forces, which is identical to that obtained in the field-free case (eqn (8.192)), there are further contributions to the virials of the forces acting on the electrons in the basin of the atoms resulting from their interaction with the external fields. The surface virial $\mathscr{V}_{\mathrm{S}}(\Omega)$ reduces to the flux in the virial of the mechanical stresses $-\mathbf{r} \cdot \overleftrightarrow{\sigma} \cdot \mathbf{n}$ in the absence of an external field, eqn (8.194). As in the case of a stationary state, only the value of the total atomic virial is independent of the choice of origin for the vector $\mathbf{r}$.

As discussed in Chapter 6, the atomic statement of the virial theorem, since it defines the atomic averages of the kinetic and potential energies, serves to define the energy of an atom in a molecule. The average electronic potential energy is equal to the virial of the forces acting over the basin of the atom. Thus, the energy of an atom in a molecule in the presence of an electromagnetic field is defined to be

$$E(\Omega) = T_\pi(\Omega) + \mathscr{V}(\Omega, \mathbf{B}, \phi) = -T_\pi(\Omega). \tag{8.249}$$

This energy possesses the characteristic, as do all atomic properties, that it yields the average value for the total system when summed over all the atoms in the molecule

$$E_e = \sum_\Omega E(\Omega) = \langle \hat{H} \rangle \qquad (8.250)$$

where the Hamiltonian operator is defined in eqn (8.226). The second equality assumes that no external mechanical forces act on the nuclei. The atomic contributions to the molecular polarizability and magnetic susceptibility are defined and discussed in the following section.

8.5.4   Atomic contributions to molecular polarizability and susceptibility

An atomic property, as discussed in Chapter 6, is determined by averaging a corresponding property density over the basin of the atom. The property density, however, has the essential characteristic that it includes the effects of all the particles in the system, as determined by the property in question. This is a result of the operator being defined for the interactions experienced by a single electron and of the mode of averaging this operator over the motions of the remaining particles in the system. The density which determines the Ehrenfest force acting on the electrons provides an example of an atomic property which illustrates all of these points. The operator is $- \mathbf{V}_1 \hat{V}$ where the gradient is taken with respect to the coordinates of a single electron, and $\hat{V}$ is the many-particle potential energy operator. As illustrated and discussed in Chapter 6, the operator $- \mathbf{V}_1 \hat{V}$ determines the total force exerted on the electron at $\mathbf{r}_1$ by all of the other particles in the system. (It is worth noting that if one sums this operator for all of the electrons, the electron–electron contribution to the force vanishes.) The averaging of this operator over all the coordinates but $\mathbf{r}_1 \equiv \mathbf{r}$ by the usual recipe, as determined by the atomic action principle, yields the force density

$$\mathbf{F}(\mathbf{r}, t) = N \int d\tau' \Psi^* (- \mathbf{V}_1 \hat{V}) \Psi. \qquad (8.251)$$

This density is N times the force exerted on the electron at $\mathbf{r}$ by the remaining particles in the system, the motions of the latter particles being averaged over the total system. Since the electrons are indistinguishable, multiplication by N gives the force exerted on the total density of electrons at the point $\mathbf{r}$ and integration of the density over the basin of the atom yields the total Ehrenfest force acting on the average electron population of atom Ω. Because of the single-particle nature of the definition of a property density, the sum of the atomic contributions over the atoms in the molecule yields the molecular value for the property. The atomic properties are additive, an essential characteristic for the atoms of chemistry. It should be emphasized that the definition of an atomic property and its consequences are a result of the existence of the atomic principle of stationary action, as opposed to being the result of necessary, but otherwise arbitrarily imposed definitions.

The determination of a property density at some point in a molecule by the total distribution of particles in the system is essential to the definition of atomic contributions to the electric and magnetic properties of a system. The densities for properties resulting from the molecule being placed in an external field must describe how the perturbed motion of the electron at **r** depends upon the field strength everywhere inside the molecule, a point that has been emphasized by others (Maaskant and Oosterhoff 1964). This requirement is met by the definition of an atomic property as determined by the theory of atoms in molecules. Property densities for a molecule in the presence of external electric and magnetic fields have been defined and discussed by Jameson and Buckingham (1980) and the present introduction follows their presentation.

In the presence of an external field, additional terms contribute to μ, the electronic dipole moment of a molecule. These contributions may be expanded in powers of the applied field

$$\mu = \mu^{(0)} + \mu^{(1)} + \mu^{(2)} + \ldots . \tag{8.252}$$

This expression may then be used to define corresponding quantities which describe the response of a system to the external field

$$\mu = \mu^{(0)} + \boldsymbol{\alpha} \cdot \mathscr{E} + \tfrac{1}{2}\boldsymbol{\beta} : \mathscr{E}^2 + \ldots . \tag{8.253}$$

Equation (8.253) defines the polarizability tensor $\boldsymbol{\alpha}$ as the quantity determining the first-order response of a molecular charge distribution to the applied field. The state function can also be expanded in terms of the field ε to give

$$\Psi = \Psi^{(0)} + \Psi^{(1)}\mathscr{E} + \Psi^{(2)}\mathscr{E}^2 \tag{8.254}$$

and, correspondingly, the charge density correct to first-order in the field is given by

$$\begin{aligned}
\rho(\mathbf{r}) &= N \int d\tau' \Psi^* \Psi \\
&= N \int d\tau' \{\Psi^{(0)*}\Psi^{(0)} + [\Psi^{(0)*}\Psi^{(1)} + \Psi^{(1)*}\Psi^{(0)}]\mathscr{E} + \ldots \\
&= \rho^{(0)}(\mathbf{r}) + \rho^{(1)}(\mathbf{r})\mathscr{E} + \ldots \tag{8.255}
\end{aligned}$$

where, as before, $\int d\tau'$ implies a summation over all spins and integration over all of the electronic coordinates with the exception of those denoted by the position vector **r**. The first-order correction to the electronic dipole moment is related to the first-order correction to the density through the equation

$$\mu = -e \int \mathbf{r}\rho(\mathbf{r})d\mathbf{r} = -e \int \mathbf{r}\rho^{(0)}(\mathbf{r})d\mathbf{r} - e\int \mathbf{r}\rho^{(1)}(\mathbf{r})d\mathbf{r} \cdot \mathscr{E}, \tag{8.256}$$

which, upon comparison with eqn (8.253), gives

$$\mu^{(1)} = -e\int \mathbf{r}\rho^{(1)}(\mathbf{r})d\mathbf{r} \cdot \mathscr{E} = \alpha \cdot \mathscr{E}. \tag{8.257}$$

One may define the electric polarizability density $\alpha(\mathbf{r})$ (Jameson and Buckingham 1980) as

$$\alpha(\mathbf{r}) = -e\mathbf{r}\rho^{(1)}(\mathbf{r}) = -eN\int d\tau'(\Psi^{(0)*}\mathbf{r}\Psi^{(1)} + \Psi^{(1)*}\mathbf{r}\Psi^{(0)}), \tag{8.258}$$

which has the property of yielding the polarizability of the molecule when integrated over all space. The integration of $\alpha(\mathbf{r})$ over the basin of an atom does not, however, yield the total atomic contribution to the polarizability tensor of a molecule. This is because the surface of the atom is also perturbed by the applied field $\mathscr{E}$ and the total atomic contribution to the atomic polarizability consists of both a basin and a surface term. The atomic contribution to the electronic dipole moment in the presence of the field is obtained by restricting the integration of eqn (8.256) to the basin of an atom. We denote the basin of the atom in the field by $\Omega_f = \Omega_0 + \delta\Omega$ where Ω_0 is the unperturbed basin. Retaining terms to first-order only yields

$$\mu(\Omega) = -e\int_{\Omega_f} \mathbf{r}\rho(\mathbf{r})d\mathbf{r}$$

$$= -e\int_{\Omega_0} \mathbf{r}\rho^{(0)}(\mathbf{r})d\mathbf{r} - e\int_{\Omega_0} \mathbf{r}\rho^{(1)}d\mathbf{r} \cdot \mathscr{E} - e\int_{\delta\Omega} \mathbf{r}\rho^{(0)}(\mathbf{r})d\mathbf{r}$$

$$= \mu^{(0)}(\Omega) + \mu^{(1)}(\Omega).$$

Thus, the first-order correction to the atomic electronic dipole moment is

$$\mu^{(1)}(\Omega) = -e\int_{\Omega_0} \mathbf{r}\rho^{(1)}(\mathbf{r})d\mathbf{r} \cdot \mathscr{E} - e\int_{\delta\Omega} \mathbf{r}\rho^{(0)}(\mathbf{r})d\mathbf{r}$$

where the final term is dependent upon the field $\mathscr{E}$ through the change in the basin resulting from the change in the atomic surface. The atomic contribution to the polarization, $\alpha(\Omega)$, is determined by a basin and a surface contribution,

$$\alpha(\Omega) = \alpha_b(\Omega) + \alpha_s(\Omega). \tag{8.259}$$

The basin contribution can be expressed in terms of the polarizability density $\alpha(\mathbf{r})$ as

$$\alpha_b(\Omega) = \int_{\Omega} \alpha(\mathbf{r})d\mathbf{r} \tag{8.260}$$

and the surface contribution

$$\alpha_s(\Omega) = -e\int_{\delta\Omega} \mathbf{r}\rho^{(0)}(\mathbf{r})\,d\mathbf{r} \cdot \mathscr{E}^{-1}.$$

The expression given in eqn (8.259) for $\alpha(\Omega)$ is equivalent to the result given by the calculus of variations for the first-order change in an atomic property. In the present case, the variation in $\mu(\Omega)$ is

$$\delta\mu(\Omega) = -e\int_\Omega \mathbf{r}\delta\rho(\mathbf{r})d\mathbf{r}\cdot\mathscr{E} - e\oint dS(\Omega, \mathbf{r})\delta S(\Omega, \mathbf{r})\mathbf{r}\rho(\mathbf{r}),$$

that is, as a variation of the property over the basin, plus a contribution from the variation of the atomic surface. The surface contributions cancel when summed over all the atoms in a molecule because the shifts in the interatomic surface between atoms Ω and Ω' are of opposite sign, that is $\delta S(\Omega|\Omega') = -\delta S(\Omega'|\Omega)$, and the polarizability of the molecule is obtained by summing just the basin contributions

$$\alpha = \sum_\Omega \alpha_b(\Omega). \tag{8.261}$$

To put the definition of this property into direct correspondence with the definition of other atomic properties, as one for which the property density at $\mathbf{r}$ is determined by the effect of the field over the entire molecule, we express the perturbed density in terms of the first-order corrections to the state function. This is done in a succinct manner by using the concept of a transition density (Longuet–Higgins 1956). The operator whose expectation value yields the total electronic charge density at the position $\mathbf{r}$ may be expressed in terms of the Dirac delta function as

$$\hat\rho(\mathbf{r}) = N\delta(\mathbf{r} - \mathbf{r}'). \tag{8.262}$$

From the definitive property of the δ function, the zero-order contribution to the dipole moment is, for example, given by (the unperturbed state functions are denoted by $|0\rangle$, $|k\rangle$)

$$\mu^{(0)} = -e\int \mathbf{r}\langle 0|\hat\rho(\mathbf{r})|0\rangle d\mathbf{r} = -e\langle 0|\int \mathbf{r}\hat\rho(\mathbf{r})d\mathbf{r}|0\rangle = -eN\langle 0|\mathbf{r}|0\rangle$$

showing that the electronic density of the unperturbed state is

$$\rho_{00}(\mathbf{r}) = \langle 0|\hat\rho(\mathbf{r})|0\rangle = \rho^{(0)}(\mathbf{r}). \tag{8.263}$$

The transition density $\rho_{0k}(\mathbf{r})$ is similarly defined as

$$\rho_{0k}(\mathbf{r}) = \langle 0|\hat\rho(\mathbf{r})|k\rangle. \tag{8.264}$$

The transition density describes a charge distribution whose dipole moment is equal to the transition moment connecting states $|0\rangle$ and $|k\rangle$, the matrix element which determines the admixture of the state $|k\rangle$ to the state function for the molecule when it is perturbed by an external electric field.

The first-order correction to the charge density for a molecule in the presence of an external electric field, as defined in eqn (8.255) and expressed in terms of the first-order corrections to the state function $|0\rangle$, is

$$\rho^{(1)}(\mathbf{r}) = 2{\sum_k}' \frac{\mathbf{D}_{0k}}{E_0 - E_k} \rho_{0k}(\mathbf{r}) \tag{8.265}$$

where $\mathbf{D}_{0k}$ is the transition moment connecting the ground state with the excited state $|k\rangle$

$$\mathbf{D}_{0k} = - e \int \mathbf{r} \rho_{0k} d\mathbf{r}. \tag{8.266}$$

The first-order correction to the charge density is independent of the choice of origin for the dipole moment operator, as the extra contribution to $\mathbf{D}_{0k}$ arising from a shift $\delta\mathbf{R}$ in origin is $- e\delta\mathbf{R}\langle 0|k \rangle$, a term which vanishes because of the orthogonality of the zero-order states. The expression for the electric polarizability density $\alpha(\mathbf{r})$ using the above expression for the perturbed density is

$$\alpha(\mathbf{r}) = - 2e{\sum_k}' \frac{\mathbf{D}_{0k}}{E_0 - E_k} \mathbf{r} \rho_{0k}(\mathbf{r}). \tag{8.267}$$

One may obtain from this equation an expression for the polarization susceptibility

$$\alpha(\mathbf{r}', \mathbf{r}'') = 2e^2 {\sum_k}' \frac{\langle 0|\mathbf{r}\delta(\mathbf{r} - \mathbf{r}'')|k \rangle \langle k|\mathbf{r}\delta(\mathbf{r} - \mathbf{r}')|0 \rangle}{E_0 - E_k}. \tag{8.268}$$

This expression, without the specific perturbations given by $- e\mathbf{r}$, is what Stone (1985) called the charge susceptibility of a system. It contains all of the information needed to evaluate any of the molecular polarizabilities. The polarizability density, as defined in eqn (8.258), is obtained by integrating the polarization susceptibility over the coordinate $\mathbf{r}''$

$$\alpha(\mathbf{r}') = \int d\mathbf{r}'' \alpha(\mathbf{r}', \mathbf{r}''), \tag{8.269}$$

an operation in complete analogy with the procedure used to obtain the definition of the Ehrenfest force density given in eqn (8.251). Thus, the contribution of the basin term to the atomic polarizability is determined by a double integration,

$$\alpha_b(\Omega) = \int_\Omega d\mathbf{r}' \int d\mathbf{r}'' \alpha(\mathbf{r}', \mathbf{r}''), \tag{8.270}$$

the first describing how the density at $\mathbf{r}'$ depends upon the effect of the field over the entire molecule, the second yielding the contribution of these effects to a particular region of space.

A parallel development of the atomic contribution to the magnetic susceptibility of a molecule is readily obtained, the charge density being replaced by the vector current density. The magnetic dipole moment $\mathbf{m}$ is defined in terms of the current density $\mathbf{J}(\mathbf{r})$ as

$$\mathbf{m} = \frac{1}{2c} \int \mathbf{r} \times \mathbf{J}(\mathbf{r})\,d\mathbf{r}, \tag{8.271}$$

and the magnetic dipole moment density $\mathbf{m}(\mathbf{r})$, in analogy with the electric dipole density given in eqn (8.256), is

$$\mathbf{m}(\mathbf{r}) = \frac{1}{2c} \mathbf{r} \times \mathbf{J}(\mathbf{r}). \tag{8.272}$$

The expansion of the magnetic moment in powers of the applied uniform magnetic field $\mathbf{B}$ yields

$$\mathbf{m} = \mathbf{m}^{(0)} + \mathbf{m}^{(1)} + \mathbf{m}^{(2)} + \ldots = \mathbf{m}^{(0)} + \chi \cdot \mathbf{B} + \ldots \tag{8.273}$$

where $\mathbf{m}^{(0)}$ is the intrinsic magnetic moment and the field-induced moment $\mathbf{m}^{(1)} = \chi \cdot \mathbf{B}$, is proportional to the magnetic susceptibility χ. Thus, a magnetization density $\chi(\mathbf{r})$ can be defined as (Jameson and Buckingham 1980)

$$\chi(\mathbf{r}) \cdot \mathbf{B} = \frac{1}{2c} \mathbf{r} \times \mathbf{J}_{\mathbf{B}}^{(1)}(\mathbf{r}) \tag{8.274}$$

where, in analogy with eqn (8.258) for the definition of the polarizability density in terms of the first-order density, $\mathbf{J}(\mathbf{r})_{\mathbf{B}}^{(1)}$ is the first-order correction to the current density in the presence of a magnetic field. The expression for this field-induced current (compare eqn (8.220)) including the field strength $\mathbf{B}$ and the sign of the electronic charge is

$$\mathbf{J}_{\mathbf{B}}^{(1)}(\mathbf{r}) = \frac{-e\hbar\mathbf{B}}{im} \int d\tau' \{\Psi^{(1)*}\nabla\Psi^{(0)} - \Psi^{(0)*}\nabla\Psi^{(1)}\} - \frac{e^2}{mc}\mathbf{A}\rho^{(0)}(\mathbf{r}). \tag{8.275}$$

By using the above two equations, the magnetization density can be expressed in terms of the zero-order and first-order corrections to the state function

$$\chi(\mathbf{r}) = \frac{-e\hbar}{2imc}(\Psi^{(1)*}\mathbf{r} \times \nabla\Psi^{(0)} - \Psi^{(0)}\mathbf{r} \times \nabla\Psi^{(1)*})$$

$$- \frac{e^2}{4mc^2}\rho^{(0)}(\mathbf{r})(\mathbf{r} \cdot \mathbf{r}\mathbf{I} - \mathbf{r}\mathbf{r}). \tag{8.276}$$

The atomic contribution to the magnetic susceptibility is obtained in analogy with the definition of $\alpha(\Omega)$, by a basin contribution obtained by integration of $\chi(\mathbf{r})$ over the basin of the atom,

$$\chi_b(\Omega) = \int_\Omega \chi(\mathbf{r})\,d\mathbf{r}, \tag{8.277}$$

but the surface contribution vanishes in this case and the molecular value is given by the sum of the basin contributions

$$\chi = \sum_\Omega \chi_b(\Omega). \tag{8.278}$$

While an electric field results in a first-order change in the electron density and causes measurable shifts in the interatomic surfaces, the first-order correction to $\rho(\mathbf{r})$ vanishes for an applied magnetic field, since $\Psi^{(1)}$ is purely imaginary. The reader is referred to recent work (Laidig and Bader 1990; Keith and Bader 1993; Bader and Keith 1993) for the application of these ideas.

Other magnetic properties, such as the nuclear magnetic shielding σ^N, can be expressed in terms of a density by using the perturbed vector current in eqn (8.275) (Jameson and Buckingham 1980). Thus, the magnetic shielding of a given nucleus in a molecule can be expressed as a sum of atomic contributions. Because of the $\mathbf{r}_\Omega^{-3}$ dependence of this property, one anticipates that the major contribution to σ^N will be from the basin of the atom containing the nucleus in question $\sigma^N(\Omega)$. The effect of a changing environment on $\sigma^N(\Omega)$ and on its transferability between molecules can be studied in a direct and quantitative manner.

8.6   The action principle—in the past and in the future

The idea that natural behaviour can be considered to be the result of the operation of certain minimum principles has a long history, beginning with Fermat's principle of least time. This principle, which arose in the study of optics, was enunciated in the seventeenth century. In 1744 Maupertuis proclaimed that nature acts in such a way as to render a certain quantity, which he termed the 'action', a minimum, but it was Lagrange who gave the first correct formulation of the principle of least action for the general case. Lagrange's formulation of the principle suffered from the necessity of imposing the law of conservation of energy from the outset and the principle languished in disuse until Hamilton enunciated the principle in what is generally recognized to be its most effectual form, Hamilton's principle (eqn (8.43)). Hamilton later established a formal analogy between mechanics and optics using the original statement of the principle of least action and it was this relationship that de Broglie and Schrödinger built upon to establish the new wave mechanics. (A reader, interested in the development of the action principle and its underlying philosophy, is referred to the book by Yourgrau and Mandelstam (1968).)

In 1933 Dirac published a paper entitled 'The Lagrangian in quantum mechanics'. After presenting a discussion as to why the Lagrangian formulation of classical mechanics could be considered to be more fundamental than the approach based on the Hamiltonian theory, Dirac went on to say 'For these reasons it would seem desirable to take up the question of what corresponds in the quantum theory to the Lagrangian method of classical mechanics.' In the opening section of this chapter, attention was drawn to the transformation function or transition amplitude. This function relates the state function at one time to that at another (eqn (8.26)) and it thus determines the dynamical behaviour of a system. Dirac was the inventor of transformation theory. Through repeated use of the multiplicative law of transformation functions (eqn (8.24)) he was able to express the function connecting states at times t_1 and t_2 by a sequence of transformation functions for times intermediate between the initial and final times. Taken to the limit of the successive intermediate times differing only infinitesimally one from the next, the multiplicative law yields a product of all the transformation functions associated with the successive infinitesimal increments of time. Dirac then stated that the transformation function associated with the time displacement from t to $t + dt$ corresponds to $\exp[(i/\hbar)L\, dt]$, where the Lagrangian L was to be considered as a function of the coordinates at time t and at time $t + dt$, rather than of the coordinates and velocities. The transformation function then becomes $\exp[(i/\hbar)W]$ where W, the action integral equal to $\int L\, dt$ between the limits t_1 and t_2, is interpreted as the sum over all the individual coordinate-dependent terms in the succession of values of t. With this construct Dirac was able to answer the question of what in quantum mechanics corresponds to the classical principle of stationary action.

Feynman built on this work and in 1948 it culminated in his path integral formulation of quantum mechanics. In the classical limit considered by Dirac, only one trajectory connects the system at time t_1 to that at time t_2 and he limited his discussion to this case. What Feynman did was to consider all the trajectories or paths that connect the states at the initial and final times, since he wished to obtain the corresponding quantum limit. Each path has its own value for the action W and all the values of $\exp[(i/\hbar)W]$ must be added together to obtain the total transition amplitude. Thus, the expression for the transition amplitude between the states $|q_{r1}\rangle$ and $|q_{r2}\rangle$ is the sum of the elementary contributions, one from each trajectory passing between q_{r1} at time t_1 and q_{r2} at time t_2. Each of these contributions has the same modulus, but its phase is the classical action integral $(1/\hbar)\int L\, dt$ for the path. This is expressed as,

$$\langle q_{r2}, t_2 | q_{r1}, t_1 \rangle = \frac{1}{N} \int \exp\left\{ (i/\hbar) \int_{t_1}^{t_2} L\, dt \right\} \delta q_r(t). \qquad (8.279)$$

The differential $\delta q_r(t)$ indicates that one must integrate over all paths connecting q_{r1} at t_1 and q_{r2} at t_2 and $1/N$ is a normalizing factor.

Schwinger's quantum action principle (1951) is developed in Section 8.2, and a statement of this principle, in which the action integral and Lagrange function appear as operators, is repeated as

$$\delta \langle q_{r2}, t_2 | q_{r1}, t_1 \rangle = (i/\hbar) \langle q_{r2}, t_2 | \delta \hat{W}_{12} | q_{r1}, t_1 \rangle$$

$$= (i/\hbar) \left\langle q_{r2}, t_2 \left| \delta \int_{t_1}^{t_2} \hat{\mathscr{L}}[t] dt \right| q_{r1}, t_1 \right\rangle. \tag{8.280}$$

It is a differential statement of Feynman's path integral formulation and, while Schwinger developed it independently, it can be obtained as a consequence of Feynman's principle (see for example, Yourgrau and Mandelstam 1968). The conceptual advantages afforded by these action principles in formulating the laws of quantum mechanics are contained in the quotation from Schwinger which opens this chapter. These principles represent more than alternative formulations of the laws of quantum mechanics. In fact, they may provide the real foundation of quantum mechanics and thus of physical theory (Gell-Mann 1989). One final quotation from Schwinger (1951) regarding the possible uses of the quantum action principle is relevant to the present work, '. . . to present a general theory of quantum field dynamics which unifies several independently developed procedures and which may provide a framework capable of admitting fundamentally new physical ideas'. The idea that matter is composed of atoms is not new. It is indeed amongst the oldest of scientific hypotheses. However, the idea of finding atoms within the framework of quantum mechanics is perhaps a new physical idea and this is what Schwinger's principle makes possible.

The same multiplicative composition law of transformation functions used by Dirac is used by Schwinger to show that the $\delta \hat{W}$, the generating operators of infinitesimal transformations, must satisfy an additive law of composition,

$$\delta \hat{W}_{13} = \delta \hat{W}_{12} + \delta \hat{W}_{23}. \tag{8.281}$$

The possibility developed in this treatise of a well-defined physical partitioning scheme $\{\Omega_i(t); i = 1, 2, \ldots, n\}$ at any time t, as afforded by the zero-flux boundary condition

$$\nabla \rho(\mathbf{r}, t) \cdot \mathbf{n}(\mathbf{r}, t) = 0 \quad \text{for all points in the boundary,} \tag{8.282}$$

ensures the spatially additive property of the action

$$\hat{W}_{12} = \sum_i \hat{W} [\Omega_i]_{12} \tag{8.283}$$

and of its variation

$$\delta \hat{W}_{12} = \sum_i \delta \hat{W} [\Omega_i]_{12} \tag{8.284}$$

where

$$\hat{W} [\Omega_i]_{12} = \int_{t_1}^{t_2} \hat{\mathscr{L}} [\Omega_i(t)] dt. \tag{8.285}$$

$\hat{\mathscr{L}}[\Omega_i(t)]$ is the projection of the Lagrangian function operator on to the subsystem, that is, on to the atom $\Omega_i(t)$. Thus, the action principle can be expressed as a sum over the changes in the action operators for all the atoms in the total system. This decomposition of the action is illustrated in eqn (8.286) for the general case of a system whose state vector is characterized by the eigenvalues α of some complete set of commuting observables

$$\delta\langle\alpha_2, t_2|\alpha_1, t_1\rangle = (i/\hbar)\sum_i\langle\alpha_2, t_2|\delta\hat{\mathscr{W}}[\Omega_i]_{12}|\alpha_1, t_1\rangle. \qquad (8.286)$$

This result, or a corresponding expression in terms of Feynman's formulation, enables one to follow a particular atom or, more generally, a particular open system, through any change in state. The open systems satisfying the boundary condition in eqn (8.282) can assume many forms, from individual atoms to linked grouping of atoms varying in size up to the total system, interacting weakly or strongly with their neighbours. The questions that can be posed in terms of the paths these fragments follow are manifold.

References

Bader, R. F. W. (1980). *J. Chem. Phys.* **73**, 2871.

Bader, R. F. W. and Keith, T. A. (1993). *J. Chem. Phys.* **99**, 3683.

Bader, R. F. W. and Nguyen-Dang, T. T. (1981). *Adv. quantum Chem.* **14**, 63.

Bader, R. F. W., Srebrenik, S., and Nguyen-Dang, T. T. (1978). *J. Chem. Phys.* **68**, 3680.

Dirac, P. A. M. (1933). *Physik. Zeits. Sowjetunion* **3**, 64.

Dirac, P. A. M. (1958). *The principles of quantum mechanics*. Oxford University Press, Oxford.

Ehrenfest, P. (1927). *Z. Phys.* **45**, 455.

Epstein, S. T. (1974). *The variation method in quantum chemistry*. Academic Press, New York.

Epstein, S. T. (1975). *J. Chem. Phys.* **63**, 3573.

Feynman, R. P. (1948). *Rev. mod. Phys.* **20**, 367.

Feynman, R. P., Leighton, R. B., and Sands, M. (1964). *The Feynman lectures on physics*, Vol. II. Addison-Wesley, Reading, Massachusetts, p. 19-8.

Gell-Mann, M. (1989). *Physics To-day* **42**(2), 50.

Goldstein, H. (1980). *Classical mechanics* (2nd edn). Addison-Wesley, Reading, Massachusetts.

Jameson, C. J. and Buckingham, A. D. (1980). *J. Chem. Phys.* **73**, 5684.

Keith, T. A. and Bader, R. F. W. (1993). *J. Chem. Phys.* **99**, 3669.

Laidig, K. E. and Bader, R. F. W. (1990). *J. Chem. Phys.* **93**, 7213.

Landau, L. D. and Lifshitz, E. M. (1975). *The classical theory of fields*, Vol. II. Pergamon Press, Oxford.

Longuet-Higgins, H. C. (1956). *Proc. r. Soc., London* **A235**, 537.

Masskant, W. J. A. and Oosterhoff, L. J. (1964). *Mol. Phys.* **8**, 319.

Malvern, L. E. (1969). *Introduction to the mechanics of a continuous medium*. Prentice-Hall, Englewood Cliffs, New Jersey.

Messiah, A. (1958). *Quantum mechanics*, Vol. I. North-Holland, Amsterdam.

Morse, P. M. and Feshbach, H. (1953). *Methods of theoretical physics*, Vol. I. McGraw-Hill, New York.

Pauli, W. (1958). In *Handbuch der Physik*, Vol. 5, Part 1 (ed. S. Fiugge), p. 1. Springer-Verlag, Berlin.

Roman, P. (1965). *Advanced quantum theory*. Addison-Wesley, Reading, Massachusetts.

Schwinger, J. (1951). *Phys. Rev.* **82**, 914.

Srebrenik, S., Bader, R. F. W., and Nguyen-Dang, T. T. (1978). *J. Chem. Phys.* **68**, 3667.

Stone, A. J. (1985). *Mol. Phys.* **56**, 1065.

Yourgrau, W. and Mandelstam, S. (1968). *Variational principles in dynamics and quantum theory*. Dover, New York.

APPENDIX

Table A1
Units

Length	$1 \text{ au} = a_0 = 0.529177 \text{ Å}$ $= 5.29177 \times 10^{-11} \text{ m}$
Elementary charge	$1 \text{ au} = e = 1.60219 \times 10^{-19} \text{ C}$ $= 4.8029 \times 10^{-10} \text{ esu}$
Charge density	$1 \text{ au} = e/a_0^3 = 6.7483 \text{ e} \text{Å}^{-3}$ $= 1.0812 \times 10^{12} \text{ C m}^{-3}$
Laplacian density	$1 \text{ au} = e/a_0^5 = 24.099 \text{ e} \text{Å}^{-5}$ $= 3.8611 \times 10^{32} \text{ C m}^{-5}$
Dipole moment	$1 \text{ au} = e a_0 = 2.542 \text{ Debyes}$ $= 8.479 \times 10^{-30} \text{ C m}$
Energy	$1 \text{ au} = e^2/a_0 = 627.51 \text{ kcal/mol}^{-1}$ $= 2.6255 \times 10^3 \text{ kJ mol}^{-1}$ $= 27.212 \text{ eV}$ $= 2.1947 \times 10^5 \text{ cm}^{-1}$

Table A2
Contour values for maps of the electronic charge density and of the Laplacian distribution function

Unless otherwise stated, the outermost charge density contour equals 0.002 au. The contours increase in order 2×10^n, 4×10^n, 8×10^n with n increasing in steps of unity from $n = -3$. In certain maps, as is stated in the figure caption, the outermost contour shown equals 0.001 au, the remaining contours increasing as above.

The contours of the Laplacian of the electronic charge density (dashed contours $\nabla^2 \rho < 0$, solid contours $\nabla^2 \rho > 0$) increase and decrease from a zero contour in steps $\pm 2 \times 10^n$, $\pm 4 \times 10^n$, $\pm 8 \times 10^n$, beginning with $n = -3$ and increasing in steps of unity. The innermost solid contour encompassing a nucleus with $Z > 1$ encloses the innermost region of charge concentration.

Table A3

Properties of ρ at $(3, -1)$ critical point together with bonded and non-bonded radii for ground-state diatomic molecules*

System	State	R	$r_b(A)$	$r_b(B)$	$r_n(A)$	$r_n(B)$	$r_p^\dagger$	ρ_b	$\nabla^2\rho_b$	$\lambda_1 = \lambda_2$	λ_3
HH	$^1\Sigma_g^+$	1.4000	0.7000	0.7000	2.54	2.54	2.93	0.2728	-1.3784	-0.9917	0.6049
LiH	$^1\Sigma^+$	3.0150	1.3444	1.6706	1.88	3.27	3.68B	0.0407	0.1571	-0.0619	0.2808
BeH	$^2\Sigma^+$	2.5380	1.0829	1.4551	4.68	2.88	3.54A	0.0965	0.1638	-0.2032	0.5703
BH	$^1\Sigma^+$	2.3360	1.0015	1.3345	4.34	2.75	3.59A	0.1843	-0.5847	-0.4622	0.3397
CH	$^2\Pi$	2.1240	1.3584	0.7656	3.86	2.52	3.54A	0.2787	-1.0389	-0.7572	0.4754
NH	$^3\Sigma^-$	1.9614	1.4227	0.5387	3.51	2.37	3.38A	0.3394	-1.6034	-1.2399	0.8765
OH	$^2\Pi$	1.8342	1.4509	0.3833	3.23	2.18	3.29A	0.3756	-2.6422	-1.8968	1.1454
FH	$^1\Sigma^+$	1.7328	1.4648	0.2680	2.97	2.01	3.18A	0.3884	-4.3872	-2.8015	1.2159
NaH	$^1\Sigma^+$	3.5660	1.8790	1.6870	2.64	3.33	3.72B	0.0337	0.1320	-0.0394	0.2108
MgH	$^2\Sigma^+$	3.2710	1.6640	1.6070	5.18	3.09	3.92A	0.0529	0.1844	-0.0695	0.3233
AlH	$^1\Sigma^+$	3.1140	1.5176	1.5964	4.97	3.03	4.13A	0.0743	0.1883	-0.1054	0.3990
SiH	$^2\Pi$	2.8740	1.3882	1.4858	4.49	2.82	4.34A	0.1134	0.1330	-0.1666	0.4663
PH	$^3\Sigma^-$	2.7080	1.3238	1.3842	4.16	2.64	4.25A	0.1590	-0.1142	-0.2222	0.3302
SH	$^2\Pi$	2.5510	1.5482	1.0028	3.86	2.46	4.14A	0.2128	-0.5663	-0.3732	0.1801
ClH	$^1\Sigma^+$	2.4087	1.6716	0.7371	3.68	2.28	3.99A	0.2540	-0.7824	-0.6035	0.4247
LiLi	$^1\Sigma_g^+$	5.0510			2.61	2.61	4.29				
LiBe	$^2\Sigma^+$	5.1662	1.6797	3.4866	3.77	4.59	4.02A	0.0130	0.0286	-0.0086	0.0439
LiB	$^1\Sigma^+$	4.5000	1.4662	3.0338	1.94	4.49	3.59B	0.0269	0.0854	-0.0275	0.1404
LiC	$^2\Pi$	3.8000	1.3143	2.4858	1.88	4.07	3.68B	0.0445	0.2097	-0.0664	0.3424
LiN	$^3\Sigma^{-1}$	3.4000	1.2241	2.1759	1.94	3.65	3.54B	0.0603	0.3656	-0.1127	0.5910
LiO	$^2\Pi$	3.1840	1.1850	1.9990	1.91	3.54	3.51B	0.0673	0.4825	-0.1428	0.7681
LiF	$^1\Sigma^+$	2.9550	1.1304	1.8246	1.91	3.12	3.36B	0.0802	0.7018	-0.1932	1.0882

Molecule	State										
BeB	$^2\Pi$	4.0000	1.3847	2.6153	4.35	3.89	4.25B	0.0409	−0.0100	−0.0303	0.0506
BeC	$^3\Sigma^-$	3.5000	1.1819	2.3181	4.17	3.63	4.05A	0.0580	0.1277	−0.0617	0.2511
BeN	$^2\Sigma^+$	3.2500	1.0999	2.1501	3.65	3.42	3.86A	0.0758	0.2918	−0.0992	0.4901
BeO	$^1\Sigma^+$	2.5149	0.8944	1.6205	1.50	3.41	3.71B	0.1842	1.5892	−0.4586	2.5064
BeF	$^2\Sigma^+$	2.5720	0.9268	1.6453	4.58	3.08	3.56A	0.1535	1.3393	−0.4925	2.3242
BB	$^3\Sigma_g^-$	3.0050	1.5025	1.5025	3.89	3.89	3.98	0.1250	−0.1983	−0.0998	0.0014
BC	$^4\Sigma^-$	2.7031	0.8852	1.8179	3.88	3.96	3.76A	0.2316	0.1101	−0.4891	1.0884
BN	$(^3\Sigma^-)$‡	2.4210	0.8166	1.6044	4.31	3.54	3.74A	0.3142	0.4378	−1.1563	2.7504
BO	$^2\Sigma^+$	2.2750	0.7765	1.4985	3.95	3.27	3.44B	0.3194	2.0866	−0.9405	3.9676
BF	$^1\Sigma^+$	2.3910	0.8093	1.5817	4.23	2.99	3.68A	0.2391	1.5852	−0.9234	3.4319
CC	$^1\Sigma_g^+$	2.3481			3.45	3.45	3.75				
CN	$^2\Sigma^+$	2.2140	0.7633	1.4507	3.51	3.47	3.48B	0.4474	0.1547	−0.7565	1.6677
CO	$^1\Sigma^+$	2.1320	0.7228	1.4092	3.77	3.21	3.42A	0.5101	0.2692	−1.7696	3.8083
CF	$^2\Pi$	2.4020	0.7783	1.6237	3.69	2.96	3.44A	0.2924	0.6047	−0.7822	2.1691
NN	$^1\Sigma_g^+$	2.0680	1.0340	1.0340	3.42	3.42	3.42	0.7219	−3.0500	−1.9337	0.8175
NO	$^2\Pi$	2.1747	0.9153	1.2594	3.33	3.14	3.38A	0.5932	−2.0353	−1.6460	1.2568
NF	$^3\Sigma^-$	2.4890	1.0211	1.4679	3.27	2.88	3.33A	0.3420	−0.4972	−0.7853	1.0733
OO	$^3\Sigma_g^-$	2.2820	1.1410	1.1410	3.08	3.08	3.23	0.5513	−1.0127	−1.4730	1.9333
OF	$^2\Pi$	2.4958	1.1698	1.3260	2.94	2.90	3.20A	0.3692	0.0302	−0.9414	1.9130
FF	$^1\Sigma_g^+$	2.6800	1.3400	1.3400	2.84	2.84	3.08	0.2945	0.2287	−0.7327	1.6941

Atomic units used throughout. Calculated from functions close to the Hartree–Fock limit using a large STO basis set: AB, Cade, P. E. and Huo, W. M. (1975). Atomic Data and Nuclear Data Tables 15, 1; AH, Cade, P. E. and Huo, W. (1973). Atomic Data and Nuclear Data Tables 12, 415; A$_2$, Cade, P. E. and Wahl, A. C. (1974). Atomic Data and Nuclear Data Tables 13, 339; OF, O'Hare, P. A. G. and Wahl, A. C. (1970). J. Chem. Phys. 53, 2469. R is the experimental value of the bond length where known. No Slater-type functions are given for LiBe and BC. The results for these molecules are for functions obtained using the 6–311G basis in ROHF calculations.

†For each system r_p is the maximum distance perpendicular to the bond axis from a nucleus to the 0.001 au contour of ρ. The letter A or B listed to the immediate right of this value denotes the nucleus of the atom giving the larger r_p value.

‡The ground state of BN is $^3\Pi$, the Slater function is for a $^3\Sigma^-$ state, the ground state presenting convergence difficulties. The atomic charges given for BN in Fig. 6.2 are for a partially converged function of $^3\Pi$ symmetry for an optimized distance of 1.286 Å (exp 1.281 Å) using the 6–311G* basis set.

Table A4

Atomic charges, moments, and volumes of ground-state diatomic molecules*

System AB	State X	q(A)	M(A)	M(B)	Q_{zz}(A)	Q_{zz}(B)	v(A)	v(B)	v(AB)†
HH	$^1\Sigma_g^+$	0.000	0.095	−0.095	0.759	0.759	60.24	60.24	120.48
LiH	$^1\Sigma^+$	0.912	−0.001	0.389	0.051	1.524	30.53	193.48	224.01
BeH	$^2\Sigma^+$	0.868	1.520	0.571	−2.878	0.885	161.03	137.67	298.70
BH	$^1\Sigma^+$	0.754	1.951	0.493	−1.534	0.352	166.79	95.19	261.98
CH	$^2\Pi$	0.032	0.806	−0.121	−0.008	0.604	169.28	54.61	223.89
NH	$^3\Sigma^-$	−0.323	0.182	−0.176	0.067	0.199	150.14	34.70	184.84
OH	$^2\Pi$	−0.585	−0.225	−0.148	0.065	−0.012	144.43	21.32	165.75
FH	$^1\Sigma^+$	−0.761	−0.453	−0.103	−0.022	−0.078	128.65	12.45	141.10
NaH	$^1\Sigma^+$	0.811	0.018	0.135	0.600	2.529	83.48	189.62	273.10
MgH	$^2\Sigma^+$	0.796	1.701	0.302	−3.159	1.636	255.10	155.18	410.28
AlH	$^1\Sigma^+$	0.825	2.274	0.359	−1.792	0.934	268.18	131.47	399.65
SiH	$^2\Pi$	0.794	1.973	0.427	3.011	1.115	272.52	110.25	382.77
PH	$^3\Sigma^-$	0.578	1.461	0.316	4.800	0.803	251.72	82.44	334.16
SH	$^2\Pi$	0.094	0.587	−0.009	4.445	0.624	249.49	54.76	304.25
ClH	$^1\Sigma^+$	−0.242	−0.007	−0.103	3.666	0.311	231.87	36.97	268.84
LiLi	$^1\Sigma_g^+$	0.000	0.391	−0.391	0.365	0.365	110.19	110.19	220.38
LiBe	$^2\Sigma^+$	0.464	−0.648	−1.088	1.130	−6.512	161.71	378.98	540.69
LiB	$^1\Sigma^+$	0.761	0.242	1.016	−0.122	−10.646	60.64	326.36	387.00
LiC	$^2\Pi$	0.884	0.044	0.645	−0.040	−4.702	33.57	257.58	291.15
LiN	$^3\Sigma^{-1}$	0.916	0.008	0.408	0.043	−1.721	28.06	204.93	232.99
LiO	$^2\Pi$	0.935	0.010	0.295	0.056	−0.332	26.56	180.34	206.90

LiF	$^1\Sigma^+$	0.938	0.017	0.281	0.061	0.726	25.45	149.30	174.75
BeB	$^2\Pi$	0.438	0.892	0.466	2.732	0.166	234.78	258.41	493.19
BeC	$^3\Sigma^-$	0.853	0.893	0.920	1.333	0.489	184.92	241.82	426.74
BeN	$^2\Sigma^+$	1.236	0.539	1.218	0.801	0.320	124.05	231.63	355.68
BeO	$^1\Sigma^+$	1.694	0.095	1.243	0.110	0.190	32.42	220.08	252.50
BeF	$^2\Sigma^+$	0.945	1.429	0.547	-2.091	0.551	157.06	142.97	300.03
BB	$^3\Sigma_g^-$	0.000	0.289	-0.289	2.914	2.914	197.51	197.51	395.02
BC	$^4\Sigma^-$	1.088	1.265	0.771	0.411	-2.761	129.71	212.36	342.07
BN	$(^3\Sigma^-)$‡	0.831	2.024	0.544	-1.132	-1.482	160.54	156.08	316.62
BO	$^2\Sigma^+$	1.553	1.219	1.162	-1.075	-0.214	105.48	160.51	265.99
BF	$^1\Sigma^+$	0.934	1.883	0.729	-0.829	0.084	154.99	125.06	280.05
CC	$^1\Sigma_g^+$	0.000	0.172	-0.172	3.583	3.583	150.20	150.20	300.40
CN	$^2\Sigma^+$	1.123	0.931	0.592	1.513	0.553	104.67	145.90	250.57
CO	$^1\Sigma^+$	1.346	1.701	1.061	-0.463	-0.060	110.18	133.90	244.08
CF	$^2\Pi$	0.780	1.470	0.521	1.087	0.267	125.91	107.65	233.56
NN	$^1\Sigma_g^+$	0.000	0.590	-0.590	1.492	1.492	115.49	115.49	230.98
NO	$^2\Pi$	0.495	0.990	-0.041	1.594	1.552	117.98	114.84	232.82
NF	$^3\Sigma^-$	0.438	0.876	0.069	1.930	1.205	112.61	98.89	211.50
OO	$^3\Sigma_g^-$	0.000	0.437	-0.437	1.725	1.725	102.40	102.40	204.80
OF	$^2\Pi$	0.201	0.518	-0.160	1.969	1.553	102.52	96.34	198.86
FF	$^1\Sigma_g^+$	0.000	0.197	-0.197	1.656	1.656	96.26	96.26	192.52

* Atomic units throughout. In all cases the nuclei are oriented such that the A nucleus is at the origin and the B atom is located at positive z. A positive value for $M(\Omega)$ implies the direction A^-B^+ for the dipole.

† $v(AB) = v(A) + v(B)$. Volumes determined by the 0.001 au density envelope.

‡ See corresponding footnote in Table A3.

Table A5

Radii of isolated ground-state spherical atoms and ions*

Atom	r_a (0.001)	r_a (0.002)
H	2.88	2.53
He	2.50	2.27
Li	4.18	3.29
Be	4.17	3.68
B	3.91	3.51
C	3.62	3.27
N	3.35	3.06
O	3.18	2.92
F	3.03	2.78
Ne	2.89	2.66
Na	4.25	3.26
Mg	4.60	4.01
Al	4.61	4.08
Si	4.45	4.01
P	4.23	3.85
S	4.07	3.72
Cl	3.91	3.58
Ar	3.72	3.42
Li^+	1.82	1.69
Na^+	2.47	2.30
F^-	3.49	3.20
Cl^-	4.36	3.92

*In atomic units. Calculated from atomic functions of Clementi, E. and Roetti, C. (1974), *Atomic Data and Nuclear Data Tables* **14**, 177.

INDEX